MIXED SURFACTANT SYSTEMS

SURFACTANT SCIENCE SERIES

ADDITIONAL VOLUMES IN PREPARATION

MIXED SURFACTANT SYSTEMS
Second Edition, Revised and Expanded

edited by

Masahiko Abe
Faculty of Science and Technology
and Institute of Colloid and Interface Science
Tokyo University of Science
Chiba, Japan

John F. Scamehorn
Institute for Applied Surfactant Research
University of Oklahoma
Norman, Oklahoma, U.S.A.

CRC Press
Taylor & Francis Group
Boca Raton London New York

CRC Press is an imprint of the
Taylor & Francis Group, an **informa** business

First published 2005 by Marcel Dekker

Published 2020 by CRC Press
Taylor & Francis Group
6000 Broken Sound Parkway NW, Suite 300
Boca Raton, FL 33487-2742

First issued in paperback 2020

© 2005 by Taylor & Francis Group, LLC
CRC Press is an imprint of Taylor & Francis Group, an Informa business

No claim to original U.S. Government works

ISBN 13: 978-0-367-57813-8 (pbk)
ISBN 13: 978-0-8247-2150-3 (hbk)

This book contains information obtained from authentic and highly regarded sources. Reasonable efforts have been made to publish reliable data and information, but the author and publisher cannot assume responsibility for the validity of all materials or the consequences of their use. The authors and publishers have attempted to trace the copyright holders of all material reproduced in this publication and apologize to copyright holders if permission to publish in this form has not been obtained. If any copyright material has not been acknowledged please write and let us know so we may rectify in any future reprint.

**Visit the Taylor & Francis Web site at
http://www.taylorandfrancis.com**

**and the CRC Press Web site at
http://www.crcpress.com**

Library of Congress Cataloging-in-Publication Data
A catalog record for this book is available from the Library of Congress.

Sherril D. Christian (1931–2000)

Dedicated to the memory of Sherril D. Christian, George Lynn Cross Professor of Chemistry at the University of Oklahoma. Sherril was a solution thermodynamicist, who spent the last twenty years of his career studying the fundamentals and applications of surfactants. He had the highest standards as scientist and educator and was a good friend.

Preface

It has been known for a long time that surfactant mixtures can exhibit substantially different properties than single surfactants. Papers on this topic have been published in the scientific literature for at least half a century. Practical applications utilizing these mixture synergisms are widespread and new uses continue to be found.

As far as we know, the first book dedicated to surfactant mixtures was published in 1986 (Scamehorn, J.F. (Editor), *Phenomena in Mixed Surfactant Systems*, ACS *Symp. Ser.*, Vol. 311). In 1992, a second book appeared (Rubingh, D.N., and Holland, P.M. (Editors), *Mixed Surfactant Systems*, ACS *Symp. Ser.*, Vol. 501). Finally, in 1993, the first edition of this book was published as Volume 46 of the Marcel Dekker Surfactant Science Series (Ogino, K., and Abe, M. (Editors), *Mixed Surfactant Systems*, Marcel Dekker, New York).

The current editors organized a session at PACIFICHEM held in Hawaii in December, 2000 and this second edition of Mixed Surfactant Systems evolved from that symposium. With the retirement of Professor Ogino, Professor Abe became senior editor and he asked Professor Scamehorn to become co-editor. In editing the book, we replaced about two thirds of the contributors in the first edition, avoided overlap of subjects with those in the first edition except where progress merited an updated chapter, and attempted to incorporate the results of the newest experimental and theoretical investigations on mixed surfactant systems as much as possible. Thus, the book contains a variety of topics including monolayers of mixed surfactants, diffusion of mixed micelles, mixed micelles of fluorinated surfactants, mixed micelles of conventional surfactants and biosurfactants, sponge-like vesicles of mixed surfactants, admicelles of mixed surfactants, liquid crystals of mixed surfactants, mixed micelles of stimulus-responsive surfactants, mixtures of surfactants and polymers, photolysis of mixed surfactants, and new measurement methods and techniques.

We earnestly hope that this book will help the reader appreciate the rapid progress made in understanding surfactant mixtures and to help them make unique inventions using mixed surfactant systems.

Masahiko Abe
John F. Scamehorn

Contents

Contributors

Nicholas L. Abbott Department of Chemical and Biological Engineering, University of Wisconsin, Madison, Wisconsin, U.S.A.

Masahiko Abe Faculty of Science and Technology and Institute of Colloid and Interface Science, Tokyo University of Science, Chiba, Japan

Paschalis Alexandridis Department of Chemical and Biological Engineering, University at Buffalo, The State University of New York, Buffalo, New York, U.S.A.

Makoto Aratono Kyushu University, Hakozaki, Fukuoka, Japan

Tsuyoshi Asakawa Kanazawa University, Kanazawa, Japan

D. M. Bloor School of Sciences, Chemistry, University of Salford, Salford, United Kingdom

Dobrin Petrov Bossev National Institute of Standards and Technology, Gaithersburg, Maryland, U.S.A.

Bret A. Coldren Advanced Encapsulation, Santa Barbara, California, U.S.A.

Gerardino D'Errico Dipartimento di Chimica, Università di Napoli "Federico II", Complesso di Monte S. Angelo, Napoli, Italy

Kunio Esumi Department of Applied Chemistry and Institute of Colloid and Interface Science, Tokyo University of Science, Tokyo, Japan

Kazuhiro Fukada Department of Biochemistry and Food Science, Faculty of Agriculture, Kagawa University, Kagawa, Japan

Jeffrey H. Harwell Institute for Applied Surfactant Research, University of Oklahoma, Norman, Oklahoma

Kathi L. Herrington W. L. Gore and Associates, Newark, Delaware, U.S.A.

H. Hidaka Frontier Research Center for the Global Environment Protection, Meisei University, Tokyo, Japan

H. Hoffmann University of Bayreuth, Physical Chemistry I, Bayreuth, Germany

Daniel J. Iampietro Merck and Company, Rahway, New Jersey, U.S.A.

Ken-Ichi Iimura Faculty of Engineering, Utsunomiya University, Utsunomiya, Japan

Hee-Tae Jung Department of Chemical and Biomolecular Engineering, Korea Advanced Institute of Science and Technology, Daejon, Korea

Eric W. Kaler Department of Chemical Engineering, University of Delaware, Newark, Delaware, U.S.A.

Teiji Kato Faculty of Engineering, Utsunomiya University, Utsunomiya, Japan

Hironobu Kunieda Graduate School of Engineering, Yokohama National University, Yokohama, Japan

Masahiro Manabe Niihama National College of Technology, Niihama, Ehime, Japan

Mutsuo Matsumoto Institute for Chemical Research, Kyoto University, Kyoto, Japan

Shigeyoshi Miyagishi Kanazawa University, Kanazawa, Japan

M. Müller University of Bayreuth, Physical Chemistry I, Bayreuth, Germany

Shigemi Nagadome Department of Chemistry, Fukuoka University, Fukuoka, Japan

Masaru Nakahara Institute for Chemical Research, Kyoto University, Kyoto, Japan

Akio Ohta Kanazawa University, Kanazawa, Japan

Ornella Ortona Dipartimento di Chimica, Università di Napoli "Federico II", Complesso di Monte S. Angelo, Napoli, Italy

Luigi Paduano Dipartimento di Chimica, Università di Napoli "Federico II", Complesso di Monte S. Angelo, Napoli, Italy

Carlos Rodriguez Graduate School of Engineering, Yokohama National University, Yokohama, Japan

Hideki Sakai Faculty of Science and Technology and Institute of Colloid and Interface Science, Tokyo University of Science, Chiba, Japan

John F. Scamehorn Institute for Applied Surfactant Research, University of Oklahoma, Norman, Oklahoma

N. Serpone Department of Chemistry and Biochemistry, Concordia University, Montreal (Quebec), Canada

Jason Y. Shin Department of Chemical and Biological Engineering, University of Wisconsin, Madison, Wisconsin, U.S.A.

Gohsuke Sugihara Department of Chemistry, Fukuoka University, Fukuoka, Japan

Kazuo Tajima Department of Chemistry, and Hightech Research Center, Faculty of Engineering, Kanagawa University, Kanagawa, Japan

Takanori Takiue Kyushu University, Hakozaki, Fukuoka, Japan

T. Thurn School of Sciences, Chemistry, University of Salford, Salford, United Kingdom

Toshihiro Tominaga Department of Applied Chemistry, Okayama University of Science, Okayama, Japan

Marina Tsianou Department of Chemical and Biological Engineering, University at Buffalo, The State University of New York, Buffalo, New York, U.S.A.

Vincenzo Vitagliano Dipartimento di Chimica, Università di Napoli "Federico II", Complesso di Monte S. Angelo, Napoli, Italy

M. S. Vohra Frontier Research Center for the Global Environment Protection, Meisei University, Tokyo, Japan

N. Watanabe Frontier Research Center for the Global Environment Protection, Meisei University, Tokyo, Japan

E. Wyn-Jones School of Sciences, Chemistry, University of Salford, Salford, United Kingdom

Shoko Yokoyama School of Pharmaceutical Sciences, Kyushu University of Health and Welfare, Miyazaki, Japan

Joseph A. Zasadzinski Department of Chemical Engineering and Materials, University of California, Santa Barbara, California, U.S.A.

1

Miscibility in Binary Mixtures of Surfactants

MAKOTO ARATONO and TAKANORI TAKIUE

Kyushu University, Hakozaki, Fukuoka, Japan

SYNOPSIS

The miscibility in binary mixtures of surfactants in the organized assemblies at the interface and in the solution is summarized on the basis of the phase diagrams of adsorption, micelle formation, and vesicle formation. In Section II, the thermodynamic equations to obtain the phase diagrams are given for some combinations of binary surfactant systems. Furthermore the criterion of an ideal mixing is proposed. In Section III, the miscibility in the adsorbed film and micelle is demonstrated for some representative systems by using the phase diagrams and the excess Gibbs free energy. In Section IV, the vesicle–micelle transition in the aqueous mixture and the phase transition in the adsorbed films at the oil/water interface are shown and examined from the viewpoint of the miscibility of the surfactants.

I. INTRODUCTION

The adsorption and micelle formation of surfactant mixtures have been studied extensively, not only by various experimental techniques, but also by theoretical considerations. In this decade several experimental techniques have been developed such as the scattering and reflection of X-rays and neutrons, electron microscopy, probe microscopy, imaging techniques, and so on. They and their developments have been introduced and reviewed in the literature [1,2]. By means of these techniques, many researchers have inquired deeply into the structures of interfaces, adsorbed films, micelles, and vesicles and their dependencies on the composition of surfactants in the bulk solution. At the same time, theoretical considerations have been improved from the one-parameter model of Rubingh et al. [3] and molecular thermodynamics have also been developed steadily, especially by Blankschtein *et al.* [4,5]. The theoretical aspects of micellization of surfactant mixtures have been reviewed very recently by Hines [6], in which the more rigorous and refined theoretical developments have been updated.

In our previous reviews on miscibility in binary mixtures of surfactant [7], the miscibilities of surfactants in the adsorbed films and micellar states have been considered by developing the thermodynamic equations for constructing the phase diagrams of adsorption and micelle formation and then by examining the phase diagrams closely. During the course, newly defined compositions of surfactants in the adsorbed film and micelle were introduced to take a dissociation of ionic surfactants into account explicitly and could describe the miscibility very adequately from the thermodynamic viewpoint: its usefulness was examined and proved in the cases of nonionic–nonionic, ionic–ionic, and nonionic–ionic surfactant mixtures.

Although the review here is basically a further extension of our previous reviews from a theoretical point of view, the criterions of an ideal mixing in the adsorbed film and micelle are newly and definitely proposed for some types of surfactant mixtures. The nonideal mixing is then expressed quantitatively either in the activity coefficients or in the excess Gibbs free energy of adsorption and micelle formation. Furthermore, the spontaneous vesicle formation of anionic–cationic surfactant mixtures, and then the vesicle–micelle transition, are introduced on the basis of our thermodynamic strategy [8]. Other new topics in this review are the phase transition of adsorbed films of the long chain hydrocarbon and fluorocarbon alcohol mixtures at the oil/water interface and the relationship between the miscibility and the phase transition of the adsorbed films [9,10].

II. THERMODYNAMIC EQUATIONS

To evaluate the composition of surfactants in the adsorbed film thermodynamically, it is indispensable that the surface tension of their aqueous solution/air interface is measured as a function of the two concentrations of the two surfactants at a given temperature and pressure. Among some combinations of the two concentration variables, the combination of the total concentration of the surfactants and the mole fraction of one surfactant is most useful to evaluate directly the composition of surfactant in the adsorbed film from the surface tension results [7,11]. For a nonionic surfactant mixture, the total concentration and the mole fraction are defined unequivocally. For a mixture comprising at least one ionic surfactant, however, the dissociation of ionic surfactants should be taken into account in their definition and thus there are some different ways to define them. Therefore, it is convenient to employ the most suitable concentration variables as the case may be and then describe the total differential of the surface tension as a function of them. A thorough derivation of the basic expression of the total differential of surface tension is given in our previous studies from this perspective [7]. Then only the very key points of the thermodynamic equations are demonstrated here. Furthermore, although the dependencies of the surface tension on temperature and pressure afford information on the entropy and volume changes associated with the adsorption, let us leave out this topic for want of space.

In the following, thermodynamic equations for a binary ionic surfactant mixture without common ions (Section II.A) are summarized from our previous papers [11], because the equations are most general in their form. The analogues for a binary ionic surfactant mixture with common ions (Section II.B) are slightly different from those of the Section II.A, but essentially the same. For ionic–nonionic (Section II.C) and nonionic–nonionic (Section II.D) surfactant mixtures, the analogues are apparently quite the same as the equations of Section II.A.

A. Ionic and Ionic Surfactant Mixtures Without Common Ions

Let us first consider the adsorption from an aqueous solution of a binary ionic surfactant mixture without common ions at their aqueous solution/air interface. Surfactants 1 and 2 are assumed to be strong electrolytes and dissociate into $\nu_{1,a}$ a ions and $\nu_{1,c}c$ ions, and $\nu_{2,b}b$ ions and $\nu_{2,d}d$ ions, respectively. Then the total differential of the surface tension γ is written as

$$\mathrm{d}\gamma = -\Gamma_a \,\mathrm{d}\mu_a - \Gamma_c \,\mathrm{d}\mu_c - \Gamma_b \,\mathrm{d}\mu_b - \Gamma_d \,\mathrm{d}\mu_d \qquad (1)$$

at constant temperature and pressure, where the electroneutrality conditions of surfactants in the bulk solution are already taken into account and Γ_i is the surface excess concentration of ion i according to the two dividing planes method [12]. By assuming the aqueous solutions are ideally dilute, the surface tension γ is expressed as a function of the total molality of the ions m and the mole fraction of the second surfactant in the bulk solution X_2 as

$$\mathrm{d}\gamma = -(RT\Gamma/m)\,\mathrm{d}m - (RT\,\Gamma/X_1X_2)(Y_2 - X_2)\,\mathrm{d}X_2 \qquad (2)$$

Here m and X_2 are defined by

$$m = m_a + m_c + m_b + m_d$$
$$= \nu_1 m_1 + \nu_2 m_2 \tag{3}$$

and

$$X_2 = (m_b + m_d)/m$$
$$= \nu_2 m_2/m \tag{4}$$

respectively. Here ν_1 and ν_2 are the number of ions dissociated from surfactants 1 and 2 defined as

$$\nu_1 = \nu_{1,a} + \nu_{1,c} \tag{5}$$

and

$$\nu_2 = \nu_{2,b} + \nu_{2,d} \tag{6}$$

The mole fraction of the second surfactant in the adsorbed film Y_2 is analogously defined in terms of the surface excess concentrations of the ions as

$$Y_2 = (\Gamma_b + \Gamma_d)/\Gamma \tag{7}$$

where Γ is the total surface excess concentration of ions given by

$$\Gamma = \Gamma_a + \Gamma_c + \Gamma_b + \Gamma_d \tag{8}$$

The theoretical background of why these definitions are appropriate is pointed out in our previous studies [7,11].

Then, from Eq. (2), the mole fraction Y_2 is evaluated by applying the equation

$$Y_2 = X_2 - (X_1 X_2/m)(\partial m/\partial X_2)_{T,p,\gamma} \tag{9}$$

to the m vs X_2 curve at a given surface tension. Plotting the m vs X_2 curve together with the m vs Y_2 curve, we have a diagram expressing the quantitative relation between the mole fractions of the two states, that is the bulk solution and the adsorbed films. We call this the phase diagram of adsorption (PDA) at the given surface tension.

To inquire more deeply into the miscibility of surfactant molecules, it is advantageous to examine the deviation of the Y_2 values from the corresponding values of the ideal mixing Y_2^{id} and then evaluate the activity coefficients of the surfactant in the adsorbed film. For this purpose, we have to derive the expression for Y_2^{id} and then know how to estimate the activity coefficients. This is performed by deriving the thermodynamic relations for the equilibrium between the adsorbed film and the aqueous solution at a given surface tension. In principle, the equilibrium condition is that the electrochemical potentials of ions in the bulk solution are equal to those in the adsorbed film. However, since the concentration of an ion cannot be changed individually without changing its counter ion concentration, it is convenient to introduce the mean chemical potential of the surfactant i in the bulk solution μ_i and that in the adsorbed film ζ_i of the binary surfactant mixture. For the first surfactant, these are defined by

$$\mu_1 = (\nu_{1,a}\tilde{\mu}_a + \nu_{1,c}\tilde{\mu}_c)/\nu_1 \tag{10}$$

and

$$\zeta_1 = (\nu_{1,a}\tilde{\zeta}_a + \nu_{1,c}\tilde{\zeta}_c)/\nu_1 \tag{11}$$

respectively. Here $\tilde{\mu}_j$ and $\tilde{\zeta}_j$ are the electrochemical potentials of ion j. Similarly, introducing the mean chemical potential μ_i^0 and the molality m_i^0 of the pure surfactant i at the given surface tension, μ_i can be expressed as [11]

$$\mu_i = \mu_i^0 + RT \ln X_i m/m_i^0 \tag{12}$$

Here the solution is assumed to be ideally dilute as shown in Eq. (1).

Even when the bulk solution is ideally dilute, the preferential adsorption of counter ions to the pair surfactant ions, i.e., $\Gamma_a/\nu_{1a} \neq \Gamma_c/\nu_{1c}$ and $\Gamma_b/\nu_{2b} \neq \Gamma_d/\nu_{2d}$, often comes about and then the ions do not mix ideally in the adsorbed film.

From this point of view, the definition of the mean chemical potentials ζ_i is rather complicated. Even in those cases, however, it has been proved that ζ_i is written in a similar form to Eq. (12) as

$$\zeta_i = \zeta_i^0 + RT \ln f_i^H Y_i \qquad (13)$$

where ζ_i^0 is the mean chemical potential of the pure surfactant i at the given γ [11]. Here it should be noted that f_i^H is the mean activity coefficient of the surfactant i and becomes unity neither when the preferential adsorption takes place nor when the interaction among species in the mixed-surfactant system is different from that in the pure surfactant system. Therefore f_i^H can elucidate quantitatively nonideal mixing of surfactants in the adsorbed film in terms of preferential adsorption and interaction between ions [11].

Now substituting the mean chemical potentials into the equilibrium conditions between the bulk solution and the adsorbed film given by

$$\mu_i = \zeta_i \qquad (14)$$

and using the equilibrium condition for the pure surfactant i at the same γ given by

$$\mu_i^0 = \zeta_i^0 \qquad (15)$$

we yield the equation describing the equilibrium relationship among the bulk concentration, the mole fractions in the bulk solution and the surface, and the activity coefficient

$$X_i m / m_i^0 = f_i^H Y_i \qquad (16)$$

Therefore, f_i^H is evaluated from the surface tension measurements since the quantities on the left-hand sides of Eq. (16) are obtained from the γ vs m curve at different X_2, and Y_i is evaluated by using Eq. (9) at a given surface tension. Once f_i^H is evaluated, the excess Gibbs free energy of adsorption per mole of surfactant mixture g^{HE} is calculated according to the equation

$$g^{HE} = RT(Y_1 \ln f_1^H + Y_2 \ln f_2^H) \tag{17}$$

The nonideal mixing of surfactants in the adsorbed film is recognized from the shape of the PDA as follows. Since $f_i^H = 1$ corresponds to the ideal mixing, Eq. (16) for the first and second surfactants with $f_i^H = 1$ provides the criterion of the ideal mixing as the straight line connecting the molality values of the pure surfactants at the given surface tension

$$m = m_1^0 + (m_2^0 - m_1^0)Y_2 \tag{18}$$

Here it should be emphasized again that the dissociation of surfactants into ions are explicitly taken into account in the definition of m, m_i^0, and Y_2 as given by Eqs (3) and (7).

Now let us turn to the micelle formation of surfactant mixtures. We have shown that the micelle formation is described by the analogous equations to those describing the adsorption by using the excess molar thermodynamic quantities of mixed micelle [13,14]. The analogue of Eq. (2) is given by

$$(RT/C)\,dC = -(RT/X_1X_2)(Z_2 - X_2)\,dX_2 \tag{19}$$

Here C is the total molality of ions at the critical micelle concentration (CMC) defined by Eq. (3) and Z_2 is the mole fraction of the second surfactant in the micelle particles defined in terms of the excess number of ions N_i by

$$Z_2 = (N_b + N_d)/N \tag{20}$$

where N is the total excess number of ions given by

$$N = N_a + N_c + N_b + N_d \tag{21}$$

Then the mole fraction Z_2 is estimated from the C vs X_2 curves by using the equation

$$Z_2 = X_2 - (X_1X_2/C)(\partial C/\partial X_2)_{T,p} \tag{22}$$

Furthermore the equilibrium relationship among the bulk monomer concentration C and the mole fractions of the surfactant i in their monomeric state X_i and that in the micelle particles Z_i, and the activity coefficients in the micelle f_i^M at the CMC are expressed by using the total molality at the CMC of the pure surfactant i, C_i^0, as

$$X_i C / C_i^0 = f_i^M Z_i \tag{23}$$

Therefore the excess Gibbs free energy of micelle formation per mole of surfactant mixture g^{ME} and the criterion of the ideal mixing in the micelle on the phase diagram of micelle formation (PDM) are respectively given by

$$g^{ME} = RT(Z_1 \ln f_1^M + Z_2 \ln f_2^M) \tag{24}$$

and

$$C = C_1^0 + (C_2^0 - C_1^0)Z_2 \tag{25}$$

At the concentrations above the CMC, micelle particles in solution are in equilibrium with the adsorbed film. Using Eqs (2) and (19), we have the equation

$$Y_2^C = Z_2 - (X_1 X_2 / RT\Gamma^C)(\partial \gamma^C / \partial X_2)_{T,p} \tag{26}$$

Then the relation between the mole fractions of the micelle Z_2 and adsorbed film Y_2^C at the CMC can be examined from the change of the surface tension at the CMC γ^C with X_2.

B. Ionic and Ionic Surfactant Mixtures with Common Ions

Let us consider that ion c is common to the two surfactants: surfactant 1 dissociates into $\nu_{1,a} a$ ions and $\nu_{1,c} c$ ions and surfactant 2 into $\nu_{2,b} b$ ions and $\nu_{2,c} c$ ions, respectively. Taking account of this dissociation, the total differential of γ is given by

$$d\gamma = -\Gamma_a \, d\mu_a - \Gamma_c \, d\mu_c - \Gamma_b \, d\mu_b \tag{27}$$

The mole fraction of surfactant 2, Y_2, in the adsorbed film is reasonably defined as the ratio of the total excess concentration of ions originated from surfactant 2 to the total excess concentration of all the ions Γ by

$$
\begin{aligned}
Y_2 &= (v_{2,b}\Gamma_2 + v_{2,c}\Gamma_2)/\Gamma' \\
&= v_2\Gamma_2/\Gamma
\end{aligned}
\tag{28}
$$

where

$$
\Gamma = \Gamma_a + \Gamma_c + \Gamma_b = v_{1,a}\Gamma_1 + v_{1,c}\Gamma_1 + v_{2,c}\Gamma_2 + v_{2,b}\Gamma_2 \tag{29}
$$

In this case, it has been shown that Eq. (2) is transformed into

$$
d\gamma = -(RT\Gamma/m)dm - \Phi(v)(RT\Gamma/X_1X_2)(Y_2 - X_2)\,dX_2 \tag{30}
$$

where $\Phi(v)$ are defined by

$$
\Phi(v) = 1 - (v_{1,c}/v_1)(v_{2,c}/v_2)/\big[(v_{1,c}/v_1)X_1 + (v_{2,c}/v_2)X_2\big]
\tag{31}
$$

Then the mole fraction Y_2 is evaluated by using the equation

$$
Y_2 = X_2 - (X_1X_2/m)(\partial m/\partial X_2)_{T,p,\gamma}/\Phi(v) \tag{32}
$$

The analogue of Eq. (16) is derived by following the same process as shown in Section II.A, but the resulting equation is found to be a rather complicated form [11]. Let us then take up a special case of $v_{1,c}/v_1 = v_{2,c}/v_2$, being of most frequent occurrence. In this case, we have the relation

$$
X_i^{1/v}m/m_i^0 = f_i^H Y_i^{1/v} \tag{33}
$$

and then the activity coefficient f_i^H and the excess Gibbs free energy of adsorption defined by Eq. (17) are calculated. Now the ideal mixing in the adsorbed film is obtained by putting $f_i^H = 1$ in Eq. (33) as

$$
m^v = (m_1^0)^v + \big[(m_2^0)^v - (m_1^0)^v\big]Y_2 \tag{34}
$$

where we put $1/v = v_{1,c}/v_1 = v_{2,c}/v_2$. Comparing Eq. (34) to Eq. (18), we note the large difference between them: the criterion of the ideal mixing for a binary mixture with common ions is not the straight line connecting m_1^0 and m_2^0, but the one connecting $(m_1^0)^v$ and $(m_2^0)^v$ at the given surface tension.

Also for the micelle formation, the mole fraction in the mixed micelle Z_2 is evaluated by

$$Z_2 = X_2 - (X_1 X_2/C)(\partial C/\partial X_2)_{T,p}/\Phi(v) \qquad (35)$$

Furthermore the activity coefficient in the micelle is calculated from the equation

$$X_i^{1/v} C/C_i^0 = f_i^M Z_i^{1/v} \qquad (36)$$

The relation between C and Z_2 for the ideal mixing on PDM is given by

$$C^v = (C_1^0)^v + [(C_2^0)^v - (C_1^0)^v]Z_2 \qquad (37)$$

for the mixtures of $1/v = v_{1,c}/v_1 = v_{2,c}/v_2$.

C. Ionic and Nonionic Surfactant Mixture

Let us consider a mixture in which surfactant 1 is ionic and dissociates into $v_{1,a} a$ ions and $v_{1,c} c$ ions and surfactant 2 is nonionic. The total differential of γ is written as

$$d\gamma = -\Gamma_a \, d\mu_a - \Gamma_c \, d\mu_c - \Gamma_2 \, d\mu_2 \qquad (38)$$

Introducing m, X_2, Γ, and Y_2 defined by

$$\begin{aligned} m = m_a + m_c + m_2, \quad X_2 = m_2/m, \\ \Gamma = \Gamma_a + \Gamma_c + \Gamma_2, \quad Y_2 = \Gamma_2/\Gamma \end{aligned} \qquad (39)$$

we obtain the same equation as Eq. (2) in its apparent form. Furthermore, introducing the analogous definitions to Eqs (39) for N and Z_2 by

$$N = N_a + N_c + N_2, \quad Z_2 = N_2/N \qquad (40)$$

for the mixed micelle formation, we also have the same equation as Eq. (22) in its apparent form. Therefore the equations derived for Section II.A are also applicable in this section. However it should be noted that m, X_2, Γ, Y_2, N, and Z_2 are defined differently in each case.

D. Nonionic and Nonionic Surfactant Mixture

The total differential of γ is simply written as

$$d\gamma = -\Gamma_1 \, d\mu_1 - \Gamma_2 \, d\mu_2 \tag{41}$$

Also in this case we obtain the same equation as Eq. (2) in its apparent form with the following definitions

$$m = m_1 + m_2, \quad X_2 = m_2/m, \quad \Gamma = \Gamma_1 + \Gamma_2, \quad Y_2 = \Gamma_2/\Gamma \tag{42}$$

For the mixed micelle formation, we also have the same equation as Eq. (22) with the definitions given by

$$N = N_1 + N_2, \quad Z_2 = N_2/N \tag{43}$$

Therefore the equations derived for Section II.A. are again applicable to this section.

III. ADSORPTION AND MICELLE FORMATION

A. Nonionic–Nonionic Surfactant Mixtures

The mixtures of analogues of nonionic surfactants often show ideal mixing in the adsorbed films and micelle when the length of hydrocarbon chains is not very different from each other [15–18]. However, even in these cases, the compositions in the adsorbed film and micelle are different from that in the bulk solution. Here we show two cases of nonionic surfactant mixture. The first one is the very typical nonionic surfactant mixture: the mixture of pentaethyleneglycol monodecyl ether (C10E5) and pentaethyleneglycol monooctyl ether (C8E5). The second one is the hydrocarbon and fluorocarbon surfactant mixture: the mixture of tetraethyleneglycol monodecyl ether (C10E4) and tetraethyleneglycol mono-1,1,7-trihydrododeca-fluoroheptyl ether (FC7E4). The former is expected to be ideally and the latter nonideally mixed in their adsorbed films and micelles [19].

1. Pentaethyleneglycol Monodecyl Ether (C10E5)–Pentaethyleneglycol Monooctyl Ether (C8E5)

The surface tension γ is shown as a function of the total molality m at constant mole fraction of C8E5 X_2 defined by Eq. (42) (Fig. 1) [20]. The shape of the curves change very regularly with X_2 and the molality at the CMC C was determined unambiguously from the break points of the curves. The mole fraction of C8E5 in the adsorbed film Y_2 was evaluated by applying Eq. (9) to the m vs X_2 curves at a given surface tension and plotted as the m vs Y_2 curves together with the corresponding m vs X_2 curves in Fig. 2. Figure 2 gives the PDA at different surface tensions. The C values are plotted against X_2 and the mole fraction of C8E5 in the micelle Z_2 calculated by applying Eq. (22) to the C vs X_2 curve are also plotted in Fig. 3. Figure 3 is the PDM at the CMC.

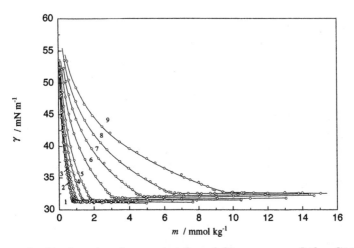

Figure 1 Surface tension vs total molality curves of the C10E5–C8E5 system at constant mole fraction: (1) $X_2 = 0$ (C10E5), (2) 0.1273, (3) 0.2777, (4) 0.4801, (5) 0.6389, (6) 0.8200, (7) 0.9117, (8) 0.9596, (9) 1 (C8E5).

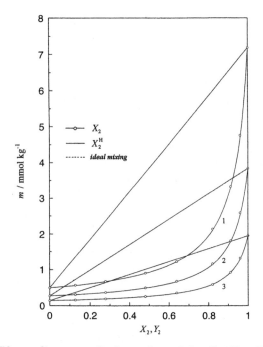

Figure 2 Phase diagram of adsorption of the C10E5–C8E5 system: (1) $\gamma = 35\,\text{mN}\,\text{m}^{-1}$, (2) 40, (3) 45.

The PDAs and PDM manifest that the adsorbed films and the micelle are richer in C10E5 molecules, being more surface active and having lower CMC, than the bulk solution. Furthermore we note the linear m vs Y_2 and C vs Z_2 relations. Thus it is realized that C10E5 and C8E5 molecules are mixed ideally in the adsorbed film and micelle on the basis of Eqs (18) and (25). The C10E5–C8E4 mixture shows the ideal mixing in the adsorbed film but the positive deviation from the ideal mixing in the micelle [20].

2. Tetraethyleneglycol Monodecyl Ether (C10E4)–Tetraethyleneglycol Mono-1,1,7-tri-hydrododecafluoroheptyl Ether (FC7E4)

It is well known that very weak interactions between hydrocarbon and fluorocarbon chains of surfactants often

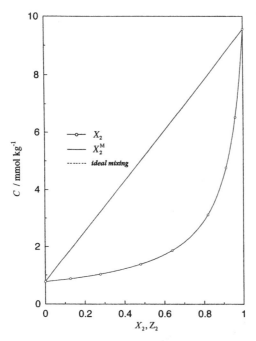

Figure 3 Phase diagram of micelle formation of the C10E5–C8E5 system.

give rise to de-mixing of the surfactants or its tendency in the adsorbed films and micelle [19,21]. However, there are only a few studies on the nonionic hydrocarbon and fluorocarbon surfactant mixtures [22]. Here we demonstrate the miscibility of fluorinated surfactant FC7E4 in the adsorbed film and micelle with hydrogenated C10E4 [23]. The latter was chosen because its surface activity is very similar to that of the former.

The γ vs m curves at constant X_2 are shown in Fig. 4. Although all the curves appear to sit very closely to each other at concentrations below the CMC, the m value at a given surface tension changes very regularly with X_2 as is shown clearly by solid lines in Fig. 5. The m values increase with the other component added and reach a maximum. Applying Eq. (9) to the solid lines, we evaluated the mole fraction of FC7E4 in the adsorbed film Y_2 (broken lines) and constructed the PDAs in Fig. 5. It is noticeable that the m vs Y_2 curves deviate positively from the ideal mixing line given by Eq. (18)

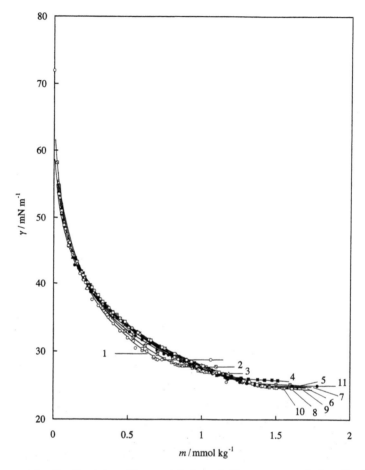

Figure 4 Surface tension vs total molality curves of the C10E4–FC7E4 system at constant mole fraction: (1) $X_2 = 0$ (C10E4), (2) 0.190, (3) 0.366, (4) 0.500, (5) 0.622, (6) 0.701, (7) 0.750, (8) 0.780, (9) 0.801, (10) 0.900, (11) 1 (FC7E4).

and always sit outside the m vs X_2 curves with the coincidence at the maximum point as expected from Eq. (9). Thus the PDA is azeotropic.

The positively azeotropic miscibility mainly comes from the weak interaction between the hydrocarbon and fluorocarbon chains. This is estimated quantitatively by the activity coefficients in the adsorbed films f_i^H and then the excess Gibbs free energy g^{HE} given by Eqs (16) and (17), respectively.

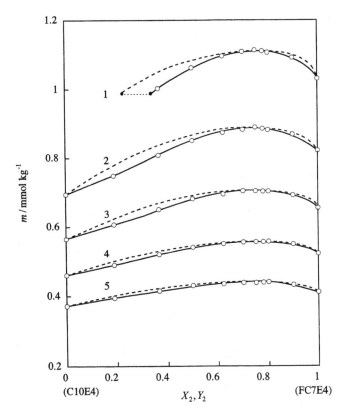

Figure 5 Phase diagram of adsorption of the C10E4–FC7E4 system: (1) $\gamma = 27\,\text{mN}\,\text{m}^{-1}$, (2) 29, (3) 31, (4) 33, (5) 35; (—) m vs X_2, ($\cdots\cdots$) m vs Y_2.

The g^{HE} values are plotted against the mole fraction in the adsorbed films at various surface tensions in Fig. 6. The positive values of g^{HE} suggest that the interaction between C10E4 and FC7E4 molecules is less attractive than that between the same species alone. This is also substantiated by the positive excess area per adsorbed molecule A^{HE}, which was evaluated by applying the equation

$$A^{\text{HE}} = -(1/N_A)(\partial g^{\text{HE}}/\partial \gamma)_{T,p,Y_2} \tag{44}$$

to the dependence of g^{HE} on the surface tension shown in Fig. 6.

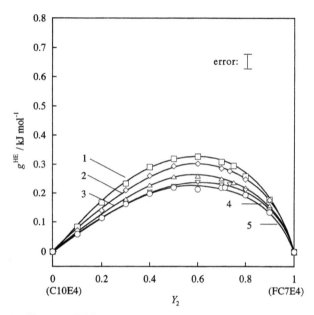

Figure 6   Excess Gibbs free energy of adsorption vs mole fraction in the adsorbed film curves of the C10E4–FC7E4 system: (1) $\gamma = 29\,\mathrm{mN\,m^{-1}}$, (2) 31, (3) 33, (4) 35, (5) 37.

The PDM demonstrates more clearly the less attractive interaction between C10E4 and FC7E4 molecules. Figure 7 is the PDM and reveals the positive azeotropy that the addition of one component to the other increases the CMC with each other. Furthermore, in Fig. 8 the excess Gibbs free energy of micelle formation g^{ME} is compared to that of adsorption at the CMC $g^{HE,C}$. The values of the latter are smaller than those of the former. The existence of the difference between g^{ME} and $g^{HE,C}$ reveals the nonideal mixing and the large difference is attributable to the different molecular orientations in the adsorbed film and the micelle; the hydrophobic chains are expected to be closer to each other in the micelle than in the adsorbed film.

B.   Nonionic–Ionic Surfactant Mixtures

Ionic–nonionic surfactant systems have been investigated for different kinds of surfactant mixtures and exhibit the strong

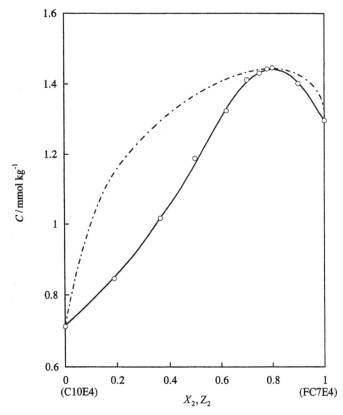

Figure 7 Phase diagram of micelle formation of the C10E4–FC7E4 system : (—) C vs X_2, ($\cdots\cdots$) C vs Z_2.

interaction between different species both in the adsorbed films and in micelle [24–29]. However, there exist several controversies and uncertainties on the type, magnitude, and mechanism of the interaction. Here we demonstrate the strong interaction observed in the anionic–nonionic and cationic–nonionic mixtures: sodium dodecyl sulfate (SDS)–tetraethyleneglycol monooctyl ether (C8E4) [29] and dodecylammonium chloride (DAC)–C8E4 systems [27]. From the results of the two mixtures, we propose a probable physical picture of these strong interactions on the basis of the thermodynamic quantities obtained.

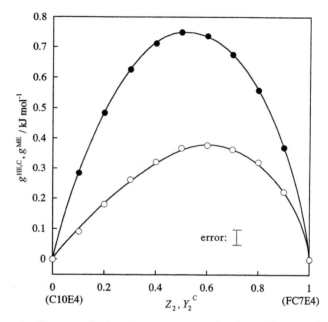

Figure 8 Excess Gibbs free energy of adsorption and micelle formation vs mole fraction curves of the C10E4–FC7E4 system: (●) $g^{HE,C}$ vs Y_2^C, (○) g^{ME} vs Z_2.

1. Sodium Dodecyl Sulfate (SDS)–Tetraethylene Glycol Monooctyl Ether (C8E4)

Figure 9 shows the surface tension vs the total molality curves at different X_2 values. The m vs X_2 curves were constructed by picking up the m values at a given surface tension from Fig. 9 and are then illustrated by solid lines in Fig. 10. The m value decreases steeply with increasing X_2 in the composition range near $X_2 = 0$ and decreases very slightly in the larger X_2 range. Judging from the existence of a minimum, although it is very shallow, on the m vs X_2 curves at the composition range around $X_2 = 0.9$, we expect a synergistic action between SDS and C8E4 molecules.

The mole fraction of C8E4 in the adsorbed films Y_2 was evaluated by using Eq. (9) and given in the form of m vs Y_2 curves by chained lines together with the m vs X_2 curves

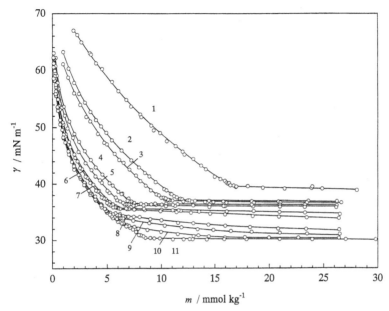

Figure 9 Surface tension vs total molality curves of the SDS–C8E4 system at constant mole fraction: (1) $X_2 = 0$ (SDS), (2) 0.030, (3) 0.050, (4) 0.200, (5) 0.300, (6) 0.500, (7) 0.600, (8) 0.800, (9) 0.875, (10) 0.9950, (11) 1 (C8E4).

in Fig. 10. Thus Fig. 10 corresponds to the PDA. It should be noted that the m vs Y_2 curves show a negative deviation from the straight line (dotted lines in Fig. 10), showing the ideal mixing in the adsorbed film as given by Eq. (18). Thus an attractive interaction between SDS and C8E4 is undoubtedly demonstrated. Figure 11 shows the PDM of the SDS–C8E4 mixture and reveals the attractive interaction more clearly; the C vs X_2 and then the C vs Z_2 curves have a minimum around $X_2 = Z_2 = 0.8$ and they deviates negatively from the straight line showing an ideal mixing. The PDM in this case shows a negative azeotrope, which is in striking contrast to the positive one demonstrated for the hydrocarbon–fluorocarbon surfactant mixture given in Fig. 7.

The estimated g^{HE} values are obviously negative and their absolute value increases with increasing surface tension as

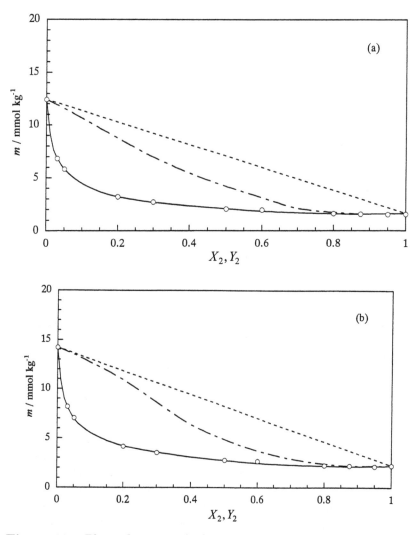

Figure 10   Phase diagram of adsorption of the SDS–C8E4 system:
(a) $\gamma = 45\,\mathrm{mN\,m^{-1}}$, (b) 42.5, (c) 40; (—) m vs X_2, (– – – –) m vs Y_2, (······)
ideal mixing.

given in Fig. 12. The former indicates that the interaction
between SDS and C8E4 molecules in the adsorbed film is more
attractive than that between SDS molecules alone or between
C8E4 molecules alone. Taking account of Eq. (44), on the other
hand, the latter is related to the packing of the molecules in

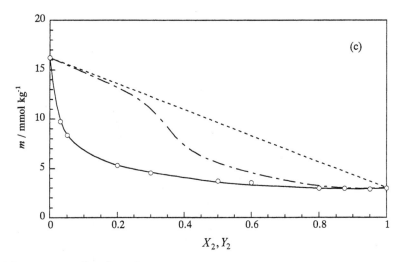

Figure 10 Continued.

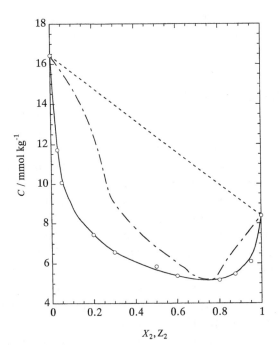

Figure 11 Phase diagram of micelle formation of the SDS–C8E4 system: (—) C vs X_2, (– – – –) C vs Z_2, ($\cdots\cdots$) ideal mixing.

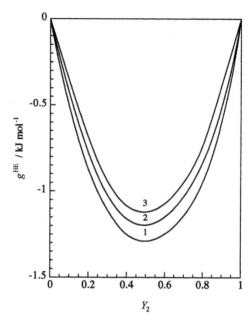

Figure 12 Excess Gibbs free energy of adsorption vs mole fraction in the adsorbed film curves of the SDS–C8E4 system: (1) $\gamma = 45\,\text{mN m}^{-1}$, (2) 42.5, (3) 40.

terms of the excess surface area A^{HE}. It is found that A^{HE} is positive and thus the adsorbed molecules tend to expand its area as compared to the ideal mixing, despite the attractive interaction. If the dispersion forces are mainly responsible for the negative g^{HE} values, the opposite situation must be true; the g^{HE} value changes from the less negative to the more negative one as the distance between hydrophobic chains decreases. We are reasoning from these findings that one of the probable interactions is a kind of anisotropic attraction between the hydrophilic parts of SDS and C8E4 molecules, which has a large optimal interaction distance as compared to the van der Waals interaction. Since this energetically more favorable configuration is expected to be prevented at least partly in the dense assembly of surfactant molecules, the absolute values of g^{HE} in the high surface density region are smaller than those in the low surface density region.

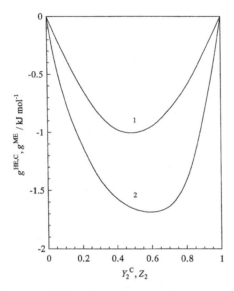

Figure 13 Excess Gibbs free energy of adsorption and micelle formation vs mole fraction curves of the SDS–C8E4 system: (1) $g^{HE,C}$ vs Y_2^C, (2) g^{ME} vs Z_2.

This is also clearly demonstrated in Fig. 13, where g^{ME} and $g^{HE,C}$ are plotted against Z_2 and Y_2 at the CMC. The negative g^{ME} value indicates the energetic stabilization accompanied by the micelle formation. Furthermore, the energetic superiority of the mixed micelle formation over the mixed adsorption confirms our idea on the interaction mentioned above: SDS and C8E4 molecules can take a more favorable conformation for the attractive interaction, acting at the long optimal interaction distance in the mixed micelle rather than at that in the mixed adsorbed film because the hydrophilic portion of the molecules and counter ions can effectively use a wedge-like space in a spherical micelle particle in contrast to a cylindrical space in a plane mixed adsorbed film.

2. Dodecylammonium Chloride (DAC)– Tetraethyleneglycol Monooctyl Ether (C8E4)

The strong attractive interaction between ionic and non-ionic surfactant was observed in the cationic–nonionic combinations as well as in the anionic-nonionic one of the previous section.

The surface tension of the DAC–C8E4 mixture was measured as a function of the total molality m and the mole fraction of C8E4 X_2. Since the C vs X_2 curve shows a minimum, some interaction between DAC and C8E4 molecules are certainly expected. The resulting PDM and PDA are given in Figs 14 and 15, respectively. The PDM has an azeotropic point at $X_2 = 0.85$ and shows that a micelle particle abounds in C8E4 compared to the bulk solution at the compositions below the azeotropic point, while in DAC above it. This fact discloses the attractive interaction between DAC and C8E4 molecules in the mixed micelle. The negative deviation of the m vs Y_2 from the straight line given by Eq. (18) also manifests the attractive interaction in the adsorbed film.

Now let us consider the attractive interaction more closely. Our previous studies on the adsorption and micelle formation of cationic surfactant and alcohol mixtures suggested that the deviation from the ideal mixing is caused mainly by

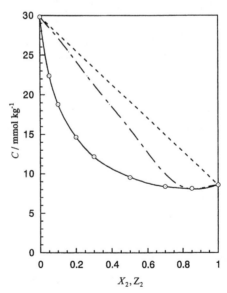

Figure 14 Phase diagram of micelle formation of the DAC–C8E4 system: (—) C vs X_2, (-- --) C vs Z_2, (······) ideal mixing.

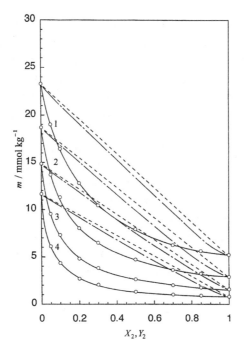

Figure 15 Phase diagram of adsorption of the DAC–C8E4 system: (1) $\gamma = 35\,\text{mN m}^{-1}$, (2) 40, (3) 45, (4) 50; (—) m vs X_2, (— · — · —·) m vs Y_2, (······) ideal mixing.

two contributions [30], that is, the chemical nature of the components and the packing of the surfactant molecules in the aggregates. Considering the chemical nature that DAC dissociates into surfactant ions and counter ions and C8E4 has a nonionic polar head group, there probably exists some kind of electrostatic interaction such as an ion-dipole one. With respect to the packing molecule, on the other hand, the positive excess surface area evaluated by using Eq. (44) suggests that the mixing of DAC and C8E4 molecules in the adsorbed film causes an increase in the area. This may imply that the interaction is not favorable from the standpoint of the occupied area. Therefore, it is said that the attractive interaction of this system comes from the electrostatic interaction.

The influence of the geometry on the interaction is clearly demonstrated by comparing the excess Gibbs free energy of micelle formation g^{ME} to that of adsorption at the CMC $g^{HE,C}$ in Fig. 16. The g^{ME} value is much more negative than that in the adsorbed film, which suggests that C8E4 molecules have a long and bulky hydrophilic part and DAC molecules probably can take a more favorable conformation for the attractive interaction in the mixed micelle than in the adsorbed film because molecules can share a wedge-like space in a spherical micelle particle as already described in the previous section.

Taking note that the attractive interactions between the ionic surfactant and the nonionic one is observed irrespective of the ionic nature of the ionic surfactants, let us propose the scheme of the attractive interaction. In our previous paper on the adsorption and micelle formation of the aqueous solutions of HCl–C8E4 and NaCl–C8E4 mixtures [27], we have examined the interaction between H^+, Na^+, and Cl^- ions with C8E4

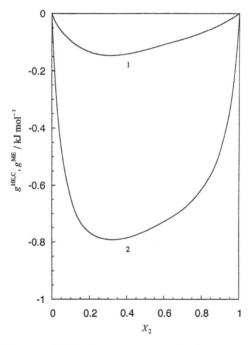

Figure 16 Excess Gibbs free energy of adsorption and micelle formation vs mole fraction curves of the DAC–C8E4 system: (1) $g^{HE,C}$ vs X_2, (2) g^{ME} vs X_2.

molecules and demonstrated that ether oxygen interacts attractively with cationic species both in the adsorbed films and in the micelles from the PDA and PDM [27]. On the basis of this view, we have concluded that the attractive interaction in the DAC–C8E4 mixtures originates from the ion–dipole interaction or hydrogen bonding between the ammonium group and oxygen atom of the ethylene oxide group because the counter ion does not play an important role for the attractive interaction.

With respect to the attractive interaction in the SDS–C8E4 system, on the other hand, an indirect interaction between DS$^-$ and the ether oxygen through Na$^+$ ions seems to be plausible because of the following findings:

1. The ether oxygens of C8E4 do not interact attractively with anionic species, but with the cationic species mentioned above.
2. The excess Gibbs energies of the SDS–C8E4 system are more negative than those of the DAC–C8E4 system as demonstrated in Fig. 17. Nevertheless, DA$^+$ ions are likely to interact rather directly with the hydrophilic part of C8E4 as mentioned in (1).
3. The positive excess area suggests that, when DS$^-$ ions and C8E4 molecules interact in the adsorbed film, they are at a longer distance compared to an ideal distance despite a strong attractive interaction between them.

Summarizing the interaction scheme, the Na$^+$ counter ions contribute to the stabilization in the SDS–C8E4 system, whereas the DS$^-$ ions interact indirectly with C8E4 through Na$^+$ ions. On the other hand, the DA$^+$ ions interact directly with C8E4 in the DAC–C8E4 system.

C. Ionic–Ionic Surfactant Mixtures

1. Decylammonium Chloride (DeAC)–Dodecylammonium Chloride (DAC)

The mole fractions Y_2 and Z_2 were evaluated by applying Eqs (32) and (35) to the surface tension data [31]. The PDA at $40 \, \text{mN} \, \text{m}^{-1}$ and PDM are given in Fig. 18 [11]. Judging from

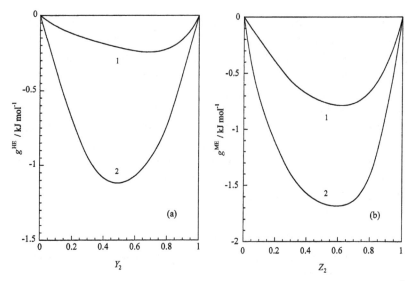

Figure 17 (a) Excess Gibbs free energy of adsorption at $40\,\mathrm{mN\,m^{-1}}$ vs mole fraction in the adsorbed film curves: (1) DAC–C8E4 system, (2) SDS–C8E4 system. (b) Excess Gibbs free energy of micelle formation vs mole fraction in the micelle curves: (1) DAC–C8E4 system, (2) SDS–C8E4 system.

that the surfactants are analogues with each other having the difference of the carbon number only by two and the same counter ion and also the shape of the diagram is similar to that of a cigar, one may expect the ideal mixing of DeAC and DAC molecules in the adsorbed film and micelle.

However, comparing the m vs Y_2 curves with the ones showing the ideal mixing of Eq. (34) with $\nu = 2$ given by

$$m^2 = (m_1^0)^2 + [(m_2^0)^2 - (m_1^0)^2]Y_2 \tag{45}$$

we note that the adsorbed film is enriched slightly in the more surface active DAC molecules than in the less surface active DeAC molecules compared to the adsorbed film of the ideal mixing. Taking note of the equation of the ideal mixing in micelle given by

$$C^2 = (C_1^0)^2 + [(C_2^0)^2 - (C_1^0)^2]Z_2 \tag{46}$$

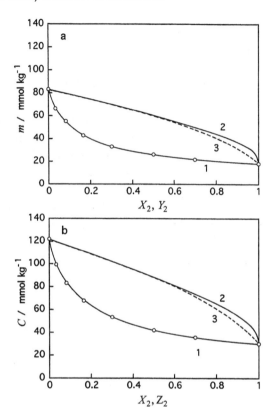

Figure 18 (a) Phase diagram of adsorption of the DeAC–DAC system at $40\,\text{mN m}^{-1}$: (1) m vs X_2, (2) m vs Y_2, (3) ideal mixing. (b) Phase diagram of micelle formation of the DeAC–DAC system: (1) C vs X_2, (2) C vs Z_2, (3) ideal mixing.

it is noted that the situation in the micelle is very similar to that in the adsorbed film. This example shows that, even in the mixture of analogues, nonideal mixing happens and is revealed only based on the correct criterion of ideal mixing given by Eqs (45) and (46).

2. Decylammonium Bromide
   (DeAB)–Dodecylammonium Chloride (DAC)

The only difference of this mixture from the DeAC–DAC is that the counter ion is not common in the DeAB–DAC but common in the DeAC–DAC system. However, we realize that this difference leads to a large nonideal mixing as follows [11].

The mole fractions Y_2 and Z_2 were evaluated by applying Eqs (9) and (22) to the surface tension data [32]. The PDA at $40\,\mathrm{mN\,m^{-1}}$ and PDM are given in Fig. 19 [11]. First it is seen that the m vs Y_2 and C vs Z_2 curves are convex downward and then the nonideal mixing in both the adsorbed film and micelle is expected. Here it should be noted that the ideal mixing is given by the straight lines of Eqs (18) and (25), in contrast to the quadratic equations given by Eqs (45) and (46). The excess Gibbs free energies were calculated from the respective PDAs and PDM and shown in Fig. 20 together with the corresponding

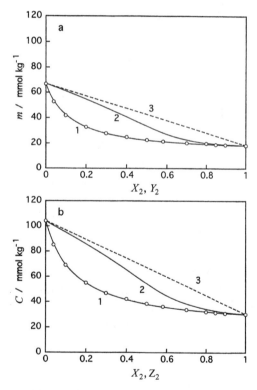

Figure 19 (a) Phase diagram of adsorption of the DeAB–DAC system at $40\,\mathrm{mN\,m^{-1}}$: (1) m vs X_2, (2) m vs Y_2, (3) ideal mixing; (b) Phase diagram of micelle formation of the DeAB–DAC system: (1) C vs X_2, (2) C vs Z_2, (3) ideal mixing.

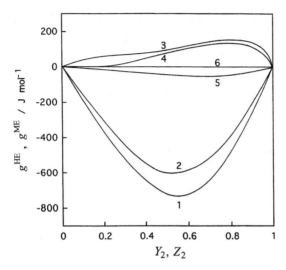

Figure 20 Excess Gibbs free energy of adsorption at $40\,\text{mN}\,\text{m}^{-1}$ and micelle formation vs mole fraction curves: (1) g^{HE} vs Y_2 of the DeAB–DAC system, (2) g^{ME} vs Z_2 of the DeAB–DAC system, (3) g^{HE} vs Y_2 of the DeAC–DAC system, (4) g^{ME} vs Z_2 of the DeAC–DAC system, (5) g^{HE} vs Y_2 of the DeAC–DeAB system, (6) g^{ME} vs Z_2 of the DeAC–DeAB system.

values of the DeAC–decylammonium bromide (DeAB) system [33]. Judging from the results of the DeAC–DeAB system where the surfactant cations are common to each other, chloride and bromide ions mix ideally in the micelles and very slightly nonideally in the adsorbed films. Since the activity coefficients arising from preferential adsorption is less than unity whenever the preferential adsorption takes place [11], the negative g^{HE} comes from the preferential adsorption of bromide ions to chloride ions and this coincides with the observation that the mole fraction Y_2 is shifted to the DeAB side from the ideal mixing line. This preferential adsorption is more pronounced in the DeAB–DAC mixtures, of which g^{HE} is largely negative compared to that of the DeAC–DeAB system. Taking account of the finding that the adsorbed film of the DeAC–DAC mixture abounds more in DA^+ than DeA^+ compared to the ideal mixing, the results of the DeAB–DAC system clearly indicates that attractive interaction, and thus negative values of g^{HE} due to the preferential adsorption of bromide ions to chloride ions, overcomes the small positive values of g^{HE} due to nonideal mixing of hydrophobic chains.

Thus the examination in terms of the excess Gibbs energy reveals the miscibility of surfactant molecules from the viewpoint of not only the surfactant ions but also counter ions such as the preferential adsorption.

IV. TRANSITION BETWEEN TWO STATES OF BINARY SURFACTANT MIXTURES

Mixing of surfactants often produces synergetic action, new functions, and new organized molecular assemblies. In this section, first the spontaneous vesicle formation followed by micelle formation is demonstrated with respect to a cationic–anionic surfactant mixture. The vesicles of cationic–anionic surfactant mixtures are often called catanionic vesicles, which were reviewed thoroughly and very recently by Tondre and Caillet [34]. Here the spontaneous formation of catanionic vesicles were studied by measuring surface tension and by constructing the phase diagram of vesicle and micelle formation by applying the thermodynamic method [8]. Second, the adsorption of long chain hydrocarbon and fluorocarbon alcohol mixtures from their oil solution to the oil/water interface will be demonstrated and their miscibility in the adsorbed film will be proved to be greatly changed by the phase transition of the adsorbed film [9,10].

Thus, the transition in the aqueous solution between micelle and vesicle for the binary water–soluble surfactant mixture and that in the adsorbed film between the expanded and condensed states for the binary oil–soluble surfactant mixture will be shown.

A. Vesicle–Micelle Transition of Sodium Decyl Sulfate (SDeS)–Decyltrimethylammonium Bromide (DeTAB) Mixture

The surface tension of the mixture was carefully measured as a function of the total molality defined by Eq. (3) with $\nu_1 = \nu_2 = 1$

at the fixed mole fraction of DeTAB [8]. The molality at the critical micelle concentration CMC, C^M, and that at the critical vesicle concentration CVC, C^V, were determined from the γ vs m curves. Then the phase diagram of aggregate formation was constructed by using the thermodynamic consideration. First the simple model of the total molality vs mole fraction diagram of aggregate formation will be described and then the experimental results followed by thermodynamic analysis will be demonstrated.

1. Simple Model of the Molality vs Mole Fraction
   Diagram of Aggregate Formation

Micelles and vesicles are not macroscopic phases in the thermodynamic sense because they never exist separately from the solution. However, when their thermodynamic quantities are defined in terms of the excess ones, the thermodynamic theory of micelle formation, in which micelle particles are treated as if they are macroscopic ones, has been successful in understanding micelle formation and properties in the solution [7,14]. Then we show a very simple way to predict the coexisting regions of vesicles and micelles in the total molality vs mole fraction of the surfactant diagram by using our thermodynamic method and the mass balance equation.

In the SDeS–DeTAB system, each pure surfactant does not form vesicles but the mixture does in a limited mole fraction range because the strong electrostatic molecular interaction between the two surfactant ions results in a kind of synergism. Thus the concentration of aggregate formation C is decreased steeply from their respective pure CMC values by adding the other component. An example of such C vs X_2 behavior is illustrated by the three curves connected at two break points at C^t and X_2^ts where the vesicle–micelle transition is assumed to take place as shown in Fig. 21, where C_1^0 and C_2^0 are the CMCs of the respective pure surfactants.

The mole fraction of the second component in the micelle Z_2^M is evaluated by applying [see Eq. (22)]

$$Z_2^M = X_2 - (X_1 X_2 / C^M)(\partial C^M / \partial X_2)_{T,p} \qquad (47)$$

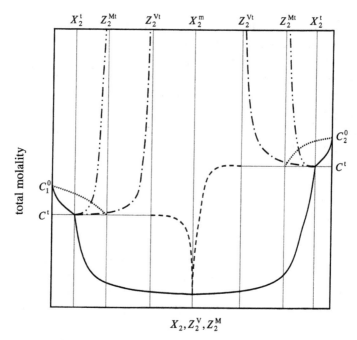

Figure 21 Total molality vs mole fraction diagram predicted: (—) C^M vs X_2 at $0 \leq X_2 \leq X_2^t$ and $X_2^t \leq X_2 \leq 1$, C^V vs X_2 at $X_2^t \leq X_2 \leq X_2^m$ and $X_2^m \leq X_2 \leq X_2^t$, (·····) C^M vs Z_2^M at $0 \leq Z_2^M \leq Z_2^{Mt}$ and $Z_2^{Mt} \leq Z_2^M \leq 1$, (----) C^V vs Z_2^V at $Z_2^{Vt} \leq Z_2^V \leq Z_2^m$ and $Z_2^m \leq Z_2^V \leq Z_2^{Vt}$, (—··—) m^I vs X_2, (—···—) m^{II} vs X_2.

to the critical micelle concentration C^M vs X_2 curves and similarly the mole fraction of the second component in the vesicle Z_2^V is evaluated by applying

$$Z_2^V = X_2 - (X_1 X_2 / C^V)(\partial C^V / \partial X_2)_{T,p} \tag{48}$$

to the critical vesicle concentration C^V vs X_2 curves. The C^M vs Z_2^M and C^V vs Z_2^V are schematically shown in Fig. 21, where Z_2^{Mt} and Z_2^{Vt} are the mole fractions in the micelle and vesicle at the vesicle–micelle transition point.

When vesicles and micelles are in equilibrium with each other in the solution, the mass balance relations for the total surfactants and surfactant 2 are respectively given by

$$m = C^t + m^M + m^V \tag{49}$$

and

$$mX_2 = C^t X_2^t + m^M Z_2^{Mt} + m^V Z_2^{Vt} \tag{50}$$

Here we assumed that the monomer concentration is equal to C^t and the mole fractions in vesicle and micelle are equal to Z_2^{Vt} and Z_2^{Mt}, respectively, and they do not change as the total concentration is further increased because vesicle and micelle are assumed to be a kind of macroscopic phase. Let us examine the right-hand part of the diagram of Fig. 21. At the composition region $Z_2^{Vt} < X_2 < Z_2^{Mt} < X_2^t$, the total molality m^I at which micelle formation starts to take place in the vesicle solution is obtained by putting $m^M = 0$ as

$$m^I = C^t (X_2^t - Z_2^{Vt})/(X_2 - Z_2^{Vt}) \tag{51}$$

At the composition region $Z_2^{Vt} < Z_2^{Mt} < X_2 < X_2^t$, on the other hand, the total molality m^{II} at which vesicle disappears is obtained by putting $m^V = 0$ as

$$m^{II} = C^t (X_2^t - Z_2^{Mt})/(X_2 - Z_2^{Mt}) \tag{52}$$

Therefore the m^I vs X_2 and m^{II} vs X_2 curves are predictable once the values of X_2^t, Z_2^{Mt}, and Z_2^{Vt} are experimentally obtained. Taking account of the relation

$$m^I - C^V = C^t (X_2^t - Z_2^{Vt})/(X_2 - Z_2^{Vt}) - C^V$$
$$> C^V (X_2^t - X_2)/(X_2 - Z_2^{Vt}) > 0 \tag{53}$$

we note that vesicle formation takes place in preference to micelle formation in the vesicle–micelle coexistence regions.

By using the diagram given in Fig. 21, the change of γ with m at a given X_2 can be predicted qualitatively as is schematically illustrated in Fig. 22 with respect to the right-hand side of Fig. 21 ($X_2^m < X_2$), where X_2^m is the mole fraction at the minimum of the C vs X_2 curve. Curve 1 (pure second surfactant $X_2 = 1$) has one break point at the CMC and the surface tension is practically constant above the CMC.

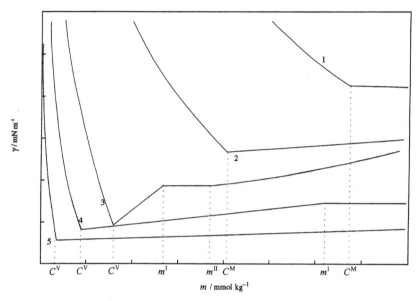

Figure 22 Surface tension vs total molality curves predicted: (1) $X_2 = 1$, (2) $X_2^t < X_2 < 1$, (3) $Z_2^{Mt} < X_2 < X_2^t$, (4) $Z_2^{Vt} < X_2 < Z_2^{Mt}$, (5) $X_2^m < X_2 < Z_2^{Vt}$.

Curve 2 ($X_2^t < X_2 < 1$) has one break at the CMC and the surface tension increases with the molality due to the change of the micelle and monomer compositions. Curve 3 ($Z_2^{Mt} < X_2 < X_2^t$) has three break points at C^V, m^I, and m^{II}. It is expected that the surface tension at the concentrations between m^I and m^{II} is constant due to the vesicle–micelle coexistence. The changes of the surface tension before and after the constant surface tension show the changes of vesicle and micelle compositions with m. On Curve 4($Z_2^{Vt} < X_2 < Z_2^{Mt}$), vesicles and micelles start to coexist at m^t and vesicles never vanish even at high m^I and therefore the surface tension is constant at concentration above m^I. Finally Curve 5 ($X_2^m < X_2 < Z_2^{Vt}$) has only one break corresponding to the CVC. The shape of the curves at $X_2^m < X_2$ is qualitatively similar to that at $X_2 < X_2^m$. The qualitative behavior predicted by this simple model will be compared to the one obtained from the thermodynamic analysis of the experimental results.

2. Vesicle Formation and Vesicle–Micelle Transition

Figure 23 shows the surface tension vs total molality curves at the mole fractions of the DeTAB-rich region [8]. We see that the variation of γ with m at low molalities and at X_2 below 0.98 looks like that of a usual surfactant mixture: the surface tension decreases very rapidly as the total molality increases and is almost constant above a break point. However, the dynamic light scattering measurement suggested the existence of much larger aggregates than micelle particles with the average radius of approximately 500 nm. The vesicle particles were observed by using the optical microscope at some X_2 and also in the TEM image of freeze-fracture replicas. Thus it is concluded that the first break points on these γ vs m curves are referred to as the critical vesicle concentration C^V.

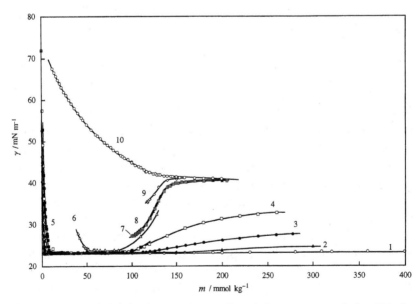

Figure 23 Surface tension vs total molality curves of the SDeS–DeTAB system at constant mole fraction: (1) $X_2 = 0.7012$, (2) 0.8004, (3) 0.9000, (4) 0.9700, (5) 0.9800, (6) 0.9990, (7) 0.9996, (8) 0.9997, (9) 0.9999, (10) 1.

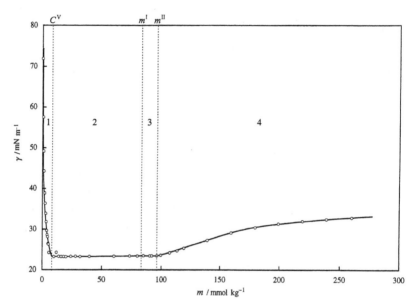

Figure 24 Surface tension vs total molality curves of the SDeS–DeTAB system at $X_2 = 0.9700$: (1) monomer, (2) monomer + vesicle, (3) monomer + vesicle + micelle, (4) monomer + micelle.

Examining more closely the curves and taking account of the visual and optical observations, we have concluded that four regions exist on each curve: the example at $X_2 = 0.9700$ is illustrated in Fig. 24. Region 1 corresponds to the monomer solution and the surface tension decreases steeply within a very narrow concentration range. In Region 2, the surface tension increases slightly but definitely with m. The solutions were turbid and slightly bluish, although the bluish color disappeared completely by freezing them followed by thawing and birefringent at some total molalites. It was expected from the turbidity that the aggregates have a size of micrometer order and in fact a donut-like shape with a size of a few micrometers was observed by differential interference microscope. Furthermore the aggregates formed in the birefringent solution may have lamellar structures because a sheet-like structure was vaguely observed with the differential interference microscope. It should be noted that the surface tension falls on the same curve despite the changes in the structures. In Region 3, the surface tension is constant and solution is observed to be bluish and transparent. It should be pointed out that the appearance of the solution and the γ values were not changed by freezing the solution followed by thawing it and the solution appeared to be bluish and transparent even at the highest total molality of this region. Judging from the appearance of the solution and the fact that vesicles were hardly observed by optical microscopy, the size of vesicles in this region is probably around 100 nm or less. In Region 4, it is important to note a new observation, which is not found in the ordinary mixed micelle systems, that the γ value starts to increase again with m. The solutions become completely clear and colorless. This suggests the absence of vesicle particles. The curves No. 1 to No. 5 in Fig. 23 are similar in their shapes.

Comparing the γ vs m curves obtained from the experiments to that from the simple theory given in Fig. 22, we note the fairly good correspondence between them. Thus, in Region 1, the surfactant monomers are dispersed in the solution and vesicles of the surfactant mixture spontaneously form at C^V. In Region 2, the vesicles and monomers are dispersed in the solution. The composition of the vesicle is different from that in monomeric states. In Region 3, micelles of the surfactant mixture appear at m^I and then the vesicles, micelles, and monomers are dispersed in the solution. Judging from the experimental finding that the surface tension is constant in this region, it is probable that the aqueous solution system behaves as if it has three kinds of macroscopic phases, that is, monomer solution, micelles, and vesicles. The molality is further increased up to m^{II}, vesicle particles disappear and then micelle particles and monomer are dispersed in the solution in Region 4.

Let us evaluate the mole fraction of DeTAB Z_2^V by applying Eq. (48) to the C^V vs X_2 curve given by the open circles on the solid line. The results are drawn by chained line in Fig. 25. The C^V vs X_2 and the C^V vs Z_2^V diagram is called the phase diagram of vesicle formation. It is very important to note that the Z_2^V values are close to 0.5 at most of bulk compositions: a vesicle particle contains almost equal numbers of surfactant cations and anions irrespective of the bulk compositions even when the monomer solution being in equilibrium with the vesicle

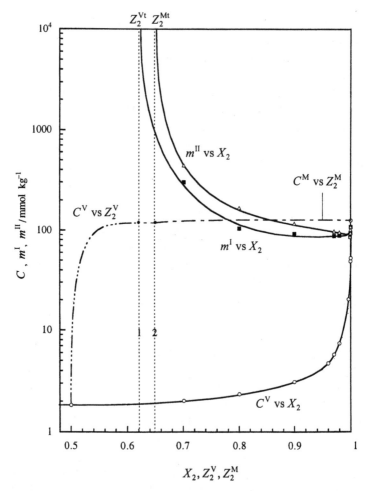

Figure 25 Total molality vs mole fraction diagram of the SDeS–DeTAB system: (1) asymptote of the m^I vs X_2 curve, (2) asymptote of the m^{II} vs X_2 curve.

particles are only 0.01 mol% surfactant anions. The m^I and m^{II} values from the experiments are plotted against X_2. Here the expected C^M vs Z_2^M curve is also shown, which was not obtained because of its experimental difficulty but is anticipated from the present results. The broken lines are the asymptotes of the m^I and m^{II} vs X_2 curves suggested by the simple theory mentioned above. We note the good correspondence of the experimental results given in Fig. 25 to the prediction from the simple theory illustrated in Fig. 21.

Thus spontaneous vesicle formation and vesicle–micelle transition are well described by our thermodynamic method developed for mixed micelle formation: the concentration vs composition diagram can predict the critical vesicle and micelle concentrations, the concentration region of the vesicle–micelle coexistence. Furthermore, the change of surface tension with total molality with the composition X_2 can be predicted by using the concentration vs composition diagram.

B. Phase Transition and Miscibility in the Adsorbed Films of 1-Icosanol (C20OH) and 1,1,2,2-Tetrahydroheptadecafluorodecanol (FC10OH)

As shown in Section III.A.2, the mutual interaction between hydrocarbon and fluorocarbon chains is very weak so that some mixtures of hydrocarbon and fluorocarbon surfactants are de-mixed or have a tendency of de-mixing in their micelle. At the aqueous solution/air (or oil) interfaces, however, even these water-soluble surfactants are miscible with each other at all proportions. This is at least partly because the water soluble surfactants have ionic or large size of hydrophilic groups and then close packing of hydrophobic chains at the interface is hardly realized due to the electrostatic or steric repulsion of the hydrophilic groups compared to that in the micelle, where a rather large distance between hydrophilic head groups is available. Therefore it is of great interest to investigate the adsorption of the oil-soluble mixtures of nonionic hydrocarbon and fluorocarbon compounds at the water/oil solution of the mixture. From this viewpoint, the adsorption of long chain hydrocarbon and fluorocarbon alcohols, and also their mixtures at oil/water interfaces have been studied systematically: the phase transition in the adsorbed film often is observed and influenced greatly by temperature, pressure, and also mixing with other compounds [35]. The phase transition observed in our studies has been proved and studied much more from the viewpoint of the molecular structure at the interface by synchrotron X-ray reflection and scattering techniques [36].

1. The Phase Transition in the Adsorbed Film of the Mixture

The interfacial tension γ of the hexane solution of FC10OH and C20OH mixture/water system was measured as a function of the total molality m and the mole fraction of FC10OH X_2 at 298.15 K under atmospheric pressure [9]. The results at the mole fractions below 0.275 and above 0.280 are shown together with those of the respective pure alcohols in Figs 26(a) and (b), respectively. It is realized that all the curves have a distinct break point at which the slope of the curve increases greatly. Furthermore it should be noted that the two break points were observed on the curves at $X_2 = 0.275$ and 0.280 as indicated by the arrows. The break points on the γ vs m curves suggest that one or two kinds of phase transition occur in the adsorbed film of the mixture. In the following, we call the break points at the higher molality at $X_2 = 0.275$ and 0.280 the second break point and the others the first one.

In Fig. 27, the interfacial tension γ^{eq} and the total molality m^{eq} at the break points are plotted against X_2. It is seen that the γ^{eq} vs X_2 curve of the first break point has a sharp-pointed minimum and the corresponding m^{eq} vs X_2 curve has a sharp-pointed maximum. The corresponding curves of the second break touch the curves of the first break at their extremum. Thus the three different states shown by A, B, and C are suggested in the adsorbed films of the FC10OH and C20OH mixture.

To identify these three states in the adsorbed films, the interfacial pressure π and the mean area per adsorbed molecule A were calculated according to the usual way and demonstrated as the π vs A curves in Fig. 28. Figure 28(a) and (b) show that the curves consist of two parts connected by one discontinuous change in A: the gradual increase and the steep one in π with decreasing A. It is noted that the A values converge at about $0.2\,\text{nm}^2$ below $X_2 = 0.250$ and at about $0.3\,\text{nm}^2$ above $X_2 = 0.300$, respectively. Judging that the cross-sectional area

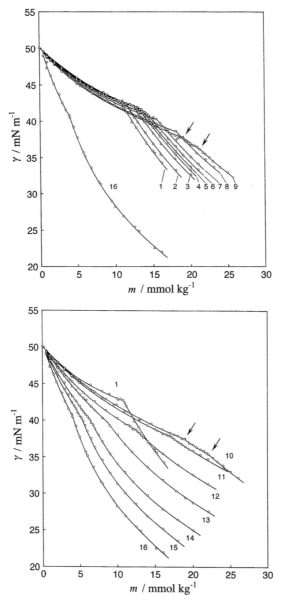

Figure 26 Interfacial tension vs total molality curves of the C20OH–FC10OH system at constant mole fraction: (1) $X_2 = 0$, (2) 0.050, (3) 0.100, (4) 0.125, (5) 0.150, (6) 0.170, (7) 0.200, (8) 0.250, (9) 0.275, (10) 0.280, (11) 0.300, (12) 0.375, (13) 0.500, (14) 0.650, (15) 0.800, (16) 1.

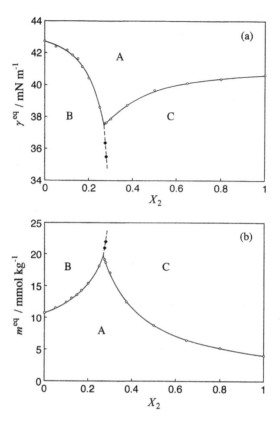

Figure 27 (a) Equilibrium interfacial tension vs mole fraction curves of the C20OH–FC10OH system: (○) first break point, (●) second break point: (b) Equilibrium total molality vs mole fraction curves of the C20OH–FC10OH system: (○) first break point, (●) second break point.

of hydrocarbon and fluorocarbon chains are very close to these A values, we have concluded that the phase transition takes place from the expanded to the condensed state at the first break points. Furthermore the convergence of A at the two different values suggests that the condensed film is constructed of C20OH molecules solely at $X_2 \leq 0.250$ and FC10OH solely at $X_2 \geq 0.300$, respectively. On the other hand, judging the finding that the A value of the expanded state varies continuously with the mole fraction X_2, the film is probably constructed by these alcohol mixtures. The composition relation between the bulk solution and the adsorbed film will be clarified by drawing the phase diagram of adsorption as demonstrated below.

Now let us look closely at the second break points at $X_2 = 0.275$ and 0.280. The π vs A curves are given together with those of the pure alcohols in Fig. 28(c). Examining these curves, we can say that the first break points correspond to the phase transition from the expanded state to the condensed film of FC10OH and the second to that from the condensed film of FC10OH to the condensed film of C20OH, respectively. Thus it is concluded that the three regions A, B, and C correspond to the expanded, C20OH condensed, and FC10OH condensed state, respectively. When the different species of these alcohol molecules are completely immiscible with each other in the C20OH and FC10OH condensed states, the thermodynamic relation can predict that the phase transition should take place from the condensed film of FC10OH to that of C20OH and that its reverse does not take place [9].

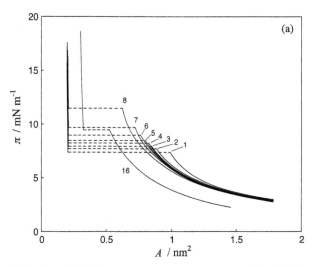

Figure 28 Interfacial pressure vs area per adsorbed molecule curves of the C20OH–FC10OH system at constant mole fraction: (1) $X_2 = 0$, (2) 0.050, (3) 0.100, (4) 0.125, (5) 0.150, (6) 0.170, (7) 0.200, (8) 0.250, (9) 0.275, (10) 0.280, (11) 0.300, (12) 0.375, (13) 0.500, (14) 0.650, (15) 0.800, (16) 1.

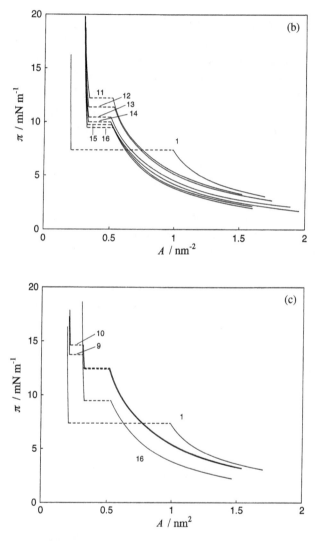

Figure 28 Continued.

2. Miscibility in the Adsorbed Film of the Mixture

To make sure the conclusion on the miscibility in the adsorbed film derived from the π vs A curves and to get more quantitative information on the miscibility, let us construct the phase diagram of adsorption PDA at different interfacial

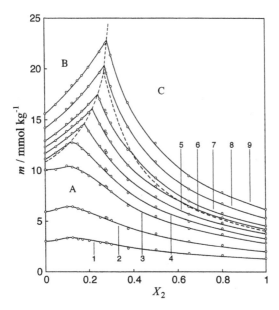

Figure 29 Total molality vs mole fraction curves of the C20OH–FC10OH system at constant interfacial tension: (1) $\gamma = 47\,\mathrm{mN\,m^{-1}}$, (2) 45, (3) 43, (4) 42, (5) 41, (6) 40, (7) 39, (8) 37, (9) 35; (- - - -) m^{eq} vs X_2.

tension values [10]. The mole fraction of FC10OH in the adsorbed film Y_2 was evaluated by applying Eq. (9) to the total molality m vs the mole fraction in the hexane solution X_2 curves given in Fig. 29, where the broken lines show the phase transition points as already given in Fig. 27(a). The PDAs are demonstrated at some selected γ values in Fig. 30.

Figure 30(a) shows the PDA at $45\,\mathrm{mNm^{-1}}$ where the adsorbed films are all in the expanded states. It is seen that C20OH and FC10OH molecules are miscible at all the proportions and their mixture forms a positive azeotrope at about $X_2 = 0.1$: the adsorbed film is richer in C20OH molecules below and in FC10OH molecules above the azeotropic point than the bulk oil solutions. Since the m vs Y_2 curve deviates positively from the straight line of the ideal mixing given by Eq. (18), the molecular interaction between different species is less attractive compared to that between the same species.

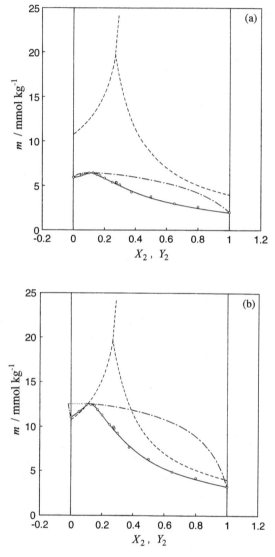

Figure 30 Phase diagram of adsorption of the C20OH–FC10OH system: (a) $\gamma = 45\,\text{mN m}^{-1}$, (b) 42.1, (c) 39, (d) 37, (e) 35; (—) m vs X_2, (– – – –) m vs Y_2, (······) m^{eq} vs X_2.

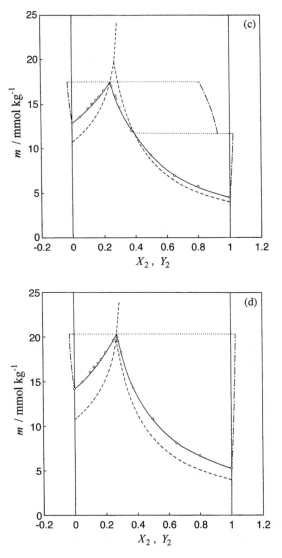

Figure 30 Continued.

As the interfacial tension decreases, the azeotropic point approaches gradually the m^{eq} vs X_2 curve and finally touches it at $42.1\,\mathrm{mN\,m^{-1}}$ as shown in Fig. 30(b). The m vs X_2 curve at lower X_2 sits in the condensed region (B in Fig. 29) and the Y_2 values evaluated are nearly zero or even very slightly negative.

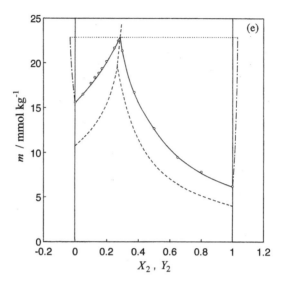

Figure 30 Continued.

This finding shows clearly that the condensed film in the B region is constructed only from C20OH molecules and thus FC10OH molecules are expelled from the interfacial region, which forms a striking contrast with the expanded film.

At $39\,\text{mN m}^{-1}$ given in Fig. 30(c), the m vs X_2 curve is divided into three parts by the two break points on the m^{eq} vs X_2 curve. At the larger X_2 (region C), the condensed film of FC10OH appears and the mole fraction Y_2 is almost equal to unity: the condensed film is composed of only FC10OH molecules and C20OH molecules are expelled from the adsorbed film. This is a totally opposite situation to the one at the smaller X_2 (region B). At the intermediate X_2 (region C), the adsorbed film is in the expanded state and the Y_2 value changes with X_2. At the interfacial tensions below $37\,\text{m N m}^{-1}$ given in Figs 30(d) and (e), the adsorbed film is in a condensed state in a whole composition range and the PDA shows the heteroazeotrope. Thus the adsorbed film is constructed from only C20OH molecules below and from only FC10OH molecules above the bulk mole fraction at the azeotropic point.

Taking note of these PDAs, we draw a conclusion on the second break points observed on the γ vs m curves at $X_2 = 0.275$

and 0.280 (curves 9 and 10 in Fig. 26) as follows. Looking at the PDAs in Figs 30(d) and (e), it is seen that the heteroazeotropic point sit below $X_2 = 0.275$ at $37\,\text{mN}\,\text{m}^{-1}$ and above $X_2 = 0.280$ at $35\,\text{mN}\,\text{m}^{-1}$. Therefore it is said that the adsorbed films at these X_2 at a concentration between the first and second break points is composed of only FC10OH molecules and those above the second break point is composed of only C20OH molecules. This is totally consistent with our conclusion that the second break point corresponds to the phase transition from the FC10OH condensed state to the C20OH condensed state.

The positive azeotrope and the complete immiscibility of C20OH and FC10OH molecules in the adsorbed film come from the rather weak interaction between hydrocarbon and fluorocarbon chains. The excess Gibbs free energy of adsorption at different interfacial tensions is given in Fig. 31: the g^{HE} values are positive at all Y_2. Therefore it is said that the mutual interaction between C20OH and FC10OH molecules in the adsorbed film is very weak compared to that between the

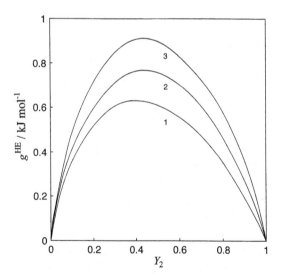

Figure 31 Excess Gibbs free energy of adsorption vs mole fraction in the adsorbed film of the C20OH–FC10OH system: (1) $\gamma = 47\,\text{mN}\,\text{m}^{-1}$, (2) 45, (3) 43.

same type. It should be noted that the g^{HE} value is much larger than that of nonionic fluorocarbon and hydrocarbon surfactant mixture given in Fig. 6. Furthermore the excess area in the adsorbed film calculated by using Eq. (44) was positive and its maximum value is about $0.1\,nm^2$ [10]. Judging from the fact that the cross-sectional areas of hydrocarbon and fluorocarbon chains are about 0.2 and $0.3\,nm^2$, respectively, and the maximum value of the corresponding excess area of the C10E4 and FC7E4 mixture is only about $0.025\,nm^2$, we realize that the mixing of C20OH and FC10OH molecules in the adsorbed film causes a great increase in the mean area per molecule. This supports strongly the weak mutual interaction between the hydrocarbon and fluorocarbon alcohols.

V. CONCLUSION

The thermodynamic equations were abstracted from our systematic studies on binary surfactant mixtures to shed light on their miscibility in the adsorbed film and micelle. The representative systems of nonionic–nonionic, ionic–nonionic, ionic–ionic with and without common ions were examined by applying the equations. It was clearly shown that the dissociation of ionic surfactants should be taken into account in the definition of the concentration variables and then in the thermodynamic equations and that the resulting equations were different from each other, depending on the combination of binary surfactants. Furthermore the criterion of the ideal mixing in the phase diagram of adsorption and that of micelle formation were proposed and examined to elucidate the deviation from the ideal mixing in terms of the activity coefficients and excess Gibbs free energy.

The methods for ordinary micelles and adsorbed films were further extended to the spontaneous vesicle formation of cationic–anionic surfactant mixture and the phase transition of hydrocarbon and fluorocarbon alcohol mixtures at the oil/water interface, respectively. In the former, it was demonstrated that the phase diagram suggested from the simple model is consistent with that predicted from the surface tension measurements and also with the visual and microscope observations. In the latter, the phase diagram of adsorption manifested that the miscibility of the alcohols were drastically changed by the phase transition in the adsorbed film.

REFERENCES

1. For example, *Curr. Opinion in Colloid Interface Sci.* **1**: 1 (1996); **2**: 243 (1997); **3**: 285 (1998); **4**: 175 (1999); **5**: **6**: 93 (2001).

2. J. R. Lu, R. K. Thomas, and J. Penfold, *Adv. Colloid Interface Sci.* **84**: 143 (2000).

3. D. N. Rubingh, *Solution Chemistry of Surfactants* (K. L. Mittal, ed.), Plenum Press, New York (1979), p.337.

4. D. Blankschtein, A. Shiloach, and N. Zoeller, *Curr. Opinion in Coll. Interface Sci.* **2**: 456 (1997).

5. I. Reif, M. Mulqueen, and D. Blankschtein, *Langmuir* **17**: 5801 (2001) (and references cited therein).

6. J. D. Hines, *Curr. Opinion in Coll. Interface Sci.* **6**: 350 (2001).

7. K. Motomura and M. Aratono, *Mixed Surfactant Systems* (K. Ogino and M. Abe, eds), *Surfactant Science Ser. 46*, Marcel Dekker, New York (1992), p. 99.

8. M. Villeneuve, S. Kaneshina, T. Imae, and M. Aratono, *Langmuir* **15**: 2029 (1999).

9. T. Takiue, T. Matsuo, N. Ikeda, K. Motomura, and M. Aratono, *J. Phys. Chem.* **102**: 4906 (1998).

10. T. Takiue, T. Matsuo, N. Ikeda, K. Motomura, and M. Aratono, *J. Phys. Chem.* **102**: 5840 (1998).

11. M. Aratono, M. Villeneuve, T. Takiue, N. Ikeda, and H. Iyota, *J. Colloid Interface Sci.* **200**: 161 (1998).

12. K. Motomura, *J. Colloid Interface Sci.* **64**: 348 (1972).

13. K. Motomura, S. Iwanaga, M. Yamanaka, M. Aratono, and R. Matuura, *J. Colloid Interface Sci.* **86**: 151 (1982).

14. K. Motomura, M. Yamanaka, and M. Aratono, *Colloid Polym.* **262**: 948 (1984).

15. J. H. Clint, *J. Chem. Soc. Faraday I* **71**: 1327 (1975).

16. B. T. Ingram, *Colloid Polymer Sci.* **258**: 191 (1980).

17. M. Aratono, T. Kanda, and K. Motomura, *Langmuir* **6**: 843 (1990).

18. Y. J. Nikas, S. Puvvada, and D. Blankschtein, *Langmuir* **8**: 2680 (1992).

19. N. Funasaki, *Mixed Surfactant Systems* (K. Ogino and M. Abe, eds), *Surfactant Science Ser. 46*, Marcel Dekker, New York (1992), p.145 (and references cited therein).

20. A. Ohta, H. Matsubara, N. Ikeda, and M. Aratono, *Colloids Surfaces A*, **183–185**: 403 (2001).

21. H. Matsuki, S. Kaneshina, N. Ikeda, M. Aratono, and K. Motomura, *J. Colloid Interface Sci.* **150**: 331 (1992).

22. J. C. Ravey, A. Gherbi, and M. Stébé, *Prog. Colloid Polym. Sci.* **79**: 272 (1989).

23. M. Villeneuve, T. Nomura, H. Matsuki, S. Kaneshina, and M. Aratono, *J. Colloid Interface Sci.* **234**: 127 (2001).

24. A. Shiloach, and D. Blankschtein, *Langmuir* **14**: 1618 (1998).

25. J. D. Hines, R. K. Thomas, P. R. Garrett, G. K. Rennie, and J. Penfold, *J. Phys. Chem. B* **101**: 9215 (1998).

26. V. M. Garamus, *Chem. Phys. Lett.* **290**: 251 (1998).

27. H. Matsubara, A. Ohta, M. Kameda, M. Villeneuve, N. Ikeda, and M. Aratono, *Langmuir* **15**: 5496 (1999).

28. H. Matsubara, A. Ohta, M. Kameda, M. Villeneuve, N. Ikeda, and M. Aratono, *Langmuir* **16**: 7589 (2000).

29. H. Matsubara, S. Muroi, M. Kameda, N. Ikeda, A. Ohta, and M. Aratono, *Langmuir* **17**: 7752 (2001).

30. M. Villeneuve, H. Sakamoto, H. Minamizawa, N. Ikeda, K. Motomura, and M. Aratono, *J. Colloid Interface Sci.* **194**: 301 (1997).

31. K. Motomura, H. Matsukiyo, and M. Aratono, In *Phenomena in Mixed Surfactant Systems* (J. F. Scamehorn, ed.), *ACS Symposium 311*, American Chemical Society, Washington, D.C. (1986), p. 163.

32. K. Motomura, N. Ando, H. Matsuki, and M. Aratono, *J. Colloid Interface Sci.* **139**: 188 (1990).

33. H. Matsuki, N. Ando, M. Aratono, and K. Motomura, *Bull. Chem. Soc. Jpn.* **62**: 2507 (1989).

34. C. Tondre, and C. Caillet, *Adv. Colloid Interface Sci.* **93**: 115 (2001).

35. T. Takiue, T. Fukuta, H. Matsubara, N. Ikeda, and M. Aratono, *J. Phys. Chem. B.* **105**: 789–795 (2001). [This is the latest paper and No. 8 of the series on the thermodynamic study on phase transition in adsorbed film of fluoroalkanol at the hexane/water interface. Other papers of this series are cited in this article.]

36. Z. Zhang, D. M. Mitrinovic, A. M. Williams, Z. Huang, and M. L. Schlossman, *J. Chem. Phys.* **110**: 7421 (1999).

2

Micro-Phase Separation in Two-Dimensional Amphiphile Systems

TEIJI KATO and KEN-ICHI IIMURA
Faculty of Engineering, Utsunomiya University,
Utsunomiya, Japan

SYNOPSIS

Micro-phase separation in binary mixed Langmuir monolayers of immiscible amphiphiles is an interesting phenomenon. In this chapter, we mainly discuss two subjects in relation to this phenomenon. The first one is the mechanism of micro-phase separation in μm size region and control of the size and shape of two-dimensional condensed phase domains. The second one is formation of surface micelles in the monolayers of partially fluorinated long-chain acids and micro-phase separation in their mixed monolayers in nm size region.

We discuss applications of these micro-phase separation structures in binary mixed monolayers to create patterned functional surfaces by transferring them to smooth solids. Applications of some of these structures and future aspects are also shown.

I. INTRODUCTION

Lateral phase-separation in insoluble (Langmuir) monolayers at the air–water interface has been a topic of long-standing interest [1–6]. Researchers are motivated in part by fundamental scientific interests in formation of domain assemblies and their shapes, and in intermolecular interactions in relation to biological phenomena, structures, and functions of cell membranes, such as a raft model [7]. Another motivation comes from the potential applications of ultra thin organic films in the field of molecular electronics and engineering. Functions can arise from defined arrangements of functionalized molecules in molecular films. In this sense, therefore, desired arrangement of film molecules in their assemblies is crucially important. However, patterns in shape, size, and distribution of micro-domain assemblies and the coexistence of contrasting properties in monolayers also have the potentiality to generate new functions.

This chapter deals with two-subjects in relation to the micro-phase separation in mixed Langmuir monolayers. The first one is microphase separation in binary mixed Langmuir monolayers of n-alkyl fatty acids and a perfluoropolyether amphiphile in μm size region. The second one is the formation of two-dimensional nanometer-size micelles in monolayers of partially fluorinated long-chain acids and micro-phase separation in their mixed Langmuir monolayers with various immiscible amphiphiles in nm size region.

We also discuss possible developments and applications of these micro-phase separated structures in Langmuir monolayers transferred to smooth solid surfaces.

II. MICRO-PHASE SEPARATION IN BINARY MIXED LANGMUIR MONOLAYERS OF *n*-ALKYL FATTY ACIDS AND A PERFLUOROPOLYETHER AMPHIPHILE [8]

Figure 1 shows the molecular structures of the film materials used in this section. These are highly purified *n*-alkyl fatty acids (Cn) (>99.9%), stearic acid (octadecanioc acid, C18), arachidic acid (eicosanoic acid, C20), behenic acid (docosanoic acid, C22), and lignoceric acid (tetracosanoic acid, C24), and a carboxyl-terminated perfluoropolyether (perfluoro-2,5,8-trifluoromethyl-3,6,9-trioxadodecanoic acid, PFPE), (>97%). Spectro-grade chloroform (Dojin Chemicals) was used as a solvent for spreading solutions. The total concentration was kept constant at 2.5 mM for all spreading solutions of mixed amphiphile systems. Monolayers were spread onto a temperature-controlled aqueous subphase surface and allowed to stand for 30 min before compression to assure complete evaporation of the solvent. Average molecular areas at spreading were 0.7, 1.5, 1.75, 2.0, 2.25, and 2.5 nm^2/molec. at mixing ratios of Cn/PFPE = 10/0, 8/2, 6/4, 4/6, 2/8, 0/10, respectively, unless otherwise specified. The subphase was 5.0×10^{-4} M aqueous solution of cadmium acetate dihydrate, adjusted to pH 7 with potassium hydrogen carbonate. Monolayers at the solution surface were

(a) $CH_3(CH_2)_{n-2}-COOH$

$$n = 18 : C18$$
$$20 : C20$$
$$22 : C22$$
$$24 : C24$$

(b) $F(CFCF_2O)_3-\overset{\displaystyle CF_3}{\underset{\displaystyle |}{C}}F-COOH$ with $\overset{\displaystyle CF_3}{\underset{\displaystyle |}{}}$

PFPE

Figure 1 Film molecules used in this section; (a) *n*-alkyl fatty acids (Cn): C18 = stearic acid (octadecanioc acid), C20 = arachidic acid (eicosanoic acid), C22 = behenic acid (docosanoic acid), C24 = lignoceric acid (tetracosanoic acid); (b) perfluoro-2,5,8-trifluoromethyl-3,6,9-trioxadodecanoic acid (PFPE).

transferred onto an atomic-flat solid surface by the horizontal scooping-up method that was devised to transfer monolayers without changing their structures [9]. The solid substrate used was a highly polished silicon wafer having a natural oxide layer. Prior to deposition, the substrate was cleaned by RCA cleaning method and then stored in ultra pure water until use. AFM observation was carried out under ambient conditions with Nanoscope III (Digital Instruments) in a tapping mode, using a $125 \times 125 \, \mu m^2$ scanning head and a silicon tip on a rectangular-shaped cantilever (129 μm long, nominal spring constant of 37–61 N/m). In all observations, the force applied to sample surface with the tip was reduced to the smallest possible value at which the surface could be imaged steadily and reproducibly. A spatial resolution of 256×256 pixels was adopted for usual imaging. However, $30 \times 30 \, \mu m^2$ images were obtained at the resolution of 512×512 pixels and binarized in order to extract information about surface structures using a digital analysis software, Win ROOF (Mitani Corp.).

A.   Monolayer Behavior at Air/Water Interface

Figure 2(a), (c) and (e) present π-A isotherms of C20/PFPE mixed monolayers at 5, 20, and 30°C, respectively. At every temperature, C20 single component monolyers underwent transition directly from the gaseous/solid co-existent state to the solid state during compression. Brewster angle microscopic (BAM) observations of the C20 monolayers indicated that bright macroscopic condensed phase domains were formed in the dark gaseous phase just after spreading. The macroscopic domains were, then, forced to fuse by compression, leading to disappearance of the gaseous phase and a simultaneous steep increase of surface pressure at around the cross-sectional molecular area of *n*-alkyl chains. Both the fully condensed characteristics in the isotherm and the spontaneous formation of huge condensed domains are indicative of a strong attractive interaction among C20 molecules. It is noteworthy that C20 exists as its cadmium salt dimmer under the present experimental conditions, referring to the previous work concerning salt formation of the fatty acids in Langmuir monolayers [10].

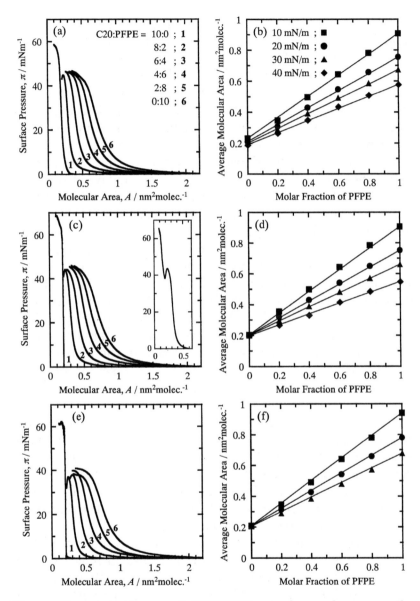

Figure 2 π-A isotherms of C20/PFPE mixed monolayers on the 0.5 mM Cd^{2+} subphase (left), and average molecular area as a function of the molar fraction of PFPE at given surface pressures (right).

In contrast to C20, PFPE formed completely expanded monolayers at the air/water interface over the temperature range examined. This extremely expanded state of PFPE monolayers can be explained by a relatively strong affinity of the hydrophilic groups to water and a weak cohesive energy certainly due to the branched perfluoropolyether groups in PFPE molecules. The broad breaks at around 40 mN/m correspond to a gradual collapse of the monolayers. Cadmium salt formation of PFPE was confirmed by measurement of X-ray photoelectron spectroscopy (XPS) of one-layer LB film on a silicon wafer. The atomic concentration ratio obtained from the spectrum is $F : Cd = 97 : 3$, in good agreement with the value expected for $Cd(PFPE)_2$ dimer.

The isotherms of mixed monolayers changed systematically depending on the mixing ratio of the two components. With increasing molar fraction of C20, the isotherm shifted to the lower molecular area side. However, it should be noted here that these isotherms do not show the whole behavior of the mixed monolayers, because the lower molecular area regions of the isotherms could not be measured due to the instrumental limitations of our Langmuir trough. An inset of Fig. 2(c) presents a π-A isotherm of a C20/PFPE (8/2) mixed monolayer at 20°C, measured while starting compression at $0.6\,nm^2$/molec. One can see a two-step collapse of the mixed monolayer, corresponding to collapses of the two pure component monolayers. This can be interpreted as independent contributions of the individual components to the monolayer properties; an expanded region of the isotherms is governed by PFPE component, and after collapse of PFPE component, the feature of C20 emerges at a lower molecular area.

Average molecular areas experimentally determined at constant surface pressures are plotted against the mole fractions of PFPE in the mixed monolayers in Fig. 2(b), (d) and (e) at 5, 20, and 30°C, respectively. These results together with the two-step collapse characters of the mixed monolayer, lead to the conclusion that C20 and PFPE are completely immiscible in the mixed monolayers in the whole composition range.

π-A isotherms were measured for all other combinations at every mixing molar ratio of Cn/PFPE at 5, 20, and 30°C.

The behavior observed for mixed monolayers of the combinations was essentially the same as that for C20/PFPE monolayers. At all temperatures examined, the Cns showed only a solid film region in the isotherm as the stable two-dimensional phase, and the isotherms of Cn/PFPE mixtures revealed regular shifts with change in the mixing ratio, resulting in a linear correlation between the average molecular area at constant surface pressure and the composition of the mixed monolayer.

B. Surface Morphologies of Mixed Monolayers of Cn/PFPE Observed by AFM

In the present monolayer systems, the experimental parameters that would affect the monolayer morphology are mixing ratio, water surface temperature, n-alkyl chain length of Cn and molecular area (surface pressure). Three parameters are varied widely and systematically, changing one variable with the other parameters fixed, in order to obtain an entire insight into the phase separation processes in our mixed monolayers.

Mixing ratio dependency of the surface morphology was examined for C20/PFPE mixed monolayers transferred at $0.7\,\text{nm}^2$/molec. at 20°C. AFM images in Fig. 3 demonstrate that the two components are in a micro-phase separated state. Apparently, surface coverage with island-like circular domains increased with an increase in the molar fraction of C20, suggesting that brighter (higher) domains and a dark (lower) surrounding continuous region are attributed to the condensed

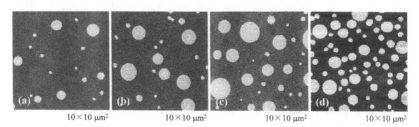

$10 \times 10\ \mu\text{m}^2$ $\quad$ $10 \times 10\ \mu\text{m}^2$ $\quad$ $10 \times 10\ \mu\text{m}^2$ $\quad$ $10 \times 10\ \mu\text{m}^2$

Figure 3 $10 \times 10\,\text{mm}^2$ AFM images of C20/PFPE mixed monolayers at 20°C; C20/PFPE = (a) 2/8, (b) 4/6, (c) 6/4, (d) 8/2. Monolayers were transferred on Si wafer at $0.7\,\text{nm}^2$/molec.

phase of C20 and the expanded phase of PFPE, respectively. The cross-sections of the images indicated that the C20 domains are 1.5 ± 0.1 nm higher than the surrounding PFPE monolayer. According to Sugi *et al.*, the bilayer long spacing, d (nm), in Langmuir–Blodgett (LB) films of cadmium salts of n-alkyl fatty acids increases with the number of carbon atoms, n, as $d = 0.53 + 0.25\,n$ [11]. Assuming the thickness of C20 domain to be a half of the bilayer long-spacing, it should be 2.76 nm, and the thickness of the surrounding PFPE phase is calculated to be about 1.2 nm.

Dependency of the morphology on water surface temperature and the chain length of Cn was examined for Cn/PFPE (2/8) mixed monolayers transferred at 0.7 nm^2/molec. AFM images are summarized in Fig. 4. It turned out that the phase-separated structures strongly depend on the two experimental parameters. Although C18/PFPE (2/8) mixture formed circular-shaped condensed phase domains irrespective of water surface temperature, irregular-shaped domains were formed in C20/PFPE (2/8) monolayer at 5°C. This shape deformation was already seen at 20°C in C22/PFPE (2/8) monolayer, and temperature lowering led to the formation of elongated narrow domains of C22. For C24/PFPE mixed monolayers, circular domains were no longer observed even at 30°C. Evidently, the branched elongated domains are formed for longer chain Cns and/or at lower temperatures, and the circular domains preferentially appear for shorter chain Cns and/or at higher temperatures. Irregular shaped domains are seen in the intermediate conditions. It is also found that the spatial frequency of occurrence varies with change in the domain shape. We can see that a straight line drawn from end to end of the image meets more times to the condensed phase domains for a branched elongated shape than for a circular one.

C. Structural Regimes of Cn/PFPE (2/8) Mixed Monolayers

The findings described above lead to an idea that the monolayer structures can be characterized by some quantities which

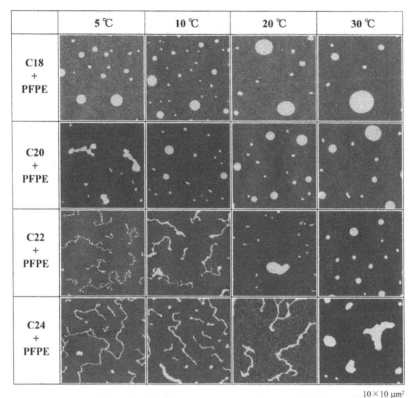

$10 \times 10 \ \mu m^2$

Figure 4 $10 \times 10 \ mm^2$ AFM images of Cn/PFPE (2/8) ($n = 18$, 20, 22, 24) mixed monolayers at 5, 10, 20, and 30°C. Monolayers were transferred on Si wafer at $0.7 \ nm^2$/molec.

reflect the distribution of phases and the shape of condensed phase domains. Characteristic length and fractal dimension, determined from binarized $30 \times 30 \ \mu m^2$ images of Cn/PFPE (2/8) mixed monolayers, are such quantities and are plotted as a function of temperature for each mixing combination in Fig. 5(A) and (B), respectively. Each of the data is the average of values obtained from at least 10 microscopically separate areas on at least two samples. Here, we define the characteristic length, λ, as the sum of the average length of white segments (Cn phase domains) and that of black segments (PFPE phase regions) on vertical and horizontal scanning lines

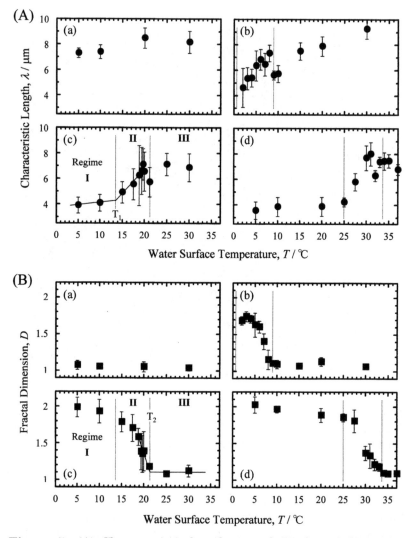

Figure 5 (A) Characteristic length, λ, and (B) fractal dimension, *D*, for Cn/PFPE (2/8) mixed monolayers as a function of water surface temperature; Cn = (a) C18, (b) C20, (c) C22, and (d) C24. Each datum was the average of values obtained from at least ten AFM images (30×30 mm^2) taken at macroscopically separated areas on separately prepared two samples. The monolayer deposition area was 0.7 nm^2/molec. The structural regimes are designated in the plots for the C22/PFPE (2/8) mixed monolayer for clarity.

(512 × 512 lines) of the binarized images. The fractal dimension, D, is determined by applying the definition [12,13]

$$S^{1/2} \propto P^{1/D} \tag{1}$$

where S and P are the area and perimeter of condensed domains, respectively. Examples of the double-logarithmic plots of S vs P are shown in Fig. 6 for (a) C18/PFPE (2/8) monolayer at 20°C and for (b) C24/PFPE (2/8) monolayer at 20°C. We find that a linear relation between log S and log P is well established for both monolayers in a wide range of the plots. As shown below, the fractal dimension takes a value close to unity for the round-shaped domains with smooth boundaries, whereas it increases with an increase of the perimeter for elongated domains of the same areas.

Figure 5 implies that the two indices reflect well the structural change of Cn/PFPE (2/8) mixed monolayers, depending on the chain length and temperature. We can divide the structural evolution into at most three regimes according to changes of λ and D values. The three regimes appear in the plots for C22/PFPE (2/8) monolayers and are designated in Fig. 5 for clarity. The regime boundary temperatures,

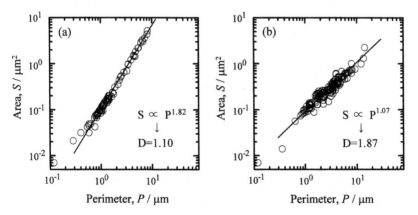

Figure 6 Double-logarithmic plots of area and perimeter of condensed phase domains for determination of the fractal dimension; (a) C18/PFPE (2/8) monolayer at 20°C, (b) C24/PFPE (2/8) monolayer at 20°C.

T_1 and T_2, are experimentally determined by taking the intersection of two approximated lines in the λ vs T plot and in the D vs T plot, respectively, as drawn in the figures. The boundary temperatures obtained for C22/PFPE (2/8) mono-layers are $T_1 = 13.4°C$ and $T_2 = 21.3°C$. The temperature range below T_1 is referred to the regime I where λ takes values as small as about 4 μm and D keeps the largest values close to 2. These reflect the characteristic surface structure of the branched narrow condensed domains distributed with a high spatial frequency. A temperature region above T_2 is defined as the regime III where a low population of round-shaped domains with smooth boundaries characterizes the structure. The regime II corresponds to a transition region between the regimes III and I. Shortening the chain length of Cn leads to a shift of the regimes to the low temperature side. For C20/PFPE (2/8) monolayers, the regime I moves to a temperature range below the range studied, and only the regimes II and III emerge. The T_1 and T_2 are estimated to be 0.6°C and 9.0°C, respectively. C18/PFPE (2/8) monolayers show only the regime III, indicating that the condensed phase domains have only a circular shape. The boundary temperatures for C24/PFPE (2/8) mixed monolayers are $T_1 = 25.8°C$ and $T_2 = 33°C$. Again, we observe a regular shift of the regimes by about 12°C with decrease in two methylene units in the alkyl chain, indicating a regular progression in the structural evolution.

D. Mechanism of the Micro-Phase Separation in Cn/PFPE Mixed Monolayers

The equilibrium size and shape of two-dimensional domains are determined by competition between the line tension at the two-phase boundary and long-range electrostatic repulsive forces within the domains [14,15]. If the line tension is high enough, the domains would take a circular shape in order to reduce their perimeter since there is a large energetic cost necessary to increase the boundary length. A relative decrease of the line tension with respect to the electrostatic forces allows an increase of the domain boundary/area ratio, leading to an elongation of the domains or formation of stripe-shaped domain structures. Such characteristic morphologies have been observed in mixed monolayers containing tension-lowering materials such as cholesterol [16,17] and an amphiphilic protein, SP-B [18], using fluorescence microscopy. The observations have clarified that the width of elongated domains is influenced by experimental parameters such as molecular area, the concentration of line tension lowering material in the mixtures, pH and ionic strength of the subphase, and temperature. A noteworthy finding is the effect of temperature observed for L-α dimirystoylphosphatidic acid (DMPA) monolayer containing 2 mol% cholesterol and 1 mol% fluorescent dye amphiphile [17]. In this monolayer system, a lowering of water temperature at a fixed molecular area causes a reversible deformation of domain texture; initially circular condensed phase domains of highly charged DMPA change to stripes via banana shape accompanying a continuous change of the domain width from about 22 to 2 μm. This elongation is ascribed to an increase in electrostatic repulsive force due to enhanced alignment of polar DMPA molecules at low temperatures. We have also observed an elongation of Cn domains at low temperatures in our mixed monolayers. This phenomenon itself is very similar to a change induced by in-plane dipole moments. However, no similarity in formation mechanisms exists as evidence by the observations for C24/PFPE (2/8) monolayers shown in Fig. 7. The monolayers in Fig. 7 were spread at 5°C and heated up to 30°C at the water surface at a heating rate of 1°C/min after compression

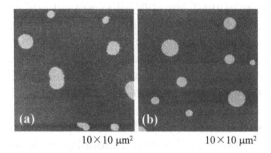

10×10 μm^2 10×10 μm^2

Figure 7 10×10 mm^2 AFM images of C22/PFPE (2/8) mixed monolayers deposited (a) after heating from 5°C to 30°C and (b) after cooling from 30°C to 5°C, at rates of $\pm$ 1.0°C/min at 0.7 nm^2/molec.

up to $0.70 \, \text{nm}^2/\text{molec}$. AFM images reveal that heating makes those originally elongated domains the round ones, but the originally circular domains at high spreading temperatures are never elongated or branched by cooling. This observation demonstrates that decrease of water temperature does not produce enough electrostatic repulsive forces for elongation of the condensed domains in our mixed monolayers. The phase-separated structures in Cn/PFPE mixed monolayers are governed predominantly by the line tension. The relatively small electrostatic contribution should be caused by charge neutralization of the carboxyl groups of the amphiphiles through cadmium salt formation.

The line tension arises from a difference in the inter-molecular interactions between two adjacent phases. At any temperatures, since the Cns and PFPE form a fully condensed phase and a fully expanded phase, respectively, there should be an essentially high line tension at the domain boundary, resulting in formation of circular domains. Additionally, at a constant temperature, the line tension is expected to increase with increasing chain length of Cn, and thus round-shaped domains should appear preferentially in binary mixed mono-layers containing longer chain Cn as one component of the film. However, it is apparent that the surface morphology observed in the present work does not always follow this theoretical expectation.

It seems reasonable to assume that the mobility of Cn molecules on the water surface is drastically reduced by complete evaporation of the spreading solvent and simultaneous incorporation of the molecules into condensed phase domains. If the Cn molecules still have a sufficient mobility in the con-densed phase domains, the domains should take a round shape to minimize the perimeters. However, under the conditions where molecular motions are extremely suppressed, the domain structure would be frozen immediately after complete evapora-tion of the spreading solvent. The most plausible mechanism of structural growth in the mixed monolayers is schematically shown in Fig. 8. At an early stage of the phase separation, a characteristic structure consisting of linearly developed branched Cn phase domains surrounded by a homogeneous

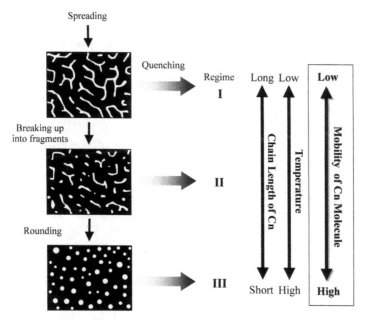

Figure 8 Schematic diagram showing the evolution of phase-separated structures in the binary mixed Langmuir monolayer of Cn/PFPE.

PFPE expanded phase is formed. If Cn molecules were sufficiently mobile at this stage, the elongated domains would become circular or be broken up into circular pieces due to the line tension, a two-dimensional version of the capillary instability. However, when the mobility is not sufficient, the structure would be frozen at a certain stage during the deformation processes because of complete solvent evaporation. The mobility of film molecules is largely dependent upon the chain length of Cn and the temperature of water surface, and increases in the order of the regime I to III.

As described above, the phase-separated structures should be developed from the elongated domains, but the genetic mechanisms for the phase separation still remains open to question. There are two possible mechanisms for the phase separation in general, the nucleation and growth (NG) mechanism and the spinodal decomposition (SD) mechanism [19]. By whichever mechanism the phase separation takes place, there exists a possibility for elongated domains to emerge with completely immiscible components. If the phase separation occurs through the NG mechanism, the only way for elongated domains to be formed is that the phase-separation starts with the first exclusion of PFPE component from the spreading solution. The affinity between a film material and a solvent can be evaluated by the solubility parameter (δ). Closer δ values of the materials indicate a better affinity of the combination. The δ values estimated for chloroform, C18, C20, C22, C24 and PFPE are 9.2, 8.9, 9.1, 9.3, 9.5 and 6.0 $(\text{cal/ml})^{1/2}$, respectively, suggesting that PFPE can be preferentially segregated from the spreading solution with solvent evaporation. When PFPE molecules are excluded from the solution first, they should start to take an expanded state while pushing the remaining solution, like a two-dimensional bubble growing, resulting in elongated branching development of Cn domains after evaporation of the solvent. On the other hand, it is also true that the characteristic structure in the regime I is very similar in appearance to the structure expected to be formed by the SD mechanism. As is well known in the field of two component polymer systems having upper or lower consolute temperatures or two-component metal alloys, the SD mechanism produces very characteristic morphologies, so-called bicontinuous structures, by temperature jump [20,21] or rapid evaporation of co-solvents [22,23]. If the SD works in mixed spread monolayer systems, it should produce a very thin layer of spreading solution during the rapid evaporation of the solvent. Cn component and PFPE component separate into a two-dimensional bicontinuous structure. Then, PFPE component expands and Cn component shrinks to form the structure in the regime I.

We cannot decide yet which mechanism works during the spreading of mixed monolayers of completely immiscible two components. Irrespective of the mechanism, however, the elongated domain structure formed in the regime I is regarded as a nonequilibrium one frozen at the initial stage of the structure development because of a low mobility of Cn molecules under the experimental conditions.

We have previously reported a micro-phase separation in binary mixed Langmuir monolayers of combinations of a fluorinated silazane derivative, which forms an expanded monolayer, and trichloro(octadecyl)silane, which forms a condensed monolayer [24]. The structural change in micro-phase separation of these systems is almost the same as that described above. Ohishi *et al.* have reported that even for combinations of alkanoic acids, their mixed Langmuir mono-layers undergo micro-phase separation when the difference of methylene number between the components is larger than 6 [25]. These are examples of micro-phase separation in mixed monolayers of combinations of condensed–condensed components.

Thus, the micro-phase separation phenomena in binary mixed monolayers are rather trivial irrespective of the combination of components wherever the components are immiscible in monolayers with each other.

III. MONOLAYERS OF PARTIALLY FLUORINATED LONG-CHAIN ACIDS

A. Formation of Surface Micelles

Existence of surface micelles in Langmuir monolayers has been repeatedly discussed since Langmuir proposed the first idea in the 1930s [26–28]. However, no one has succeeded in observing surface micelles because of the experimental difficulties. Mann and his collaborators have reported formation of surfactant hemi-micelles on the hydrophobic solid surfaces in aqueous solutions of amphiphiles based on their AFM observations [29]. Ducker and his collaborators have also observed hemi-cylindrical sodium dodecyl sulfate micelles adsorbed onto a graphite surface from aqueous solutions, using AFM [30,31]. Eisenberg and his group have reported surface micelle formation at the air/water interface from diblock copolymers of hydrophilic and hydrophobic components. They investigated aggregation numbers of surface micelles by analyzing TEM observed images [32–34]. Chi and her collaborators have found that stearic acid monolayers exhibit a phase transition from an expanded to a condensed state on the subphase containing a water-soluble polycation, polyimine. They observed that many micro-grains of 30–50 nm and macro-grains of a few μm coexist in the transition region [35,36].

We have found that some partially fluorinated long-chain acids form surface micelles in the Langmuir monolayers [37,38]. Table 1 lists the molecular structures, chemical names and abbreviated names of the materials used in this study. The abbreviated names of the materials are based on the total carbon number and total fluorine number in the molecules. The structure of these materials changes systematically. Materials 1, 2, and 3 have the same fluorocarbonic group, whereas the number of methylene groups in the hydrocarbon chain increases by six from 1 to 2 and from 2 to 3. For materials 3, 4 and 5, fluoromethylene number decreases by two from 3 to 4 and from 4 to 5 whereas methylene number is the same. Figure 9 displays molecular models of the amphiphiles calculated as the energetically stable structures.

Figure 10 shows π-A isotherms of C19F17 monolayers on the subphases containing (a) K^+, (b) Cd^{2+} and (c) La^{3+} ions at

Table 1 Materials used in Langmuir monolayer studies [37,38].

Materials	Chemical structures	M.W.	m.p./°C	Abbreviation
Heptadecafluorononanedecanoic acid	$CF_3(CF_2)_7(CH_2)_{10}COOH$	604	90	C19F17
Heptadecafluoropentadecanoic acid	$CF_3(CF_2)_7(CH_2)_{16}COOH$	689	98	C25F17
Heptadecafluorotriacontanoic acid	$CF_3(CF_2)_7(CH_2)_{22}COOH$	773	105	C31F17
Tridecafluorononacosanoic acid	$CF_3(CF_2)_5(CH_2)_{22}COOH$	673	95	C29F13
Nonafluoroheptacosanoic acid	$CF_3(CF_2)_3(CH_2)_{22}COOH$	573	87	C27F9
Perfluoroundecanoic acid	$CF_3(CF_2)_9COOH$	454	98	C11F21

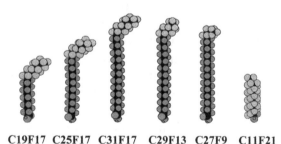

C19F17 C25F17 C31F17 C29F13 C27F9 C11F21

Figure 9 Molecular models of the partially fluorinated acids and perfluoroundecanoic acid.

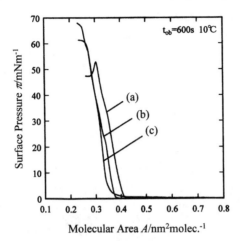

Figure 10 π- A isotherms of C19F17 on the subphase containing (a) K^+ ions, (b) Cd^{2+} ions, (c) La^{3+} ions, at pH 7.

the concentration of $5.0 \times 10^{-4}\,M$ at pH 7.2 and 10°C. With increasing valency of the metal ion in the subphase, the molecular area at the condensed state decreased due to formation of metal salt of the amphiphile. Figure 11(A), (B) and (C) show AFM images of the corresponding monolayers in Fig. 10, transferred at 35 mN/m. There is almost no contrast in structure in image (A), while we can see formation of uniformed-sized surface micelles in images (B) and (C). However, the monolayer condenses with K^+ at pH 11, and we can observe surface micelles of almost the same size in image (B), the shape of which is rather irregular. Figure 11(D) is a power spectrum of the signal along a line in image (C) that gives a sharp peak corresponding to a uniform size of the micelles, 18 nm. Image (E) is a two-dimensional Fourier transform spectrum of the image data of (C), showing a ring that corresponds to a Debye–Scherrer ring of powder X-ray diffraction. This also means a uniform size of the micelles, but they have no long-distance order. Image (F) is the result of an auto-correlation analysis of AFM image (C). This image suggests that positional correlation of micelles extends over only a few at most. Figure 12(A) is an AFM image of C19F17 monolayer on the lanthanum ion subphase transferred at

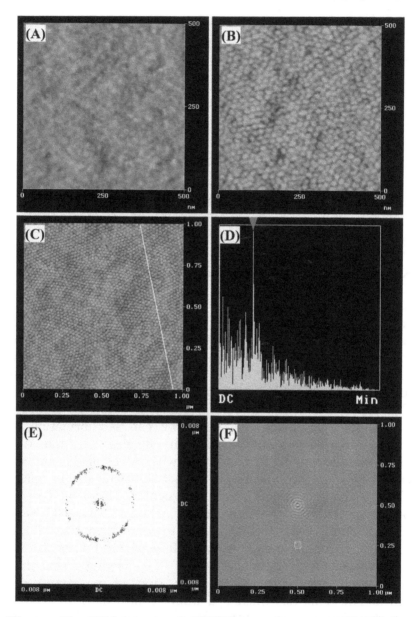

Figure 11 AFM images of C19F17 monolayers on (A) K$^+$ ion subphase, (B) Cd^{2+} ion subphase, (C) La^{3+} ion subphase, transferred at 35 mN/m, (D) a power spectrum along a line in (C), (E) a two-dimensional Fourier transform spectrum of image (C), (F) an autocorrelation analysis of the image (C).

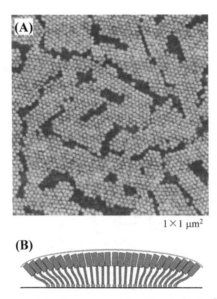

Figure 12   (A) An AFM image of C19F17 on La^{3+} ion subphase, transferred at 0.45 nm^2/molec. (B) A schematic model of a surface micelle.

0.45 nm^2/molec. (Refer the π-A isotherm c in Fig. 10.) At this point, the micelles gather to form islands of assembly where the micelles are arranged in a rather ordered manner. (B) is a schematic model of a micelle of C19F17.

Figure 13 shows AFM images of one layer LB films of (A) C19F17, (B) C25F17, (C) C31F17, (D) C29F13, and (E) C27F9, transferred at the points just before the pressure starts rising on the isotherm. With increasing number of methylene groups in hydrocarbon chain from C19F17 to C31F17 through C25F17, the size of micelles increases from 18 nm to about 70 nm and the shape changes from circular to irregular. C29F13 and C27F9 also form surface micelles of the size near 100 nm. The most characteristic feature of micelles of C29F13 and C27F9 is that they have a hole of the order of 10 to 20 nm at the center. Since we used etched silicon cantilevers of the nominal radius of tip-top curvature of 20 nm, there is a possibility that the hole size might actually be larger than that imaged. Cross section

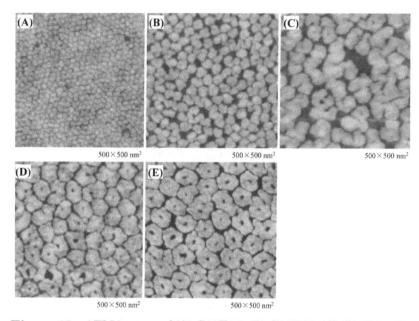

Figure 13 AFM images of (A) C19F17, (B) C25F17, (C) C31F17, (D) C29F13, (E) C27F9 monolayers on the Cd^{2+} ion subphase at 10°C, transferred at the points just before pressure rising.

profiles of the images revealed that the thickness of surface micelles is in all cases less than the fully extended molecular length of the amphiphile. This means that molecules are tilted to some degree in the micelles on average. We have measured ex situ and in situ FT-IR spectra of one-layer LB films on the smooth solid substrate and on the water surface. Polarization modulation infrared reflection absorption spectrum measurements revealed that the hydrocarbon part of amphiphile takes *trans* planar zigzag conformation and tilted on average around 25° in surface micelles against the normal to the water surface [39,40].

The mechanism of surface micelle formation for partially fluorinated long-chain acids in Langmuir monolayer is as follows. There are strong attractive interactions among the planar zigzag hydrocarbon chains and the molecules gather to form a condensed phase aided with the condensation effect of salt formation with multi-valent metal ions, while the bulky fluorocarbon groups at the top of molecules inhibits the amphiphiles from approaching the distance of strong van der Waals attractive interactions. The balance of these antagonistic effects determines the size of surface micelles. Thus, with lengthening hydrocarbon part from C19F17 to C31F17 through C25F17 with the same fluorocarbon top, the size of surface micelles increases due to an increased attractive interaction among the hydrocarbon chains. C29F27 and C27F9 form rather large micelles because of the weak inhibiting effect of their shorter fluorocarbon groups. We do not know yet why there is a hole in the center of a micelle for C29F13 and C27F9, but forming central hole may be one possible mechanism to reduce the inhibiting effect of the bulky fluorocarbon heads in surface micelles.

B. Micro-Phase Separation in Mixed Monolayers

Figure 14 shows AFM images $(3 \times 3\,\mu m^2)$ of mixed mono-layers of C19F17 and alkanoic acids (1/1) on the subphase containing Cd^{2+} ions at 10°C; (A) behenic acid, (B) arachidic acid, (C) stearic acid, and (D) palmitic acid. These are examples of condensed–condensed combination monolayers. Figure 14(E) is an enlarged AFM image of (B) $(1 \times 1\,\mu m^2)$. In all cases, the condensed phase micro-domains of alkanoic acids are surrounded by C19F17 surface micelles. The size and shape of the micro-domains change with the length of alkyl groups. Thus, the domains of behenic acid take an irregular but angular shape, those of arachidic acid a hexagonal shape, and those of stearic and palmitic acid a circular shape. The size of domains increases with decreasing alkyl chain length from less than 200 nm for behenic acid to larger than 500 nm for larger domain of palmitic acid. Z. F. Liu and his collaborators have reported micro-phase separation in mixed monolayers of sodium octadecanesulfate (C18S) and perfluorononanoic acid (C9F) [41]. Their system also exhibits hexagonal domains of C18S surrounded by C9F phase due to mutual phobicity of the components and segregation of the condensed component from the expanded one. It is reasonable to consider that

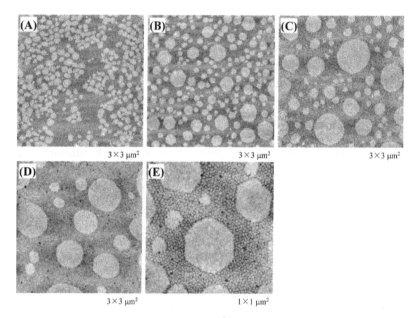

Figure 14   AFM images ($3 \times 3\ \mu m^2$) of mixed monolayers of C19F17 with alkanoic acids (1/1) on the subphase containing Cd^{2+} ions at 10°C: (A) behenic acid, (B) arachidic acid (C) stearic acid and (D) palmitic acid. (E) is an enlarged AFM image of (B) ($1 \times 1\ \mu m^2$).

two-dimensional crystals grow in a hexagonal shape when the alkyl groups of the amphiphile take a hexagonal crystal lattice in two dimensions as a rotator. Whether crystal growth yields two-dimensional lattices of orthorhombic, rectangular parallel, or hexatic system depends on the chain length of the alkanoic acid, temperature, and surface pressure [42]. Growth of hexagonal micro-crystals of arachidic acid shown in Fig. 14(B) or (E) is permitted in a very narrow window of the experimental conditions.

When the partially fluorinated acid is mixed with a suitable amphiphile which forms an expanded type monolayer and is immiscible in monolayer with the main component, micro-phase separation in nm size and an ordered arrangement of micelles in long distance over several hundreds nm can be achieved [43].

Figure 15(A) shows an AFM image of a mixed monolayer of C19F17 and perfluoro-undecanoic acid (C11F21) (9 : 1)

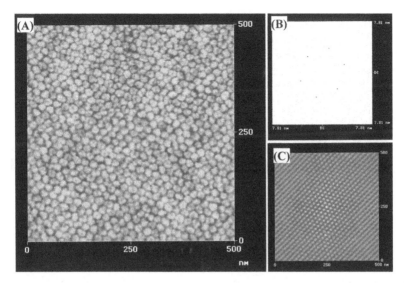

Figure 15 (A) an AFM image of a mixed monolayer of C19F17 and C11F21 (9 : 1) on the subphase containing 5.0×10^{-4} M cadmium acetate at pH 7 and 10°C, (B) a two-dimensional Fourier transform spectrum of image data of (A), (C) an auto-correlation analysis of the image (A).

on the subphase containing 5.0×10^{-4} M cadmium acetate at pH 7 and 10°C. Surface micelles of C19F17 are arranged in two-dimensional hexagonal arrays. This is confirmed more clearly in (B), a two-dimensional Fourier transform spectrum of the image data of (A). There are clear six points on the hexagon, suggesting a hexagonal array of micelles and the sharp monodispersity of their size. Figure 15(C) is the result of an auto-correlation analysis of the image (A), which indicates a positional order of micelles from the centered one extends over nearby twenty micelles.

Figure 16(A) is a schematic model of the micro-phase separation mechanism in mixed monolayers of C19F17 and an alkanoic acid that forms a condensed monolayer. With evaporation of the spreading solvent, C19F17 forms surface micelles and, at the same time, the alkanoic acid is segregated to form condensed micro-domains due to the mutual phobicity of the two components. Figure 16(B) shows a schematic model of the

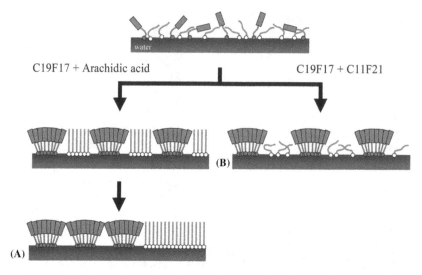

Figure 16 A schematic formation mechanism of micro-phase separation structures in mixed monolayers of C19F17 with alkanoic acids (condensed-condensed combinations) and with C11F21 (a condensed-expanded combination).

ordering of C19F17 micelles through mixing with an immiscible amphiphile such as C11F21 that forms an expanded monolayer. The expanded monolayer disturbs coagulation of micelles and facilitates ordering by weakening the interaction and endowing micelles with thermal mobility. When we use a chemically reactable amphiphile as the surrounding phase, we can create a functional surface of regular pattern in nm size. This leads to the development of molecular nano-technology.

IV. FUTURE ASPECTS OF THE MICRO-PHASE SEPARATED STRUCTURES IN MIXED MONOLAYERS AND THEIR APPLICATIONS

Controlled construction of supra-molecular architectures through self-organization processes in molecular films has been a challenging object in surface science. Several approaches have been attempted by utilizing phase separations in Langmuir monolayers. For example, Frommer *et al.* used a phase-separated surface of circular hydrocarbon domains surrounded by a fluorocarbon sea in the one layer Langmuir–Blodgett (LB) film for adsorption of tobacco mosaic virus [44]. They found the virus preferentially adsorbs to the hydrocarbon domains or the side wall of the domains where the methylene groups are exposed. This suggests a possibility of patterned arrangement of nano-particles, macromolecules, proteins or cells by the use of the phase-separated monolayers transferred to smooth solid surfaces. Duschl *et al.* reported a combined fabrication route using a mixed Langmuir monolayer of palmitic acid and a thio-lipid [45]. Upon compression, the monolayer starts to de-mix into condensed phase domains predominantly composed of palmitic acid and an expanded phase of the thio-lipid. The average size and surface density of condensed domains depend on the experimental conditions such as mixing ratio, surface pressure, temperature, and ionic strength. By deposition onto a gold substrate, the palmitic acid component in condensed domains physisorbs, but the thio-lipid component covalently binds to the gold surface. After removal of palmitic acid by rinsing with an organic solvent, newly exposed gold surface can be used for chemisorption of suitable thiols having pendant functional groups to catch proteins, nano-size metal clusters, metal complexes, and so on. Another construction rout using phase-separated structures in Langmuir monolayers was reported by Fang and Knobler [46]. They showed formation of condensed phase micro-domains of *n*-alkyl silanes in a LB monolayer transferred on mica from a liquid-condensed state of a spread monolayer. The domain formation in this case is ascribed to dewetting of water film after monolayer deposition, and the domain structure is altered by changing deposition conditions like transfer speed or surface pressure. The domain structures are then exposed to chemisorption of a fluorinated silane compound on the gaps among the hydrocarbon islands. Inversely, monolayers of fluorinated silanes were also used as precursors for self-assembly of the alkyl-silane. Proteins, such as bovine serum albumin adsorbed preferentially to methyl-terminated regions of the phase-separated surfaces.

Figure 17 schematically shows an example of a rout for formation of reactive template surfaces using the micro-phase separation processes [47]. Figure 17(A) shows the formation of a micro-phase separated monolayer by mixing a reactive component such as octadecyl-triethoxy-silane (ODTES) that forms a condensed monolayer and PFPE which forms an expanded monolayer. After transfer the micro-phase separated monolayer to the smooth solid surface such as silicon wafer, ODTES reacts with surface hydroxyl groups and fixed by chemical bondings (B). Then, the surrounding phase of PFPE is washed out by rinsing with a suitable organic solvent (C). Another reactive functional molecules such as aminopropyl-triethoxysilane (APTES) are adsorbed to the vacant surface of the solid from a solution, and APTES reacts with the solid surface (D) by the thermal treatment. Then, this template reactive surface is exposed to a solution of nanometer size gold clusters. The gold clusters are captured by the pendant amino groups through strong chemical interactions among them (D). Figure 18 shows the resultant surface with adsorbed gold clusters having the average size of 20 nm on the template structure. As clearly shown, gold clusters adsorb to the amino group region very densely as a mono-particulate layer and the boundary is very sharp.

Thus, micro-phase separated structures having controlled shape and size described in this chapter can also be used to develop functionalized surfaces if we use chemically reactable amphiphiles such as alkyl chlorosilanes, methoxy- or ethoxy-silanes or alkyl-thiols, as one component of immiscible combinations in mixed Langmuir monolayers. This will lead to the controlled construction of two-dimensional supra-molecular architectures in nm to μm size.

V. SUMMARY

In all cases described above, the driving force of the micro-phase separation in binary mixed Langmuir monolayers is the generation of the mutual phobicity of the components accompanied by the evaporation of the solvent during

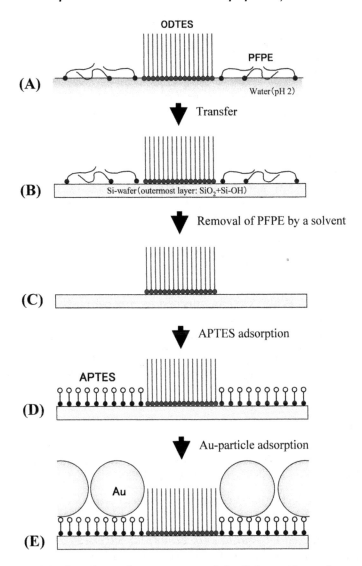

Figure 17 A schematic process model of formation of reactive template surfaces using the micro-phase separation processes. (A) Formation of a micro-phase separation structure. (B) Transfer and formation of chemical bonding to a solid surface. (C) Washing of the non-bonded component with an organic solvent. (D) Adsorption of a functionalized component, (E) Adsorption of gold nano-clusters to the APTES region.

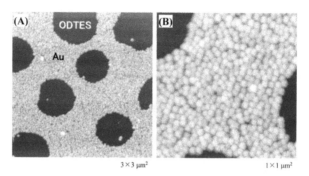

3 × 3 μm²   1 × 1 μm²

Figure 18 AFM images of a surface adsorbed by gold clusters having average size of 20 nm to the template structure of APTES.

spreading and/or segregation of one component from another during spreading when the difference in cohesive energy density between the components is larger than a certain critical value. When we can succeed in controling the shape and size of micro-phase separated structures freely, we can use the structures for construction of functionalized surfaces in mesoscopic to nm size. These functionalized surfaces will be very effective for construction of supra-molecular architectures.

REFERENCES

1. G. L. Gains, *Insoluble Monolayers at Liquid-Gas Interfaces*, Interscience Publishers, New York (1969).

2. G. G. Roberts (ed.), *Langmuir-Blodgett Films*, Plenum Press, New York (1990).

3. A. Ulman (ed.), *Organic Thin Films and Surfaces—Directions for the Nineties*. Academic Press, San Diego (1995).

4. C. M. Petty, *Langmui-Blodgett Films, An Introduction*, Cambridge University Press, Cambridge (1996).

5. K. S. Birdi, *Self-Assembly Monolayer Structures of Lipids and Macromolecules at Interfaces*, Kluwer Academic/Plenum Publishers, New York (1999).

6. M. Seul, and D. Andelman, *Science* **267**: 476–483 (1995).

7. T. Harder, and K. Simon, *Curr. Opin. Cell Biol.* **9**: 534 (1997).

8. K. Iimura, T. Shiraku, and T. Kato, *Langmuir*, **18**: 10183–10190.

9. T. Kato, N. Matsumoto, and M. Kawano, *Thin Solid Films* **242**: 223–228 (1994); **273**: 232–235 (1996).

10. K. Kobayashi, T. Takaoka, and S. Ochiai, *Thin Solid Films* **159**: 267–273 (1988).

11. A. Matsuda, M. Sugi, T. Fukui, S. Iizima, M. Miyahara, and Y. Otsubo, *J. Appl. Phys.* **48**: 771–774 (1977).

12. S. Lovejoy, *Science* **216**: 185–187 (1982).

13. Z.-Y.,Wang, M. Konno, and S. Saito, *J. Chem. Eng. Jpn.* **24**: 256–258 (1991).

14. H. M. McConnel, and V. T. Moy, *J. Phys. Chem.* **92**: 4520–4525 (1988).

15. H. M. McConnel, and R. De Koker, *J. Phys. Chem.* **96**: 7101–7103 (1992).

16. M. R. Weis, and H. M. McConnel, *J. Phys. Chem.* **89**: 4453–4459 (1985).

17. W. M. Heckel, and H. M. Moehwald, *Ber Bunsenges. Phys. Chem.* **90**: 1159–1167 (1986).

18. M. M. Lipp, C. Y. K. Lee, A. J. Zasadzinski, and J. A. Waring, *Science* **273**: 1196–1199 (1996).

19. R. G. Strobl, *The Physics of Polymers*, Springer, Berlin (1996).

20. A. Nakai, T. Shiwaku, W. Wang, H. Hasegawa, and T. Hashimoto, *Macromol.* **29**: 5990–6001 (1996).

21. A. Nakai, T. Shiwaku, W. Wang, H. Hasegawa, and T. Hashimoto, *Macromol.* **31**: 5391–5398 (1998).

22. T. Inoue, T. Ougizawa, O. Yasuda, and K. Miyasaka, *Macromol.* **18**: 57–63 (1985).

23. Y. Miyake, Y. Sekiguchi, and S. Kohjiya, *J. Chem. Eng. Jpn.* **26**: 543–550 (1993).

24. K. Iimura, and T. Kato, *Mol. Cryst. Liq. Cryst.* **322**: 117–122 (1998); *Colloids and Surfaces A* **143**: 491–501(1998).

25. M. Kuramori, N. Uchida, K. Suehiro, and Y. Ohishi, *Bull. Chem. Soc. Jpn.* **73**: 829–835 (2000).

26. K. S. Birdi, *Lipid and Biopolymer Monolayers at Liquid Interfaces*, Plenum, New York and London (1989), Section 4.7.1.

27. B. I. Yue, C. M. Jackson, J. A. G. Talor, J. Mingins, and B. A. Pethica, *J. Chem. Soc., Faraday Trans. 1*, **72**: 2685 (1976); **78**: 523 (1982).

28. O. Albrecht, H. Gruler, and E. Sackmann, *J. Phys. (Paris)* **39**: 301 (1978).

29. M. Jaschke, H.-J. Butt, H. E. Gaub, and S. Mann, *Langmuir* **13**: 1381–1384 (1997).

30. E. J. Wanless, and W. A. Ducker, *J. Phys. Chem.* **100**: 3207 (1996).

31. W. A. Ducker, and E. J. Wanless, *Langmuir* **12**: 5915–5920 (1996).

32. J. Zhu, J. Eisenberg, and R. B. Lennox, *Langmuir* **7**: 1579–1584 (1991); **9**: 2243–2246 (1993).

33. J. Zhu, J. Eisenberg, and R. B. Lennox, *J. Am. Chem. Soc.* **113**: 5583–5588 (1991).

34. J. Zhu, J. Eisenberg, and R. B. Lennox, *Macromol.* **96**: 4727 (1992).

35. L. F. Chi, M. Anders, H. Fuchs, R. R. Johnston, and H. Ringsdorf, *Science* **259**: 213 (1993).

36. L. F. Chi, M. Anders, H. Fuchs, R. R. Johnston, and H. Ringsdorf, *Thin Solid Films* **242**: 151–156 (1994).

37. T. Kato, M. Kameyama, and M. Kawano, *Thin Solid Films* **273**: 232–235 (1996).

38. T. Kato, M. Kameyama, M. Ehara, and K. Iimura, *Langmuir* **14**: 1786–1798 (1998).

39. Y. Ren, K. Iimura, and T. Kato, *J. Phys. Chem.* **105**: 4305–4312 (2001).

40. Y. Ren, K. Iimura, and T. Kato, *J. Phys. Chem.* **106**: 1327–1333 (2002).

41. B. Y. Zhu, P. Zhang, L. Wang, and Z.-F. Liu, *J. Colloid Interface Sci.* **185**: 551–553 (1997).

42. Y. Ren, K. Iimura, and T. Kato, *J. Chem. Phys.* **114**: 1949–1951 (2001); **114**: 6502–6504 (2001).

43. T. Kato, M. Ehara, and K. Iimura, *Mol. Cryst. Liq., Cryst.* **294**: 167–170 (1997).

44. J. Frommer, R Lüthi, E. Meyer, D. Anselmetti, M. Dreier, R. Overney, H. J. Güntherrodt, and M. Fujihira, *Nature* **364**: 198 (1993).

45. C. Duschl, M. Liley, O. Corradin, and H. Vogel, *Biophysical J.* **67**: 1229–1237 (1994).

46. J. Fang, and M. C. Knobler, *Langmuir* **12**: 1368–1374 (1996).

47. K. Iimura, and T. Kato, *Langmuir* (submitted).

3

The Differential Conductivity Technique and Its Application to Mixed Surfactant Solutions for Determining Ionic Constants

MASAHIRO MANABE

Niihama National College of Technology,
Niihama, Ehime, Japan

I. SIGNIFICANCE OF DIFFERENTIAL CONDUCTIVITY

A. Introduction

Conductivity is a physical quantity we can measure with high precision and good reproducibility. In the conductometric studies of ionic solutions, the two conductivities have been traditionally employed: specific conductivity (κ) and equivalent conductivity (Λ) or molar conductivity defined as a quotient of κ divided by concentration (C). In the area of colloid science of ionic surfactant solution, these conductivities have been extensively used mainly for determining the critical micelle concentration (CMC) of ionic surfactant, at which the concentration dependence of each conductivity breaks.

In addition to these conductivities, another conductivity was used for the CMC determination. This is the "differential conductivity", the slope of κ–C curve at any C denoted by $d\kappa/dC$ or $\Delta\kappa/\Delta C$ and calculated as $(\kappa_2-\kappa_1)/(C_2-C_1)$. The differential conductivity was first used by Mukerjee [1] who determined the CMC of hydrocarbon–fluorocarbon surfactant mixture, the differential conductivity curve of which has two kinks involving the coexistence of two kinds of immiscible micelles. Carcia-Mateos *et al.* [2] adopted the second differential conductivity $d^2\kappa/dC^2$ for determining the CMC of ionic surfactants, along the definition of CMC by Phillips [3] in which the CMC is considered as the concentration at which the slope of the plot of a physical property of the solution against concentration is changing most rapidly.

As mentioned above, the conductivity has been mostly used only for the purpose of determining CMC, except for such rare case of Evance [4] who used not only the break but also the slopes of κ–C curve (corresponding to the differential conductivity) below and above CMC, in order to evaluate the degree of counter-ion dissociation of micelles (α) and their aggregation number. Recently, Manabe *et al.* [5–11] discovered the great usefulness of the differential conductivity. The conductivity allows us to determine not only CMC but also ionic constants of aggregates or complexes, such as equilibrium constant, amount of charge, aggregation number, and ionization degree. The present chapter will describe the novel method developed by Manabe *et al.* and named "differential conductivity method" for determining ionic constants in surfactant solution.

B.   Differential Conductivity Curve of SDS

The differential conductivity ($d\kappa/dC$) and the equivalent conductivity ($\Lambda = 1000\,\kappa/C$) of sodium dodecylsulfate (SDS) at

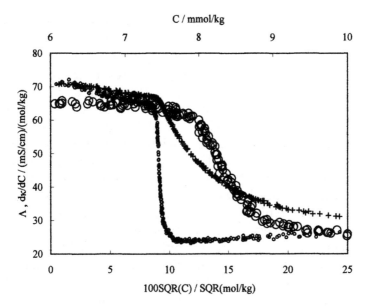

Figure 1 Concentration dependence of dκ/dC and Λ of SDS at 25°C. ○, dκ/dC − $SQR(C)$; +, Λ − $SQR(C)$; ◯, dκ/dC − C.

25°C are plotted together in Fig. 1 against the square root (SQR) of concentration [6]. This is a representative aspect of ordinary ionic surfactants.

In this chapter, different types of differential conductivities will be adopted. Whenever the differential conductivity is plotted against concentration, remember that the concentration implies a mean concentration calculated as $(C_1 + C_2)/2$, although it is denoted by C for simplicity. Furthermore, the temperature is 25°C for all systems treated in the present chapter.

The dκ/dC curve of SDS has some characteristics. At low concentrations, dκ/dC decreases linearly with a little steeper slope than Λ. Once attaining a certain concentration, dκ/dC decreases very rapidly in a narrow concentration region in which the κ–C curve has a curvature. So far, a lot of CMC values for surfactants have been determined by various methods. Kresheck [12] and Mukerjee and Mysels [13] summarized CMCs of SDS around room temperature, most of which locate between ca. 7.2 and 8.8 mM, except some abnormally low values. It should be noted that the location agrees well with the $d\kappa/dC$ dropping region, its expanded view being given in Fig. 1. This agreement indicates that micelles are formed in a concentration region, rather than at a particular concentration, and that the quoted CMC is an average value in the region. Then it is named "CMC region," here. In the summarized data, the surface tension method tends to provide the lowest value and the light scattering and solubilization methods do the highest ones. The values from conductivity are rather intermediate between them. The CMC obtained from surface tension, which is not a bulk property but a surface one, might exhibit a very initial step of micelle formation as an embryo or infant aggregate. In contrast, the CMC by the latter techniques might involve the final step, above which well defined micelles form and their size is large enough to reflect light and to solubilize an additive in them. In the CMC region, with increasing concentration, micelles might grow up rapidly from small embryo micelles to well-defined ones with a definite aggregation number. It should then be emphasized that the differential conductivity curve definitely shows a CMC region in which respective CMC values determined by various methods must be all involved. The CMC region is meaningful and the curve in the region would offer a new theory of micelle formation, different from those based on a definite point of CMC. In the present chapter, the CMC of SDS (7.80 mmol/kg, denoted by CMC_{io}) is taken to be the concentration at which $d\kappa/dC$ starts dropping, the initial point in the CMC region, because of the lowest arbitrarity for determination.

The other characteristic in Fig. 1 is that the $d\kappa/dC$ above the CMC region remains almost constant, whereas Λ tends to decrease monotonically with an increase in C. The distinction arises from the mathematical forms, a differential and a quotient, respectively. Their correlation is compared to that of partial molar quantity and apparent molar one in thermodynamics. Then, it is acceptable that $d\kappa/dC$ reflects the property of an ionic species itself existing at any concentration, and the constancy above the CMC region implies explicitly that the property of micelles remains unaltered over a wide

concentration region, although it is difficult to obtain information from Λ. Along this consideration, the following form of
linear combination can be assumed, not only for micellar
solutions but also for general electrolyte solutions in which
some ionic association occurs

$$\kappa = \sum (d\kappa/dC_j)C_j \tag{1}$$

where C_j and $(d\kappa/dC_j)$ refer to the concentration and the
differential conductivity of each ionic species $(_j)$, e.g., free monomer and micellar species in a surfactant solution. The relation
enable to estimate the concentration of each ionic species, that
is, concentration analysis at any composition is possible based on
the conductivity data, not just on the break point such as CMC.
Equation (1) is the basic equation of the differential conductivity
method for determining equilibrium constants of coexisting
ionic species. The application will be mentioned below.

II. AGGREGATION NUMBER OF SMALL
SELF-AGGREGATES OF SURFACTANT
ION: PREMICELLE FORMATION

A. Introduction

The differential conductivity method is applied first to the
system of single component ionic surfactant solution, in order
to determine the aggregation number of small self-aggregate of
the surfactant ion. Ordinary ionic surfactants aggregate to
form micelles whose aggregation number is around several tens
and the degree of dissociation of counter-ion (α) is about 0.3
[14], a behavior regarded as weak electrolytes. In contrast, if
surfactant ions form small aggregates, such as dimmer, trimer,
tetramer, and so on, they must behave as strong electrolytes
dissolved in completely ionized states, without any counter-ion
condensation, different from ordinary large micelles.

   Such a polyvalent ion of oligo-aggregates must be higher
in ionic mobility than monomer and the aggregation should
lead to an increase in the differential conductivity, since
the differential of κ with C is equivalent to the ionic mobility.

So, the magnitude of $d\kappa/dC$ provides the amount of charge (i.e., aggregation number) of the aggregates bearing no counter-ion.

Here, the way of determining aggregation number by differential conductivity will be illustrated for the three ionic univalent surfactants at 25°C: AOT (Aerosol OT: sodium di(2-ethylhexyl) sulfosuccinate) bearing a double-chain of hydrophobic moiety, NaTDC (sodium taurodeoxycholate), and NaTUDC (sodium tauroursodeoxychaolate), the latter two of which are derivatives of a bile salt bearing a bulky structure of hydrophobic moiety.

B. Differential Conductivity Curve of Self-association Systems

Figure 2 shows the differential conductivity curves for these surfactants. In the initial low concentration region, $d\kappa/dC$ decreases linearly and the slope as well as the intercept of each surfactant is close to each other, in line with the aspect of SDS

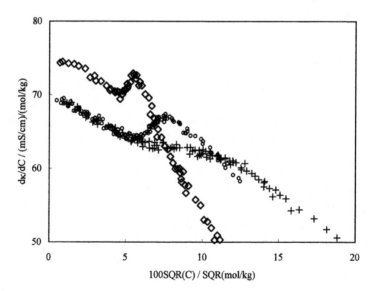

Figure 2 Concentration dependence of $d\kappa/dC$ of double-chain surfactant and bile salt derivatives at 25°C. ◇, AOT; ○, NaTDC; +, NaTUDC.

below the CMC region in Fig. 1. The agreement reflects monomerical dissolution of univalent ion. In particular, for both NaTDC and NaTUDC, the plots lie on the same line, because of little difference in size and shape of the univalent ions. Then, extrapolation to $C \to 0$ leads to the differential conductivity, which is equal to the equivalent conductivity (Λ_{fo}) of free monomer surfactants at infinite dilution, since, in mathematical form, the equivalent conductivity (κ/C) is equal to the differential conductivity ($d\kappa/dC$) at infinite dilution.

A unique pattern of the differential conductivity curve in Fig. 2 appears after the initial linear decrease. The $d\kappa/dC$ increases to the maximum and then another linear decrease is followed. ESR measurements of an amphiphilic probe dissolved in the AOT solution evidenced that at a concentration around the maximum, the mobility of the prove was restricted, indicating the formation of some aggregates to solubilize the proof. Taking the aggregation into account, we can attribute the increase in $d\kappa/dC$ to the formation of multi-charged aggregates, in terms of self-association of surfactant ions without any counter-ion condensation. The increase in $d\kappa/dC$ during the aggregation of the present surfactants (Fig. 2) is in contrast to the decrease observed for ordinary ones such as SDS (Fig. 1): the difference is due to the extent of counter-ion condensation.

The linear decrease over the maximum, consistent with Kohlrausch's square root law, can be regarded as the concentration dependence of the aggregates with a definite aggregation number. In other words, the decrease is due to the behavior of a $1 : Z$ electrolyte, Z referring to aggregation number or charge valence. It is supposed that another maximum appears when further aggregation occurs at higher concentration.

C. Aggregation Number of Oligo-aggregate Ion

By extrapolating the respective linear relations to $C \to 0$ below the minimum and above the maximum in Fig. 2, we can estimated the differential conductivities, i.e., the equivalent conductivities of the monomer and aggregate species at infinite dilution, denoted by Λ_{fo} and Λ_{ao}. Note that $d\kappa/dC$ in the present unit in Fig. 2 is numerically equal to Λ in the traditional unit ($S\,cm^2/eq$). By applying Kohlrausch's law of the independent migration of ions, we can divided Λ_{fo} and Λ_{ao} into the contributions of the free monomer surfactant ion (λ_{fo}) or the aggregate-ion (λ_{ao}) and their counter-ion (λ_{go}).

$$\Lambda_{fo} = \lambda_{fo} + \lambda_{go}, \quad \Lambda_{ao} = \lambda_{ao} + \lambda_{go} \tag{2}$$

Then, λ_{ao} and λ_{fo} can be calculated as

$$\lambda_{ao} = \Lambda_{ao} - \lambda_{go}, \quad \lambda_{fo} = \Lambda_{fo} - \lambda_{go} \tag{3}$$

In the numerical calculation, $\lambda_{go} = 50.1$ [15], the equivalent conductivity of Na^+ ion at limiting dilution at $25°C$ is taken.

Assuming that the surfactant ion and its aggregate are spheres with radius, r_f and r_a, respectively, the aggregation number (N) can be calculated as

$$N = (r_a/r_f)^3 \tag{4}$$

where the volume change on aggregation of the surfactant ion is ignored. For the mobilities of respective spherical ions, U_a and U_f, Stokes' law gives the relation

$$U_a = Ze/6\pi\eta r_a, \quad U_f = e/6\pi\eta r_f \tag{5}$$

where, η and e are the viscosity of solvent and the elementary charge. Then, taking into account the relation between the equivalent conductivity and mobility of an ion, $\lambda = FU$, ($F = $ Faraday constant), the conductivities are related with r_a, r_f, and N

$$\lambda_{ao}/\lambda_{fo} = N/(r_a/r_f) \tag{6}$$

Finally, from Eqs (4), (5), and (6), N which is identical with charge (Z) of the self-aggregate is estimated from the conductivity ratio

$$N = Z = (\lambda_{ao}/\lambda_{fo})^{3/2} \tag{7}$$

Table 1 Aggregation number and charge of self-associated surfactant ion at 25°C.

Surfactant	AOT	NaTDC	NaTUDC
Λ_{fo}	76.2	70.0	69.9
Λ_{ao}	102.0	81.4	79.9
λ_{fo}	26.0	19.9	19.8
λ_{ao}	51.9	31.3	29.8
$N(Z)$	2.8	2.0	1.9

The equivalent conductivity (Λ, λ) in $S\,cm^2\,eq$ is numerically equal in the present unit of $(mS/cm)/(mol/kg)$. $\lambda_{go} = 50.1$ was used for calculating λ.
The aggregation number (N) and the charge (Z) are in entirely ionized state of aggregates.

Table 1 lists the determined values of N, together with the concerning conductivity data. The resultant value of aggregation number is around 3 for AOT and 2 for NaTDC and NaTUDC. Such small aggregates can be considered to be the ones formed in the just initial stage of aggregation at a lowest concentration, and they are supposed to grow up with increasing concentration. Such small aggregates correspond to the premicelle formed just before the ordinary micelles, proposed by Mukerjee [16].

So far, the aggregation number of bile salt derivatives has been determined mainly by light scattering technique. In most cases, the studies were carried out in the presence of extra salt. In salt free water, few studies [17] on the aggregation number of bile salts was performed by ultracentrifugation and light scattering: the formation of dimmer $(N = 2)$ was detected at 20°C for the sodium salts of the following bile acids: cholic acid, glycocholic acid, and glycodeoxycholic acid. The present results for the bile salts are in good agreement with the reported ones of related bile salts. The present technique is much simpler than the ultracentrifugation and light scattering. The differential conductivity method is advantageous in extra salt-free solutions, for determining the aggregation number of such small self-aggregates.

III. CRITICAL COMPOSITION FOR COUNTERION CONDENSATION IN IONIC–NONIONIC MIXED MICELLES [8,11]

A. Introduction

The degree of dissociation of counterion (α) of micelles is known to be around 0.3 for ordinary ionic surfactants [14]. There has been some observations that the counterion dissociation is enhanced when a nonionic amphiphile, such as 1-alkanols [18] or nonionic surfactants [19], is solubilized in ionic micelles to form mixed micelles, the decrease of surface charge density (σ) on which enhances ionization. So far, it has been accepted that α is a continuous function of σ on the surface of micelles. On the contrary, there exists a critical surface charge density below which the counterion is completely released ($\alpha = 1$), according to Manning [20] who concerned with polyelectrolytes in linear rod-like form. It is therefore reasonably supposed that even on a plane surface of micelles, there must exist such a critical surface charge density. If a trace of ionic surfactant is solubilized in nonionic micelles, the counterion must be completely released. Few studies [21,22] have so far been carried out to confirm and determine the critical density.

In the present section, a differential conductivity method will be mentioned how to ascertain the existence of such critical surface charge density (σ_c) and to estimate its numerical value. The following mixed surfactant systems were adopted: an ionic surfactant (SDS) mixed with each of the three nonionic surfactants, hexaoxyethylene-decylether ($R_{10}E_6$), -dodecylether ($R_{12}E_6$), and MEGA10 (n-decanoyl-N-methylglucamide). On conductivity measurements, small amounts of ionic surfactant (SDS) were successively added in a micellar solution of the nonionic surfactants at 25°C.

B. Differential Conductivity Curve of Critical Mixed Micelle Formation

As shown in Fig. 3, when SDS (concentration $= C_i$) is added to a nonionic surfactant solution at a concentration (C_n) far lower

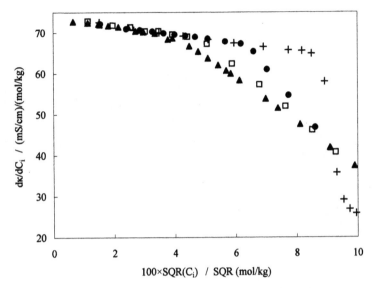

Figure 3 Differential conductivity curve of SDS in the presence of MEGA10. C_n/mmol/kg: +, 0; ●, 0.6943; □, 1.397; ▲, 1.9998.

than the CMC of nonionic surfactant itself denoted by CMC_{no}, the differential conductivity of SDS, $d\kappa/dC_i$, lies initially on the straight line for SDS alone in water below CMC, and then goes down toward the low constant value given in Fig. 1, through a CMC region which becomes wider with increasing C_n. The concentration of SDS at the break point decreases with increasing C_n. The concentration is taken as the CMC of SDS in the presence of the nonionic surfactants. The aspect is similar to that for the conductivity curve of SDS in the presence of alkanols [5].

Figure 4 illustrates the differential conductivity curves of SDS added into a micellar solution of the nonionic surfactants. $d\kappa/dC_i$ starts at a lower value than that for pure SDS in water and goes up to a maximum, higher than that for pure SDS. The higher value than that for the uni-valence monomer suggests the formation of multi-charged ionic species, as in the self-association of surfactant ions in Section II. The maximum value of $d\kappa/dC_i$, denoted by $(d\kappa/dC_i)_c$, decreases and the C_i at

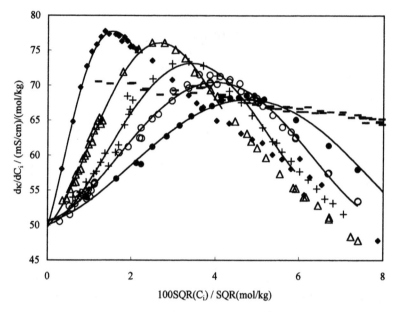

Figure 4   Differential conductivity curve of SDS in micellar solution of $R_{10}E_6$. C_n/mmol/kg: —, 0; ◆, 1.998; △, 4.999; +, 7.990; ○, 11.99; ●, 17.96.

the maximum point, denoted by C_{ic}, shifts toward higher C_i with an increase in C_n. At C_n around CMC_{no}, the linearly extrapolated value of $d\kappa/dC_i$ to $C_i \to 0$ attains about 50 in the present unit. It is significant that this value coincides with the equivalent conductivity of Na^+ ion at limiting dilution $(\lambda_{go} = 50.10\,S\,cm^2/eq)$[15]. A plausible explanation for the agreement is as follows. As long as C_i is very low, the surfactant ion with strong hydrophobicity is entirely incorporated in nonionic surfactant micelles with much bigger size than the monomeric ion, resulting in an extremely low ionic mobility (i.e., the lowest possible transport number) of the surfactant ion, and at the same time, only the counter ion left in the bulk water contributes to the conductivity of λ_{go}. The validity of the explanation was experimentally confirmed by using homologous sodium alkylsulfates (from hexyl to dodecyl) and some other ionic surfactants with different counter-ions (Na^+

and K^+) dissolved in the micellar solution of $R_{10}E_6$ and $R_{12}E_6$ [11].

Based on the mechanism, the increase in $d\kappa/dC_i$ to the maximum with increasing C_i is ascribed to a gradual increase in the net charge of mixed micelles, caused by successive solubilization of surfactant ions. The increase in charge leads to an increase in the surface charge density up to a critical value above which, once it is attained, counter-ion condensation occurs. Then, the charge increase produced by successive solubilization of surfactant ions might be cancelled by the counter-ion binding. The decreasing tendency over the maximum seems therefore to be the concentration dependence of mixed micelles with a certain amount of charge, as discussed below.

According to the model of completely ionized mixed micelle formation, it seems to be reasonable that the C_i at which $d\kappa/dC_i$ equals to λ_{go} (=50.1) is not limited to $C_i \to 0$ but increases up to higher C_i with increasing C_n, since the infinite dilution region of ionic surfactant widens with increasing C_n. Namely, the increasing tendency of $d\kappa/dC_i$ to the maximum cannot generally be regarded as linear but sigmoidal. Then, quadratic equations are applicable for curve fitting in Fig. 4, in order to determine the values of $(d\kappa/dC_i)_c$ and C_{ic}. The quadratic regressive analysis was carried out under such condition that the constant term of the quadratic equation becomes as close to λ_{go} as possible in a suitable C_i region, by the successive calculation from the measured lowest C_i to the highest C_i, above which the constant deviates from λ_{go}. Figure 4 shows the resultant best fitting curve (solid line).

Furthermore, it is possible to say that at the maximum at different C_n, the mixed micelles have the same mole fraction of ionic surfactant (X_{ic}), i.e., the mixed micelle with X_{ic} can be regarded as a stoichiometric complex. Consequently, the relation between $(d\kappa/dC_i)_c$ and C_{ic} can be considered to be the concentration dependence of conductivity for the critical complex with X_{ic} at $\alpha = 1$. The $(d\kappa/dC_i)_c$ determined by the quadratic fitting curve is plotted against $SQR(C_{ic})$, according to the Kohlrausch square-root law, as seen in Fig. 5. It is apparent that the relation is linear for each non-ionic surfactant

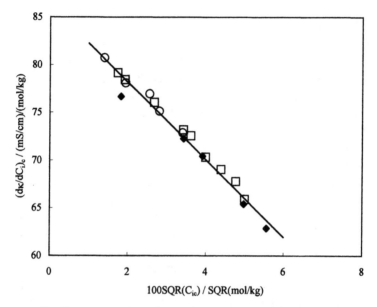

Figure 5 Concentration dependence of $(d\kappa/dC_i)_c$ in non-ionic micellar solution at 25°C. Non-ionic surfactant: $\blacklozenge$, MEGA10; $\square$, $R_{10}F_6$; $\bigcirc$, $R_{12}E_6$.

(MEGA10, $R_{10}E_6$, and $R_{12}E_6$), and that, inexplicably, these relations are indistinguishable. Then, they are regarded as lying on a single straight line. The least mean squares method for the straight line gives the intercept of 86.20 and the slope of -4.008 in the unit in Fig. 5. The slope is about four times larger than that for monomerical SDS below CMC_{io} in Fig. 1, due to the multi-charged particles. The intercept can be taken as the equivalent conductivity of the critical complex at infinite dilution, denoted by Λ_{co}. The Λ_{co} provides the characteristic parameters concerning the complex with X_{ic}, as mentioned below.

C. Critical Mole Fraction for Counter-ion Binding (X_{ic})

In the same manner as that used for determining the equivalent conductivity of monomeric surfactant ion λ_{io}, as described in Section II, the equivalent conductivity of the complex ion at limiting dilution denoted by λ_{co} is calculated as

$$\lambda_{co} = \Lambda_{co} - \lambda_{go} \tag{8}$$

Here, one critical mixed micelle consists of Z surfactant ions and N nonionic surfactant molecules. The ratio of the radius of aggregate to that of monomer, (R/r), is related with the aggregation numbers, Z and N, assuming that the volumes of respective surfactants in free monomer and aggregated states are approximately identical with each other, as

$$R/r = (N + Z)^{1/3} \tag{9}$$

Applying the Stokes' law to the equivalent conductivities, the ratio of mobility of complex ion (U_{co}) to free monomer ion (U_{io}) is

$$U_{co}/U_{io} = Z/(R/r) \tag{10}$$

Taking the relation $\lambda_{co}/\lambda_{io} = U_{co}/U_{io}$ into account, the amount of charge, Z, of the complex is derived from these equations

$$Z = (\lambda_{co}/\lambda_{io})^{3/2}/X_{ic}^{1/2} \tag{11}$$

where $X_{ic} = Z/(Z+N)$ for the critical micelles.

In order to estimate Z from Eq. (11), the value of X_{ic} is essential. The X_{ic} can be determined as follows. The C_{ic} is plotted against C_n in Fig. 6. For each surfactant system, the relation is apparent to be linear. Linear regressive analysis provides the relations with the slopes, 0.0822, 0.1258, and 0.1154, and the intercepts, -0.0648, -0.0444, 0.0089 mmol/kg, for MEGA10, $R_{10}E_6$, and $R_{12}E_6$, respectively. The relation can be taken as proportional since each intercept is negligibly small. The proportional relation in Eq. (12) therefore holds in the micellar solution

$$C_{ic} = qC_n \tag{12}$$

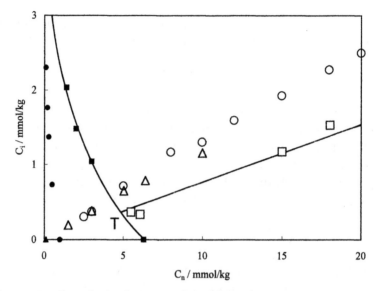

Figure 6   Correlation between C_{ic} and CMC of SDS in the presence of nonionic surfactant. Open and closed marks indicate C_{ic} and CMC, respectively. Solid lines with MEGA10 cross at T. Non-ionic surfactant: □, MEGA10; ○, $R_{10}E_6$; △, $R_{12}E_6$.

where q is a characteristic constant, the physical meaning of which is described later. The proportional straight line is now called the critical line of counter-ion condensation (CLC).

For comparison with the CLC, the CMC of SDS determined at the break point in Fig. 3 is also plotted in Fig. 6. According to the pseudo phase separation model of mixed micelle formation [23], the CMC curve is the boundary between the monomerically dissolved phase and the micellar one, providing a phase diagram. So, the CLC must cross the CMC curve at the triple point denoted by T, representatively in MEGA10 system in Fig. 6. It should be emphasized that the CLC undercuts a new phase from the traditional micellar phase. In the novel phase, the counter-ion of mixed micelles is completely released ($\alpha = 1$).

If an ionic surfactant is, in general, successively added in a micellar solution of nonionic surfactant with a given C_n, the whole composition in the phase diagram goes up vertically from a point on the abscissa, above the CMC_{no}, and the composition of monomer species coexisting with the mixed micelles changes along the CMC curve from the CMC_{no}. The CMC_{no} values of the present nonionic surfactants shown in Fig. 6 were determined by the surface tension method. Once the whole composition attains any point on the CLC, the monomer composition must locate at the point T. Now, the concentrations of ionic surfactant and nonionic one at T are denoted by C_{iT} and C_{nT}, respectively. Then, X_{ic} for the critical mixed micelles can be expressed by the concentrations on the proportional line

$$X_{ic} = (C_{ic} - C_{iT})/(C_{ic} - C_{iT} + C_n - C_{nT}) \qquad (13)$$

Then, C_{ic} is calculated as a function of X_{ic}

$$C_{ic} = X_{ic}/(1 - X_{ic})C_n + C_{iT} - X_{ic}/(1 - X_{ic})C_{nT} \qquad (14)$$

This relation is another form of Eq. (12), where the last two terms should yield zero, judging from the proportional relation of CLC. Then, the relation is derived from Eq. (14).

$$X_{ic}/(1 - X_{ic}) = C_{iT}/C_{nT} \qquad (15)$$

From Eqs (12) and (15)

$$q = X_{ic}/(1 - X_{ic}) \qquad (16)$$

Then, q means the ratio of mole fractions in the critical mixed micelles. Finally, X_{ic} can be evaluated from q

$$X_{ic} = q/(1 + q) \qquad (17)$$

and the values are given in Table 2.

The values of X_{ic} are around 0.1 in the present surfactant systems studied. So far, few studies were carried out on the determination of X_{ic}. Trainer *et al.* [21,22] determined X_{ic} by the EMF measurements on counter-ion selective electrodes: X_{ic} of SDS in $R_{12}E_4$ and $R_{12}E_{23}$ are 0.08 and 0.15, respectively [21], and of copper dodecyl sulfate in Triton X-100 and Brij 35 are 0.02 and 0.05, respectively [22]. The reported values concerning SDS are consistent with the present ones. It should be

Table 2 Critical mole fraction (X_{ic}) of SDS in mixed micelles with nonionic surfactant and relating quantities at 25°C.

Surfactant	MEGA10	R10E6	R12E6
q^a	0.0794	0.129	0.120
X_{ic}	0.0736	0.114	0.107
Z^b	5.75	5.81	5.82
N^c	73	45	49
AN^d	78(20°C)[e], 75(40°C)[e]	73[f]	400[f]

[a]Slope of proportional relation through original point in Fig. 6 through Eq. (12).
[b]In calculating for Z by Eq. (11), the equivalent conductivities of SDS and the critical complex at limiting dilution are used. $\Lambda_{io} = 73.2$, $\Lambda_{co} = 86.2$ (mS/cm)/(mol/kg).
[c]Aggregation number of nonionic surfactant in the critical mixed micelles.
[d]Aggregation number of nonionic surfactant alone in water.
[e]From Ref. 24.
[f]From Ref. 25.

mentioned that under the condition of such low concentrations of ions, conductivity is much more sensitive and reliable than EMF, in the measurements.

The value of Z and then N can be calculated with Eq. (11) (Table 2). As a result, the values of Z are found to be around 6 with little difference, and the values of N differ a little from each other for respective systems. The N is reasonably compared with the aggregation number of nonionic surfactant itself for MEGA10 [24] and $R_{10}E_6$ [25] in water (Table 2).

D. Critical Surface Charge Density for Counter-ion Binding (σ_c)

The present result enables to estimate the critical surface charge density (σ_c) at X_{ic} above which counter-ion condensation occurs on the surface of mixed micelles. Considering a spherical micelle, the radius of its hydrocarbon core, denoted by L, is regarded as a hydrocarbon chain length of surfactant, in Angstrom unit [26]

$$L = 1.5 + 1.265m \tag{18}$$

where m is the carbon number in the alkyl chain of a surfactant. When m is taken to be 12 for a tentative calculation, L is 1.67 nm, and then σ_c in the surface area S is calculated. When $Z = 6$ from Table 2

$$\sigma_c = Ze/S = 0.027 \, \text{C/m}^2 \tag{19}$$

where e refers to the elementary charge, 1.6×10^{-19} C. Huang *et al.* [27] reported the surface charge density for fully ionized carboxyl-terminated dendrimer with different generations in a buffer solution. When hydrodynamic radius is 1.7 nm for the second generation dendrimer, close to the present tentative length, the charge density of $0.16 \, \text{C/m}^2$ was estimated. The latter is ca. 6 times greater than the present density. The difference between them, however, can be considered to be not so serious but rather reasonably comparable to each other, in the present stage, since the determination of the critical surface charge density on fully ionized surface of colloid particles is still in its infancy. It is then concluded that there exists a critical surface charge density on the colloid particle of mixed micelles with $\alpha = 1$.

IV. BINDING CONSTANT OF IONIC SURFACTANT-CYCLODEXTRIN INCLUSION COMPLEX FORMATION [7]

A. Introduction

Cyclodextrins (cycloamyloses), abbreviated as CD, consist of several α-D-glucopyranose units to form a cylindrical cavity, which is hydrophobic inside and hydrophilic outside. The unique molecular structure leads to inclusion complex formation by including a hydrophobic guest in the cavity. When a free surfactant ion is accommodated in a CD molecule to become bigger in its ion-size, the complexation must reduce its ionic mobility. The lowering of mobility causes a decrease of differential conductivity and allows to determine the concentration of complex-ion in equilibrium with free surfactant-ions, thereby permitting to determine the equilibrium constant K_{CD}

for the complex formation. The present section will introduce
the application of the differential conductivity method to the
determination of the complex formation constant of sodium
perfluorooctanoate (SPFO) with β- and γ-CD.

B. Differential Conductivity Curve of Inclusion Complex Formation

The κ is measured on successive addition of β- or γ-CD
(concentration $= C_d$) into a solution at a certain concentration
of SPFO (concentration: $C_i = 1.000$ mmol/kg), lower than its
critical micelle concentration (32 mmol/kg). The κ–C_d curve
provides C_d dependence of the differential conductivity, $d\kappa/dC_d$,
as is seen in Fig. 7. The stepwise, sigmoid curve in β-CD/SPFO
system comes from the linear decrease of κ and a nearly
constant value with increasing C_d. On the other hand, in γ-CD/
SPFO, the asymptotical curve of $d\kappa/dC_d$ arises from the
monotorically decreasing κ with C_d.

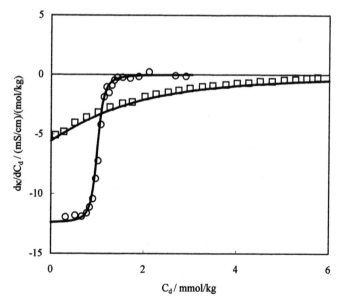

Figure 7 Differential conductivity curve of CD/SPFO system at
25°C. $C_i = 1.000$ mmol/kg. Solid line indicates calculated value with
K_{CD}. ○, β-CD, $K_{CD}=19\,900$; □, γ-CD, $K_{CD}=590$ kg/mol.

In the β-CD/SPFO, it is obvious that $d\kappa/dC_d$ goes up suddenly in a narrow C_d region just around 1 mmol/kg of C_d, from a negative low constant value to a high one (zero). The sigmoid curve suggests the formation of 1:1 complex, in the following respects: (1) C_d at the inflection point of the curve is in good agreement with C_i (=1 mmol/kg), (2) the negative constant of $d\kappa/dC_d$, i.e., the decrease in κ in the initial addition of CD, can be attributed to the formation of complex with a lower ionic mobility (bigger ion) than free monomer surfactant ion, and (3) the $d\kappa/dC_d$ being zero at high C_d implies that the added CD dissolves independently of the surfactant ions after saturation. For γ-CD, the increasing tendency of $d\kappa/dC_d$ gives an asymptotically monotonous curve, differing from the sigmoid curve for β-CD. This difference may be ascribed to the difference of the magnitude of K_{CD}, which is determined in the following manner, based on the function of $d\kappa/dC_d$ with C_d in Fig. 7.

C. Binding Constant of CD-surfactant Ion Complex

Provided that a free monomer surfactant ion (concentration: C_{if}) and uncomplexed free CD (concentration: C_{df}) form 1:1 complex (concentration: C_X), the concentrations are related as

$$C_i = C_{if} + C_X, \quad C_d = C_{df} + C_x \tag{20}$$

where C_i indicates the concentration of ionic surfactant. Then, the equilibrium constant (K_{CD}) of 1:1 complex is expressed as

$$K_{CD} = C_x/(C_{if} \cdot C_{df}) \tag{21}$$

Combination of Eqs (20) and (21) yields C_x as

$$C_x = (1/2)[(C_i + C_d + 1/K) - SQR\{(C_i + C_d + 1/K)^2 \\ - 4C_i \cdot C_d\}] \tag{22}$$

where SQR indicates square root.

If κ is a function of the concentrations of ionic species, C_{if} and C_X, the total differential of κ is given by

$$d\kappa = (d\kappa/dC_{if})\, dC_{if} + (d\kappa/dC_X)\, dC_X \tag{23}$$

Dividing $d\kappa$ by dC_d, at a certain C_i, in line with the present experimental condition, gives Eq. (24)

$$d\kappa/dC_d = \bar{\kappa}_{if}(dC_{if}/dC_d) + \bar{\kappa}_X(dC_X/dC_d) \tag{24}$$

where $\bar{\kappa}_{if}$ and $\bar{\kappa}_x$ are the differential conductivities of monomer surfactant and complex, in Eq. (23), respectively. Inserting C_{if} of Eq. (20) into Eq. (24) allows to relate $d\kappa/dC_d$ with dC_X/dC_d

$$d\kappa/dC_d = (\bar{\kappa}_x - \bar{\kappa}_{if})(dC_X/dC_d) \tag{25}$$

where dC_X/dC_d means the complexation fraction of added CD and is derived by differentiating Eq. (22) with C_d at a constant C_i

$$dC_X/dC_d = (C_i - C_X)/(C_i + C_d + 1/K_{CD} - 2C_X) \tag{26}$$

If κ is measured as a function of C_i at a constant C_d, the following equations corresponding to the above ones are derived

$$d\kappa/dC_i = \bar{\kappa}_{if} + (\bar{\kappa}_X - \bar{\kappa}_{if})(dC_X/dC_i) \tag{27}$$

$$dC_X/dC_i = (C_d - C_X)/(C_i + C_d + 1/K_{CD} - 2C_X) \tag{28}$$

where dC_X/dC_i is also the fraction of complexation of added surfactant.

When the added CD is entirely complexed, $dC_X/dC_d = 1$, i.e., $d\kappa/dC_d = (\bar{\kappa}_x - \bar{\kappa}_{if})$ in Eq. (25). This value can be approximately estimated in Fig. 7 by extrapolating $d\kappa/dC_d$ to $C_d \to 0$. On the other hand, when $dC_X/dC_d = 0$, the added CD does not participate in complex formation at all and $d\kappa/dC_d = 0$ in Eq. (25). This $d\kappa/dC_d$ should correspond to the value obtained by extrapolating the conductivities at high C_d in Fig. 7 to $C_d \to \infty$. As a result, experimentally measured $d\kappa/dC_d$ at a constant C_i can be related with C_d through Eqs (25) and (26).

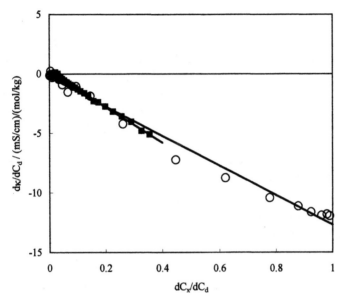

Figure 8 Determination of K_{CD} in CD/SPFO system at 25°C. $C_i = 1.000$ mmol/kg. ○, β-CD, $K_{CD} = 19\,900$; ■, γ-CD, $K_{CD} = 590$ kg/mol.

Equation (25) indicates that the experimentally determined $d\kappa/dC_d$ is a linear and proportional function of (dC_X/dC_d) calculated at any C_d through K_{CD} as an adjustable parameter. The linear regressive analysis involving the adjustable parameter, K_{CD}, provides therefore the best fitting linear relation given in Fig. 8. Satake *et al.* [28] developed a conductivity method for the determination of K_{CD} in ionic surfactant–CD complex, in which they used the traditional equivalent conductivity. The equations of the equivalent conductivities are, in principle, in consistent with the present ones, the latter being more definite in the physical meaning and graphical analysis.

The binding constants obtained are $K_{CD} = 1.99 \times 10^4$ kg/mol in β-CD/SPFO and 5.90×10^2 kg/mol in γ-CD/SPFO systems (Fig. 8). It is apparent that K_{CD} of β-CD/SPFO is about 35 times as high as that of γ-CD/SPFO, which means that β-CD includes fluorocarbon-chain of SPFO more tightly through the strong van der Waals interaction between host and guest molecules in the inclusion state, compared to γ-CD. Palepu and Reinsborough [29] obtained $K_{CD} = 4500\,\text{dm}^3/\text{mol}$ for β-CD/SPFO and $K_{CD} = 640\,\text{dm}^3/\text{mol}$ for γ-CD/SPFO at $C_i = 2\,\text{mmol/dm}^3$, by using Satake's method [28]. Comparing these data, K_{CD} of γ-CD/SPFO is in good agreement with the present one, but for β-CD/SPFO, K_{CD} is ca. 4.5 times as low as the present one. The difference might be due to the difficulty of curve fitting to such a sharp break on the conductivity curve for β-CD. The linear analysis along Eq. (25) shown in Fig. 8 seems to provide a more reliable K_{CD} value than the polynomial regression [28]. The resultant best fitting curves are shown by the solid line in Fig. 7, too, for comparison with experimental data.

The linear relation in Fig. 8 gives information concerning the differential conductivity (i.e., mobility) of each ionic species. For β- and γ-CD, the respective straight lines have the intercepts -0.27 and $+0.23$, and the slopes -12.35 and -15.04, in the present unit. The intercept can be taken to be negligibly small, consistent with the proportional relation of Eq. (25). According to Eq. (25), the slope should be equal to $d\kappa/dC_d$ at $dC_x/dC_d = 1$. In fact, in β-CD/SPFO, the slope of -12.35 is close to the measured value of $d\kappa/dC_d$ ($= -11.95$) the lowest C_d at which dC_x/dC_d is almost unity, seen in Fig. 8. In addition, the slopes in both CD systems can be well compared with each other. The consistency indicates the validity of the present analysis.

The differential conductivity data, $\bar{\kappa}_x - \bar{\kappa}_{if}$, evaluated as the slope in Eq. (26) provide the conductometric information of the complex ion itself. The differential conductivities at infinite dilution, equal to the equivalent conductivity, of the complex and free surfactant, denoted by Λ_{xo} and Λ_{ifo}, can be separated into the contributions of component ionic species in the same manner as above.

$$\Lambda_{ifo} = \lambda_{ifo} + \lambda_{go}, \quad \Lambda_{xo} = \lambda_{xo} + \lambda_{go} \qquad (29)$$

where λ_{ifo}, λ_{xo}, and λ_{go} stand for the differential conductivities (i.e., equivalent conductivities) of surfactant ion, complex ion, and counter-ion, respectively, and $\lambda_{go} = 50$ for Na^+ ion. Extrapolating the differential conductivity curve of SPFO itself into non-micellar region gives $\Lambda_{ifo} = 75.0$. Assuming that the concentration dependence of differential conductivity for each species at any concentration is identical, the difference, $\bar{\kappa}_x - \bar{\kappa}_{if}$ should be equal to $(\Lambda_{xo} - \Lambda_{ifo})$. In the following approximate numerical calculations, the mean value of the slopes in Fig. 8 in β- and γ-CD, $\bar{\kappa}_x - \bar{\kappa}_{if} = -13.9$, is used, neglecting the difference of complex size between β- and γ-CD. Then, λ_{xo} can be calculated as

$$\lambda_{xo} = \Lambda_{xo} - \lambda_{go} = \Lambda_{ifo} - 13.9 - \lambda_{go}$$
$$= 75.0 - 13.9 - 50.1 = 11.0 \tag{30}$$

which is less than a half of $\lambda_{ifo} = 75.0 - 50.1 = 24.9$. According to Stokes' law, the ratio of $\lambda_{ifo}/\lambda_{xo} = 2.26$ corresponds to the ionic radius ratio of complex/monomer ion. The ratio is converted to the ionic volume ratio, $(2.26)^3 = 11.54$ (see, Section II.C). The other estimation of volume ratio is made from the molecular structures. The volume calculated from the height and diameter of a cylindrical molecule is 1430(β-CD) and 1750(γ-CD) in A^3 [30]. Based on the relation from Tanford [26], the volume of hydrocarbon octane is $216 A^3$, instead of fluorocarbon. Then the volume ratios of CD/octane are 6.6 and 8.1 for β- and γ-CD, respectively. These ratios can be fairly well compared with that from conductivity, 11.54, indicating the validity of the differential conductivity method.

V. THE PARTITION COEFFICIENT AND THE COUNTERION RELEASING EFFECT OF NONIONIC HOMOLOGOUS AMPHIPHILES IN IONIC MICELLAR SOLUTION [5,6,9,10]

A. Introduction

When a nonionic amphiphile is dissolved in a micellar solution of an ionic surfactant, it is distributed between the bulk water and the micelles. In the bulk water, the monomerically dissolved amphiphile drives the singly dispersed surfactant ion to micellize by the hydrophobic interaction, judging from the well-known fact that CMC is depressed by the addition of a small amount of organic additives [5,23,31]. The micellization decreases the conductivity of micellar solution.

On the other hand, the other portion of the additive is solubilized in micelles and the polar head group is located in the surface of micelles. And the groups wedged among the ionic head groups of surfactant reduces the surface charge density of micelles, resulting in the acceleration of counterion release. This factor increases the conductivity. Then, the conductivity change controlled by the two competing factors will enable to evaluate the quantities described below.

So far, many methods have been used to determine the partition coefficient (K_x) between the bulk water and the micelles [32]. Here, a novel method developed by Manabe *et al.* [5] based on the differential conductivity will be mentioned. The method enables to estimate the two quantities, $d\alpha/dX_{am}$ and dC_{sf}/dC_{af}, in addition to K_x, at infinite dilution of the non-ionic amphiphile solubilized in a ionic surfactant micellar solution. The $d\alpha/dX_{am}$ implies the acceleration rate of the degree of counterion dissociation of micelles against the mole fraction (X_{am}) of solubilized additive, and dC_{sf}/dC_{af} is the decreasing rate of the monomer surfactant concentration (C_{sf}) against the monomer additive concentration (C_{af}) in the bulk water of micellar solution, which corresponds to the decreasing rate of CMC $(\Delta CMC/\Delta C_a)$ on addition of a non-ionic additive just at CMC. Here, the results will be mentioned on the systems in which some homologous series of nonionic amphiphiles with different polar groups are dissolved in SDS solution.

B. Differential Conductivity Curve of Partitioning Systems

In order to determine partition coefficient and related quantities, the conductivity of a micellar solution of SDS was measured as a function of the concentration of

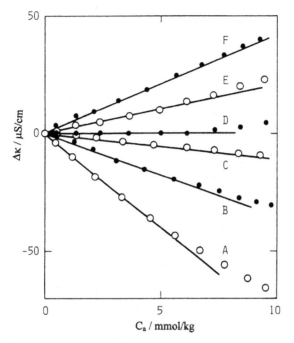

Figure 9 Change in κ with C_a of hexanol added to SDS micellar solution at 25°C. Cs/mmol/kg: 8.53 (A); 14.9 (B); 20.0 (C); 27.1 (D); 37.0 (E); 50.0 (F).

non-ionic additives. On the other hand, the conductivity of an aqueous solution of additive was measured as a function of SDS to determined the CMC of SDS in the presence of the additive.

When a nonionic amphiphile is added in a micellar solution of the ionic surfactant (SDS) with a concentration of C_s, the specific conductivity (κ) changes almost linearly as long as the concentration of additive (C_a) is very low, as illustrated with 1-hexanol in Fig. 9. The slope at infinite dilution denoted by $(d\kappa/dC_a)_o$ is determined from the κ–C_a curve, and the differential conductivity is plotted in Fig. 10 against C_s above CMC_{io} (CMC of SDS, itself). With a given amphiphile, the $(d\kappa/dC_a)_o$ increases asymptotically from negative to positive. The differential conductivity curves can be correlated with the ionic constants as follows.

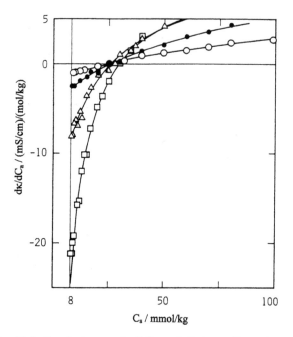

Figure 10   Relation between $d\kappa/dC_a$ and C_s in 1-alkanol-SDS system. 1-Alkanol: butanol (○); pentanol (●); hexanol (△); heptanol (□).

C. K_x, $d\alpha/dX_{am}$, dC_{sf}/dC_{af}, and $\Delta CMC/\Delta C_a$

The two competing factors mentioned above are taken into account for evaluating K_x, $d\alpha/dX_{am}$, and dC_{sf}/dC_{af}, in the following procedure. With the surfactant $(_s)$ and additive $(_a)$ in free monomer $(_f)$ and micellar $(_m)$ states, their concentrations (C) are related as

$$C_s = C_{sf} + C_{sm}; \quad C_a = C_{af} + C_{am} \tag{31}$$

For the ionic surfactant, the differential conductivities $(\bar{\kappa})$ of free monomer and micellar surfactant species are defined as

$$\bar{\kappa}_f = d\kappa/dC_{sf}, \quad \bar{\kappa}_m = d\kappa/dC_{sm} \tag{32}$$

Here, $\bar{\kappa}_f$ and $\bar{\kappa}_m$ correspond to the differential conductivities of the ionic surfactant itself below and above CMC_{io}, respectively. Then the κ of a micellar solution can be expressed as a linear combination of the differential conductivities and concentrations of the respective species, as seen in Eq. (1).

$$\kappa = \bar{\kappa}_f C_{sf} + \bar{\kappa}_m C_{sm} \tag{33}$$

When a small amount of a nonionic substance is added in a micellar solution with a given C_s, the rate of conductivity change represented in the form of $(d\kappa/dC_a)$ can be derived by differentiating Eq. (33) with C_a under the condition of constant C_s

$$d\kappa/dC_a = \bar{\kappa}_f(dC_{sf}/dC_a) + \bar{\kappa}_m(dC_{sm}/dC_a) + (d\bar{\kappa}_m/dC_a)C_{sm} \tag{34}$$

In the derivation, the term $(d\bar{\kappa}_f/dC_a)$ can be neglected from the experimental fact that no influence of the additive was observed on the conductivity of SDS in non-micellar region. Further, $\bar{\kappa}_m$ is correlated with the degree of counter-ion dissociation of micelles (α) as

$$\bar{\kappa}_m = \alpha\bar{\kappa}_m^* \tag{35}$$

where $\bar{\kappa}_m^*$ is a hypothetical differential conductivity of completely ionized micelles with $\alpha = 1$, for the ionic surfactant.

Now, the following quantities are defined: the partition coefficient (K_x) of additive in mole fraction unit, the decreasing rate of monomer surfactant (k), and the solubilization fraction (J)

$$K_x = x_{af}/X_{am}; \quad k = dC_{sf}/dC_{af}; \quad J = dC_{am}/dC_a \tag{36}$$

The fraction J varies from 0 at $C_s = CMC_{io}$ to 1 at $C_s \rightarrow \infty$. Under the condition of $C_a \rightarrow 0$, the approximations can be made to yield [5]

$$J = (K_x/n_w)(C_s - CMC_{io})/(1 + (K_x/n_w)(C_s - CMC_{io})) \tag{37}$$

where n_w stands for the mole number of water in 1 kg of solution. Equation (37) allows us to calculate J at any C_s when

K_x can be estimated. As a result, Eq. (34) can be rewritten using the above quantities at infinite dilution of additive

$$(d\kappa/dC_a) = (\bar{\kappa}_f - \bar{\kappa}_m)k + \{\bar{\kappa}_m^*(d\alpha/dX_{am}) - (\bar{\kappa}_f - \bar{\kappa}_m)k\}J$$

$$(38)$$

Here, $(d\alpha/dX_{am})$ implies the acceleration rate of ionization of micelles by a solubilized additive. On the right-hand side, only J is a variable. Equation (38) therefore indicates that the experimentally determined $(d\kappa/dC_a)$ at infinite dilution of additive is a linear function of J. The linear relation can be determined by a regressive analysis for the best fitting K_x as an adjustable parameter, as illustrated for 1-alkanols in Fig. 11. Respective intercepts at $J = 0$ and $J = 1$ of the straight line provide k and $(d\alpha/dX_{am})$. More detailed derivation of the equations can be found in our paper [5]. In numerical analysis, the following parameters of SDS itself taken from its

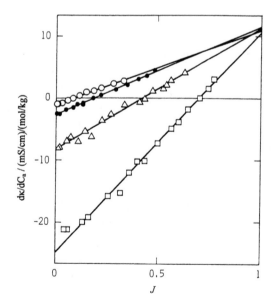

Figure 11 Relation between $d\kappa/dC_a$ and solubilization fraction (J) in 1-alkanol-SDS system. 1-Alkanol: butanol (O); pentanol (●); hexanol ($\triangle$); heptanol ($\square$).

differential conductivity curve are used:$\bar{\kappa}_f = 65.0$ and $\bar{\kappa}_m = 25.0$ in the unit of differential conductivity, just at CMC_{io} and above CMC region (the constant value), respectively, and $CMC_{io} = 7.8$ mmol/kg as is seen in Fig. 1. In addition, $\bar{\kappa}_m^*$ is postulated to be identical with $\bar{\kappa}_f$. The determined linear relations in Fig. 11 are also shown by the solid line in Fig. 10 for comparison with experimental data.

In addition to K_x, k and $(d\alpha/dX_{am})$ can be obtained from the respective intercepts at $J = 0$ and 1, in Fig. 11. Table 3 lists these values, together with the CMC decreasing rate, $\Delta CMC/\Delta C_a$, estimated as the limiting slope of CMC – Ca plot, and other related quantities.

D. Dependence of K_x on m

In each series of homologues, $\log(K_x)$ is plotted against the carbon number of the hydrophobic part of alkyl-chain in the additives, m (Fig. 12). The relation has the following characteristics. The relation seems to be linear in each homologous series. The free energy of transfer of an additive from the bulk water to micelles is calculated using the relation $\Delta G^\circ = -RT\ln(K_x)$, where R and T are the gas constant and absolute temperature, respectively. The slope of the linear relation in Fig. 12 gives the free energy change per CH_2 group, denoted by $\Delta G^\circ(CH_2)$, as a measure of hydrophobicity of the group (Table 3). $\Delta G^\circ(CH_2)$ in 1-alkanols is estimated to be -2.36 kJ/mol as a mean value, which is in good agreement with the value in the corresponding processes [33].

In the homologues of diols, the line lies below that for alkanols caused presumably by the stronger hydrophilicity of two OH groups. The relation for the diols is almost parallel to that of 1-alkanols, i.e., $\Delta G^\circ(CH_2)$ can be fairly well compared with that of 1-alkanols, consistent with the finding that the longer-chain diols than hexane behave as a hydrophobic substance like 1-alkanols [34].

In the case of branched alkanols (i-ROH), the slope becomes less steep with increasing i, the OH group shifting from an end towards the center of a hydrocarbon chain. It means

Table 3 Partitioning of nonionic homologues in SDS micellar system at 25°C.

Homologue[a]	m^d	Mwt[e]	K_x	$-\Delta G°(CH_2)^f$	$d\alpha/dX_{am}$	$(d\alpha/dX_{am})_{av}^g$	$-k^h$	$-\Delta CMC/\Delta C_a$
1-ROH	4	17	3.61×10^2		0.151		0.0293	0.028
	5	17	9.44×10^2		0.187		0.0796	0.078
	6	17	2.44×10^3		0.172		0.240	0.21
	7	17	6.44×10^3		0.159		0.669	0.53
	8	17	1.63×10^4	2.36	0.133		1.65	1.65
2-ROH	5	17	9.99×10^2		0.151		0.0733	0.073
	6	17	2.22×10^3		0.182		0.181	0.16
	7	17	4.61×10^3		0.184		0.470	0.43
	8	17	1.04×10^4	1.94	0.181		1.24	1.19
3-ROH	5	17	8.88×10^2		0.132		0.0454	0.054
	6	17	1.83×10^3		0.164		0.133	0.13
	7	17	4.11×10^3		0.195		0.413	0.34
	8	17	8.44×10^3	1.86	0.181		0.928	0.87
	9	17	(1.48×10^4)		(0.293)		2.23	3.58
4-ROH	7	17	3.61×10^3		0.191		0.368	0.32
	8	17	7.10×10^3	1.68	0.193		0.720	0.73
	9	17	(1.23×10^4)		0.175		1.31	2.44

Mark	Carbon no.	Mol. wt.		ΔG°				
5-ROH	9	17	(1.16×10^4)		0.201	0.171	1.29	2.3
cAc	8	17	3.60×10^3		0.246	0.246	0.472	0.56
Dio	8	34	2.29×10^3		0.421		0.249	0.39
	9	34	6.79×10^3		0.401		0.802	0.56
	10	34	2.55×10^4	2.99	0.409	0.410	2.60	1.97
β-Glu	6	179	2.45×10^3		0.286		0.216	0.27
	7	179	6.62×10^3		0.290		0.608	0.60
	8	179	1.84×10^4	2.50	0.280		1.84	1.56
α-Glu	8	179	1.40×10^4		0.314	0.293	1.71	1.8
MEG	7[b]	222	7.43×10^3		0.441		0.751	0.95
	8[c]	222	1.34×10^4	1.46	0.417	0.429	1.96	3.0

[a]Mark of homologues: *i*-ROH: isomeric alkanols with OH at the *i*th carbon atom; cAc: cycloalcohol, Dio: α,ω-alkanediol; α- and β-Glu: α- and β-D-alkylglucoside; MEG: MEGA(*n*-alkanoyl-N-methylglucamide).

[b]For MEGA8.

[c]For MEGA9.

[d]Carbon number in hydrocarbon chain of hydrophobic moiety.

[e]Molecular weight of polar head group.

[f]Average $\Delta G^\circ (CH_2)$ for each series of homologues.

[g]Average of $d\alpha/dX_{am}$ for each polar group. The values in parentheses are omitted on averaging.

[h]$k = dC_{sb}/dC_{af}$.

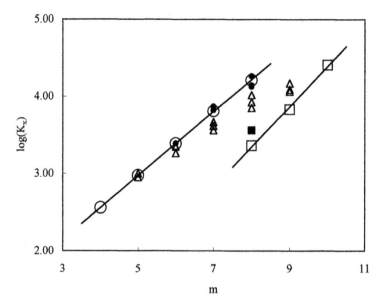

Figure 12 The relation between $\log(K_x)$ and carbon number in hydrophobic alkyl-chain in homologous amphiphiles. ○, 1-ROH; △, *i*-ROH; ●, monopolar aliphatics; ■, cyclooctanol; □, diol.

that the hydrophobicity is weakened by branching (OH shifting), as seen from $\Delta G^\circ(CH_2)$ in Table 3.

K_x of cyclooctanol is far lower than those of 1-octanol and *i*-octanols, and rather near that of octanediol, which suggests that cyclicity weakens the hydrophobicity of the hydrocarbon chain more effectively than branching.

It is surprising that for α- and β-glucosides and MEGAs, K_x is located on the line of 1-alkanols. They have a common molecular structure in the respect that an alkyl chain carries one polar group on its one end, called "monopolar aliphatics" here. Judging from the agreement, it can be said that the partition coefficient depends only on the hydrophobic alkyl-chain length (m), and independent of the type of polar groups, against the usual expectation that the polar group with stronger hydrophilicity suppresses more effectively the transfer of amphiphiles from the bulk to micelles. The contradiction will be discussed later.

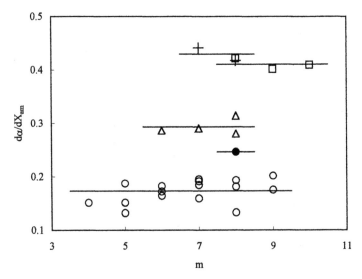

Figure 13 The relation between $d\alpha/dX_{am}$ and carbon number in hydrophobic alkyl-chain in amphiphiles. $\bigcirc$, aliphatic alkanol; $\bullet$, cyclooctanol; $\triangle$, glucoside; $\square$, diol; $+$, MEGA.

E.   Dependence of $d\alpha/dX_{am}$ on Molecular Weight of Polar Group

The $(d\alpha/dX_{am})$ is plotted against m in Fig. 13. The plot clearly shows that homologues of each series have a common value. For all alkanols studied (i-alkanols including 1-alkanols), the values are equal to one another, although the hydrophobicity is influenced by branching, as mentioned before, on $\Delta G^{\circ}(CH_2)$. The $(d\alpha/dX_{am})$ for β-glucosides and α-glucoside cannot be distinguished from each other although they are different in molecular conformation in the solid state [35]. The agreement in glucosides may be due to the free rotation of the polar group in the molecule in the liquid state. It is therefore concluded that $(d\alpha/dX_{am})$ is an intrinsic parameter of the polar head group, independent of the hydrophobic moiety.

The fact that $(d\alpha/dX_{am})$ is positive for each polar group deduces reasonably that the polar groups incorporated in the surface region of ionic micelles reduce the surface charge

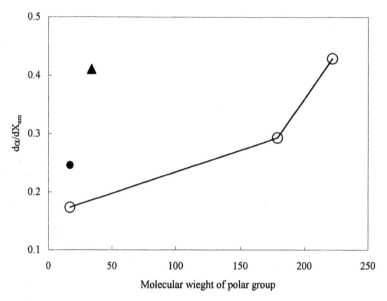

Figure 14 The relation between $d\alpha/dX_{am}$ and molecular weight of polar group in homologous amphiphiles. -⊖-, Monopolar aliphatics; ●, Cyclooctanol; ▲, Diols.

density, resulting in the acceleration of counter-ion dissociation. The cross-sectional area of the polar group should then be taken as a parameter of the polar group. It is difficult, however, to assign a consistently reasonable size to each group, especially in such solubilized state. So, the molecular weight of polar head group is inevitably adopted as a parameter for size, and the relation between $(d\alpha/dX_{am})$ and the molecular weight is shown in Fig. 14.

The monopolar aliphatics give a consistent relation in which $(d\alpha/dX_{am})$ tends to increase monotonically with increasing molecular weight, suggesting that the larger size of polar group accelerates counter-ion releasing more effectively. The monotonous relation infers that α is a function of surface charge density.

With cyclooctanol, however, $d\alpha/dX_{am}$ is higher than those for 1- and i-alkanols, whereas K_x is far lower than those for the other octanol derivatives in Fig. 12. The finding suggests that cyclization might weaken the hydrophobicity of alkyl chain and force a portion of the chain to behave hydrophilically, resulting in the enlarging of the effective size of OH group, as well as the weakening of hydrophobicity.

It should be emphasized that $(d\alpha/dX_{am})$ of diols is about two times higher than that of mono-alkanols, and even close to MEGAs. An acceptable explanation for such a high value is as follows. The two OH groups might behave not independently but cooperatively to form a stretched umbrella-shape by molecular rotating, where the central part of hydrophobic alkyl chain might be varied in nonpolar region of micelles, as supposed from the $\Delta G(CH_2)$ mentioned before.

These results lead to the conclusion that the larger size of polar group dilutes more effectively the surface charge density of micelles to accelerate counter-ion dissociation. In other words, the reduction of charge density caused by solubilized polar group is cancelled by the counter-ion releasing. After all, the surface charge density must be unaltered by the solubilization of nonionic amphiphiles. This is the reason why K_x is independent of the polar group structure of monopolar aliphatics in Fig. 12. As a result, the electrostatic term in free energy change would remain unaltered.

F. The Relation between dC_{sf}/dC_{af} and $\Delta CMC/\Delta C_a$

Table 3 lists the values of k calculated from the intercept at $J = 0$ in Fig. 11 based on Eq. (38) for all the additives studied here. From the definition, k implies the relation between the concentrations of monomerically dissolved surfactant (C_{sf}) and additive (C_{af}) in the bulk water. The negative value of k indicates that C_{sf} is reduced by the addition of an additive dissolved in monomerical form in the bulk water. The decrease in C_{sf} is in line with the well-known fact that CMC decreases with an increase in the concentration of organic additives, owing to the hydrophobic interaction [5,23,31]. In order to compare k with the CMC decreasing rate, the CMC in the presence of all additives studied here was determined by the differential conductivity curve given in Fig. 3. It is obvious that for each additive, k and $\Delta CMC/\Delta C_a$ are in good agreement with each other. Remember that the former is determined above CMC in micellar solution, and the latter is just at CMC. The agreement reveals that the hydrophobic interaction between monomer species of SDS and an additive remains the same at infinite dilution of additive because C_{sf} remains substantially constant at low concentrations close to CMC_{io}, over the wide C_s region. It is concluded that $\Delta CMC/\Delta C_a$ is usable, instead of k, in order to evaluate the relation between the concentrations of surfactant and additive in monomeric form in micellar solution.

Finally, a simple relation was reported between $\Delta CMC/\Delta C_a$ and K_x [36]

$$-\Delta CMC/\Delta C_a = \theta CMC_o/n_w K_x \tag{39}$$

where θ is a constant involving all activity coefficients. The relation between $\Delta CMC/\Delta C_a$ and k with K_x in Table 3 gives the constant value around 0.8 for θ. Except for some short-chain substances, q is possible to be unity reflecting ideal behavior [10].

VI. SUMMARY

The applications of the differential conductivity method have been introduced. The differential conductivity corresponds to the ionic mobility, which is a function of the charge and size of ionic species. This method is then applicable to the system where an ionic species with a different charge and/or size is formed with increasing concentration of a solute. The measurements should be made at low concentrations (lower than about 0.1 mol/kg) of electrolytes, in order to obtain precise differential conductivity data.

We hope the method will be applied to various systems of electrolyte solutions.

REFERENCES

1. P. Mukerjee, *J. Phys. Chem.* **80**: 1388 (1976).

2. I. Garcia-Mateos, M. M. Velazquez, and L. J. Rodriguez, *Langmuir* **6**: 1078 (1990).

3. J. N. Phillips, *Trans. Faraday Sci.* **51**: 561 (1955).

4. H. C. Evans, *J. Chem. Soc.* 579 (1956).

5. M. Manabe, H. Kawamura, A. Yamashita, and S. Tokunaga, *J. Colloid Interface Sci.* **115**: 147 (1987).

6. M. Manabe, H. Kawamura, A. Yamashita, and S. Tokunaga, *The Structure, Dynamics and Equilibrium Properties of Colloidal Systems* (D. M. Bloorand E. Wyn-Jones, eds.), NATO ASI Series–Vol. 324, Kluwer, Dortrecht (1990), p. 63.

7. M. Manabe, H. Kawamura, H. Katsuura, and M. Shiomi, *Proceedings of the Ninth International Symposium on Cyclodextrins* Ed. by (J. J. Torres Labandeira, and J. L. Vila-Jato, eds.), Kluwer, Dordrecht (1999), p. 653.

8. M. Manabe, H. Kawamura, H. Katsuura, and M. Shiomi, *Proceedings of the International Conference on Colloid and Surface Science*, Studies in Surface Science and Catalysis, Vol. 132, (Y. Iwasawa, N. Oyama, and H. Kunieda, eds.), Elsevier, Amsterdam (2001), p. 97.

9. M. Manabe, A. Tokunaga, H. Kawamura, H. Katuura, M. Shiomi, and K. Hiramatsu, *Colloid Polym. Sci.* **280**: 929 (2002).

10. M. Manabe, M. Kaneko, T. Miura, C. Akiyama, H. Kawamura, H. Katsuura, and M. Shiomi, *Bull. Chem. Soc. Jpn.* **75**: 1967 (2002).

11. M. Manabe, M. Funamoto, F. Kohgami, H. Kawamura, and H. Katsuura, *Colloid Polym. Sci.* **281**: 239 (2003).

12. G. C. Kresheck, *Water: A Comprehensive Treaties*, Vol. 4, (F. Franks, ed.), Plenum, New York (1975), p. 96.

13. P. Mukerjee and K. J. Mysels, *Critical Micelle Concentration of Aqueous Surfactant Systems*, Natl. Stand. Ref. Data Ser. (U.S. Natl. Bur. Stand) No.36 (1971), p. 17.

14. Y. Moroi, *Micelles—Theoretical and Applied Aspects*, Plenum, New York (1992), p. 62.

15. R. A. Robinson and R. H. Stokes, *Electrolyte Solutions*, Butterworths, London, (1965), p. 465.

16. P. Mukerjee, *Adv. Colloid Interface Sci.* **1**: 241 (1967).

17. D. M. Small, *Molecular Association in Biological and Related System*, Advances in Chemistry, Vol. 84, American Chemical Society (1984), p. 31.

18. M. Manabe, H. Kawamura, S. Kondo, M. Kojima, and S. Tokunaga, *Langmuir* **6**: 1596 (1990).

19. J. F. Rathman, and J. F. Scamehon, *J. Phys. Chem.* **88**: 5807 (1984).

20. G. S. Manning, *J. Phys. Chem.* **79**: 263 (1975).

21. C. Trainer, A. A. Khodja, and M. Fromon, *J. Colloid Interface Sci.* **128**: 416 (1989).

22. C. Tainer, M. Fromon, and M. H. Mannebach, *Langmuir* **5**: 283 (1989).

23. K. Shinoda, T. Nakagawa, B. Tamamushi, and T. Isemura, *Colloidal Surfactants*, Academic Press, New York (1963).

24. M. Okawauchi, M. Hagio, Y. Ikawa, G. Sugihara, Y. Murata, and M. Tanaka, *Bull. Chem. Soc. Jpn.* **60**: 2718 (1987).

25. M. J. Schick (ed.), *Nonionic Surfactants*, Surfactant Science Series, Vol. 1, Marcel Dekker, New York (1967), p. 496.

26. C. Tanford, *The Hydrophobic Effects*, Wiley, New York (1973), p. 75.

27. Q. R. Huang, P. L. Dubin, C. N. Moorefield, and G. R. Newkome, *J. Phys. Chem.* **104**: 88 (2000).

28. I. Satake, S. Yoshida, K. Maeda, and Y. Kusumoto, *Bull. Chem. Soc. Jpn.* **59**: 3991 (1986).

29. R. Palepu, and J. E. Richardson, *Langmuir* **5**: 218 (1989).

30. J. Szejtli, *Cyclodextrin Technology*, Kluwer, Dordrecht (1988), p. 4.

31. M. Manabe, H. Kawamura, G. Sugihara, and M. Tanaka, *Bull. Chem. Soc. Jpn.* **61**: 1551 (1988).

32. C. Trainer, *Solubilization in Surfactant Aggregation*, Surfactant Science Series, Vol. 55 (S. D. Christian, and J. F. Scamehorn, eds.), Dekker, New York (1995), p. 383.

33. In Ref. 26, p. 51.

34. M. Manabe and M. Koda, *Bull. Chem. Soc. Jpn.* **51**: 1599 (1978).

35. S. Bogusz, R. M. Venable, and R. W. Pastor, *J. Phys. Chem. B*, **105**: 8312 (2001).

36. M. Manabe, M. Koda, and K. Shirahama, *J. Colloid Interface Sci.* **77**: 189 (1980).

4

Diffusion Processes in Mixed Surfactant Systems

TOSHIHIRO TOMINAGA

Department of Applied Chemistry, Okayama
University of Science, Okayama, Japan

SYNOPSIS

In the Introduction, differences between self-diffusion coefficient and mutual diffusion coefficient are described and methods
to measure the coefficients are briefly reviewed. The self-
diffusion coefficient of counterions reflects the degree of counterion binding to micelles as well as the micellar size, shape, and
morphology. Some examples are shown in Section II. From the
self-diffusion coefficient of each surfactant in mixed surfactant
solution, the fraction of each surfactant in the micelle, and then
the composition of mixed micelles can be obtained, with some

examples shown in Section III. Section IV describes the self-diffusion coefficient of mixed micelles, which reflects the size and charge of micelles and ionic strength of the medium. In Section V, the mutual diffusion coefficients of mixed surfactants are described. In the case of zwitterionic and long-chain ionic mixed surfactants, where the concentration of monomeric surfactant ion is extremely low, the mutual diffusion coefficient behaves like that of a spherical polyion. In the case of mixed surfactants, however, cross-term diffusion coefficients can be large.

I. INTRODUCTION

Elucidation of diffusion processes in mixed surfactant systems is important both in fundamental and practical aspects such as detergency, emulsification, digestion, gallstone formation/dissolution, and drug delivery systems [1]. There are two types of diffusion coefficients. One is the self-diffusion coefficient, also called intra-diffusion coefficient or tracer diffusion coefficient, which is defined for each chemical species in the absence of concentration gradient. To measure the self-diffusion coefficient, we tag some chemical species of interest either by radioactive isotope or by some spectroscopic techniques. The most typical modern method to measure self-diffusion is the pulsed field gradient spin-echo NMR (PGSE NMR) method [2,3]. The other type is the mutual diffusion coefficient or inter-diffusion coefficient, for which the solute diffuses with concentration gradient. There are a number of classical methods to measure the diffusion coefficients under macroscopic concentration gradient [2,4–6]. A modern technique to measure the mutual diffusion coefficient is the quasi-elastic or dynamic light scattering method (QELS or DLS), in which the decay of naturally occurring microscopic concentration gradient (fluctuation) is measured [7, 8]. A schematic representation of solute distributions before and after the diffusion is shown in Fig. 1.

For a binary system consisting of a surfactant (component 1) and solvent (component 0), the flux of the surfactant,

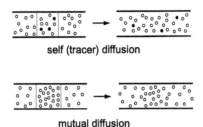

self (tracer) diffusion

mutual diffusion

Figure 1 Schematic representation of solute distributions before and after diffusion.

J, i.e., the number of moles of the surfactant that passes through unit cross sectional area per unit time is given by

$$J = -D\nabla C \tag{1}$$

For a ternary system consisting of two surfactants and solvent, the flux of surfactant i is given by

$$J_1 = -D_{11}\nabla C_1 - D_{12}\nabla C_2 \tag{2}$$

$$J_2 = -D_{21}\nabla C_1 - D_{22}\nabla C_2 \tag{3}$$

where ∇C_is are the concentration gradient for component i. D_{11} and D_{22} are main diffusion coefficients, which relate the fluxes of components 1 and 2, respectively, to their own concentration gradient. D_{12} and D_{21} are cross-term diffusion coefficients, which relate the fluxes of components 1 and 2, respectively, to the concentration gradients of the other component. Although D_{ij} $(i \neq j)$ values can be zero if there is no coupling between the flows of components i and j, they can be large positive and negative values when there is a strong coupling between the flows. Although the treatment can be extended to systems containing three or more solute components, mutual diffusion in such systems is more difficult to study than self-diffusion, where as far as the species of interest can be detected by some means its self-diffusivity can be measured even in the presence of many other components.

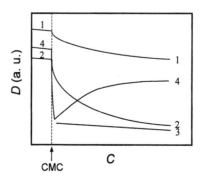

Figure 2 Typical examples of self- and mutual diffusion coefficients for an ionic surfactant as a function of concentration. 1, Self-diffusion coefficient of counterions; 2, self-diffusion coefficient of surfactant ions; 3, self-diffusion coefficient of solubilized molecules in the micelle; 4, mutual diffusion coefficient of the surfactant.

Figure 2 shows self- and mutual diffusion coefficients in a one-component ionic surfactant solution as a function of the total surfactant concentration, C. The self-diffusion coefficient of counterions (curve 1) decreases above the critical micelle concentration (CMC) because counterions are bound to micelles. The self-diffusion coefficient of surfactant ions (curve 2) is a time-averaged value of self-diffusion coefficients of monomer ions and micelles, and it decreases and approaches the self- or tracer-diffusion coefficient of micelles with increasing surfactant concentration. The self- or tracer-diffusion coefficient of micelles (curve 3) is obtained by measuring the diffusion coefficients of probe molecules such as dyes and tetramethylsilane solubilized in micelles. The mutual diffusion coefficient of the total surfactant (curve 4) drops suddenly at the CMC and then increases with increasing surfactant concentration.

II. SELF-DIFFUSION OF COUNTERIONS

The self-diffusion coefficient of counterions reflects the degree of their binding to micelles as well as the size, shape, and morphology of micelles. It has been measured for a number of systems by the PGSE NMR method and others such as the open-ended capillary tube method using radioactive isotopes. According to the two-state model, the experimentally observed self-diffusion coefficient, D_s^{obs}, is the averaged value of the self-diffusion coefficient of dissociated free counterions, D_s^{free}, and that of micelle-bound counterions, which is the same as the self-diffusion coefficient of micelles, D_s^{mic}, and can be expressed as

$$D_s^{obs} = (1 - \beta)D_s^{free} + \beta D_s^{mic} \tag{4}$$

where β is the fraction of micelle-bound counterions. As the D_s values of counterions are sensitive to the degree of their binding to micelles, the latter values can be obtained from the measured D_s^{obs} values.

Kamenka *et al.* [9] measured self-diffusion coefficients for Na^+ and Cl^- ions using ^{22}Na-enriched NaCl and ^{36}Cl-labeled NaCl, respectively. The results are shown in Fig. 3. When dodecyltrimethylammonium bromide ($C_{12}TABr$) was added to a sodium dodecyl sulfate (NaDS) solution, keeping the total surfactant concentration constant, i.e., 50 mM ($1 M = 1 mol/dm^3$ throughout this chapter), the sodium ion self-diffusion coefficient, D_s^{Na+}, increased until the $C_{12}TABr$ molar fraction, X, reached 0.32, reflecting the release of micelle-bound Na^+ ions. In this range ($0 < X < 0.32$), the chloride ion self-diffusion coefficient, D_s^{Cl-}, does not change because Cl^- ions do not bind to the micelles. In the range $0.32 < X < 0.38$, both D_s^{Na+} and D_s^{Cl-} values decrease with increasing molar fraction of $C_{12}TABr$. These unexpected decreases were found to be due to the vesicle formation, where both Cl^- and Na^+ ions are (partially) entrapped in the closed assemblies even if they do not bind to the aggregates.

When lithium perfluorooctylsulfonate (LiFOS) was added to tetraethylammonium perfluorooctylsulfonate (TEAFOS), keeping the total surfactant concentration constant ($c = 100 mM$), the self-diffusion coefficient of TEA^+ ion, D_s^{TEA+}, decreased with increasing molar fraction of LiFOS, X_{Li}, until X_{Li} reaches 0.6 [10]. This was interpreted in terms of the preferential binding of TEA^+ ions to the micelles. In the range $0.6 < X_{Li}$, the D_s^{TEA+} value increases with increasing X_{Li},

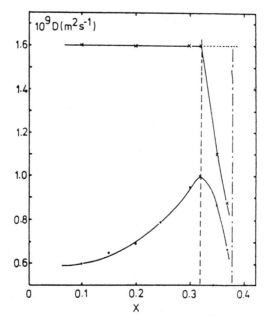

Figure 3 Self-diffusion coefficients of Na^+ (●)and Cl^- (x) ions in $C_{12}TABr/NaDS$ mixtures as a function of the $C_{12}TABr$ mole fraction X. Total concentration is 0.050 M. The turbid systems are observed between the vertical broken and broken/dotted lines. (Adapted from Ref. 9.)

reflecting the micellar shape change from threadlike ($X_{Li} < 0.6$) to spherical ($0.6 < X_{Li}$). The D_s^{TEA+} values are shown later (Fig. 5) together with the surfactant self-diffusion coefficients.

III.   SELF-DIFFUSION OF SURFACTANTS

As was the case for counterions, the experimentally observed self-diffusion coefficient of a surfactant is the average value of the surfactant diffusion coefficients in micellar and monomeric states when the micelle-monomer exchange rate is rapid, which is the case for many micellar systems.

$$D_s^{obs} = (1 - \beta)D_s^{mon} + \beta D_s^{mic} \qquad (5)$$

where D_s^{mon} and D_s^{mic} are the self-diffusion coefficients of the surfactant in monomeric and micellar states, respectively, and β is the fraction of the surfactant in micellar state. Equation (5) is used to obtain the fraction of surfactant in the micelle from the D_s values. For multi-component surfactant systems, as far as the three D_s values in Eq. (5) can be obtained for each surfactant, the composition of mixed micelles can be obtained. Stilbs *et al.* have proposed the "component-resolved (CORE)" NMR method and applied to measure D_s values in multi-component surfactant systems [11].

A. Nonionic–Ionic Surfactant Mixtures

Figure 4 shows the self-diffusion coefficients of sodium hexanesulfonate (C_6SNa), pentaethylene glycol monohexyl ether (C_6E_5), and tetramethylsilane (TMS) in D_2O at 25°C as a function of the C_6E_5 molality, m_{C6E5} [12]. The molality of C_6SNa, m_{C6SNa}, is kept constant at $0.08\,mol/kg_{D2O}$ (Fig. 4a), which is below the CMC of neat C_6SNa ($m_{C6SNa} = 0.56mol/kg_{D2O}$), and D_s values correspond to the respective monomeric surfactants at $m_{C6E5} < 0.081\,mol/kg_{D2O}$. The break point of the D_s curve for C_6E_5 indicates that the two surfactants form mixed micelles at concentrations $m_{C6E5} = 0.081$ and $m_{C6SNa} = 0.08\,mol/kg_{D2O}$. Hydrophobic TMS molecules are solubilized in the mixed micelles, and the D_s values for TMS can be taken as the self-diffusion coefficients of the mixed micelles as long as TMS molecules do not change the shape and size of micelles. Figure 4(b) shows D_s values for the same system with the C_6SNa concentration ($m_{C6SNa} = 0.7\,mol/kg_{D2O}$) higher than its CMC ($m_{C6SNa} = 0.56\,mol/kg_{D2O}$). In this case, D_s values for both surfactants decrease from the beginning when C_6E_5 is added since both surfactants are incorporated in the micelles. At high C_6E_5 concentrations ($m_{C6E5} = 0.4$–$0.6\,mol/kg_{D2O}$), D_s^{C6E5} values are close to D_s^{TMS} values but D_s^{C6SNa} values are still higher than D_s^{TMS} values. This fact reflects that almost all C_6E_5 molecules are in the micellar state but some fraction of C_6SNa is in the monomeric state. In fact, these authors have calculated mixed micelle compositions using Eq. (5).

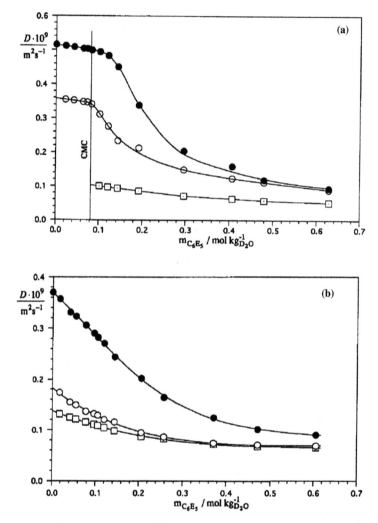

Figure 4 Self-diffusion coefficients of C_6SNa (●), C_6E_5 (○), and TMS (□), in the mixed system, $D_2O + C_6SNa + C_6E_5$, at 298.2 K. (a) $m_{C6SNa} = 0.08$ mol/kg$_{D2O}$; (b) $m_{C6SNa} = 0.7$ mol/kg$_{D2O}$. (Adapted from Ref. 12.)

For mixed solutions of NaDS and a sugar-based nonionic surfactant, dodecylmalonobis(N-methylglucamide) (DBNMG), Griffiths *et al.* have determined the self-diffusion coefficients of both surfactants [13] applying the new component-resolved

(CORE) NMR method [11], in addition to the conventional PGSE NMR method. When NaDS is added to DBNMG (CMC = 0.44 mM) of fixed concentrations (9.1, 15.0, and 25.0 mM), D_s^{DS-} values increase with increasing NaDS concentration (0–30 mM), reflecting the increasing fraction of monomeric DS$^-$ ions. The D_s^{DBNMG} values, which can be regarded as the D_s values of the mixed micelles because the monomer contribution is low, are the same as those in the absence of NaDS, suggesting that "the inclusion of NaDS into the mixed micelles does not significantly alter the size of the micelles." Using Eq. (5), the authors calculated the fractions and concentrations of monomeric DS$^-$ ions, and they reported that the two surfactants mix noncooperatively and ideally.

A similar method has been applied to the system, pentaethylene glycol monodecyl ether $(C_{10}E_5)$ + NaDS, to obtain the fractions of each surfactant in the mixed micelles and, hence, the composition of the mixed micelles as a function of the total surfactant concentration [14]. The micellar composition could be explained in terms of the regular solution theory, and the two surfactants were found to interact moderately. Dodecyl amido propyl betaine (LAPB) and NaDS were found to interact very strongly, and "the regular solution theory fails to account for the experimental data at high NaDS molar fractions" [14].

B. Anionic–Anionic Surfactant Mixtures

For mixed micellar systems of TEAFOS + LiFOS, Nakahara *et al.* [10] measured the self-diffusion coefficient of FOS$^-$ surfactant ions in addition to that of TEA$^+$ counterions as described in Section II. The results are shown in Fig. 5. The D_s^{FOS-} value is low in the molar fraction range $0 < X_{Li} < 0.6$, where micelles are threadlike in structure and both the monomer fraction and the D_s value of the micelles are low. The D_s^{FOS-} value increases with increasing X_{Li} in the range $0.6 < X_{Li}$, where the micelle shape becomes spherical and the both monomer fraction and the D_s value of the micelles increase.

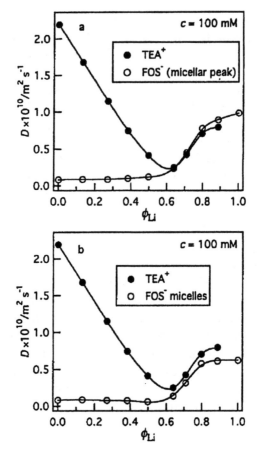

Figure 5 Self-diffusion coefficients as a function of the Li^+ mole fraction ($\Phi_{Li} = n_{Li}/(n_{Li} + n_{TEA})$) for the mixture, TEAFOS + LiFOS, in D_2O at 303.2 K. (a) D_s^{FOS} values obtained from the "micellar" peak of FOS^- ions; (b) D_s^{FOS} values calculated for the micellized FOS^- ions according to Eq. (5). The D_s^{TEA} values are average values measured both for (a) and (b). (From Ref. 10.)

C. Anionic–Cationic Surfactant Mixtures

For NaDS + octyltrimethylammonium bromide (C_8TABr) mixed surfactant solutions in D_2O, the self-diffusion coefficients of the respective surfactant ions, DS^- and C_8TA^+, have been measured at the total surfactant concentration range 10–35 mM and the molar fraction of C_8TABr, X_2, being 0.11, 0.22, and 0.28 [15]. The D_s^{C8TA+} values are smaller than D_s^{DS-} values, reflecting that the fraction of C_8TABr in the micelle is larger than that of NaDS in the micelle. Similar results have been found for sodium decanesulfonate $(NaC_{10}SO_3) + C_8TABr$ and sodium octanesulfonate $(NaC_8SO_3) + C_8TABr$ mixed surfactant systems [16].

IV. SELF-DIFFUSION OF MICELLES

As can be seen from Eq. (5) and Fig. 4, D_s values for surfactants are the average values for monomeric and micellar states. When the monomer contribution is negligible, D_s values for surfactants can be taken as those for micelles. If hydrophobic molecules are solubilized in micelles without changing micellar size and shape, D_s values for the solubilized molecules can be taken as those for micelles. Although diffusion coefficients obtained by QELS are mutual diffusion coefficients, they can be considered as the self-diffusion coefficients of micelles if medium ionic strength is sufficiently high, where micelle-micelle and micelle-counterion interactions are negligible (see Section V, however).

A. Effect of Size and Charge on the Self-Diffusion of Micelles

According to the Stokes–Einstein equation, the self-diffusion coefficient of a sphere of radius r_S in a continuum medium of the viscosity η is given by

$$D_s = \kappa_B T/(6\pi\eta r_S) \tag{6}$$

where κ_B is the Boltzmann constant, T is the temperature in K. For ionic micelles, their hydrodynamic radii calculated from Eq. (6) using measured D_s values are in many cases larger than those estimated for hydrated micelles, suggesting a drag due to the electrostatic interactions between micelles and the medium including counterions [17,18]. Figure 6 shows D_s values for

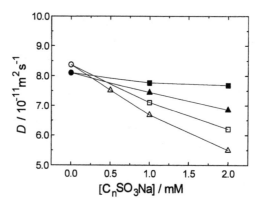

Figure 6 Tracer diffusion coefficients for 10 mM C_{14}DAPS as a function of the C_nSO_3Na concentration at 298.2 K. O, C_{14}DAPS in water; △, C_{14}DAPS + $C_{18}SO_3Na$ in water; □, C_{14}DAPS + $C_{10}SO_3Na$ in water; ●, C_{14}DAPS in 0.1 M NaBr; ▲, C_{14}DAPS + $C_{18}SO_3Na$ in 0.1 M NaBr; ■, C_{14}DAPS + $C_{10}SO_3Na$ in 0.1 M NaBr. (From Ref. 19.)

mixed micelles of a zwitterionic surfactant, 3-(N-tetradecyl-N,N-dimethylammonio)propanesulfonate (C_{14}DAPS), and an ionic surfactant, sodium octadecanesulfonate ($C_{18}SO_3Na$) or sodium decanesulfonate ($C_{10}SO_3Na$), which were obtained by measuring the diffusion coefficient of pyrene molecules solubilized in the micelles using the Taylor dispersion method [19]. Essentially all pyrene molecules are solubilized in the micelles, and the D values can be taken as D_s values for the micelles. When the respective ionic surfactants are added to 10 mM C_{14}DAPS (CMC = 0.3 mM), i.e., the charge of the micelles is increased, the D values decrease significantly, i.e., 34% and 26% decreases when 2 mM $C_{18}SO_3Na$ and $C_{10}SO_3Na$ are added, respectively. Addition of salt (0.1 M NaBr) decreases the D value of neat C_{14}DAPS micelles only a little (3%), but increases the D values appreciably for the charged mixed micelles. In the case of 10 mM C_{14}DAPS + 2 mM $C_{18}SO_3Na$ system, the D value increases 22% when 0.01 M NaBr is added, and remains essentially constant over the concentration range 0.01–0.5 M NaBr (only the result in 0.1 M NaBr is shown in Fig. 6). The constant D value in the wide salt concentration range suggests that the drag due to the electrostatic interaction between the micelles and the medium becomes negligible and the size of the micelles remains constant in this NaBr concentration range. In the case of $10\,mM$ $C_{14}DAPS + 2\,mM$ $C_{10}SO_3Na$ system, the D value increases by 24% when $0.1\,M$ NaBr is added. The very similar degree of increase in the D value for the two systems suggests that essentially all $C_nSO_3^-$ ions are solubilized in the micelles and the 22–24% increase (18–19% decrease from NaBr solution to water) is ascribed to the electrostatic interaction. Even in $0.1\,M$ NaBr, when $2\,mM$ $C_{10}SO_3Na$ and $C_{18}SO_3Na$ are separately added to $10\,mM$ $C_{14}DAPS$, the D value decrease 5% and 17%, respectively. This fact suggests the increase in micellar size by incorporation of the ionic surfactants. The increase in the aggregation number by incorporation of $C_{18}SO_3Na$ to $C_{14}DAPS$ micelles was confirmed by fluorescence decay measurements for pyrene solubilized in the micelles [20]. Thus, the decrease in D value with increasing concentration of C_nSO_3Na (Fig. 6) reflects both electrostatic interactions and size change in water, but only micellar size change in $0.1\,M$ NaBr.

In Fig. 7, D_s values are plotted against κa, where a is the micellar radius for which r_S values at an appropriately high

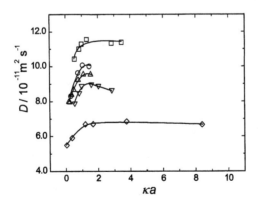

Figure 7 Tracer diffusion coefficients as a function of κa. $\diamondsuit$, $10\,mM$ $C_{14}DAPS + 2\,mM$ $C_{18}SO_3Na$ (298.2 K); $\triangledown$, $C_{14}TAB$ (298.2 K); $\triangle$, $C_{18}TAC$ (308.2 K); $\bigcirc$, $C_{16}TAB$ (308.2 K); $\square$, $C_{14}TAB$ (308.2 K). (From Ref. 19.)

ionic strength are used and κ^{-1} is the Debye screening length defined as

$$\kappa^2 = e^2 \sum_i n_i z_i^2 / (\varepsilon_0 \varepsilon_r k_B T) \tag{7}$$

where e is the electronic charge, n_i is the concentration of the ith ion, z_i is the valency of the ith ion, ε_0 is the vacuum permittivity, and ε_r is the relative permittivity of the medium [19]. In addition to the mixed micelles of 10 mM $C_{14}DAPS$ + 2 mM $C_{18}SO_3Na$, the results for one component ionic micelles are also shown. In all cases, D_s value decreases with decreasing κa when $\kappa a < 1$. This means that caution has to be taken if we want to obtain the size of colloidal particles from D_s value at low ionic strength. For one order of magnitude larger colloidal particles, D_s value has also been reported to decrease with decreasing κa, i.e., for polystyrene latex spheres of $a = 18$ nm down to $\kappa a = 0.6$ [21] and of $a = 60$ nm down to $\kappa a = 0.2$ [22]. Some theories predict that D_s value passes through a minimum and increases again with decreasing ionic strength (κa) [23–25], and there are experimental studies for gold particles ($a = 10$ nm) and polystyrene latex spheres ($a = 20$ nm) [26] as well as for tobacco mosaic virus [27] that show D_s value has a minimum at $\kappa a \approx 1$ and increases again with decreasing κa. Diffusion processes of charged particles at low ionic strength are not completely understood.

V. MUTUAL DIFFUSION OF SURFACTANTS

A. One-Component Ionic Surfactants in Water with and without Added Salts

Before we go into mixed surfactant systems, it is worth summarizing the results for one-component ionic surfactant systems. In water, the measurements have been made by various methods such as QELS [28–35], Taylor dispersion [36–38], boundary spreading [39], and Harned's conductivity cell [40]. In contrast to self-diffusion coefficient, mutual diffusion coefficient increases with increasing surfactant concentration above its CMC (see Fig. 2, curve 4) except when a salt of high concentration is added, where the opposite concentration dependence has been observed.

One interpretation of the positive concentration dependence of mutual diffusion coefficient is that, because of the intermicellar repulsion, micelles diffuse faster (from more concentrated region to less concentrated region) than expected from the concentration gradient. Mutual diffusion coefficients for NaDS in aqueous NaCl solutions were fitted by

$$D = D_0(1 + \kappa_D c_m) \tag{8}$$

where D_0 is the mutual diffusion coefficient at the CMC (extrapolated value), c_m is the micellar concentration (total surfactant concentration minus CMC), and κ_D is a constant [29,39]. The κ_D value reflects the interaction due to interparticle forces and the hydrodynamic interaction resulting from the fact that the movement of one particle through the fluid generates a velocity field which affects the motion of neighboring particles. The κ_D value takes the largest positive value when no salt is added, reflecting a large repulsive interparticle interactions; it decreases with increasing concentration of added salt and finally becomes negative, reflecting attractive interparticle interactions. Corti and Degiorgio [29] obtained, as the best-fit parameters, a micellar charge $q = 37$ and the Hamaker constant $A = 11.3 \kappa_B T \cong 4.5 \times 10^{-20}$ J, by which they could explain positive and negative κ_D values for salt concentrations lower than 0.4 M and higher than 0.5 M, respectively. However, the micellar charge $q = 37$ corresponds to the degree of counterion dissociation $1 - \beta = 0.6$, which is significantly larger than those obtained by other methods, i.e., 0.2–0.4 [41].

Another interpretation of the increase in mutual diffusion coefficient with increasing surfactant concentration is the electrostatic coupling between micelles, counterions, and monomeric surfactant ions. In the simplest case of a polyion having charge q and q monovalent counterions (q-uni-valent electrolyte), counterions diffuse faster than the polyion, but because of the charge neutrality, the polyion and counterions have to

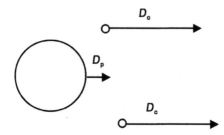

Figure 8 Schematic representation of the electrostatic coupling between a polyion and counterions. Only two counterions are shown.

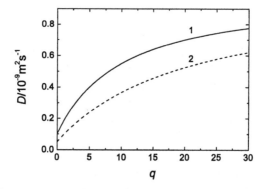

Figure 9 The calculated mutual diffusion coefficient according to Eq. (9) as a function of the polyion charge q 1, $D_p = 0.1 \times 10^{-9}$ m^2/s; 2, $D_p = 0.5 \times 10^{-9}$ m^2/s. $D_c = 1 \times 10^{-9}$ m^2/s for both cases.

diffuse together (see Fig. 8). In this case, the mutual diffusion coefficient of the polyelectrolyte is given by [4]

$$1/D = \left(\frac{1}{D_p} + \frac{q}{D_c}\right)/(1 + q) \tag{9}$$

where D_p and D_c are the self-diffusion coefficients of the polyion and counterions, respectively. Figure 9 shows the D values as a function of q calculated according to Eq. (9). In the case of micellar systems, there are monomeric surfactant ions in addition to micelles and counterions. This makes the expression of D for the surfactant quite complicated [36,37,40]. Leaist [40] has obtained the following equation for the mutual diffusion coefficient of the one-component ionic surfactant when no salt is added

$$D = \frac{c_1 c_2 D_1 D_2 + (n-q)^2 c_1 c_m D_1 D_m + n^2 c_2 c_m D_2 D_m}{c_1 D_1 + c_2 D_2 + q^2 c_m D_m}$$
$$\times \frac{c_1 + c_2 + q^2 c_m}{c_1 c_2 + (n-q)^2 c_1 c_m + n^2 c_2 c_m} \qquad (10)$$

where c_1, c_2, c_m, D_1, D_2, D_m are the equilibrium concentrations and self-diffusion coefficients of the monomeric surfactant, counterion, and micelle, respectively, n is the aggregation number of the micelle, and q is the charge of the micelle.

B. Mixed Surfactant Systems at Low Monomeric Surfactant Ion Concentrations

In the case of one-component ionic surfactant, when surfactant concentration, C, is not sufficiently high compared to CMC, there are significant fractions of monomeric surfactant ions, which interrupt the electrostatic coupling between micelles and counterions and reduce the mutual diffusion coefficient. If we add a long-chain ionic surfactant to a nonionic or zwitterionic surfactant micellar solution, most fraction of the ionic surfactant is expected to be solubilized in micelles, making the monomer-ion concentration very low. This means $c_1 \cong 0$ and $c_2 \cong q c_m$, and Eq. (10) reduces to Eq. (9) by substitution of $D_2 = D_c$ and $D_m = D_p$. In a one-component surfactant system, this situation is only met at high C ($>>$CMC), but in mixed micelle system it would be met at relatively low concentrations where the hydrodynamic interaction is not significant.

The above-mentioned idea was tested by adding sodium octadecanesulfonate ($C_{18}SO_3Na$) or octadecyltrimethylammonium chloride ($C_{18}TACl$) to a solution of a zwitterionic surfactant, N-tetradecyl-N,N-dimethyl-3-ammonio-1-propane-sulfonate ($C_{14}DAPS$) [20,42]. Figure 10 shows the D values when $C_{18}TACl$, $C_{18}SO_3Na$, and $C_{18}SO_3K$ are respectively added to 10 mM solution of $C_{14}DAPS$ [42]. Also shown are the tracer (self-) diffusion coefficients of the corresponding micelles. When

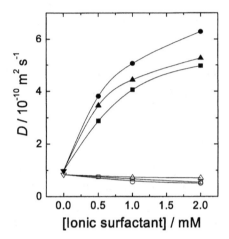

Figure 10 Mutual diffusion coefficients for mixed surfactants (D_m) and tracer diffusion coefficients for mixed micelles (D_t) for 10 mM $C_{14}DAPS + x$ mM ionic surfactants in water at 298.2 K. ●, D_m for $C_{18}SO_3K$; ▲, D_m for $C_{18}TACl$; ■, D_m for $C_{18}SO_3Na$; ○, D_t for $C_{18}SO_3K$; △, D_t for $C_{18}TACl$; □, D_t for $C_{18}SO_3Na$. (From Ref. 42.)

no ionic surfactant is added, the mutual diffusion coefficient for $C_{14}DAPS$ surfactant and the tracer diffusion coefficient for $C_{14}DAPS$ micelle are very close to each other. When the ionic surfactants are added, the tracer diffusion coefficient decreases as described in Section IV, while the mutual diffusion coefficient increases significantly. D values are larger when $C_{18}SO_3K$ is added than when $C_{18}SO_3Na$ is added. At least qualitatively, this is consistent with the expectation from Eq. (9) because the D_c value is larger for K^+ $(1.96 \times 10^{-9}\,m^2/s$ at infinite dilution) than for Na^+ $(1.33 \times 10^{-9}\,m^2/s$ at infinite dilution). Although the D_c value for Cl^- $(2.03 \times 10^{-9}\,m^2/s$ at infinite dilution) is even larger than that for K^+, the D values are smaller when $C_{18}TACl$ is added than when $C_{18}SO_3K$ is added. This turned out to be larger aggregation numbers for $C_{14}DAPS + C_{18}SO_3M$ $(M^+ = Na^+$ or $K^+)$ micelles than those for $C_{14}DAPS + C_{18}TACl$ micelles, i.e., larger q values for the anionic mixed micelles [Eq. (9) and Fig. 9].

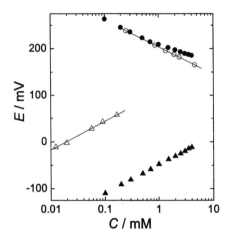

Figure 11 EMF values as a function of the $C_{18}TACl$ or NaCl concentration at 298.2 K. $\triangle$, $E(C_{18}TA^+)$ for $C_{18}TACl$ in water; $\blacktriangle$, $E(C_{18}TA^+)$ for $C_{18}TACl$ in 10 mM $C_{14}DAPS$; $\bigcirc$, $E(Cl^-)$ for NaCl in water; $\bullet$, $E(Cl^-)$ for $C_{18}TACl$ in 10 mM $C_{14}DAPS$. (Adapted from Ref. 42.)

Figure 11 shows EMF values for the $C_{14}DAPS + C_{18}TACl$ system obtained with $C_{18}TA^+$ and Cl^- ion selective electrodes. The E_{C18TA+} values show that 99.7% of $C_{18}TA^+$ ions are solubilized in the micelles and only 0.3% is in the monomeric state in the aqueous phase. From the E_{Cl^-} values, the degree of counterion dissociation, α, was calculated. Table 1 summarizes the results for the $C_{14}DAPS + C_{18}TACl$ and $C_{14}DAPS + C_{18}SO_3Na$ systems. The agreement between the observed and calculated values by Eq. (9) are within 10% for the 10 mM $C_{14}DAPS + 1$ mM $C_{18}TACl$ and 10 mM $C_{14}DAPS + 1$ mM $C_{18}SO_3Na$ systems. Strictly speaking, Eq. (9) is valid only at infinite dilution. The larger deviation for the 2 and 4 mM $C_{18}TACl$ systems needs further study to find a reasonable explanation for it.

Figure 12 shows the mutual diffusion coefficient for the $C_{14}DAPS + C_{18}SO_3Na$ system as a function of the concentration [20]. Not only the zwitterionic micelles that have no net charge but also the charged mixed micelles show no positive concentration dependence. This is in marked contrast with

Table 1 Aggregation number of mixed micelles, n, degree of dissociation, α, micellar charge, q, calculated mutual diffusion coefficient, D_{calc}, and the ratio D_{obs}/D_{calc}.

[C$_{14}$DAPS] (mM)	[Ionic surfactant] (mM)	n	α	q	D_{calc} (10^{-10}m^2/s)	D_{obs}/D_{calc}
C$_{18}$SO$_3$Na[a]						
10	0	87				
10	1	104	0.92	8.7	4.50	0.90
10	2	120	0.81	16.2	5.65	0.88
C$_{18}$TACl[b]						
10	1	94	0.87	7.4	4.83	0.92
10	2	96	0.69	11.0	6.17	0.85
10	4	95	0.56	15.2	6.94	0.77

[a]From Ref. 20.
[b]From Ref. 42.

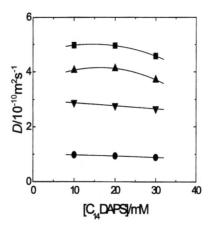

Figure 12 Mutual diffusion coefficients for the C$_{18}$SO$_3$Na + C$_{14}$DAPS system. C$_{18}$SO$_3$Na : C$_{14}$DAPS (mol : mol) = 0 : 1 (●), 0.05:1 (▼), 0.1 : 1 (▲), and 0.2 : 1 (■). (Adapted from Ref. 20.)

the results for one-component ionic micellar systems that show large positive concentration dependence. The result suggests that the factor that increases the D value is not the intermicellar repulsion but the electrostatic coupling between micelles and counterions, at least in this case.

C. Cross-Term Diffusion Coefficients for Ionic Surfactant and Salts

QELS (DLS) is a widely used technique to study small particles, especially macromolecules and micelles. Although the theory for QELS diffusion measurement is well established for binary solutions [7,8], it is not the case for multi-component solutions. In practice, ionic micelles are studied in salt solutions. When no salt is added, experimental results reflect electrostatic interactions very strongly and scattered light intensity is low. If one wants to obtain the hydrodynamic radius of micelles, the measured values have to be extrapolated back to CMC (infinite dilution of micelles), where there is no intermicelle interaction; the extrapolation process can accompany large uncertainty. With increasing salt concentration, the slope of D vs C plot decreases and the uncertainty in determining the hydrodynamic radius becomes smaller unless the aggregation number of micelles is altered.

Using the Taylor dispersion method, Least and Hao [38] obtained D_{ij} values for NaDS (component 1) in NaCl (component 2) solutions, and compared the D_{11} values with the most reliable QELS values [29]. Figure 13 shows their results. Surprisingly, the QELS values were found to be larger than the D_{11} values. Furthermore, the QELS values were found to be close to the lower eigenvalue, $\mathscr{D}_L$, of the mutual diffusion coefficient matrix. The higher and lower eigenvalues are expressed as

$$\mathscr{D}_L = \tfrac{1}{2}\left(D_{11} + D_{22} + \sqrt{(D_{11} - D_{22})^2 + 4D_{12}D_{21}}\right) \qquad (11)$$

$$\mathscr{D}_H = \tfrac{1}{2}\left(D_{11} + D_{22} - \sqrt{(D_{11} - D_{22})^2 + 4D_{12}D_{21}}\right) \qquad (12)$$

As can be seen from Eqs. (11) and (12), if either of the cross-terms D_{ij} $(i \neq j)$ is zero, $\mathscr{D}_H$ and $\mathscr{D}_L$ values are equal to D_{ii} values. In this case of NaDS + NaCl, the lower eigenvalue was found to be 7–14% larger than the D_{11} value. Although the discrepancy between Corti and Degiorgio's QELS values [29]

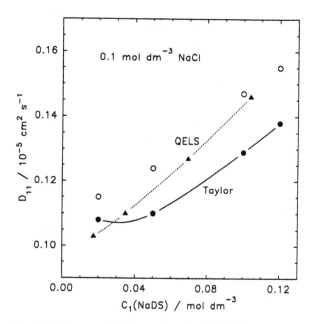

Figure 13 Diffusion coefficients for NaDS (1) in 0.100 M aqueous NaCl (2) solutions at 298.2 K. ▲, QELS; ●, D_{11} from Taylor dispersion; ○, lower eigenvalue calculated from Taylor dispersion data. (From Ref. 38.)

and Leaist and Hao's D_{11} values [38] is not very large, the latter authors state that "it does seem to contradict the interpretation that QELS gives the diffusion coefficient of the micelle species."

D. Nonionic–Ionic Surfactant Mixtures with Cross-Term Diffusion

Vitagliano *et al.* [43] measured main and cross-term diffusion coefficients for the ternary system, C_6E_5 (component 1) + C_6SNa (component 2) + water, using a Gouy diffusiometer. The five average compositions at which D_{ij} were measured are shown in Fig. 14 and the D_{ij} values obtained are summarized in Table 2. At point 1, where the surfactant concentration is below the CMC, both main diffusion coefficients D_{11} and D_{22} are close to the binary system values, i.e., 0.464×10^{-9} and

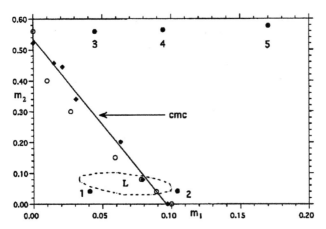

Figure 14 Average composition in molality of diffusion runs (●) for the system, C_6E_5 (1) $+ C_6SNa$ (2). ◆, Measured critical micelle composition; ○, critical micelle composition from self-diffusion measurements in D_2O solutions. L is approximate range of miscibility gap. (From Ref. 43.)

$0.910 \times 10^{-9} \, m^2/s$ for C_6E_5–water and C_6SNa–water systems, respectively. Although a cross-term diffusion coefficient D_{12} is close to zero, the other cross-term value D_{21} is large and positive. This means that even without initial concentration gradient of itself, C_6SNa diffuses with C_6E_5 that diffuses according to its own concentration gradient. This can be interpreted as due to "the tendency of C_6SNa to diffuse away from the cloud area"; note that region L in Fig. 14 shows the miscibility gap. At point 2, which is a little above the CMC and where mixed micelles exist, D_{11} value drops drastically. The sharp drop of D_{11} value after passing through the CMC is in contrast with the self-diffusion coefficient of C_6E_5 that decreases gradually after passing through the CMC (see Fig. 4a). D_{22} value also decreases but not so drastically, reflecting that most of the micelle constituents are C_6E_5 and most of C_6SNa is in the monomeric state. D_{12} value is positive while D_{21} value is negative. These signs are in accord with the tendency of C_6E_5 (toward a region of less concentrated C_6SNa) and C_6SNa (toward a region of more concentrated C_6E_5) to

Table 2 Main and cross-term diffusion coefficients for the system C_6E_5 (1) $+ C_6SNa$ (2) $+$ water measured at five different average concentrations C_1^0 and C_2^0 and 298.15 K[a].

Point[b]	C_1^0	C_2^0	D_{11}	D_{12}	D_{21}	D_{22}
1	0.0399	0.0399	0.439 ± 0.005	0.006 ± 0.005	0.119 ± 0.024	0.883 ± 0.017
2	0.1002	0.0401	0.189 ± 0.005	0.049 ± 0.002	-0.038 ± 0.003	0.748 ± 0.013
3	0.0409	0.5117	0.197 ± 0.004	0.006 ± 0.001	-0.224 ± 0.015	0.581 ± 0.020
4	0.0850	0.5090	0.162 ± 0.002	0.008 ± 0.001	-0.210 ± 0.009	0.490 ± 0.012
5	0.1500	0.5088	0.159 ± 0.009	0.022 ± 0.004	-0.118 ± 0.024	0.487 ± 0.030

[a]From Ref. 43. Unit: C_1^0, mol/dm^{-3}; D_{ij}, 10^{-9} m^2/s.
[b]For the location, see Fig. 14.

move apart from the phase separation area. At points 3, 4, and 5, where the C_6E_5 concentration is varied keeping the C_6SNa concentration constant, D_{11} value decreases slightly with increasing C_6E_5 concentration and approaches the C_6E_5 micelle (self-) diffusion coefficient ($0.125 \times 10^{-9}\,m^2/s$). D_{22} value also decreases slightly with increasing C_6E_5 concentration but is still much larger than D_{11} value, indicating that only small part of C_6SNa is in the mixed micelles. While the D_{12} values are positive but small, the D_{21} values are large and negative. The latter fact can be interpreted as follows. Although the total C_6SNa concentration is constant, the C_6SNa concentration in the mixed micellar state is higher and that in the monomeric state is lower when the C_6E_5 concentration is higher. As a result, C_6SNa diffuses with C_6E_5 micelles according to the C_6E_5 gradient for some period of time, but it diffuses as a monomer against the C_6E_5 concentration gradient for most period of time, thus giving the net counterflow.

E. Ionic–Ionic Surfactant Mixtures with Cross-Term diffusion

Leist and MacEvan [44] measured main and cross-term diffusion coefficients for the mixed surfactant system, NaDS (component 1) + sodium octanoate (NaOct, component 2) in water. The D_{ij} values for the $C_1 + C_2 = 0.1\,M$ system are shown in Fig. 15 as a function of the NaDS mole fraction. The reason for the increase in D_{11} with increasing NaDS fraction is similar to that for the case of one-component NaDS. The decreasing tendency of D_{22} reflects the tendency for NaOct to diffuse in the mixed micelles. The most remarkable point is the large value of D_{12}, particularly, at large NaDS fractions. When there is an initial concentration gradient for NaOct but not for NaDS, the diffusion of free Na^+ and Oct^- produces an electric field, i.e., diffusion potential [4], and accelerates the diffusion/migration of the mixed micelles. The mobility or mass transfer efficiency is much higher when surfactant ions diffuse as a micelle than when they diffuse as the same number of monomers [40]. In sharp contrast, the D_{21} values are small. When there is an initial concentration gradient for NaDS but not for NaOct, as a

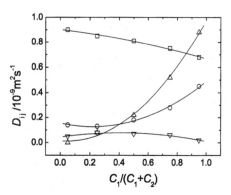

Figure 15 Main and cross-term diffusion coefficients of aqueous NaDS (C_1) + NaOct (C_2) solutions for $C_1 + C_2 = 0.100\,\text{M}$ at 298.2 K. $\bigcirc$, D_{11}; $\square$, D_{22}; $\triangle$, D_{12}; ∇, D_{21}. (Adapted from Ref. 44.)

result of stronger micelle forming tendency of NaDS, relatively small gradients in free Na^+ and DS^- ions are produced. Furthermore, the micelles are highly enriched in NaDS and only small fraction of NaOct is transported in the micellar form, resulting in small D_{21} values.

In summary, because of the usefulness of PGSE-NMR technique, quite a number of active researches are going on for the self-diffusivity measurement. The QELS method is also popular for the study of macromolecules. Although this is a powerful technique, it uses naturally occurring concentration gradient/fluctuation, and is not necessarily suitable for the study of cross-term diffusion at the moment. When people found the increase in mutual diffusivity with increasing concentration, it was a kind of surprise. Now we know what is going on there. To understand interacting flows in mixed surfactant systems, we have to wait for future study to be done.

REFERENCES

1. D. F. Evans, and H. Wennerstrom, *The Colloidal Domain*, 2nd edn. Wiley-VCH, New York (1999), pp. 33–37, 181–183.

2. H. J. V. Tyrrell, and K. R. Harris, *Diffusion in Liquids. A Theoretical and Experimental Study*. Butterworths, London (1984).

3. P. Stilbs, *Prog. NMR Spectrosc* **9**: 1–45 (1987).

4. R. A. Robinson, and R. H. Stokes, *Electrolyte Solutions*, 2nd edn. Butterworths, London (1970).

5. E. L. Cussler, *Diffusion: Mass Transfer in Fluid Systems.* Cambridge University Press, Cambridge (1984).

6. E. L. Cussler, *Multicomponent Diffusion.* Elsevier, Amsterdam (1976).

7. B. J. Berne, and R. Pecora, *Dynamic Light Scattering.* Wiley, New York (1976).

8. R. Pecora (ed.), *Dynamic Light Scattering. Applications Photon Correlation Spectroscopy.* Plenum Press, New York (1985).

9. N. Kamenka, M. Chorro, Y. Talmon, and R. Zana, *Colloids Surf.* **67**: 213–222 (1992).

10. D. P. Bossev, M. Matsumoto, T. Sato, H. Watanabe, and M. Nakahara, *J. Phys. Chem. B.* **103**: 8259–8266 (1999).

11. P. Stilbs, K. Paulsen, and P. C. Griffiths, *J. Phys. Chem.* **100**: 8180–8189 (1996).

12. D. Ciccarelli, L. Costantino, G. D'Errico, L. Paduano, and V. Vitagliano, *Langmuir* **14**: 7130–7139 (1998).

13. P. C. Griffiths, P. Stilbs, K. Paulsen, A. M. Howe, and A. R. Pitt, *J. Phys. Chem. B*, **101**: 915–918 (1997).

14. A. M. Misselyn-Bauduin, A. Thibaut, J. Grandjean, G. Broze, and R. Jérôme, *Langmuir* **16**: 4430–4435 (2000).

15. T. Kato, H. Takeuchi, and T. Semiya, *J. Colloid. Interface. Sci.* **140**: 253–257 (1990).

16. T. Kato, H. Takeuchi, and T. Seimiya, *J. Phys. Chem.* **96**: 6839–6843 (1992).

17. T. Tominaga, and M. Nishinaka, *J. Chem. Soc. Faraday Trans.* **89**: 3459–3464 (1993).

18. T. Tominaga, and M. Nishinaka, *J. Mol. Liq.* **65/66**: 333–336 (1995).

19. T. Tominaga, and T. Nakamura, *J. Mol. Liq.* **73/74**: 413–417 (1997).

20. T. Tominaga, T. Nakamura, and H. Saiki, *Chem. Lett.* **1997**: 979–980 (1997).

21. S. Gorti, L. Plank, and B. R. Ware, *J. Chem. Phys.* **81**: 909–914 (1984).

22. T. Okubo, *J. Phys. Chem.* **93**: 4352–4354 (1989).

23. J. M. Schurr, *Chem. Phys.* **45**: 119–132 (1980).

24. F. Booth, *J. Chem. Phys.* **22**: 1956 (1968).

25. H. Ohshima, T. W. Healy, L. R. White, and R. W. O'Brien, *J. Chem. Soc. Faraday Trans. 2* **80**: 1299–1317 (1984).

26. G. A. Schumacher, and T. G. M. van de Ven, *Faraday Discuss. Chem. Soc.* **83**: 75–85 (1987).

27. G. A. Schumacher, and T. G. M. van de Ven, *J. Chem. Soc. Faraday Trans.* **87**: 971–976 (1991).

28. A. Rohde, and E. Sackmann, *J. Phys. Chem.* **84**: 1598–1602 (1980).

29. M. Corti, V. Degiorgio, *J. Phys. Chem.* **85**: 711–717 (1981).

30. J. Briggs, R. B. Dorshow, C. A. Bunton, and D. F. Nicoli, *J. Chem. Phys.* **76**: 775–779 (1982).

31. R. B. Dorshow, J. Briggs, C. A. Bunton, and D. F. Nicoli, *J. Phys. Chem.* **86**: 2388–2395 (1982).

32. R. B. Dorshow, C. A. Bunton, and D. F. Nicoli, *J. Phys. Chem.* **87**: 1409–1416 (1983).

33. P. J. Missel, N. A. Mazer, G. B. Benedek, and M. C. Carey, *J. Phys. Chem.* **87**: 1264–1277 (1983).

34. S. J. Candau, E. Hirsch, and R. Zana, *J. Phys.* **45**: 1263–1270 (1984).

35. M. Drifford, L. Belloni, J. P. Dalbiez, and A. K. Chattopadhyay, *J. Colloid Interface Sci.* **105**: 587–604 (1985).

36. R. M. Weinheimer, D. F. Evans, and E. L. Cussler, *J. Colloid Interface Sci.* **80**: 357–368 (1981).

37. D. F. Evans, S. Mukherjee, D. J. Mitchell, and B. W. Ninham, *J. Colloid Interface Sci.* **93**: 184–204 (1983).

38. D. G. Leaist, and L. Hao, *J. Phys. Chem.* **97**: 7763–7768 (1993).

39. J. P. Kratohvil, and T. M. Aminabhavi, *J. Phys. Chem.* **86**: 1254–1256 (1982).

40. D. G. Leaist, *J. Colloid Interface Sci.* **111**: 230–239 (1986).

41. S. Ikeda, *Colloid Polym. Sci.* **269**: 49–61 (1991).

42. Y. Nogami, H. Watanabe, H. Ohtaka-Saiki, and T. Tominaga, *Colloids Surf. A* **169**: 227–232 (2000).

43. M. Castaldi, L. Costantino, O. Ortona, L. Paduano, and V. Vitagliano, *Langmuir* **14**: 5994–5998 (1998).

44. D. G. Leaist, and K. MacEwan, *J. Phys. Chem.* **105**: 690–695 (2001).

5

Mixed Micellar Aggregates of Nonionic Surfactants with Short Hydrophobic Tails

VINCENZO VITAGLIANO,
GERARDINO D'ERRICO, ORNELLA ORTONA,
and LUIGI PADUANO

Dipartimento di Chimica, Università di Napoli
"Federico II", Complesso di Monte S. Angelo,
Napoli, Italy

SYNOPSIS

A review (67 references) on the theoretical and experimental investigation of the properties of aqueous mixtures containing at least one ethoxylated surfactant with short hydrophobic tail is presented. For short hydrophobic tail, hydrocarbon chains with less than 12 carbon atoms are intended.

Theoretical models of micellization and mixed micellization are summarized and discussed in relation to their applicability in the interpretation of experimental results. Peculiarities of short-tailed surfactants, arising from the looseness of the aggregates they form, are highlighted. Besides classical models, such as the pseudo-phase separation, the mass action, and the multistep equilibrium models, a more recent molecular thermodynamic model is discussed which shows the last one to be more realistic and reliable. However, because of its large employment in the literature, the pseudo-phase separation model is widely discussed, introducing the ideal and regular approach to describe the mixing of surfactants into the aggregates.

Experimental results, presented in the literature, on mixtures of two ethoxylated surfactants and on mixtures of one ethoxylic and one ionic surfactant are overviewed, to extract some general information.

In the case of nonionic–nonionic surfactant mixtures, the hydrophilic interactions among ethoxylic chains are more responsible for deviation from ideal behavior than the hydrophobic interaction among the hydrocarbon tails. The effect of polydispersity of ethoxylic chains is discussed. Mixtures of surfactants having fluorinated and nonfluorinated hydrocarbon tails show considerable deviations from ideality even when the ethoxylic moiety is the same.

In the case of nonionic–ionic mixtures, the mixed micellar systems deviate only slightly from ideality and they can be handled by the simple regular solution approach.

I. INTRODUCTION

In the last two to three decades, technological exploitation of surfactant's mixtures has grown dramatically and continues to increase steadily. Commercial surfactants typically consist of different compounds, since industrial syntheses produce mixtures that, for practical purposes, are used without further purification. Moreover, in many cases mixtures of surfactants are intentionally formulated since their properties can be easily optimized for specific applications. For instance, textile processing, detergency, metal processing, agriculture, polymer synthesis, cosmetics, pharmaceuticals, food preparation, paper, and leather processing currently employ such systems [1–3].

Mixtures containing, or exclusively composed of, nonionic surfactants of the ethoxylated type $[CH_3(CH_2)_{r-1}(OCH_2CH_2)_m OH, C_rE_m]$, have appeared to be particularly appealing to the formulation technologist. This is due to the properties of these surfactants, especially their low foaming, superior fiber cleaning, and tolerance to water hardness. In addition, nonionic surfactants are regarded as non-toxic and non-pollutant [4].

The development of the applications of surfactant mixtures has promoted a wide scientific interest in the argument. In general, mixtures of surfactants present an aggregation behavior substantially different from that of each single constituent. These differences have to be ascribed to specific interactions (synergistic or antagonistic) between the surfactants present in the mixture [5–7]. However, the subject is far from being completely understood, information present in the literature being extremely scattered. In fact, even though different techniques generally give concordant qualitative pictures of a system, quantitative analyses of the parameters related to the mixtures aggregation process often largely disagree. On the other hand, theoretical studies on mixed micellar solutions have produced simplified models that, despite some predictive ability, still seem not completely adequate to fully describe the system reality [8,9]. Alternatively some complex and more comprehensive models have been proposed, but often their formulation includes a large number of adjustable parameters, thus making comparison with experimental results quite difficult or unreliable [10,11].

Most of the articles present in the literature deal with mixtures of surfactants with quite long hydrophobic tails (number of carbon atoms equal to or higher than 12). Much less attention has been paid to mixtures of surfactants with short hydrophobic tails (number of carbon atoms lower than 12).

Nevertheless, from a practical point of view, mixtures of these surfactants are of great importance because of their use as wettability agents [12].

Furthermore, from a scientific point of view, surfactants with short hydrophobic tails show interesting peculiarities with respect to the homologue terms with longer tails [13,14]. They self-aggregate at a relatively high concentration, so that micelles are dipped in a quite concentrated aqueous solution of monomers. In the case of ionic surfactants, this implies high ionic strength of the medium surrounding the micelles that depresses the electrostatic interactions among them. The lower aggregation number of short hydrophobic tail surfactants in comparison with that shown by the longer ones, is responsible for a looser, more open structure of the aggregates that can be penetrated by solvent molecules. As a consequence, the hydrophobic core of these surfactants shows low efficiency in terms of solubilization ability.

Surfactants with short hydrophobic tails allow an experimental investigation also in premicellar solutions making possible the study of the solvent–surfactant interaction [15].

The aim of this review is to give a brief overview of the theoretical and experimental investigation on the properties of surfactants' mixtures in aqueous solutions. In particular, the focus is on mixtures of surfactants with short hydrophobic tails.

II. WATER–SURFACTANT BINARY SYSTEMS

Surfactants are compounds having in their molecules a hydrophilic part (generally called head) that makes them soluble in water, and a hydrophobic moiety (so called tail) that favors the aggregation of such molecules in supramolecular structures, the micelles. The tail is commonly made of an aliphatic chain. The aggregation tendency of these molecules increases with increasing length of the hydrophobic tail.

The hydrophobic character is empirically measured with a parameter, the hydrophilic–lipophilic balance (HLB), that weights the relative extension of heads and tails [16–18]. However, it is not straightforward to extract an operative definition of the HLB parameter. From a thermodynamic point of view, it can be defined as a function of the μ_o/μ_w ratio where μ_o and μ_w are the reference chemical potentials of a surfactant in an oil (i.e., hydrocarbon solvent) and water, respectively. The μ_o/μ_w ratio can be determined from the partition coefficient of the surfactant between the oil and water phases.

This review is focused on the aggregation behavior of surfactants with short hydrophobic tails that usually present high values of the HLB. In this case, molecules are able to strongly interact with the solvent, thereby dissolving easily in the bulk as monomers. Nevertheless, even in this context, the aliphatic chains at moderate concentration are responsible for hydrophobic interactions among each other that could lead to a self-aggregation process.

In the following, we briefly discuss models for micellization of nonionic surfactants, highlighting the peculiarities of the self-aggregation of surfactants with short hydrophobic tails. We note that the same models apply to ionic surfactants: however, in this case the fraction of counterions condensed on the micelle surface should also be considered.

The aggregation process is well represented by a multistep equilibrium model

$$S + S \leftrightarrows S_2$$
$$S_2 + S \leftrightarrows S_3$$
$$S_3 + S \leftrightarrows S_4$$
$$S_{i-1} + S \leftrightarrows S_i$$
$$\dots\dots\dots\dots\dots$$
$$S_{n-1} + S \leftrightarrows S_n \tag{1}$$

where n is the upper limit of the aggregation number. This model assumes the presence of monomers, dimers, trimers, ... The presence of these adducts is well established by the literature on short surfactants [19–27].

Ignoring activity coefficients, the multiple equilibrium (1) is described by the following set of equations

$$[S_2] = K_2[S]^2 \tag{2a}$$

$$[S_3] = K_3[S_2][S] \tag{2b}$$

$$\dots\dots\dots\dots\dots$$

$$[S_i] = K_i[S_{i-1}][S] \tag{2c}$$

$$[S_n] = K_n[S_{n-1}][S] \tag{2d}$$

where $[S_i]$ is the concentration of a generic species with aggregation number i, $[S]$ being the monomer species concentration.

In terms of monomer units concentration, the set of equations (2) can be written as

$$[S_o] = [S] + 2[S_2] + 3[S_3] + \cdots i[S_i] \cdots + n[S_n] \tag{3}$$

or

$$1 = \sum_{i=1}^{n} \alpha_i \tag{4}$$

where $[S_o]$ is the stoichiometric concentration of the surfactant and α_i is the fraction of surfactant in the associated state i.

$$[S_o] = [S] + 2K_2[S]^2 + 3K_2K_3[S]^3 + \cdots iK_2K_3 \cdots K_i[S]^i \cdots$$
$$+ nK_2K_3 \cdots K_n[S]^n \tag{5}$$

and, defining

$$\mathbf{K_i} = \prod_{j=1}^{i} K_j \tag{6}$$

where $K_1 = 1$ by definition

$$[S_o] = \sum_{i=1}^{n} i\, \mathbf{K_i}[S]^i \tag{7}$$

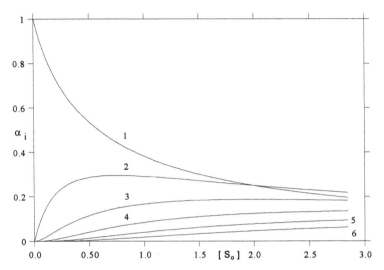

Figure 1 Fraction of surfactant S in various association states, on the assumption that all constants of Eq. (4) are equal (arbitrary self-consistent units for K_i and $[S_o]$).

The relative values of the various K_i define the degree of cooperativity of each association step. Various conditions can be simulated by assigning different distributions of the K_i values. In the absence of cooperativity, all K_i have the same value (isodesmic model). In Fig. 1 this case is shown assuming all $K_i = 1/(\text{volume units} \times \text{mol}^{-1})$, concentrations being given in arbitrary units self-consistent with the choice $K_i = 1$.

It can be seen that up to $[S_o] = 3$ only species with $n \leq 8$ are present in an appreciable amount.

The opposite condition of full cooperativity corresponds to the presence of the only equilibrium

$$nS \leftrightharpoons S_n \tag{8}$$

In this case the equilibrium constant can be written as

$$K_n = \frac{1 - \alpha}{n\, \alpha^n [S_o]^{(n-1)}} \tag{9}$$

29

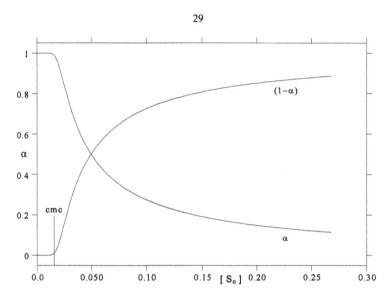

Figure 2 Fractions of monomer and associated surfactant for the equilibrium: 11 $S \leftrightharpoons S_{11}$ with $K_{11} = 10^{15}$ (arbitrary self-consistent units for K_{11} and $[S_o]$).

where α is the fraction of monomer surfactant.

Figure 2 shows the fractions α and $1 - \alpha$ in the case of a degree of association $n = 11$ and an association constant, $K_{11} = 10^{15}$ (arbitrary units consistent with those of concentration).

In this case the micellization process appears in a very short range of concentration around a value defined as *critical micelle concentration*, CMC.

As the degree of association becomes very large, the association process approaches an all-or-nothing process and the equilibrium *monomer* $\leftrightharpoons$ *micelle* approaches a first-order transition, like any phase separation process. As a consequence, micellization of surfactants with long hydrophobic tails is sufficiently well described in terms of a pseudo-phase separation, occurring at a unique surfactant concentration, the CMC. In contrast, it should be stressed that this model is questionable when applied to surfactants with short hydrophobic tails for which the CMC definition is ambiguous.

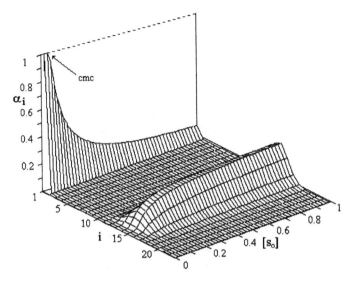

Figure 3 Fractions of various associated species in the case of a gaussian distribution of the equilibrium constants around $i=15$: $\mathbf{K}_i = 10^i \exp\{-[0.5\ (i-15)]^2\}$ (arbitrary self-consistent units for $\mathbf{K}_i$ and $[S_o]$).

An intermediate condition between the full cooperativity represented by Eq. (8) and the noncooperativity of Eqs. (1), with $K_1 = K_2 = \cdots K_n = $ constant, is probably more realistic in the case of micellization processes.

An example of intermediate condition is shown in Fig. 3, where a gaussian distribution of the constants $\mathbf{K}_i$ [Eq. (5)] has been chosen around the association degree $i = 15$. In this case species having different values of n are also present, whose amount depends on the width of the gaussian distribution.

It is interesting to note that, in consideration of the low n and $\mathbf{K}_i$ chosen values, which are realistic for short hydrophobic tail surfactants, the fraction of monomer is never negligible, even at high surfactant concentrations.

Finally, let us remember here the very general approach to the treatment of surfactant solutions proposed by Blankschtein and his school [28–30]. Although a discussion of their molecular-thermodynamic model is beyond the purpose of this review and the reader is addressed to the original papers, we would like to outline here the thermodynamic basis of these authors' treatment.

Using the principle of multiple equilibrium between micelles of different sizes and monomer (namely the equality of chemical potentials, $\mu_n = n\mu_1$, where n is the aggregation number), the following expression can be written

$$X_n = X_1^n \exp(-n[\Delta g_{\mathrm{mic}}(n)/k_{\mathrm{B}}T]) \tag{11}$$

where X_n is the mole fraction of micelles with aggregation number n, X_1 is the mole fraction of monomer surfactant (mole fractions being computed including the solvent contribution), k_{B} is the Boltzman constant, and $\Delta g_{\mathrm{mic}}(n)$ is the Gibbs energy of micellization of the micelles with aggregation number n. Δg_{mic} accounts for the formation of the micelles from monomers in their standard states.

Equation (11) is just an extension of Eqs. (2) which account for the nonideality of the surfactant solution. In fact, the authors split the Gibbs energy term, Δg_{mic}, into five contributions giving expressions for each of them:

1. *Hydrophobic contribution*: associated with the interactions between the nonpolar hydrocarbon tails and water. It represents the Gibbs energy change associated with the transfer of surfactants molecules from water to a hydrocarbon medium. This is the major contribution to the Gibbs energy which favors the micellization process.
2. *Interfacial contribution*: a repulsive contribution which opposes micellization due to formation of a micellar core/water interface.
3. *Packing contribution*: in the micelles the hydrocarbon tail ends, linked to the heads, are restricted to lie in the vicinity of the micellar core/water interface. The result is a loss of conformational degree of freedom.
4. *Steric contribution*: due to the repulsion among the surfactant heads at the core/water interface.

5. *Electrostatic contribution*: present in solutions of ionic or zwitterionic surfactants.

III. MIXED SURFACTANT SOLUTIONS

When mixed nonionic micelles are formed, the equilibrium model can be written as

$$pA + qB \leftrightarrows A_pB_q \tag{12}$$

where A and B represent the two surfactants and p and q the number of their monomers in the micelle. If ionic surfactants are present, the contribution of condensed counterions must also be considered in Eq. (12.)

The model of the chemical equilibrium, as well as that of the multistep equilibrium, are conceptually the more appropriate to interpret mixed micellization, particularly when surfactant with short hydrophobic tails are involved. However, the application of these models is often very cumbersome and is not justified by the quality of the results in comparison with those obtained from the application of simpler models such as the phase separation approach.

Another problem connected with the theoretical description of mixed surfactant systems is imputable to the fact that, almost generally, synthesis of these compounds does not lead to a single molecular species but to a wide spectrum of surfactants of different molecular weight distributed around a mean value. This is the case of nonionic ethoxylated surfactants for which a polydispersity in the length of the hydrophilic moiety is always present. This effect adds a further difficulty to a general treatment of these systems.

We are aware that the application of the phase separation model to short hydrophobic chain surfactants is a very daring approximation. However, since it is still able to give general information in this case too, we consider here the possibility of applying it to short mixed micellar systems.

If the micellar phase can be regarded as ideal, which is often the case for the ethoxylated mixed micelles, thermodynamics states simple relationships for the Gibbs energy and entropy of mixing of surfactants inside the micelle, Δg_{mix} and Δs_{mix}, and for the CMC_{mix} [31] of the mixed solution, namely

$$\Delta g_{mix} = RT[y_A \ln y_A + (1 - y_A) \ln (1 - y_A)] \tag{13a}$$

$$\Delta s_{mix} = -R[y_A \ln y_A + (1 - y_A) \ln (1 - y_A)] \tag{13b}$$

$$1/CMC_{mix} = x_A/CMC_A + (1 - x_A)/CMC_B \tag{14}$$

where y_A is the mole fraction of component A in the micelle and x_A is its stoichiometric mole fraction, computed excluding the solvent contribution. Δg_{mix} accounts for the formation of the mixed micelles from the pure ones.

Equation (14) shows that the value of CMC_{mix} depends mainly on the lower CMC. This is more evident for high ratios CMC_B/CMC_A as can be seen in Fig. 4.

If it is not possible to assume ideal behavior in the mixed micelles, as for the mixing of nonionic and ionic surfactants, it is necessary to take into account the activity coefficients

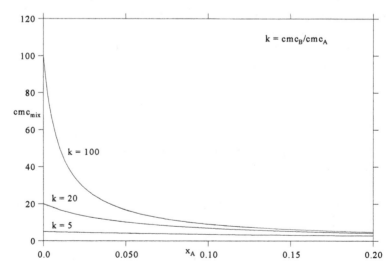

Figure 4 Critical micelle concentration of mixed surfactants solution, CMC_{mix} (arbitrary units), for different values of the ratio $k = CMC_B/CMC_A$ [from Eq. (14)].

of the two surfactants, at least in the micellar phase, while it is still reasonably valid to consider their activity coefficients unitary in the monomeric phase.

The first ones who faced this problem were Holland and Rubingh [32] who assumed a regular solution behavior for the micellar phase. In this case

$$\Delta g_{mix} = RT(y_A \ln y_A + y_B \ln y_B) + \beta y_A y_B \qquad (15a)$$

or

$$\Delta g_{mix}^{exc} = \beta y_A y_B \qquad (15b)$$

In the regular solution theory, β is independent of temperature and composition. A negative value of β indicates synergism in mixed micelle formation, corresponding to a positive heat of mixing, while a positive value of β stays for antagonism.

However, there is experimental evidence that β is often dependent on temperature [33], leading to a symmetrical non-zero excess entropy of mixing, and it may depend also on micelles composition which leads to an asymmetric shape of the excess entropy of mixing [34,35] (see Fig. 5).

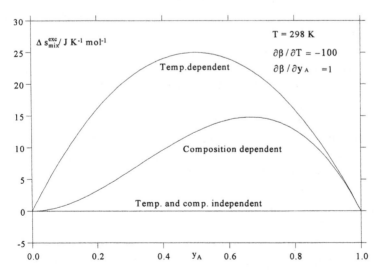

Figure 5 Excess entropy of mixing for various expressions for the parameter β.

It is worthwhile pointing out here that, according to the regular solutions theory, for a positive value of β larger than 2 a phase separation may occur between micelles that split into micelles richer in surfactant A and micelles richer in surfactant B.

Finally, let us note that the Blankschtein school [36–39] also extended its theoretical treatment to solutions of mixed surfactants. Also, in this case we shall outline only the thermodynamic basis of the treatment by these authors, referring the reader to the original papers.

The treatment starts from the general equilibrium expression.

$$X_{n,y_A} = X_{A1}^{ny_A} X_{B1}^{n(1-y_A)} \exp\left(-n[\Delta g_{\text{mic}}(y_A)k_B T - 1] + 1\right) \quad (16)$$

where X_{A1} and X_{B1} are the mole fractions of surfactants A and B in their monomeric state, X_{n,y_A} the mole fraction of micelles composed of n monomers of composition y_A (y_A being the fraction of species A in the micelle). The mole fractions are computed including the number of solvent molecules; Δg_{mic} is the Gibbs energy of mixed micellization.

Also in this case the model accounts for the presence of a distribution of micelles having different sizes.

The knowledge of the terms Δg_{mic} allows to compute any properties of the micelles present in solution (i.e., sizes, size distribution, CMC) on the basis of only structural information about the surfactants involved and solution conditions, such as temperature and presence of salts.

As in the case of a single surfactant solution, the term Δg_{mic} is split into five terms as described in the previous section, and the authors give expressions for each term.

This theory has also been used along with experimental data to evaluate the influence of temperature on the hydration number of the hydrophilic head of nonionic surfactant molecules forming the outer shell of the mixed micelle. The method proposed by Garamus [40], within the framework of the theory developed by Blankschtein and collaborators, uses some experimental parameters measured through small angle neutron scattering (SANS) such as charge, diameter, and aggregation number of micelle. The results obtained show that the hydration number decreases as the temperature is raised. It is to note that in several cases a dehydration of the ethylene oxide unit has been invoked to explain the weakening of interactions observed in nonionic–ionic systems [41].

Although the theory of Blankschtein and co-workers is the only one to have a molecular thermodynamic basis, the Rubingh's approach is, for the present, the more widely used in the literature for describing mixed surfactants solutions.

Haque *et al.* [42] presented a report in which the performances of Rubingh's approach and those of the molecular thermodynamic theory of Blankschtein's school have been compared. These authors analyzed several combinations of surfactants with different lengths of the hydrophobic tails. Both approaches have been found to be successful in describing the mixed micellar properties of nonionic–anionic surfactant solutions. However, as mentioned above, Rubingh's theory cannot thoroughly account for the interactional features of surfactants in mixed micelles, in fact only rarely Haque *et al.* found the interaction parameter β to be constant over the entire range of composition. For this reason the authors conclude that the Blankschtein theory is more reliable to extract information on mixed surfactant systems above and below the micellar compositions.

IV. MIXED MICELLAR AGGREGATES OF NONIONIC–NONIONIC SURFACTANTS WITH SHORT HYDROPHOBIC TAILS

Synthesis of nonionic surfactants produces mixtures of poly-dispersed compounds with respect to both hydrocarbon chain length and polar group dimension. These mixtures are used without further fractionation for practical purposes. Peculiarly, ethoxylated surfactants present a polydispersed length of the ethoxylic head [43].

Some efforts have been made in order to connect the physico-chemical properties of these commercial mixtures to those of the binary aqueous solution of the different components. However, the experimental investigation on micellization of this kind of surfactant mixture is difficult, because almost all techniques do not distinguish among the contributions of the various components to the observed property.

The papers present in the literature furnish quite scattered information on the subject, so that it is extremely difficult to obtain a unifying picture of such mixtures. In particular, very few articles have been published on mixtures of nonionic surfactants with short hydrophobic tails.

Recently, D'Errico et al. [44] presented an experimental investigation on mixed micellar systems formed by two nonionic ethoxylated surfactants with a short hydrophobic tail (C_6E_5 and C_6E_2) in aqueous solution. C_6E_5 micellizes in aqueous solution [45]. In contrast, no micellization process occurs in the binary mixtures C_6E_2–water [46,47], because of the surfactant having too short a hydrophilic head. At high C_6E_2 compositions, the binary mixture segregates in a water-rich and in a surfactant-rich phase. The temperature-composition phase diagram for this system is shown in Fig. 6 [44,48]. At 25°C

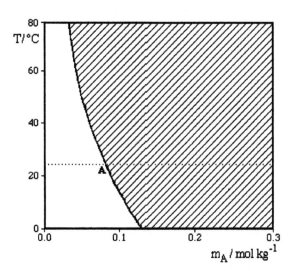

Figure 6 Temperature-composition phase diagram for the system C_6E_2–water (reproduced with permission from Ref. [44]).

and $m \cong 0.08\,\mathrm{mol\,kg^{-1}}$, the C_6E_2-water system undergoes a macroscopic phase separation (point A in Fig. 6). With increasing temperature, C_6E_2 solubility decreases because the interactions among the aqueous medium and the surfactant's polyoxyethylene chains become worse [49].

D'Errico *et al.* found that a low percentage of C_6E_5 is sufficient to allow mixed micellization of C_6E_5 and C_6E_2. These authors obtained the temperature-composition phase diagrams for the system C_6E_2–C_6E_5–water at various C_6E_5 molalities (see Fig. 7). The presence of C_6E_5 at constant molality causes the appearance of a region (L_1) in which C_6E_2 and C_6E_5 mixed micellar aggregates form. In these aggregates, the larger C_6E_5 heads balance the smaller C_6E_2 ones.

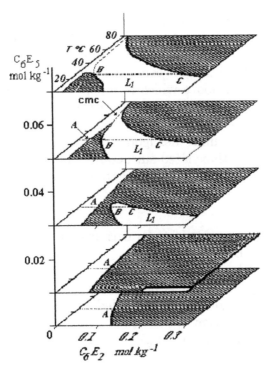

Figure 7 Temperature-composition phase diagram for the system C_6E_2–C_6E_5–water. L_1, micellar solution.

By following the horizontal dotted lines in Fig. 7 we can analyze the behavior of a C_6E_5 solution to which an increasing amount of C_6E_2 is added, at 25°C. Initially an aqueous solution of both C_6E_2 and C_6E_5 as free monomers appears. As the composition corresponding to point A is reached, a phase separation occurs. Above point A two water–surfactant mixtures coexist; one of these is a homogeneous aqueous solution of free surfactants with the composition corresponding to point A, the other one, with the composition corresponding to point B, is a solution where an equilibrium is present between micelles and monomers. It must be noted that the composition at point B corresponds to the CMC.

For C_6E_2 compositions above point B, a single phase containing mixed micelles appears, so that the mixture becomes homogeneous once again.

It is interesting to note that the binary mixture C_6E_3–water at 25°C shows a turbidity region [45] in a composition range very close to the range of phase separation found for the ternary system C_6E_2–C_6E_5–water.

Finally, the ratio between the molalities of C_6E_2 and C_6E_5 reaches the point above which a new phase separation occurs (point C). In this second phase separation range the micellar mixture presumably coexists with more complex water–surfactant structures (cubic, hexagonal, lamellar, etc), as already found for other ethoxylated surfactant solutions [50].

By comparing the phase diagrams obtained at different C_6E_5 molalities, D'Errico *et al.* found that, with increasing the C_6E_5 molality:

1. the molalities of C_6E_2 corresponding to point A and point B decrease.
2. at both points A and B the sum of the molalities of C_6E_2 plus C_6E_5 is nearly constant, showing that the aggregation is mainly ruled by the total surfactant concentration.
3. the molality of C_6E_2 corresponding to point C increases.

Some information on the mixed aggregates that form in the composition range between point B and point C at 25°C was obtained by analyzing the micelle intradiffusion coefficient, $\mathbf{D}_{mic}$, measured through the pulsed gradient spin echo (PGSE)–NMR technique [51], using tetramethylsilane (TMS) as the molecular label. Since TMS is a strongly hydrophobic molecule, it solubilizes into the micellar core, so that its intradiffusion coefficient is the same as that of the micelles [52]. By using the Stokes–Einstein equation, $\mathbf{D}_{mic}$ can be related to the hydrodynamic size of the mixed aggregates:

$$R_{app} = \frac{k_B T}{6\pi \mathbf{D}_{mic} \eta} \tag{17}$$

where k_B is the Boltzmann constant, T is the absolute temperature, and η is the viscosity of the medium in which the micelles diffuse. It was found that, in the ternary system C_6E_2–C_6E_5–water, R_{app} is nearly constant (mean value $(18.8 \pm 0.2) \times 10^{-10}$ m) and independent of the surfactant molalities. This value is approximately equal to the radius of the micelles formed by C_6E_3 (18.9×10^{-10} m [45]), so that it can be inferred that in C_6E_2–C_6E_5 mixed aggregates the ethoxylic chain of C_6E_5 molecules balances the short hydrophilic heads of C_6E_2 in such a way that aggregates similar to C_6E_3 micelles form. This experimental evidence suggests that the aggregation behavior in mixtures of ethoxylated surfactants is ruled by the mean number of ethoxylic units per surfactant molecule, while the distribution of the length of the hydrophilic heads is quite indifferent.

Other researchers investigated surfactants mixtures in which both constituents micellize also in the binary water–surfactant solutions. Differently from mixtures of ionic and nonionic surfactants or those of oppositely charged ionic surfactants, mixtures of nonionic ethoxylated surfactants do not present strong interaction among the hydrophilic heads. For this reason, it is generally assumed that the physico-chemical properties of these mixtures present small departures from those predicted by the ideal or regular solution models of mixed micellization. In contrast, recent experimental investigations have shown that differences in the length of both the

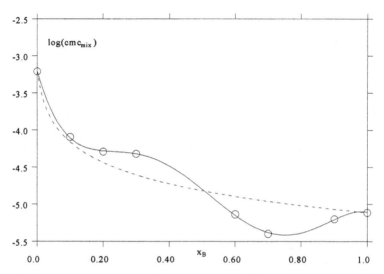

Figure 8 Variation of the critical micelle concentration with surfactant relative mole fraction for the mixed $C_{10}E_6(A)$–$C_{14}E_6$ (B)–water system. The dotted line represents the ideal behavior (reproduced with permission from Ref. [53]).

hydrophobic and the hydrophilic chains of the surfactants can cause a mixing behavior far from ideality.

In this context Portet *et al.* [53] studied, using surface tension measurements, $C_{10}E_6(A)$–$C_{14}E_6(B)$ aqueous mixtures. Figure 8 shows the CMC values as a function of the stoichiometric surfactant composition; strong differences from the prediction of the ideal solution model for mixed micellization [Eq. (14)], represented by the dashed line, can be detected.

For $x_B > 0.5$ the experimental CMC values are lower than those predicted; this evidence indicates a favorable interaction between the two surfactants when they form mixed micelles. The insertion of $C_{10}E_6$ molecules stabilizes the large $C_{14}E_6$ aggregates, that probably present an oblate shape. For $0.1 < x_B < 0.5$ the experimental CMC values are higher than those obtained through the ideal solution model; this evidence indicates unfavorable interactions between the two surfactants that could lead to a partial demixing of the micelles, i.e., to the coexistence of $C_{10}E_6$-rich and $C_{14}E_6$-rich micelles. Finally, for $x_B < 0.1$ the experimental CMCs nearly coincide with the ideal values.

The experimental CMC trend cannot be explained in terms of the regular solution approach because the experimental data are not fitted with a unique β value; this is a further confirmation of the presence of strong interactions in the surfactant mixture.

The difference in the hydrophilic chain length is also important in determining the mixing behavior of ethoxylated surfactants. Ohta *et al.* [33] investigated, using microcalorimetric titration, the $C_{10}E_5$–C_8E_5 and the $C_{10}E_5$–C_8E_4 aqueous mixtures and computed the excess molar enthalpy, Δh_{mic}^{exc}, and the excess Gibbs energy, Δg_{mic}^{exc}, defined as

$$\Delta h_{mic}^{exc} = y_A(h_{A,\,mic} - h_{A,\,mic}^o) + y_B(h_{B,\,mic} - h_{B,\,mic}^o) \qquad (18)$$

and

$$\Delta g_{mic}^{exc} = y_A(g_{A,\,mic} - g_{A,\,mic}^o) + y_B(g_{B,\,mic} - g_{B,\,mic}^o) \qquad (19)$$

where y_i is the mole fraction of surfactant i in the micelles; $h_{i,\,mic}$ is its partial molar enthalpy in the mixed micelles and $h_{i,\,mic}^o$ is its molar enthalpy in single surfactant micelles. $g_{i,\,mic}$ and $g_{i,\,mic}^o$ are the corresponding Gibbs energies.

In the case of $C_{10}E_5$–C_8E_5 mixtures, Δh_{mic}^{exc} as well as Δg_{mic}^{exc} are equal to zero in the whole micelle composition range (see Fig. 9), meaning that $C_{10}E_5$ and C_8E_5 are ideally mixed in the micelles. In contrast, in the case of $C_{10}E_5$–C_8E_4 mixtures both Δh_{mic}^{exc} and Δg_{mic}^{exc} assume positive values, showing that $C_{10}E_5$ and C_8E_4 molecules mix in a slightly nonideal fashion. The authors interpreted these experimental findings in terms of the lack of size balance which makes the packing of the hydrophilic heads less favorable in $C_{10}E_5$–C_8E_4 than in $C_{10}E_5$–C_8E_5 mixtures.

Much stronger deviations from ideality are found in mixtures of ethoxylated surfactants with hydrocarbon and fluorocarbon tails. Villeneuve *et al.* [54] studied, using surface tension measurements, the mixture of the hydrocarbon

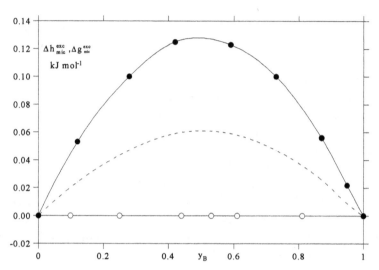

Figure 9 Molar excess enthalpy, Δh_{mic}^{exc}, and Gibbs energy, Δg_{mic}^{exc}, as a function of micelles composition. (—○—) Δh_{mic}^{exc} and Δg_{mic}^{exc} or the system $C_{10}E_5(A)$–$C_8E_5(B)$–water; (—●—) Δh_{mic}^{exc} and (----)Δg_{mic}^{exc} for the system $C_{10}E_5(A)$–$C_8E_4(B)$–water (reproduced with permission from Ref. [33]).

surfactant $C_{10}E_4$ and the fluorocarbon surfactant FC_7E_4 [$H(CF_2)_6(CH_2)(OCH_2CH_2)_4OH$] at the total surfactant concentration both below and above the mixture CMC. The composition of the film of surfactant molecules adsorbed at the water–air interface could be determined by analyzing the mixture surface tension in premicellar solutions. This film presents a composition different from that in the solution bulk, as can be seen in Fig. 10, where the total surfactant molality, m, is given as a function of the compositions of both the solution bulk, x_B, and the adsorbed film, z_B, at constant surface tension below the CMC value. All curves show positive deviations from the prediction of the ideal solution model, which would be the straight lines connecting the m values for $x_B = 0$ and $x_B = 1$.

A sort of *azeotropic point* is present, i.e., if the bulk composition of surfactant B is small, it will be smaller at the air–water interface, while if the bulk composition is high, it will be higher at the air–water interface.

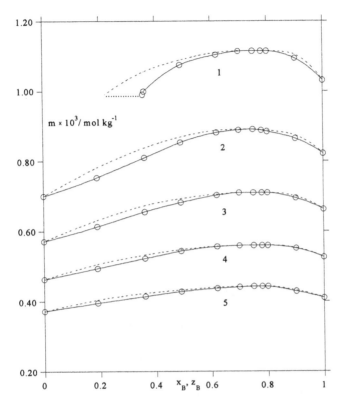

Figure 10 Phase diagram of adsorption in the system $C_{10}E_4(A)$–$FC_7E_4(B)$–water at constant surface tension: (—O—) m vs x_B, (- - - -) m vs z_B; $\gamma = (1)$ 27, (2) 29, (3) 31, (4) 33, and (5) 35 mN/m (reproduced with permission from Ref. [54]).

The mixing behavior of $C_{10}E_4$ and FC_7E_4 in the micelles was investigated by plotting the CMC value as a function of both the composition of the solution bulk, x_B, and that of the micelles, y_B (see Fig. 11). Also in this graph, an *azeotrope* can be detected, delimiting two composition ranges in which the micelles are richer in the surfactant more concentrated in the solution bulk. The authors found that the deviation from ideality is greater in the micelles than in the adsorbed film. These experimental findings indicate that the interaction between $C_{10}E_4$ and FC_7E_4 molecules is less attractive than that between the same species alone. However, the two nonionic surfactants are still miscible in the whole composition

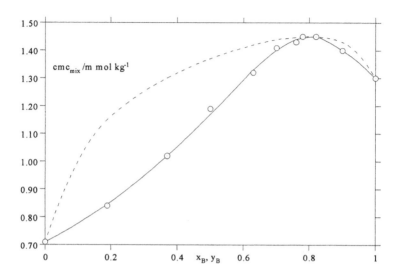

Figure 11 Phase diagram of micelle formation in the system $C_{10}E_4$(A)–FC_7E_4(B)–water (—O—) CMC vs x_B, (– – – –) CMC vs y_B (reproduced with permission from Ref. [54]).

range. This result can be compared with the findings concerning mixtures of ionic hydrocarbon and fluorocarbon surfactants, in which demixing is very often found, as a result of much stronger repulsive interactions [55].

V. MIXED MICELLAR AGGREGATES OF NONIONIC-ANIONIC SURFACTANTS WITH SHORT HYDROPHOBIC TAILS

The pentaethyleneglycol-1-hexylether (C_6E_5, A)–sodium-1-hexanesulfonate (C_6SO_3Na, B)–water is the only mixture formed by short chain surfactants present in the literature on which an extensive experimental study has been performed.

    This binary nonionic–anionic surfactants mixture has been widely investigated through mutual and intra-diffusion measurements [56,57], as well as through calorimetric measurements [58] by Vitagliano and his school.

    Castaldi *et al.* [56] measured the CMC of this system at several ratios of the two surfactants through a solubilization

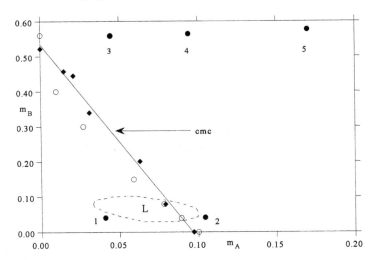

Figure 12 Ternary system $C_6E_5(A)$–$C_6SO_3Na(B)$–water, m_i/mol kg^{-1} molality of component i; ●, average composition of diffusion runs; ◆, measured critical micelle composition from Orange OT solubilization; ○, critical micelle composition from intradiffusion measurements taken in D_2O solutions [57]. L is the approximate range of miscibility gap (reprinted with permission from Ref. [56]).

technique using Orange OT as a dye [59,60]. The CMC was found to range from that of pure C_6SO_3Na to that of pure C_6E_5 depending on the surfactants ratio (see Fig. 12). Moreover, at low C_6SO_3Na concentration, a composition range was found where the system shows some turbidity, due to a phase separation. The C_6H_5–water system does not show a similar behavior [45,61] and this suggests that the miscibility gap is due to the presence of C_6SO_3Na.

These authors performed also mutual diffusion measurement on the same ternary system, at several mean concentrations using the Gouy interferometric technique. The correct approach, through the Fick's phenomenological equations, was used [51]

$$-J_A = D_{AA}\nabla C_A + D_{AB}\nabla C_B \tag{20a}$$

$$-J_B = D_{BA}\nabla C_A + D_{BB}\nabla C_B \tag{20b}$$

In Eqs (20), J_i and ∇C_i are the flow and the concentration gradient of component i, respectively, whereas D_{ii} are the main-term diffusion coefficients which account for the flux of each component on its own concentration gradient, and the D_{ij} are the cross-terms which account for the flux of each component due to the concentration gradient of the other component.

At low solutes concentrations (Point 1, see Fig. 12), the C_6E_5–C_6SO_3Na system is dominated by the presence of the critical mixing region. The values of the main terms are close to the value of the corresponding binary system, i.e., no mixed micelle are formed. The cross-term D_{BA} is large and positive which implies that the system is approaching a region where a phase separation occurs [62] (see Table 1).

In contrast, the overall picture of the diffusion data measured at the mean concentrations above CMC (points 2–5) shows the presence of mixed micelles. In fact, due to the formation of a slow moving micelle, the main diffusion coefficients are much lower than the corresponding diffusion coefficients in the binary systems. Moreover, one of the cross-term diffusion coefficients is negative as expected in systems

Table 1 Diffusion coefficients of the ternary system C_6E_5 (A)–C_6SO_3Na (B)–H_2O in $cm^2\,s^{-1} \times 10^5$. Reprinted from Ref. [56]. C_i/mol dm^{-3} average concentrations of each point. D_{ij}, measured diffusion coefficients of the ternary system. D_i diffusion coefficient of the corresponding binary systems [61]. $D_{i,\mathrm{mic}}$ diffusion coefficient of micelles in the binary systems [61] (see Fig. 12).

	Point 1		Point 2	Point 3	Point 4	Point 5
C_A	0.0399		0.1002	0.0409	0.0850	0.1500
C_B	0.0399		0.0401	0.5117	0.5090	0.5088
D_{AA}	0.4393		0.1893	0.1973	0.1618	0.1590
D_{AB}	0.0059		0.0486	0.0055	0.0076	0.0218
D_{BA}	0.1189		−0.0379	−0.2240	−0.2096	−0.1183
D_{BB}	0.8834		0.7475	0.5806	0.4898	0.4870
D_A	0.478		0.325			
D_B	0.912		0.910			
$D_{A,\mathrm{mic}}$		0.125				
$D_{B,\mathrm{mic}}$		0.215				

where a chemical equilibrium takes place and the complex formed has a smaller diffusivity with respect to the parents components [63].

The careful analysis of the values of the mutual diffusion coefficients in the micellar region suggests also that the mixed micelles change in stoichiometry and dimension as the concentration of C_6E_5 is raised.

A parallel study on the same C_6E_5–C_6SO_3Na mixture by Ciccarelli *et al.* [57] gives a further insight into this system's behavior. They measured the intradiffusion coefficients of both the nonionic and anionic surfactants, by the PGSE-NMR method, at various constant C_6SO_3Na molalities, varying the C_6E_5 composition.

The CMC of the system, according to the binary surfactant–water mixture treatment [45], was evaluated at the concentration where the trend of the intradiffusion coefficients shows a slope change. The computed values are in agreement with those measured by the solubilitation method shown in Fig. 12.

In the micellar composition range, the intradiffusion coefficients were analyzed according to a two-site model in which the molecules of both surfactants are present in free monomeric or micellar form. On this assumption, the experimental intradiffusion coefficients of both surfactants $\mathbf{D}_B$ and $\mathbf{D}_A$ are an average between the intradiffusion coefficient of monomeric surfactant molecules $\mathbf{D}_{B1}$ and $\mathbf{D}_{A1}$ and that of mixed micellar aggregates, $\mathbf{D}_{mic}$

$$\mathbf{D}_B = \frac{m_{B1}}{m_B}\mathbf{D}_{B1} + \frac{m_{B,\,mic}}{m_B}\mathbf{D}_{mic} \tag{21a}$$

$$\mathbf{D}_A = \frac{m_{A1}}{m_A}\mathbf{D}_{A1} + \frac{m_{A,\,mic}}{m_A}\mathbf{D}_{mic} \tag{21b}$$

m_{i1}, $m_{i,mic}$, and m_i are the molalities of surfactant i in the monomeric and micellizated form, and its stoichiometric molality in solution, respectively.

Ciccarelli *et al.* through Eqs. (21) calculated the micellar phase composition and monomer phase composition at the

CMC, at several constant C_6SO_3Na molalites, and analyzed the results on the basis of the Holland and Rubingh theory.

In Eqs. (21), $\mathbf{D}_{mic}$ was determined experimentally with addition of TMS to the system, whereas D_{B1} and D_{A1} were assumed to be those measured at the CMC composition corrected for the obstruction effect due to the micelles [64]

$$\mathbf{D}_1 = \mathbf{D}_1^{CMC}(1 + 0.5\phi)^{-1} \tag{22}$$

where ϕ is the volume fraction of the micelles. From the knowledge of the intradiffusion coefficients of the species, it was possible to evaluate the concentration of each species in the system (see Fig. 13).

Ciccarelli *et al.* described the mixed micellization process in terms of the phase separation model and computed the

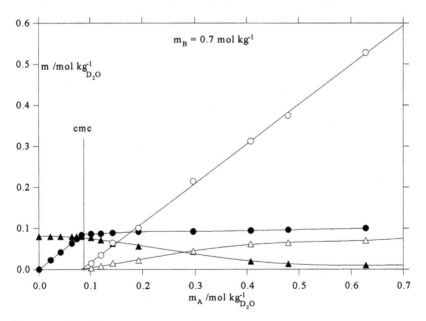

Figure 13 Ternary system $C_6E_5(A)$–$C_6SO_3Na(B)$–D_2O. Free and micellized surfactants molalities vs C_6E_5 molality: free (▲) and micellized (△) C_6SO_3Na, free (●) and micellized (○) C_6E_5 (reprinted with permission from Ref. [57]).

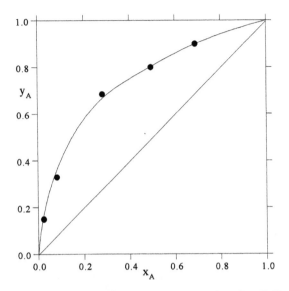

Figure 14 Micellar phase composition, y_A, for the $C_6E_5(A)$–C_6SO_3 Na(B)–D_2O system at micellar infinite dilution (i.e., at the CMC) as a function of stoichiometric solution composition, x_A. The plotted points are experimental data, the solid line is the prediction for the regular solution approximation with $\beta = -0.8$ (reprinted with permission from Ref. [57]).

surfactant mole fractions in the aqueous phase, x_A, and micellar pseudo-phase, y_A

$$x_A = \frac{m_{A1}}{m_{A1} + m_{B1}} ; \quad y_A = \frac{m_{A,mic}}{m_{A,mic} + m_{B,mic}} \qquad (23)$$

The y_A and x_A values extrapolated to the CMC line are plotted in Fig. 14.

Analysis of the CMC and y_A values according to the Rubingh's theory, allowed the authors to compute, through a best fitting procedure, the value of the β parameter for the C_6E_5–C_6SO_3Na–D_2O system ($\beta = -0.8$).

The negative value indicates favorable interaction among unlike hydrophilic heads on the micellar surface. Moreover, the good agreement between the calculated values of y_A and the

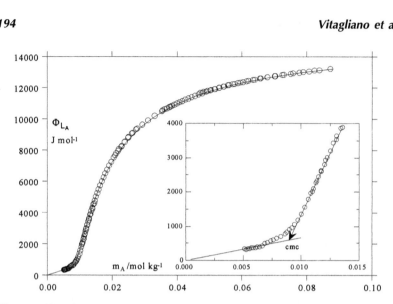

Figure 15 Apparent molar relative enthalpy of $C_8E_5(A)$ in aqueous solution as a function of surfactant molality. The line corresponds to the values computed for an association number $n = 17$ and an association constant $K_n = 10^{30}\,\mathrm{kg^{29}/mol^{29}}$, and $\Delta h_{mic} = 14\,800\,\mathrm{J/mol}$. Insert shows a possible choice of the CMC (reproduced with permission from Ref. [65]).

experimental ones suggests that in this system entropy changes on mixing are due mainly to a randomizing of the two surfactants in the micelle.

Before discussing the results of the microcalorimetric study performed on the system $C_6E_5(A)$–$C_6SO_3Na(B)$–water [58], let us summarize the calorimetric behavior of a set of nonionic surfactant–water binary systems, namely aqueous solutions of $C_{10}E_5$, C_8E_5, C_6E_5, and C_6E_3 [65]. In the paper, the molar relative enthalpies of the surfactant solutions have been determined via dilution experiments. In Fig. 15 the apparent molar enthalpy of C_8E_5 is shown as an example. All the examined surfactants show the same general sigmoid behavior. The dilution of their aqueous solutions is always an exothermic process.

The positioning of the inflexion and the steepness of the curve depend on the cooperativity level of the micellization process.

From the apparent molar enthalpy data, collected as a function of surfactant composition, some general conclusion can be extracted.

In the premicellar composition range, hydrophobic interactions among the aliphatic surfactant chains are already present, so that the surfactants molecules tend to organize themselves in structures favoring the subsequent micellization. This process is ruled by the solvent molecules. In this composition range, the apparent molar enthalpy is extrapolated to $m_A \to 0$ as a linear function of surfactant molality, as it can be seen in the insert of Fig. 15

$$\Phi_{L_A} = A_E m_A \tag{24}$$

where A_E is the first interaction parameter of the ethoxylated surfactant.

Concerning the post micellar composition range, the enthalpy of micellization is substantially constant in spite of the extension of both the hydrophilic head and the hydrocarbon chain. The micellization process is endothermic.

The micellization entropies are always positive confirming that the process is ruled by the hydrophobic interactions among the hydrocarbon chains of the surfactants molecules. The entropy gain is approximately $10 \, \text{J K}^{-1} \text{mol}^{-1}$ per methylene unit in the aliphatic chain.

In addition, dilution calorimetric runs on C_8E_5 were performed starting from the pure liquid surfactant and reaching the lowest aqueous compositions compatible with the instrumental sensibility. These experiments verified the presence of various transitions besides the micellization one. In Fig. 16 the numerical derivative of the apparent molar enthalpy is shown as a function of composition. Transitions appearing in the higher composition range can be ascribed to the saturation of different hydration layers surrounding the ethoxylic groups. They correspond to different organizations of surfactant molecules that may be arranged in various structures, hexagonal, cubic, etc. [50].

The same titration calorimetric technique was also applied to the study of the ternary system C_6E_5–C_6SO_3Na–water by

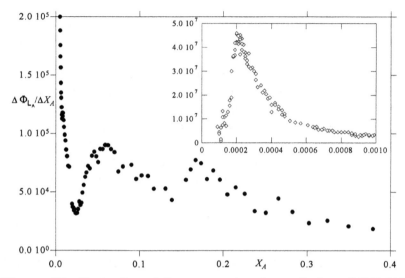

Figure 16   Derivative of the apparent molar enthalpy of $C_8E_5(A)$ plotted as a function of mole fraction. Three discontinuities are evident, corresponding to the micelle formation (insert) and to two other possible transitions (reproduced with permission from Ref. [65]).

diluting C_6E_5 in the titrating cell at constant C_6SO_3Na composition. In Fig. 17 the apparent molar relative enthalpies of C_6E_5 for various C_6SO_3Na molalities are shown.

   The curves of Fig. 17 are similar to that of Fig. 15. From these curves it is possible to evaluate the CMC and the premicellar interaction parameter, A_E. Both are function of C_6SO_3Na molality.

   The CMC values agree with those shown in Fig. 12. The A_E parameter increases smoothly with C_6SO_3Na molality (see Fig. 18), meaning a smooth increase of the interactions between aqueous C_6E_5 monomers. It is interesting to note that for $m_B = 0.587$ mol/kg a drastic decrease of the A_E parameter can be detected. At this C_6SO_3Na molality the micelles of the ionic surfactant are already present in solution, so that C_6E_5 is solubilized directly into the pre-existing micelles. As a consequence, the parameter A_E as well as the corresponding full curve (curve 6 of Fig. 17) account for the mixing behavior of C_6E_5 inside C_6SO_3Na micelles.

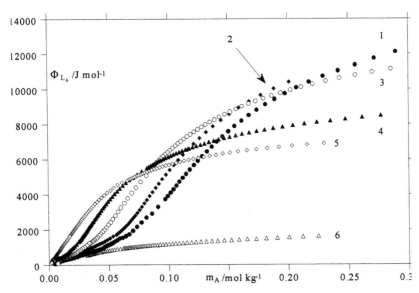

Figure 17 Ternary system $C_6E_5(A)-C_6SO_3Na(B)-H_2O$. Apparent molar relative enthalpy of C_6E_5 for various molalities of C_6SO_3Na: 1, $m_B = 0.105$; 2, $m_B = 0.199$; 3, $m_B = 0.300$; 4, $m_B = 0.429$; 5, $m_B = 0.495$; 6, $m_B = 0.587$ (reproduced with permission from Ref. [58]).

Calorimetric dilutions permit also the determination of the enthalpy of mixed micellization, provided that the distribution of both C_6E_5 and C_6SO_3Na between monomers and mixed micelles is known from intradiffusion data [57].

In Fig. 19 the micellization enthalpy is shown as a function of micelles composition, the solid and dotted lines correspond to the trend predicted through the ideal and regular solution approximations, respectively. It is not possible to discriminate between these two possibilities. The system investigated here is probably the first one for which the experimental Δh_{mic} values show that surfactants form effectively a regular or almost ideal solution in a quite wide micellar composition range. However, when the C_6E_5 molar fraction is less than 0.5, the calorimetric experiments show a behavior deviating from both theories.

Finally, with the aim to distinguish between the contribution of the hydrophobic tail and that of the hydrophilic

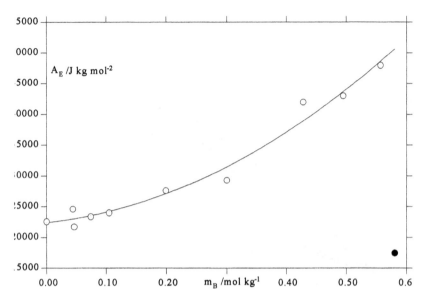

Figure 18 C_6E_5–C_6E_5 first interaction parameter as a function of C_6SO_3Na molality: o, below C_6SO_3Na CMC; •, above C_6SO_3Na CMC (reproduced with permission from Ref. [58]).

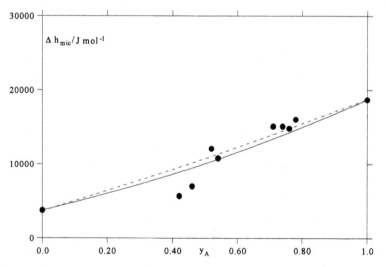

Figure 19 Molar enthalpy of micellization as a function of the mole fraction of C_6E_5 into the micelles. Solid line: prediction of the ideal solution approximation. Dotted line: prediction of the regular solutions approximation (reproduced with permission from Ref. [58]).

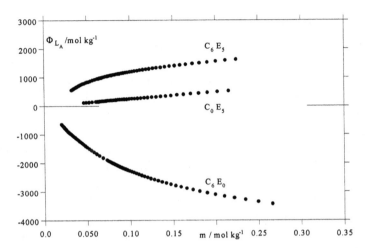

Figure 20 Comparison between the apparent molar relative enthalpies of 1-hexanol, C_6E_0, and pentaethylene glycol, C_0E_5, with that of C_6E_5 solubilized into the C_6SO_3Na micelles (molality of $C_6SO_3Na \sim 0.6\,\mathrm{mol/kg}$) (reproduced with permission from Ref. [58]).

head to the Φ_{L_A} value of C_6E_5 solubilized inside C_6SO_3Na micelles, calorimetric runs were performed on the ternary systems C_6E_0(1-hexanol)–C_6SO_3Na–water and C_0E_5(pentaethyleneglycol)–C_6SO_3Na–water at a C_6SO_3Na composition higher than its CMC in the binary system. In fact, C_6E_0 mimes the hydrophobic tail of C_6E_5 while C_0E_5 mimes its polar head.

As it can be seen in Fig. 20, the mixing of C_6E_0 is exothermic while that of C_0E_5 is endothermic as that of C_6E_5. This fact shows that, at least in a qualitative way, the interactions between the hydrophilic heads prevail over the contribution of the hydrophobic tails in ruling the enthalpic properties of the system. Recently, D'Errico et al. [66] found that mixtures of nonionic–cationic surfactants present a behavior similar to that found for nonionic–anionic surfactant mixtures.

VI. CONCLUSIONS

The aim of this paper has been to give up-to-date, general and specific information on surfactant mixtures containing at least one ethoxylated surfactant with a short hydrophobic tail, within the limits of a lack of literature on the subject.

To describe the properties of the mixtures, a trustworthy approach is the molecular-thermodynamic model proposed by Blankschtein and his school [28–30,36–39], being applicable to almost every kind of mixture. However, it is necessary to note that in the case of short-tailed surfactants some discrepancy between the theory and experimental data may exist [52]. This effect could be probably overcome taking into account the free energy contributions: hydrophobic, interfacial, packing, steric, and electrostatic factors have to be defined in more detail because micelles are loose, i.e., the surface separating the aqueous phase from the micellar core is deeply penetrated by water molecules.

The present paper analyzes the experimental results of mixtures of two ethoxylated surfactants and of mixtures of one ethoxylic and one ionic surfactant.

Some general information can be extracted through an overall analysis of the experimental results.

In particular for nonionic–nonionic mixtures, it has been observed that the presence of a surfactant, able to micellize, induces aggregation behavior in a second one that is not able to micellize, leading to mixed micelles [44].

The hydrophilic interactions are more responsible for deviation from ideal behavior than the hydrophobic ones [33]. Recently, Inoue et al. [67] found the trend of the cloud point temperature for mixtures of nonionic surfactants with the same hydrophobic tail to be well described by the Flory-Huggins model. However, mixtures of surfactants having fluorinated and non-fluorinated hydrocarbon tails show appreciable deviations from ideality even when the ethoxylic moiety is the same [53]. In this case the different flexibility of the molecular backbones and unfavorable packing effects can probably be involved.

In the case of nonionic–ionic mixtures, diffusion measurements [56,57] have shown that mixed micelles change not only their composition but also their dimensions as a function of stoichiometric mixtures composition. In the case of surfactants with short hydrophobic tails, the mixed micellar systems deviate only slightly from ideality [57] and they can be handled by the simple regular solution approach. This finding has been confirmed by calorimetry [58].

REFERENCES

1. P. M. Holland, and D. N. Rubingh, *ACS Symp. Ser.* **501**: 2–30 (1992).

2. K. Ogino, and M. Abe (eds), *Mixed Surfactant Systems.* Marcel Dekker, New York (1993).

3. B. Kronberg, *Curr. Opin. Colloid Interface Sci.* **2**: 456–463 (1997).

4. M. Abe, and K. Ogino, *Surfactant Sci. Ser.* **46**: 1–21 (1992).

5. M. J. Rosen, *Prog. Colloid Polym. Sci.* **109**: 35–41 (1998).

6. P. Somasundaran, and L. Huang, *Surfactants* 204–224 (1999).

7. A. Khan, and E. F. Marques, *Curr. Opin. Colloid Interface Sci.* **4**: 402–410 (2000).

8. A. P. Gracia, J. Lachaise, and R. S. Schechter, *Surfactant Sci. Ser.* **46**: 63–97 (1992).

9. N. Nishikido, *Surfactant Sci. Ser.* **46**: 23–61 (1992).

10. R. Aveyard, and D. Blankschtein, *Curr. Opin. Colloid Interface Sci.* **6**: 338–341 (2001).

11. J. D. Hines, *Curr. Opin. Colloid Interface Sci.* **6**: 350–356 (2001).

12. P. Somasundaran, and L. Huang, *Adv. Colloid Interface Sci.* **88**: 179–208 (2000).

13. B. Joensson, O. Edholm, and O. Teleman, *J. Chem. Phys.* **85**: 2259–2271 (1986).

14. L. Laaksonen, and J. B. Rosenholm, *Chem. Phys. Lett.* **216**: 429–434 (1993).

15. G. D'Errico, O. Ortona, L. Paduano, and V. Vitagliano, *J. Colloid Interface Sci.* **239**: 264–271 (2001).

16. P. M. Kruglikov, and A. F. Koretskii, *USSR Opyt Primen. Nov. Moyushch. Sredstv* (1973), pp. 76–83.

17. K. Shinoda, and H. Kunieda, In *Encyclopedia of Emulsion Technology* Vol. 1 (P. Becher, ed.), Marcel Dekker, NewYork (1983), p. 337.

18. R. M. Ljubic, and S. M. Gasic, *J. Serb. Chem. Soc.* **54**: 115–117 (1989).

19. P. Mukerjee, *J. Phys. Chem.* **76**: 565–570 (1972).

20. M. J. Rosen, and L. Liu, *J. Am. Oil Chem. Soc.* **73**: 885–890 (1996).

21. N. Hioka, M. J. Politi, and H. Chaimovich, *Tetrahedron Lett.* **30**: 1051–1054 (1989).

22. T. Kawai, J. Umemura, and T. Takenaka, *Colloid Polym. Sci.* **262**: 61–66 (1984).

23. J. Umemura, H. H. Mantsch, and D. G. Cameron, *J. Colloid Interface Sci.* **83**: 558–568 (1981).

24. B. Lindman, and B. Brun, *J. Colloid Interface Sci.* **42**: 388–399 (1973).

25. T. Telgmann, and U. Kaatze, *J. Phys. Chem. B* **101**: 7766–7772 (1997).

26. D. Attwood, P. H. Elworthy, and M. J. Lawrence, *J. Chem. Soc. Faraday Trans. I* **82**: 1903–1910 (1986).

27. Y. Zimmels, and I. J. Lin, *Colloid Polym. Sci.* **252**: 594–612 (1974).

28. S. Puvvada, and D. Blankschtein, *J. Chem. Phys.* **92**: 3710–3723 (1990).

29. G. Briganti, S. Puvvada, and D. Blankschtein, *J. Phys. Chem.* **95**: 8989–8995 (1991).

30. A. Naor, S. Puvvada, and D. Blankschtein, *J. Phys. Chem.* **96**: 7830–7832 (1992).

31. J. H. Clint, *J. Chem. Soc., Faraday Trans I* **71**: 1327–1334 (1975).

32. P. M. Holland, and D. N. Rubingh, *J. Phys. Chem.* **87**: 1984–1990 (1983).

33. A. Ohta, S. Miyagishi, and M. Aratono, *J. Phys. Chem. B* **105**: 2826–2832 (2001).

34. C. D. Eads, and L. C. Robosky, *Langmuir* **15**: 2661–2668 (1999).

35. T. R. Desai, and S. G. Dixit, *J. Colloid Interface Sci.* **177**: 471–477 (1996).

36. C. Sarmoria, S. Puvvada, and D. Blankschtein, *Langmuir* **8**: 2690–2697 (1992).

37. S. Puvvada, and D. Blankschtein, *J. Phys. Chem.* **96**: 5567–5579 (1992).

38. S. Puvvada, and D. Blankschtein, *J. Phys. Chem.* **96**: 5579–5592 (1992).

39. A. Shiloach, and D. Blankschtein, *Langmuir* **14**: 1618–1636 (1998).

40. V. M. Garamus, *Chem. Phys. Lett.* **290**: 251–254 (1998).

41. V. Sambhav, G. Alex, D. Hemangi, and B. Pratap, *J. Surfactants Deterg.* **2**: 213–221 (1999).

42. M. E. Haque, A. R. Das, A. K. Rakshit, and S. P. Moulik, *Langmuir* **12**: 4084 (1996).

43. J. Cross, *Nonionic Surfactants, Chemical Analysis* (J. Cross, ed.), Marcel Dekker, New York (1987).

44. G. D'Errico, D. Ciccarelli, O.Ortona, and V. Vitagliano, *J. Mol. Liq.* **100**: 241–253 (2002).

45. L. Ambrosone, L. Costantino, G. D'Errico, and V. Vitagliano, *J. Colloid Interface Sci.* **190**: 286–293 (1997).

46. L. Ambrosone, L. Costantino, G. D'Errico, and V. Vitagliano, *J. Solution Chem.* **25**: 757–772 (1996).

47. L. Ambrosone, L. Costantino, G. D'Errico, and V. Vitagliano, *J. Solution Chem.* **26**: 735–748 (1997).

48. K. H. Lim, J. S. Reckley, and D. H. Smith, *J. Colloid Interface Sci.* **161**: 465–470 (1993).

49. R. Kjellander, and E. Florin, *Chem. Soc. Trans. Faraday Soc.* **77**: 2053–2077 (1981).

50. D. J. Mitchell, G. J. T. Tiddy, L. Waring, T. Bostock, and M. P. McDonald, *J. Chem. Soc., Faraday Trans. 1* **79**: 975–1000 (1983).

51. V. Vitagliano, G. D'Errico, O. Ortona, and L. Paduano, *Handbook of Surfaces and Interfaces of Material* (H. S. Nalwa, ed.), Academic Press, San Diego (2001), Vol. 1 ch. 10, pp. 545–611.

52. L. Costantino, G. D'Errico, P. Roscigno, and V. Vitagliano, *J. Phys. Chem.* **104**: 5469–5473 (2000).

53. F. Portet, P. L. Desbene, and C. Treiner, *J. Colloid Interface Sci.* **184**: 216–226 (1996).

54. M. Villeneuve, T. Nomura, H. Matsuki, S. Kaneshina, and M. Aratono, *J. Colloid Interface Sci.* **234**: 127–136 (2001).

55. H. Matsuki, N. Ikeda, M. Aratono, S. Kaneshina, and K. Motomura, *J. Colloid Interface Sci.* **154**: 454–460 (1992).

56. M. Castaldi, L Costantino, O. Ortona, L. Paduano, and V. Vitagliano, *Langmuir* **14**: 5994–5998 (1998).

57. D. Ciccarelli, L. Costantino, G. D'Errico, L. Paduano, and V. Vitagliano, *Langmuir* **14**: 7130–7139 (1998).

58. O. Ortona, G. D'Errico, V. Vitagliano, and L. Costantino, *J. Colloid Interface Sci.* **249**: 481–488 (2002).

59. H. Schott, *J. Phys. Chem.* **68**: 3612–3619 (1964).

60. H. Schott, *J. Phys. Chem.* **70**: 2966–2973 (1966).

61. L. Paduano, R. Sartorio, V. Vitagliano, and L. Costantino, *J. Colloid Interface Sci.* **189**: 189–198 (1997).

62. V. Vitagliano, R. Sartorio, S. Scala, and D. Spaduzzi, *J. Solution Chem.* **7**: 605–621 (1978).

63. L. Paduano, R. Sartorio, and V. Vitagliano, *J. Phys. Chem. B* **102**: 5023–5028 (1998).

64. G. M. Bell, *Trans. Faraday Soc.* **60**: 1752–1759 (1964).

65. O. Ortona, V. Vitagliano, L. Paduano, and L. Costantino, *J. Colloid Interface Sci.* **203**: 477–484 (1998).

66. G. D'Errico, O. Ortona, L. Paduano, A. M. Tedeschi, and V. Vitagliano, *Phys. Chem.: Chem. Phys.* **4**: 5317–5324 (2002).

67. T. Inoue, H. Ohmura, and D. Murata, *J. Colloid Interface Sci.* **258**: 374–382 (2003).

6

The Effect of Mixed Counterions on the Micelle Structure of Perfluorinated Anionic Surfactants

DOBRIN PETROV BOSSEV

National Institute of Standards and Technology,
Gaithersburg, Maryland, U.S.A.

MUTSUO MATSUMOTO and MASARU NAKAHARA

Institute for Chemical Research, Kyoto University,
Kyoto, Japan

SYNOPSIS

In this chapter we have reviewed studies on ionic perfluorinated surfactant systems that include two different counterions, inorganic and organic. Strong deviation from the ideal mixing is observed in the counterion atmosphere due to the preferential binding of the organic counterions. This finding implies that the counterion atmosphere can be conveniently "tuned" to create micelles with different shapes and sizes and to regulate the inter-micellar interaction.

I. INTRODUCTION

There is a large number of studies on fluorocarbon/hydrocarbon surfactant mixtures where the emphasis is placed on the immiscibility of the fluoro- and hydrocarbons [1–16]. As a result of this phenomenon, fluorocarbon/hydrocarbon surfactant mixtures deviate from the ideal mixing trend and form two or more kinds of hydrocarbon-rich or fluorocarbon-rich micelles, which is called micellar demixing. However, very few studies have been done on mixed ionic surfactant systems that include identical surfactant ions but different counterions [17–29]. These mixtures are expected to form a single type of micelles and the nonideality of mixing takes place not in the micellar structures but in the formation of the counterion atmosphere surrounding the micelles. Such phenomenon leads to preferential counterion binding that affects the shape, size, and interaction of the micelles. Consequently, the macroscopic properties of the solution, such as the viscosity, are altered [28,29]. Recent studies that involve anionic perfluorinated surfactant show that mixing of two identical surfactants with different counterions may allow "fine tuning" of the counterion atmosphere that leads to a strong synergetic effect on the viscosity of the solution [17,28]. Mixture of two counterions surrounding the micelles can be obtained easily if we simply add a particular electrolyte to a single ionic surfactant. However, in this case the ionic strength of the solution also increases. In this chapter, we will focus on mixtures of two surfactants with different counterions at a constant ionic strength.

Ionic surfactants may form a variety of micellar structures depending on the nature of the counterion [30–35]. As a rule, smaller and more hydrophilic counterions induce formation of spherical micelles. This is thought to be due to the weak binding of these counterions to the micelles which leads to an effective decrease of the electrostatic repulsion between the surfactant head groups. Ionic surfactants with organic counterions form thread-like structures that result in viscoelastic behavior of the solution. For example salicylate counterion is known to induce formation of cylindrical micelles with a number of cationic surfactants that normally form spherical micelles with Cl$^-$ counterions [36,37]. This effect is attributed to the much stronger binding of the counterion to the micelles and even to intercalation between the oppositely charged surfactant head groups. As a result, the electrostatic repulsion between the surfactant head groups is screened more effectively which reduces the curvature of the surfactant membrane and leads to formation of cylindrical micelles. In a series of perfluorinated sulfonic and carboxylic acids, this transition to thread-like structures has been observed with alkylammonium counterions [14,29,38]. These solutions possess strong viscoelastic behavior due to formation of a living network of micelles [28].

The theory of the counterion binding to micelles has been steadily improved throughout the past decades [39]. Together with the refinement of the classical or continuum theory of the electric double layer [40–44] a number of computer simulations have recently been done to get an adequate molecular picture [45,46]. However, the experimental analyzes employ rather macroscopic models to shed light on the counterion binding because it is difficult to distinguish between the different regions of the counterion atmosphere. Atomic force microscopy and ξ-potential measurements are applicable only to fixed macroscopic surfaces [41] and the conductivity method gives rather collective information [38,47]. It is the NMR that is indispensable for studies on counterion binding to micelles. NMR is selectively sensitive to the different species present in the solution and can simultaneously monitor the self-diffusion coefficients of several components.

II. BINDING OF MIXED COUNTERIONS

A. NMR Method to Determine the Counterion Binding

The counterion binding to micelles can be determined by monitoring the counterion mobility in the solution. Although the conductivity is the most available method [38,47], this method is not selective to the different species in the solution. The self-diffusion coefficients of the counterions can be directly and selectively measured by radioactive-isotope labeling of the counterion under investigation [23,48]. Although very precise, this method is very unpopular because it relies upon radioactive isotopes.

Determination of the self-diffusion coefficients in multicomponent systems has been made much easier after the development of the Fourier transform pulse-gradient-spin-echo (FT PGSE) NMR technique. This method is based on the focusing/defocusing of the transverse magnetization of the atomic spins diffusing in a magnetic field gradient [49,50]. The species under investigation do not require labeling since most of the alkali, halides, and the proton have magnetic moments.

Traditionally, the counterion binding is measured on the basis of the self-diffusion coefficient [31,51–53]:

$$D^+ = (1 - f_D)D^+_{\text{free}} + f_D D^+_{\text{bound}} \qquad (1a)$$

$$f_D = \frac{D^+_{\text{free}} - D^+}{D^+_{\text{free}} - D^+_{\text{bound}}} \qquad (1b)$$

where D^+_{free} and D^+_{bound} are the self-diffusion coefficients of the free and bound counterions, respectively, and f_D is the fraction of the bound counterions due to the diffusion measurement. The value of D^+_{free} can be measured in a dilute solution of a simple electrolyte that contains the respective counterion. D^+_{bound} can be assumed to be equal to the self-diffusion coefficient of the micelles, D_{mic}. The D_{mic} value is obtained either by measuring the self-diffusion coefficient of a hydrophobic probe molecule solubilized into the micelles [51–53] or by plotting the self-diffusion coefficient of the surfactant molecule against the inverse concentration [33]. Since the micellar formation of ionic surfactants is weakly affected by the temperature and deuteration of the solvent, the D_{mic} values can be easily converted to those at other temperatures through the Stokes–Einstein hydrodynamic expression [48] (see Table 1.) Thus, f_D can be evaluated according to Eq. (1b). However, f_D, based on the self-diffusion coefficient, provides an average binding coefficient for each counterion, somehow comparable to those obtained from conductivity method.

Table 1 Self-diffusion coefficients of the micelles, D_{mic}, formed by some ionic surfactants[a] (in $10^{-10}\,m^2/s$ units).

Surfactant	D_{mic}	$T/°C$	Solvent	CMC/mM	c/mM	D_{30}[b]	Ref.
$C_8F_{17}SO_3^-Li^+$	0.57	30	D_2O	7.1	> cmc	0.57	[33]
$C_8H_{17}SO_3^-Na^+$	~1.0	25	H_2O	~150	250	0.92	[48]
$C_8H_{17}C_6H_4SO_3^-Na^+$	~1.0	33	H_2O	12.8	100	0.77	[53]
$C_{10}H_{21}NH_3^+CH_3COO^-$	1.5	38	D_2O	61	111	1.3	[24]
$C_{10}H_{21}NH_3^+CH_2ClCOO^-$	1.14	38	D_2O	43	111	0.96	[24]
$C_{10}H_{21}NH_3^+CHCl_2COO^-$	0.48	38	D_2O	27	108	0.41	[24]
$C_{12}H_{25}SO_4^-Li^+$	~0.76	25	D_2O	~8.5	92	0.85	[48]
$C_{12}H_{25}SO_4^-Na^+$	~0.65	25	H_2O	~8.0	100	0.59	[48]
$C_{16}H_{33}N(CH_3)_3^+Cl^-$	~0.54	33	D_2O	1.1	100	0.51	[48]
$C_{16}H_{33}N(CH_3)_3^+Br$	~0.50	33	D_2O	0.9	102	0.47	[48]

[a]The enlisted surfactants are supposed to form spherical micelles.
[b]Recalculated values at 30°C in D_2O on the basis of the viscosity of pure water.

The NMR chemical shift is especially sensitive to the short-range interactions between the counterions and the micellar surface. Thus the chemical shift can be used as a selective probe for the species adsorbed onto the micellar surface. The counterion binding may be determined from the chemical shifts in a way similar to that for the self-diffusion coefficients [54–56]

$$\delta = (1 - f_\delta)\delta_{free} + f_\delta \delta_{bound} \tag{2a}$$

$$f_\delta = \frac{\delta_{free} - \delta}{\delta_{free} - \delta_{bound}} \tag{2b}$$

where δ_{free} and δ_{bound} are the chemical shifts of the free and bound counterions, and f_δ is the fraction of the bound counterions due to the chemical shift. The value of δ_{free} is easily measurable in a simple solution of the respective electrolyte but there is a technical difficulty to obtain δ_{bound}. Thus this method has not been used until recently [17].

In normal circumstances only a fraction of the counterions are bound to the micelles. Hence δ_{bound} cannot be obtained unless all of the counterions are forced to bind the micelles. Bossev *et al.* [17,33] have solved this problem by combining

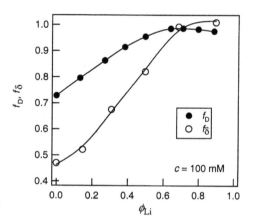

Figure 1 The fractions of bound TEA$^+$ ions evaluated through the self-diffusion coefficient (f_D) and through the chemical shift (f_δ) as a function of ϕ_{Li} in D$_2$O at 30°C [see Eqs (1) and (2)]. [From Ref. 17.] (Reproduced by permission of the American Chemical Society.)

two kinds of counterions with very different binding coefficients. The binding coefficient of the tetraethylammonium counterion (TEA$^+$) to perfluorooctylsulfonate micelles (FOS$^-$) is known to be 0.80 while that of the lithium ion (Li$^+$) is as small as 0.17. When TEAFOS and LiFOS are mixed at a constant total concentration of 100 mM, these two counterions compete to bind to the FOS surface. Thus at a high fraction of the hydrophilic Li$^+$ (ϕ_{Li}) we can expect all of the hydrophobic TEA$^+$ counterion are bound to the micellar surface. This is proven by plotting the f_D values for the TEA$^+$ counterion as a function of ϕ_{Li} (see Fig. 1). At sufficiently high ϕ_{Li} the f_D values are independent of the fraction of the lithium ions, as expected. In these conditions we can separately obtain δ_{bound}.

The fraction f_δ is smaller than the dynamical f_D because of the difference in the detection characteristic. It is proposed that the difference between f_δ and f_D originates from the fine structure of the electric double layer which consists of the two parts, Stern and diffuse layers [44]. The *Stern layer* is assumed to be a monolayer of counterions firmly bound to the micellar surface by not only Coulombic but also some additional interactions. The chemical shift probes exclusively the short-range interactions within the *Stern layer*. On the other hand, the fraction f_D includes the counterions in the *diffuse double layer* as well as those in the *Stern layer*. Thus, the counterions located within the *Stern layer* are distinguished from those in the *diffuse double layer*.

B. Competitive Counterion Binding

The competitive or preferential counterion binding in mixed ionic surfactant systems has been initially studied on hydrocarbon surfactants with small monoatomic counterions. An example for this is the early work by Fabre *et al.* [23] that employs tracer and self-diffusion and NMR methods to investigate CTAC/CTAB mixtures (cetyltrimethylammonium chloride and bromide) as a function of the molar ratio between these two surfactants. This study involves radio-isotope labeled chemicals to determine the self-diffusion coefficient of the counterions. The bulkier Br^- ion is found to be associated to the CTA micelles with a binding coefficient, β, of about 0.75, while the Cl^- ion appears less associated, $\beta \approx 0.55$. Although the Br^- and the Cl^- ions differ in their affinity to the CTA micelles, they are rather similar counterions and no significant preference has been observed when the two surfactants have been mixed. Another early study on monoatomic counterions has been conducted by Koshinuma [25] using electromotive force cells. He has determined the activity coefficients of Na^+ and K^+ ions at different molar rations in the presence of decylsulfate micelles. The larger K^+ has been found to be preferentially bound. Jansson and Stilbs [24] have used NMR to monitor β on the basis of the self-diffusion coefficients of the counterions and micelles for decylammonium acetate, chloroacetate, and dichloroacetate. The counterion binding has been found to decrease in the order, acetate < chloroacetate < dichloroacetate, and to be independent of the surfactant concentration when studied in neat systems. They have also performed experiments to study the competitive effects on the association behavior of these counterions. For this purpose, two kinds of surfactants are mixed in pairs at a 50/50 molar ratio and β is determined for each counterion separately and compared to that in neat systems, respectively. For example, the counterion binding of acetate and dichloroacetate ions is found to be 0.60 and 0.78 in neat systems but these values change to 0.44 and 0.92, respectively, when the two surfactants are mixed together. This has laid grounds for the study of the competitive counterion binding to micelles but stopped short of investigating this process in details. Another study on the mixture of chlorinated aromatic counterions employs ^{1}H and ^{13}C NMR chemical shifts to probe the intercalation of these counterions on the surface of the micelles [26]. The findings are that 3,5-dichlorobenzoate counterions insert further into the interface of the rod-like micelles of CTA than 2,6-dichloro-benzoate ones.

The non-ideal mixing of counterions have been thoroughly investigated in a recent paper [17] for two very different counterions, the hydrophilic lithium (Li^+) and the hydrophobic tetraethylammonium (TEA^+). These two counterions are combined with perfluorooctylsulfonate surfactant (FOS^-) to form LiFOS and TEAFOS, respectively. Although these surfactants consist of an identical hydrophobic chain part and headgroup, thread-like micelles are grown by TEAFOS, in contrast to the spherical micelles formed by LiFOS. The marked differences in the structure between TEAFOS and LiFOS are attributed solely to the antagonism in the counterion nature. It is the hydrophobic TEA^+ counterions that are more strongly bound to the micelles so as to effectively reduce the electrostatic surface charge density of the micelles [38,47]. Thus, the counterion binding strongly affects the ionic micellar structure as well as the double layer structure. The effect of nonideal mixing of these counterions has been investigated by varying the fraction of LiFOS (ϕ_{Li}) in the mixtures of TEAFOS and LiFOS at a constant total concentration of 100 mM.

The non-ideality of the counterion mixing becomes apparent when the viscosity of LiFOS/TEAFOS mixtures is measured (see Fig. 2). The large size of the thread-like micelles formed by TEAFOS explains the high viscosity at $\phi_{Li} = 0$. On the other corner of this plot at $\phi_{Li} = 1$, the viscosity is roughly close to that of water because of the ordinary spherical micelles formed by LiFOS. When these two surfactants are mixed upon

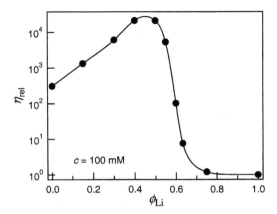

Figure 2 Viscosity of the TEAFOS/LiFOS mixtures in H_2O at a total concentration (c) of 100 mM and temperature of 26°C as a function of the LiFOS fraction (ϕ_{Li}). [From Ref. 17.] (Reproduced by permission of the American Chemical Society.)

condition that the total surfactant concentration is kept at 100 mM, the viscosity increases exponentially by 2 orders of magnitude with increasing ϕ_{Li}. At $\phi_{Li} = 0.55$, the viscosity of the mixture exhibits a clear maximum, as a result of the nonideal mixing of the two counterions. At higher fractions of the Li^+ ions, the viscosity drops sharply by about 4 orders of magnitude. A similar synergistic increase in the viscosity has been observed by Hoffmann *et al.* in the mixtures of TEAFOS and perfluorooctylsulfonic acid (HFOS) at the same total concentration of 100 mM [29]. For the explanation of the synergetic phenomenon, the viscosity (η) of this system can be expressed as in the case of the so-called living polymers organized in a transient network [57]

$$\eta = G\tau \qquad (3)$$

where G is the shear modulus of the network and τ is the lifetime of the network strands. The parameter G is a static (equilibrium) characteristic of the network, whereas the lifetime τ accounts for the network dynamics. As discussed in the next chapter, this model well describes the rheology of an aqueous solution of neat TEAFOS [58]. According to the NMR data, G increases within 20% when ϕ_{Li} is varied from 0 to 0.55.

However, this increase in G is too small to explain the increment in η (two orders of magnitude). The main factor for the non-additive effect on the viscosity originates from the dynamical behavior (the relaxation time τ) of the TEAFOS/LiFOS systems [28].

Once f_δ is determined, the concentration of TEA$^+$ counterions localized within the *Stern layer* (c_S) can be calculated as a function of ϕ_{Li} [17]

$$c_S = c f_\delta (1 - \phi_{Li}) \qquad (4)$$

where c (=100 mM) is the total concentration of the FOS$^-$ ions and $c(1 - \phi_{Li})$ is the concentration of the TEA$^+$ ions in the TEAFOS/LiFOS mixture. The remaining portion of the TEA$^+$ ions (c_D) which are located within the *diffuse double layer* are calculated simply as

$$c_D = c - c_S \qquad (5)$$

where c_S from Eq. (4) is inserted. Figure 3 shows how c_S and c_D depend on ϕ_{Li}. The concentration of the TEA$^+$ ions in the *Stern*

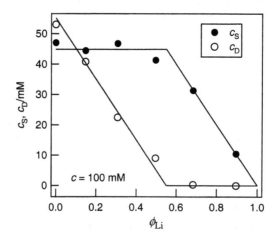

Figure 3 Concentration of the TEA$^+$ counterions bound in the *Stern layer* (c_S) evaluated through the chemical shift as a function of ϕ_{Li}. The concentration of the remaining portion of TEA$^+$ counterions (c_D) is also shown. [From Ref. 17.] (Reproduced by permission of the American Chemical Society.)

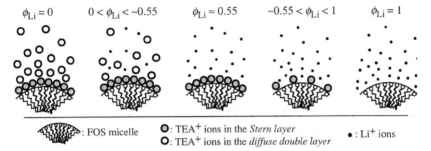

$\phi_{Li} = 0$ $0 < \phi_{Li} < {\sim}0.55$ $\phi_{Li} \approx 0.55$ ${\sim}0.55 < \phi_{Li} < 1$ $\phi_{Li} = 1$

: FOS micelle   ○ : TEA$^+$ ions in the *Stern layer*   • : Li$^+$ ions
◯ : TEA$^+$ ions in the *diffuse double layer*

Figure 4 Schematic representation of the changes in the double layer structure at different ϕ_{Li} values. Note the preferential binding of TEA$^+$ counterions. At high ϕ_{Li} values the micellar surface is incompletely covered by TEA$^+$ counterions; this is a reason for the disintegration of the threadlike network. [From Ref. 17.] (Reproduced by permission of the American Chemical Society.)

layer is $\sim$45 mM and independent of ϕ_{Li} at $0 < \phi_{Li} < 0.55$. This value of c_S shows that one TEA$^+$ counterion is bound to roughly two FOS$^-$ ions in the micellar state; i.e., this hydrophobic counterion bridges two micellized FOS$^-$ ions. At a fixed value of c, the c_D value *decreases* linearly with ϕ_{Li}. This implies that the TEA$^+$ ions in the *diffuse double layer* are proportionally replaced by Li$^+$ ions with increasing ϕ_{Li} and that the *Stern layer* remains unaffected at this stage. At $\phi_{Li} > 0.55$, all of the TEA$^+$ counterions are bound within the *Stern layer* and the concentration of the *Stern layer* decreases with increasing ϕ_{Li} due to the shortage of TEA$^+$ ions in the micellar solution. These structural changes in the counterion atmosphere are illustrated in Fig. 4. Undoubtedly, the driving force for the preferential binding of the TEA$^+$ ions in the *Stern layer* originates from their hydrophobicity.

The viscosity of the TEAFOS/LiFOS system is correlated to the structure of the electric double layer described above. The region of ϕ_{Li} where the viscosity is high ($0 < \phi_{Li} < 0.55$) exactly corresponds to unruffled *Stern layer*; compare Figs 2 and 4. The sudden drop in the viscosity at $\phi_{Li} > 0.55$ correlates with the deficiency of TEA$^+$ ions to form a completed *Stern layer*. The conclusion is that the complete coverage of the micellar surface by the TEA^+ ions is essential for the formation of threadlike structure by the FOS^- surfactant. It seems that the surplus of TEA^+ ions beyond the *Stern layer* which is given by c_D in Fig. 3 at $0 < \phi_{Li} < 0.55$, reduces the viscosity.

III. VISCOELASTICITY WITH MIXED COUNTERIONS

Interesting viscoelasticity and rheological properties are exhibited by dilute and semi-dilute micellar systems that involve thread-like micelles entangled in a network [29,36,37,59,60]. As discussed above, such systems involve at least one organic counterion. On the other hand, if the same surfactant is combined with a monoatomic counterion, it forms spherical micelles without any viscoelastic behavior. Thus mixing identical surfactants with two counterions that induce formation of thread-like and spherical micelles, respectively, may be of significant interest regarding the nonideality of the counterion binding discussed above.

Hoffmann and Würtz [29] found that the structural relaxation time τ as well as the viscosity η first increase and then decrease on increasing fraction ϕ_H of the H^+ ions in the counterion mixture (TEA^+ and H^+) in TEAFOS/HFOS solutions (see Fig. 5). They attributed these changes to a change in the surface charge density of the FOS threads. Much more detailed investigation was done by Watanabe *et al.* [28] on similar TEAFOS/LiFOS mixtures. The TEA^+/Li^+ mixture has certain advantage over the TEA^+/H^+ one because one can expect constant pH of the micellar solution when the ratio between the counterions is varied.

The stress of the thread-like micelles is related to their orientation/stretch, as in the case of unbreakable polymer chains. However, these micelles are reversibly scissored and reformed due to thermal motions of unimers (either individual surfactant molecules or their spherical micelles) and the stress can relax upon the thread scission. Thus the thread-like micelles are classified as living threads and their rheological features are different, in some respects, from those of polymer

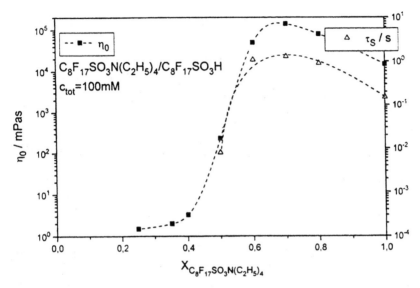

Figure 5 Zero-shear viscosity and structural relaxation time for different charge densities at the surface of the rodlike micelles. The charge density on the surface of the rodlike micelle can be varied by variation of the $C_8F_{17}SO_3H$–content. A maximum of the zero-shear viscosity and the structural relaxation time is passed for $X = 0.7$. [From Ref. 29.] (Reproduced by permission of Elsevier Science B.V.)

chains. For example, the relaxation of long threads is completed via the thermal scission before these threads exhibit global motion in their unscissored form. This terminal relaxation is well described by the single-mode Maxwell model, as found for threads of a 1 : 1 complex of cetyltrimethylammonium bromide (CTAB) and sodium salicylate (NaSal) and for the threads of cetylpyridinium chloride (CPyCl) and NaSal [36,37,60]. These CTAB : NaSal and CPyCl : NaSal threads have a cylindrical (or rod-like) shape. A different type of thread is formed in aqueous solutions of TEAFOS. As shown by cryogenic transmission electron microscopy (cryo-TEM) FOS molecules aggregate into spherical micelles and these micelles further associate with each other to form threads of a pearl-necklace shape [14,58]. As discussed in the previous section, the TEA^+ ions are tightly bound to the surfaces of the FOS micelles to form well-packed

Stern layers which means the electrostatic repulsion is more effectively reduced in comparison to the Li^+ ions at shorter distances to the micellar interface. This implies that FOS micelles should associate only weakly through hydrophobic attraction (due to the tightly bound hydrophobic TEA ions) that is somewhat stronger than the remaining electrostatic repulsion [14]. The FOSTEA threads of such weakly associating micelles should behave as feeble threads. In contrast, the CTAB:NaSal cylindrical threads behave as tough threads because of their high stability due to strong interactions of the surfactant molecules at every cross-section of the threads [58].

For the FOSTEA/FOSLi solutions at a temperature of 15°C, the storage and loss moduli (G' and G'') are plotted in Fig. 6 against the angular frequency ω at selected ϕ_{Li}. In the linear regime, the FOSTEA threads commonly exhibited the Maxwell-type terminal relaxation attributable to thermal scission [28]

$$G'(\omega) = \frac{1}{J}\frac{\omega^2\tau^2}{1+\omega^2\tau^2} \qquad G''(\omega) = \frac{1}{J}\frac{\omega\tau}{1+\omega^2\tau^2} \qquad (6)$$

where J is the compliance. The zero-frequency shear viscosity, $\eta\,(=[G''/\omega]_{\omega\to0})$, the terminal relaxation time, $\tau\,(=[G'/\omega G'']_{\omega\to0})$, and the steady-state recoverable compliance, $J\,(=[G'/(G'')^2]_{\omega\to0})$, are plotted against ϕ_{Li} in Fig. 7(a–c), respectively, at three temperatures. It is evident from these plots that J remains constant as a function of ϕ_{Li}. The main reason for the maximum of the solution viscosity is the increase in the relaxation time, τ, which can be related to the dynamics of the thread-like network. Taking into account the mechanism of the counterion bonding discussed above, the authors propose that the TEA^+ counterions not bound in the Stern layer serve as a catalyst in the thermal scission of the FOS thread. The scission is more difficult in the absence of nonbound TEA^+ counterions. To verify this hypothesis, they have conducted an additional experiment where the amount of the TEA^+ counterions not bound in the Stern layer is increased artificially by addition of TEACl. The amount of the TEA^+ counterions not

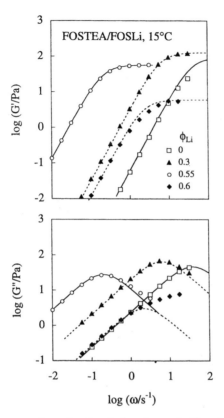

Figure 6 Linear viscoelastic behavior of the FOSTEA/FOSLi aqueous solutions at 15°C. These solutions have the same C_{FOS} (0.1 mol/l) and various ϕ_{Li} as indicated. The curves show the behavior of the Maxwell model [Eq. (1)] calculated from the J and τ data. (a) Storage moduli $G'(\omega)$; (b) loss moduli $G''(\omega)$. [From Ref. 28.] (Reproduced by permission of Springer-Verlag.)

bound in the Stern layer in the TEAFOS/LiFOS mixtures at a total concentration of 0.1 mol/l can be expressed as

$$C^*_{TEA} = 0.1[1 - \phi_{Li}] - 0.1 f_{bound} \quad (mol/l) \qquad (7)$$

Figures 8(a–c) show the η, τ, and J data, respectively, for solutions of TEAFOS/LiFOS with addition of TEACl; the

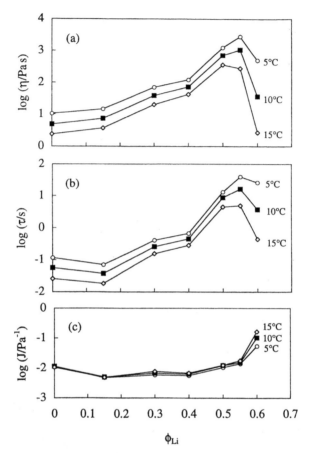

Figure 7 Dependence of (a) viscosity, (b) relaxation time, and (c) compliance of the FOSTEA/FOSLi aqueous solutions having the same C_{FOS} (0.1 mol/l) on the Li^+ fraction ϕ_{Li} in the counterions (Li^+ and TEA^+). [From Ref. 28.] (Reproduced by permission of Springer-Verlag.)

properties are plotted against the amount of the added TEACl, C^*_{TEA} (unfilled squares). The thick solid line in Fig. 8(a) indicates the C^*_{TEA} dependence of τ for the pure FOSTEA solutions. These results confirm that the nonbound TEA^+ ions catalyze the thermal scission of the well-developed FOS network to govern the slow relaxation in the TEAFOS/LiFOS solutions with $\phi_{Li} < 0.5$.

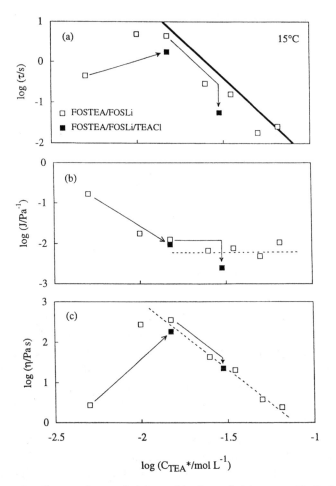

Figure 8 Comparison of (a) τ, (b) J, and (c) η at 15°C for the FOSTEA/FOSLi solutions (unfilled squares), the pure FOSTEA solutions (thick solid line in a), and FOSTEA/FOSLi/TEACl solutions (filled squares in a–c). The data are plotted against the concentration C_{TEA}^* of the TEA ions not bound to the FOS thread. [From Ref. 28.] (Reproduced by permission of Springer-Verlag.)

IV. KRAFFT POINTS WITH MIXED COUNTERIONS

The Krafft point of dodecylpoly(oxyethylene) sulfates with Na^+ and Ca^{++} counterions in pure and mixed systems has been

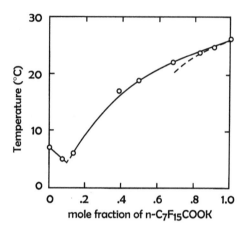

Figure 9 The Krafft point of binary surfactant mixture: n-C_7F_{15}COONa $+ n$-C_7F_{15}COOK [From Ref. 20.] (Reproduced by permission of the American Chemical Society.)

measured in order to find out the usefulness of these surfactants in hard water [21,27]. The Krafft point exhibits a clear minimum at certain Na^+/Ca^{2+} ratios. Such measurements have been done on perfluorinated surfactants such as carboxylates and sulfates with NH_4^+, K^+, H^+, and Na^+ counterions [19,20]. The binary mixtures of perfluorosurfactants with different counterions also exhibit such a minimum (see Fig. 9).

The Krafft point may be related to the phase diagram of melting points of binary mixtures since this is connected to the temperature at which a solid phase appears from the liquid mixtures [20,21]. If the mole fraction of one of the components is small, then the activity of the other component is estimated by Raoult's law. The change in the Krafft point of surfactant mixtures may be expressed just as the well-known equation for the freezing point depression

$$\ln X_1^m = -\frac{\Delta H_1^f}{R}\left(\frac{1}{T} - \frac{1}{T_f^0}\right) \tag{8}$$

where X_1^m is the mole fraction of the first component in the micellar phase, ΔH_1^f is the heat of fusion of hydrated solid surfactant (component 1), T is the absolute temperature, and T_f^0 is the Krafft point of surfactant (component 1). Putting the heat of fusion, $\Delta H_1^f = 9.6\,\text{kcal/mol}$ [61], and the Krafft point 25.6°C of $C_7F_{15}COOK$ into Eq. (8), the changes in the Krafft point of the first component of the mixture can be evaluated as shown in Fig. 9 by a dotted curve. The initial slope agrees well with the experimental values, but this equation does not explain the change in the slope at intermediate values.

V. CONCLUDING REMARKS

Mixing two identical surfactants with very different counterions may be of significant benefit for studies on the counterion atmosphere. Large deviations from the ideal mixing rule occur due to the preferential binding of one of the counterions as well as the complex structure of the electric double layer. Generally, the organic counterions bind strongly onto the perfluorinated micellar surface within the *Stern layer* while small alkali ions are less bound. This *salt* effect can be used to enforce complete binding of the organic counterion to the micellar surface.

The formation of thread-like micelles of anionic perfluorosurfactants can be correlated to the integrity of the *Stern layer*. Collapse of the *Stern layer* results in the disintegration of the thread-like micelles into isolated spherical ones. The viscoelastic behavior of the FOSTEA/FOSLi solutions is explained in terms of the dynamic network of threadlike micelles. It can be concluded that the excess of TEA$^+$ counterions, not bound to the Stern layer, acts as a catalyst to catalyze the scission and reduce the relaxation time.

REFERENCES

1. N. Funasaki, In K. Ogino, and M. Abe (eds), *Mixed Surfactant Systems*. Marcel Dekker, New York (1993), pp. 145–188.

2. R. De Lisi, A. Inglese, S. Milioto, and A. Pellerito, *Langmuir* **13**: 192–202 (1997).

3. L. Pedone, D. Chillura Martino, E. Caponetti, M. A. Floriano, and R. Triolo, *J. Phys. Chem. B* **101**: 9525–9531 (1997).

4. M. Almgren, P. Hansson, and K. Wang, *Langmuir* **12**: 3855–3858 (1996).

5. T. Asakawa, H. Hisamatsu, and S. Miyagishi, *Langmuir* **12**: 1204–1207 (1996).

6. A. V. Barzykin, and M. Almgren, *Langmuir* **12**: 4672–4680 (1996).

7. M. Almgren, K. Wang, and T. Asakawa, *Langmuir* **13**: 4535–4544 (1997).

8. S. Rossi, G. Karlsson, S. Ristori, G. Martini, and K. Edwards, *Langmuir* **17**: 2340–2345 (2001).

9. M. B. Ghoulam, N. Moatadid, A. Graciaa, G. Marion, and J. Lachaise, *Langmuir* **12**: 5048–5051 (1996).

10. P. Mukerjee, and A. Y. S. Yang, *J. Phys. Chem.* **80**: 1388–1390 (1976).

11. T. Asakawa, S. Miyagishi, and M. Nishida, *J. Colloid Interface Sci.* **104**: 279–281 (1985).

12. J. Calfors, and P. Stilbs, *J. Phys. Chem.* **88**: 4410–4414 (1984).

13. T. Asakawa, K. Johten, S. Miyagishi, and M. Nishida, *Langmuir* **1**: 347–351 (1985).

14. A. Knoblich, M. Matsumoto, K. Murata, and Y. Fujiyoshi, *Langmuir* **11**: 2361–2366 (1995).

15. T. Asakawa, S. Miyagishi, and M. Nishida, *Langmuir* **3**: 821–827 (1987).

16. R. M. Clapperton, R. H. Ottewill, and B. T. Ingram, *Langmuir* **10**: 51–56 (1994).

17. D. P. Bossev, M. Matsumoto, T. Sato, H. Watanabe, and M. Nakahara, *J. Phys. Chem. B* **10**: 8259–8266 (1999).

18. T. Wolff, and G. Bünau, *Ber. Bunsenges Phys. Chem.* **86**: 225–228 (1982).

19. M. Hato, and K. Shinoda, *Nippon Kagaku Zasshi* **91**: 27–31 (1970).

20. K. Shinoda, M. Hato, and T. Hayashi, *J. Phys. Chem.* **76**: 909–914 (1980).

21. M. Hato, and K. Shinoda, *J. Phys. Chem.* **77**: 378–381 (1973).

22. M. Almgren, J. E. Löfroth, and R. Rydholm, *Chem. Phys. Lett.* **63**: 265–268 (1979).

23. H. Fabre, N. Kamenka, A. Khan, G. Lindblom, B. Lindman, and G. J. T. Tiddy, *J. Phys. Chem.* **84**: 3428–3433 (1980).

24. M. Jansson, and P. Stilbs, *J. Phys. Chem.* **89**: 4868–4873 (1985).

25. M. Koshinuma, *Bull. Chem. Soc. Jpn.* **56**: 2341–2347 (1983).

26. P. J. Kreke, L. J. Magid, and J. C. Gee, *Langmuir* **12**: 699–705 (1996).

27. M. Hato, *Nippon Kagaku Zasshi* **92**: 496–500 (1971).

28. H. Watanabe, T. Sato, K. Osaki, M. Matsumoto, D. P. Bossev, C. E. McNamee, and M. Nakahara, *Rheol Acta* **39**: 110–121 (2000).

29. H. Hoffmann, and J. Würtz, *J. Mol. Liq.* **72**: 191–230 (1997).

30. O. Regev, M. S. Leaver, R. Zhou, and S. Puntambekar, *Langmuir* **17**: 5141–5149 (2001).

31. M. Jansson, and P. Stilbs, *J. Phys. Chem.* **91**: 113–116 (1987).

32. J. E. Brady, D. F. Evans, G. G. Warr, F. Grieser, B. W. Ninham, *J. Phys. Chem.* **90**: 1853–1859 (1986).

33. D. P. Bossev, M. Matsumoto, and M. Nakahara, *J. Phys. Chem. B* **10**: 8251–8258 (1999).

34. E. W. Anacker, and A. L. Underwood, *J. Phys. Chem.* **90**: 2463–2466 (1981).

35. K. Fontell, and B. Lindman, *J. Phys. Chem.* **87**: 3289–3297 (1983).

36. T. Shikata, H. Hirata, and T. Kotaka, *Langmuir* **3**: 1081–1086 (1987).

37. H. Rehage, and H. Hoffmann, *J. Phys. Chem.* **92**: 4712–4719 (1988).

38. H. Hoffmann, and B. Tagesson, *Z. Phys. Chem. Neue. Folge* **110**: 113–134 (1978).

39. J. Israelachvili, *Intermolecular and Surface Forces*. Academic Press, New York (1994), Chapter 12.

40. G. S. Manning, *Annual Rev. Phys. Chem.* **23**: 117–140 (1972).

41. A. Adamson, *Physical Chemistry of Surfaces*, 3rd edn. Wiley, New York (1976), Chapter IV.

42. J. F. Scamehorn, R. S. Schechter, and W. H. Wade, *J. Dispersion Sci. Technol.* **3**: 261–278 (1982).

43. J. F. Rathman, and J. F. Scamehorn, *Langmuir* **2**: 354–361 (1986).

44. J. F. Rathman, and J. F. Scamehorn, *J. Phys. Chem.* **88**: 5807–5816 (1984).

45. J. C. Shelley, M. Sprik, and M. L. Klein, *Langmuir* **9**: 916–926 (1993).

46. A. D. MacKerell Jr., *J. Phys. Chem.* **99**: 1846–1855 (1995).

47. H. Hoffmann, and W. Ulbricht, *Z. Phys. Chem. Neue Folge* **106**: 167–184 (1977).

48. B. Lindman, M. C. Puyal, N. Kamenka, R. Rymden, and P. Stilbs, *J. Phys. Chem.* **88**: 5048–5057 (1984).

49. E. O. Stejskal, and J. E. Tanner, *J. Chem. Phys.* **42**: 288–292 (1965).

50. P. Stilbs, *Prog. Nucl. Magn. Reson. Spectrosc.* **19**: 1–45 (1987).

51. P. Li, M. Jansson, P. Bahadur, and P. Stilbs, *J. Phys. Chem.* **93**: 6458–6463 (1989).

52. P. Stilbs, and B. Lindman, *J. Phys. Chem.* **85**: 2587–2589 (1981).

53. B. Lindman, M. C. Puyal, N. Kamenka, B. Brun, and G. Gunnarsson, *J. Phys. Chem.* **86**: 1702–1711 (1982).

54. H. Gustavsson, and B. Lindman, *J. Chem. Soc. Chem. Commun.* 93–94 (1973).

55. H. Gustavsson, and B. Lindman, *J. Am. Chem. Soc.* **97**: 3923–3930 (1975).

56. H. Gustavsson, and B. Lindman, *J. Am. Chem. Soc.* **100**: 4647–4654 (1978).

57. M. E. Cates, *Macromolecules,* **20**: 2289–2296 (1997).

58. H. Watanabe, T. Sato, K. Osaki, M. Matsumoto, D. P. Bossev, C. E. McNamee, M. Nakahara, M. L. Yao, *Rheol. Acta* **37**: 470–485 (1998).

59. H. Hoffmann, H. Rehage, G. Platz, W. Schorr, H. Thurn, and W. Ulbricht, *Colloid Polym. Sci.* **260**: 1042–1056 (1982).

60. P. T. Callaghan, M. E. Cates, C. J. Rofe, and J. B. A. F. Smeulders, *J. Phys. II France* **6**: 375–393 (1996).

61. K. Shinoda, S. Hiruta, and K. Amaya, *J. Colloid Interface Sci.* **21**: 102–106 (1966).

7

Delayed Degradation of Drugs by Mixed Micellization with Biosurfactants

SHOKO YOKOYAMA

School of Pharmaceutical Sciences, Kyushu
University of Health and Welfare,
Miyazaki, Japan

SYNOPSIS

It is important to suppress the hydrolysis and/or the oxidation of drugs. However mixed surfactant systems containing biosurfactants have not been studied so much. In this chapter, we report the effects of mixed micellization of biosurfactants on the degradation of drugs.

I. INTRODUCTION

Mixed surfactant systems containing synthetic ionic and non-ionic surfactants have been widely studied [1], while mixed surfactant systems containing biosurfactants have not been studied so much except for bile salt systems [2–9]. On the other hand, stabilization of drugs included in micelles have been studied [10–12] in mixed micellar systems. For the stabilization of drugs, biosurfactants are desirable compared with synthetic surfactants, since biosurfactants are biodegradable and harmless.

In this chapter, we report on the delayed degradation of drugs by mixed micellization with biosurfactants, where the biosurfactants are arachidonic acid, which is the precursor of prostaglandins, lysophosphatidylcholine, and ganglioside G_{M1}, which is a subclass of glycosphingolipids. Prostaglandins and epinephrine were chosen as model drugs.

II. DELAYED DEGRADATION OF DRUGS BY MIXED MICELLIZATION WITH BIOSURFACTANTS

A. Effect of Arachidonic Acid on Base-Catalyzed Degradation of Prostaglandin E_2 [13]

Prostaglandins (PGs) are biosynthesized from arachidonic acid (AA) via the arachidonic acid cascade [14]. Prostaglandin E_2 (PGE_2) has high pharmacological potency, particularly in uterine muscle contraction, peripheral blood vessel extension, bronchus muscle slackness, secretion suppression of gastric juice, and so on. However, PGE_2 is extremely unstable in aqueous solutions [15–21]. PGE_2 undergoes a conversion into PGA_2 and PGB_2 by dehydration and isomerization reactions [15–21]. The instability of PGEs limits the development of dosage formulas. The kinetics of dehydration and isomerization of PGE_2 have been studied [17]. However, kinetic analysis on the degradation of PGs in the presence of surfactant has not yet been studied except for our previous study [22].

The base-catalyzed degradation of PGA_2 decreased by the micellization of PGA_2 or mixed micellization of PGA_2 with a nonionic surfactant [22]. The degradation reaction of PGA_2 was previously discussed, taking into account the critical micelle concentration (CMC) [23] and the micellar surface potential of PGA_2 [24]. In this section, we report on the kinetics of the base-catalyzed consecutive degradation of PGE_2 ($PGE_2 \rightarrow PGA_2 \rightarrow PGB_2$) and the effect of arachidonic acid, which is the precursor of PGE_2, on the degradation.

1. Degradation (Dehydration and Isomerization) of Free PGE_2

The structures of PGE_2, PGA_2 and PGB_2 along with dehydration and isomerization reactions are presented in Scheme 1.

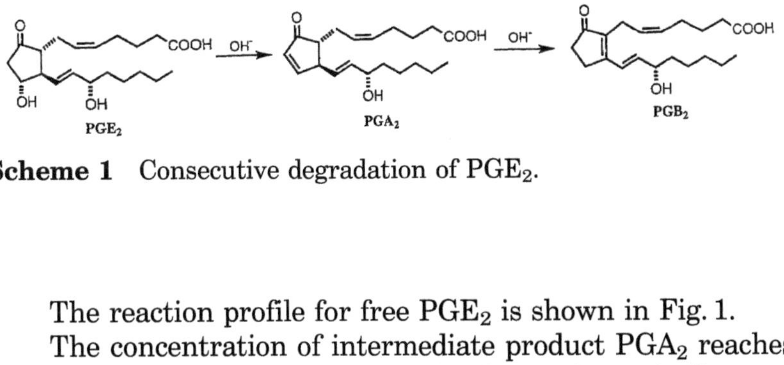

Scheme 1 Consecutive degradation of PGE_2.

The reaction profile for free PGE_2 is shown in Fig. 1.

The concentration of intermediate product PGA_2 reaches a maximum and then decreases. The thermodynamically more stable product PGB_2 soon thereafter increases with the completion of isomerization.

The degradation reaction of PGE_2 is of the first-order above pH > 4 [17]. The reaction of PGE_2 to PGB_2 is

$$PGE_2 \xrightarrow{k_E} PGA_2 \xrightarrow{k_A} PGB_2$$

where k_E and k_A are the rate constants for the dehydration reaction $PGE_2 \rightarrow PGA_2$ and for the isomerization reaction $PGA_2 \rightarrow PGB_2$, respectively.

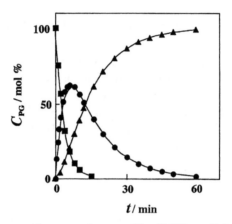

Figure 1 Consecutive reaction profile ($PGE_2 \rightarrow PGA_2 \rightarrow PGB_2$) of free PGE_2 at pH 12 and 60°C. PGs: ■, PGE_2; ●, PGA_2; ▲, PGB_2. Initial concentration of PGE_2: 9×10^{-4} M.

The concentrations of PGE_2, PGA_2, and PGB_2 change with time as

$$-\frac{d[PGE_2]}{dt} = k_E[PGE_2] \tag{1}$$

$$\frac{d[PGA_2]}{dt} = k_E[PGE_2] - k_A[PGA_2] \tag{2}$$

$$\frac{d[PGB_2]}{dt} = k_A[PGA_2] \tag{3}$$

The integration of Eq. (2) leads to

$$[PGA_2] = \frac{k_E[PGE_2]_0}{k_E - k_A}(e^{-k_A t} - e^{-k_E t}) \tag{4}$$

where $[PGE_2]_0$ is the initial concentration of PGE_2 at time 0, and $k_E > k_A$.

By taking logarithms of each side of Eq. (4) it is rewritten as

$$\ln[PGA_2] = \ln\frac{k_E[PGE_2]_0}{k_E - k_A} - k_A t \tag{5}$$

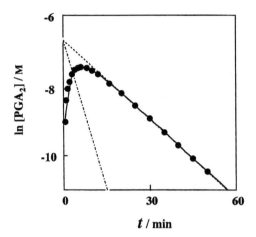

Figure 2 Plots of $\ln[\text{PGA}_2]$ vs t for $\text{PGE}_2 \to \text{PGA}_2 \to \text{PGB}_2$ of free PGE_2.

The value of $\ln[\text{PGA}_2]$ is plotted against time in Fig. 2.

The plots in the figure indicate linearity at $t > t_{max}$. The rate constant for the isomerization of PGA_2, k_A, is obtained from the slope of the linear part: k_A is $1.242 \times 10^{-3} \text{ s}^{-1}$. The rate constant for the dehydration of PGE_2, k_E, is obtained from the slope of the straight broken line in Fig. 2, by subtracting the concentrations of PGA_2 from the extrapolated dotted line values. The value of k_E for free PGE_2 is $4.844 \times 10^{-3} \text{ s}^{-1}$, approximately four times k_A. The value of PGE_2 at time t, $[\text{PGE}_2]_t$, is obtained by subtracting $[\text{PGA}_2]_t$ and $[\text{PGB}_2]_t$ from total $(=[\text{PGE}_2]_0)$, and is indicated by the closed squares in Fig. 1. The PGE_2 in the free state disappeared in 16 min and was converted to the final degradation product PGB_2 in 1 h.

2. Effect of AA on Degradation of PGE_2

Figures 3 and 4 show the degradation of PGE_2 in the presence of AA above the CMC [25]. The ordinate indicates the changes in mole percent of the relevant species with time. The plots of $\ln[\text{PGA}_2]$ vs time are shown in Figs. 5 and 6.

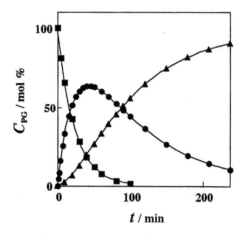

Figure 3   Consecutive reaction profile of PGE_2 at pH 12 and 60°C in the presence of 1×10^{-3} M AA. Symbols are the same as in Fig. 1.

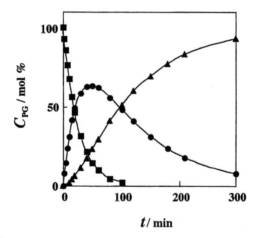

Figure 4   Consecutive reaction profile of PGE_2 at pH 12 and 60°C in the presence of 5×10^{-3} M AA. Symbols are the same as in Fig. 1.

The values of k_E and k_A are summarized in Table 1. The degradation rate of PGE_2 decreased to about 1/7 in the presence of AA micelles, while the degradation rate of PGE_2 in the presence of free AA (5×10^{-5} M below the CMC [25]) was nearly equal to that of free PGE_2 in the absence of AA. The rate

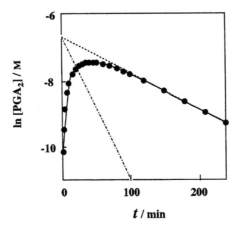

Figure 5 Plots of ln [PGA$_2$] vs t for PGE$_2$ → PGA$_2$ → PGB$_2$ in the presence of 1×10^{-3} M AA.

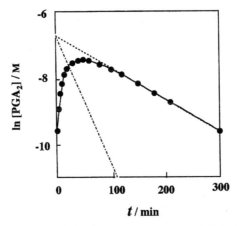

Figure 6 Plots of ln [PGA$_2$] vs t for PGE$_2$ → PGA$_2$ → PGB$_2$ in the presence of 5×10^{-3} M AA.

constant for the degradation of PGE$_2$ in the presence of 5×10^{-3} M of AA was smaller than that of 1×10^{-3} M of AA. This is considered to be due to the fact that the approach of OH$^-$ to PGE$_2$ in AA micelles is suppressed by the electrostatic repulsion between the negatively charged AA micellar surfaces and OH$^-$ and that the molar ratio of AA in the mixed micelles

Table 1 Rate constants, k_E and k_A, for consecutive degradation.

	k_E ($10^{-4}\,s^{-1}$)	k_A ($10^{-4}\,s^{-1}$)
Free PGE$_2$	47.8	12.6
PGE$_2$ + AA 1 mM	7.17	1.80
PGE$_2$ + AA 5 mM	6.52	1.61
PGE$_2$ + HED 1 mM	10.8	2.81

The initial concentration of PGE$_2$ is 9×10^{-4} M in all cases.

increases with increasing concentration of AA and that the negative surface potential of AA micelles [26] is larger than that of PGE$_2$ micelles. Hirayama *et al.* [27] reported that the degradation rate of PGE$_2$ decreased to 1/3 or 1/1.4 by including PGE$_2$ in 2,6-di-*o*-methyl-β-cyclodextrin or 2,3,6-tri-*o*-methyl-β-cyclodextrin, respectively. The efficiency of AA micelles to stabilize PGE$_2$ is about 2–2.5-fold superior to that of methylated-β-cyclodextrins.

The effect of AA micelles was compared with that of nonionic surfactant, heptaethyleneglycol dodecyl ether (HED), micelles. The result is shown in Table 1. The value of k_E in the presence of HED micelles was $1.083 \times 10^{-3}\,s^{-1}$, which is larger than that in the presence of AA micelles. The CMC of HED [28] is smaller than that of AA and the amount of HED micelles with 1×10^{-3} M is larger than that of AA with the same concentration. The superior effect of AA micelles to stabilize PGE$_2$ could be related to the electrostatically repulsive effect between negatively charged micellar surfaces and OH$^-$.

Based on k_E in Table 1, the stability and shelf-life of PGE$_2$ are evaluated. The shelf-life of a medicine is defined as the period during which biological activity is maintained at more than 90 or 95%. The result obtained by a severe test is converted into the value at pH 7 and 10°C using the Arrhenius equation, activation energy for the degradation of PGE$_2$ [17] and the log k-pH profile [17,19]. The PGE$_2$ shelf-life in AA micelles at pH 7 and 10°C is 10 days, while that of free PGE$_2$ is 1.5 days. AA is a precursor of PGs in the living body, and is considered to be safe.

B. Effect of Lysophosphatidylcholine on Acid-Catalyzed Dehydration of Prostaglandin E_1 [29]

It is important to obtain information on the acid-catalyzed degradation of PGs when PGs are orally administered or intravenously injected with acidic medicines. In this section, we report on the stabilization of PGE_1 using lysophospholipid micelles. Lysophospholipids have different properties from the corresponding diacylphospholipid [30,31]: their solubilities in water are higher than those of the respective diacyl compound and they form micelles, in contrast with diacylphospholipids, which form liposomal bilayer membranes. 1-Myristoyl-*sn*-glycero-3-phosphocholine (My-lysoPC) as a lysophospholipid is reported in this section. The CMC of My-lysoPC in water is 5.9×10^{-5} M [32] at 30°C and $4.3–7.0 \times 10^{-5}$ M [33–35] at room temperature. Only the dehydration reaction, $PGE_1 \rightarrow PGA_1$, occurs below pH 3, although the dehydration of PGE_1 prodeeds as $PGE_1 \rightarrow PGA_1 \rightarrow PGB_1$ above pH 4 [17,36].

The dehydration of PGE_1 in the absence and in the presence of My-lysoPC at pH 1.2, which corresponds to the pH of gastric juice, is shown in Fig. 7.

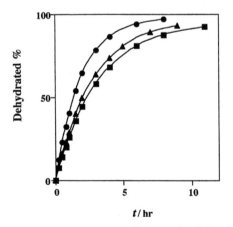

Figure 7 Effect of My-lysoPC micelles on the dehydration of PGE_1 at pH 1.2 and 60°C, Percentage of PGE_1 dehydrated vs time. Concentration of My-lysoPC: ●, 0 (free PGE_1); ▲, 1×10^{-3} M; ■, 3×10^{-3} M. Initial concentration of PGE_1: 9×10^{-4} M.

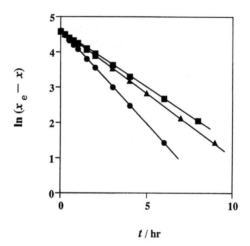

Figure 8 Plots of ln $(x_e - x)$ vs t. Symbols are the same as in Fig. 7.

Table 2 Rate constants for the dehydration of PGE_1.

	$k\,(10^{-5}\,s^{-1})$
Free PGE_1	14.72
$PGE_1 + $ My-lysoPC 0.02 mM	15.11
$PGE_1 + $ My-lysoPC 1 mM	9.79
$PGE_1 + $ My-lysoPC 3 mM	8.73
$PGE_1 + $ My-lysoPC 1 mM + SA($X_{SA} = 0.1$)	7.38
$PGE_1 + $ My-lysoPC 1 mM + SA($X_{SA} = 0.2$)	5.07

The initial concentration of PGE_1 is 9×10^{-4} M in all cases.

The rate constant for the dehydration of PGE_1, k, is defined as

$$\ln (x_e - x) = \ln x_e - kt \qquad (6)$$

where x and x_e are the concentrations of PGE_1 dehydrated during time t and the equilibrium value of the dehydration reaction, respectively. Plots of ln (x_e-x) vs t are linear as shown in Fig. 8. The values of k obtained from Fig. 8 are summarized in Table 2.

The dehydration rate of PGE_1 decreased to approximately $1/1.5$–$1/1.7$ in the presence of My-lysoPC micelles. However, the effect of My-lysoPC micelles was not so significant. This is considered to be due to the lack of micellar catalysis: namely, the lack of an electrostatic repulsion between H^+ and micellar surfaces. Next, the effects of positively-charged My-lysoPC mixed micelles containing stearyl amine (SA) on the dehydration of PGE_1 are shown in Figs. 9 and 10 and the rate constants are summarized in Table 2.

The rate constant for the dehydration of PGE_1 in My-lysoPC micelles effectively decreased by the addition of SA. The dehydration rate of PGE_1 decreased to about $1/3$ in the presence of My-lysoPC–SA mixed micelles whose mole fraction of SA was 0.2.

Based on the rate constants presented in Table 2, the stability and shelf-life of PGE_1 are evaluated as mentioned in the previous section. The results are summarized in Table 3.

The shelf-life of free PGE_1 at pH 1.2 and 37°C and at pH 3.0 and 37°C is about 1.4 h and 5 days, respectively. If PGE_1 is used together with an antacid or basic medicine,

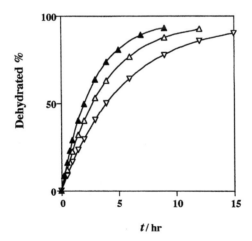

Figure 9 Effect of My-lysoPC–SA micelles on the dehydration of PGE_1 at pH 1.2 and 60°C, Percentage of PGE_1 dehydrated vs time. Mole fraction of SA: ▲, 0 (my-lysoPC micelles); △, 0.1; ▽, 0.2. Total concentration of PGE_1: 9×10^{-4} M.

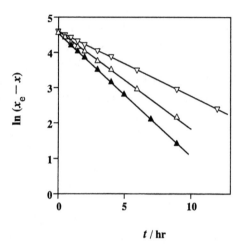

t / hr

Figure 10   Plots of $\ln(x_e - x)$ vs t. Symbols are the same as in Fig. 9.

Table 3   Stability and shelf-life of PGE_1 ($[PGE_1]_t > 90\%$).

	pH	Temperature	Shelf-life
Free PGE_1	pH 1.2	60°C	0.19 h (11 min)
		37°C	1.4 h
		5°C	43.2 h
PGE_1 + My-lysoPC 3 mM	pH 1.2	60°C	0.33 h (20 min)
		37°C	2.5 h
		5°C	72.9 h (3 day)
PGE_1 + My-lysoPC 1 mM + SA ($X_{SA} = 0.2$)	pH 1.2	60°C	0.57 h (34 min)
		37°C	4.3 h
		5°C	125 h (5 day)
Free PGE_1	pH 3.0	60°C	12.5 h
		37°C	94.4 h (4 day)
		5°C	2728 h (113 day)
PGE_1 + My-lysoPC 3 mM	pH 3.0	60°C	21.1 h
		37°C	159 h (6 day)
		5°C	4599 h (191 day)
PGE_1 + My-lysoPC 1 mM + SA ($X_{SA} = 0.2$)	pH 3.0	60°C	36.4 h
		37°C	274 h (11 day)
		5°C	7920 h (330 day)

oral administration of PGE_1 may be possible. The shelf-life of PGE_1 in the presence of My-lysoPC micelles at pH 3.0 and 5°C is 191–330 days. PGE_1 included in lysophosphatidylcholine micelles seems to be able to endure long-term cold storage even in solution. When PGE_1 is necessitated, an administration (intravenous injection) of PGE_1 stock solution may be possible using a PGE_1-lysophosphatidylcholine solution (or injection) stored in a refrigerator, since lysophosphatidylcholine, which is a constituent of biological membranes, is considered to be safe.

C. Inhibition Effects of Gangliosides G_{M1}, G_{D1a}, and G_{T1b} on Base-Catalyzed Isomerization of Prostaglandin A_2 [37]

Glycosphingolipids are localized exclusively in the external surface of mammalian membranes [38]. Gangliosides, a subclass of glycosphingolipids, are double-tailed amphiphilic molecules, in which a ceramide lipid portion, constituted by a sphingosine and a fatty acid with roughly 20 carbons each, carries a rather bulky headgroup made of several sugar rings, some of which are N-acetylneuraminic acid (sialic acid) residues. Gangliosides bear a net negative charge conferred by one or more sialic acid residues. Their highest concentration occurs in the white matter of the central nervous system [39]. Gangliosides form micelles [40]: their CMCs are relatively low and their micellar aggregation numbers are relatively large compared with those of general anionic surfactants.

In this section, the micellar effects of gangliosides on the base-catalyzed isomerization (degradation) of PGA_2 as a model experiment are evaluated. The effects of gangliosides having various numbers of sialic acid residues on the isomerization of PGA_2 are also investigated using ganglioside G_{M1} (GM1), G_{D1a} (GD1a) and G_{T1b} (GT1b). The CMCs of GM1, GD1a and GT1b are $(2 \pm 1) \times 10^{-8}$, $(2 \pm 1) \times 10^{-6}$ and $(1 \pm 0.5) \times 10^{-5} \, \mathrm{mol \, L^{-1}}$, respectively, at pH 7.4 and 20°C [40]. The chemical structures of GM1, GD1a and GT1b are shown in Scheme 2 as abbreviated expressions.

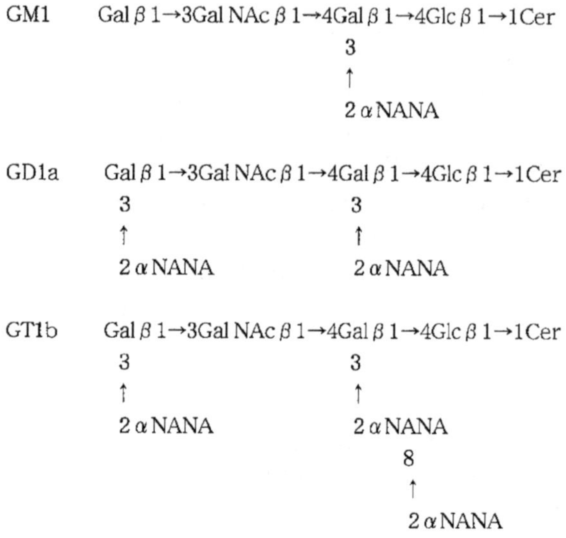

GM1   Gal β 1→3Gal NAc β 1→4Gal β 1→4Glc β 1→1Cer
                                        3
                                        ↑
                                   2 α NANA

GD1a   Gal β 1→3Gal NAc β 1→4Gal β 1→4Glc β 1→1Cer
                      3   3
                      ↑   ↑
                 2 α NANA   2 α NANA

GT1b   Gal β 1→3Gal NAc β 1→4Gal β 1→4Glc β 1→1Cer
                      3   3
                      ↑   ↑
                 2 α NANA   2 α NANA
                                             8
                                             ↑
                                        2 α NANA

NANA: *N*-acetylneuraminic acid (sialic acid)

Scheme 2   Chemical structures of gangliosides.

1. Inhibition Effects of GM1, GD1a and GT1b Micelles

The isomerization of PGA_2 in the absence and in the presence of gangliosides is shown in Fig. 11. The rate constant for the isomerization of PGA_2, k, is defined as in Eq. 6. Plots of ln $(x_e - x)$ vs t are shown in Fig. 12.

The inhibition effect of gangliosides could be due to the micellar shielding effect and electrostatic repulsion between sialic acid residues and OH^-. The inhibition effect of GT1b having three sialic acid residues was larger than that of GD1a having two sialic acid residues. In the initial isomerization reaction of PGA_2, the inhibition effect of GM1 having only one sialic acid residue was smaller than that of GD1a or GT1b. However, the inhibition effect of GM1 became larger than that of GT1b in the latter half of the isomerization reaction of PGA_2. These findings are likely due to the position of the sialic acid residue and the specific micellar shielding effect of GM1. GD1a and GT1b have an additional terminal sialic acid residue so

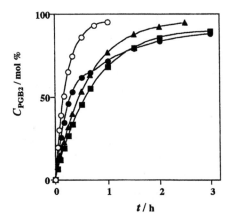

Figure 11 Relationship between percentage of isomerized PGA_2 and time in the absence and in the presence of $3 \times 10^{-4}\,\mathrm{mol\,L^{-1}}$ gangliosides at pH 12 and 60°C. $\bigcirc$, free PGA_2; $\bullet$, in the presence of GM1 micelles; $\blacktriangle$, in the presence of GD1a micelles; $\blacksquare$, in the presence of GT1b micelles. Initial concentration of PGA_2: $5 \times 10^{-5}\,\mathrm{mol\,L^{-1}}$.

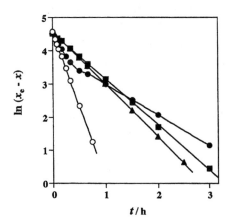

Figure 12 Plots of $\ln(x_e - x)$ vs t. Symbols are the same as in Fig. 11.

that the electrostatic shielding effect of GD1a or GT1b from OH^- is larger than that of GM1 without the terminal sialic acid residue. The absolute value of negative surface potential $|-\Delta\Psi|$ of GT1b micelles should be larger than that of GD1a micelles. However, the effect of an additional sialic acid residue at an inner site is unlikely to be so significant. On the other hand, the micellar size and micellar amount of GM1 is larger [40] than that of the other gangliosides: the CMC of GM1 is 1000 and 100 times lower than GT1b and GD1a, respectively; the micellar aggregation number of GM1 is 4 and 2 times higher than GT1b and GD1a, respectively. The motion of polar groups in micelles of GM1 is somewhat restricted due to the interaction of ganglioside oligosaccharide chains [41]. Thus, the steric shielding effect of GM1 micelles is larger than that of the other ganglioside micelles.

The plots of $\ln(x_e - x)$ vs t for the isomerization of PGA$_2$ in the presence of gangliosides are shown in Fig. 12. A single straight line was obtained in the presence of GD1a or GT1b micelles. The rate constants for the isomerization of PGA$_2$ were obtained from the slopes of the straight lines and are summarized in Table 4. The isomerization rate of PGA$_2$ decreased to approximately 1/3 in the presence of GT1b or GD1a micelles compared with that of free PGA$_2$.

In the case of GM1, the plots of $\ln(x_e - x)$ vs t indicated two linear parts as can be seen in Fig. 12. The rate constants are obtained from the first and the second linear regions and are summarized in Table 4. The first and the second linear regions are the isomerization reaction of PGA$_2$ whose reaction

Table 4 Rate constants for the isomerization of PGA$_2$ in the absence and in the presence of ganglioside micelles.

	$k/10^{-4}\,\mathrm{s}^{-1}$	
Free PGA$_2$	12.37	
PGA$_2$/GT1b	3.825	
PGA$_2$/GD1a	4.403	
PGA$_2$/GM1	7.833[a]	2.564[b]
PGA$_2$/GM1 – GT1b ($X_{GT1b} = 0.1$)	2.637[a]	0.488[b]
PGA$_2$/GM1 – GD1a ($X_{GD1a} = 0.1$)	3.343[a]	0.711[b]

[a]First linear region.
[b]Second linear region in Figs. 12 and 14.
The initial concentration of PGA$_2$ is $5 \times 10^{-5}\,\mathrm{mol\,L^{-1}}$. The concentration of ganglioside is $3 \times 10^{-4}\,\mathrm{mol\,L^{-1}}$.

site, the five-membered ring moiety of PGA_2, is located at the outer and inner regions from the position of the sialic acid residue of GM1, respectively. The rate constant for the first linear region of the GM1 system was larger than that for the GT1b and GD1a systems, while the rate constant for the second linear region of the GM1 system was smaller than that for the GT1b and GD1a systems.

If the percentage of PGA_2 whose reaction site protrudes from the position of the sialic acid residue of GM1 micelles is calculated from the inflection point shown in Fig. 12, the existence percentage is estimated as 54%.

2. Inhibition Effects of GM1–GD1a and GM1–GT1b Mixed Micelles

Next, the isomerization of PGA_2 in the presence of GM1–GD1a or GM1–GT1b mixed micelles, whose mole fraction of GD1a or GT1b is 0.1, is investigated. The relationship between isomerized PGA_2 and time and the plots of $\ln(x_e - x)$ vs t are shown in Figs. 13 and 14, respectively, where the data in the presence of GM1 alone micelles is also shown as the reference.

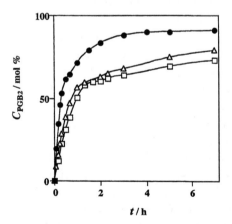

Figure 13 Effects of GM1–GD1a and GM1–GT1b mixed micelles on the isomerization of PGA_2, percentage of isomerized PGA_2 vs time. Micelles: $\triangle$, mixed GM1–GD1a micelles ($X_{GD1a} = 0.1$); $\square$, mixed GM1–GT1b micelles ($X_{GT1b} = 0.1$); $\bullet$, GM1 alone micelles. Total concentration of gangliosides: 3×10^{-4} mol L^{-1}.

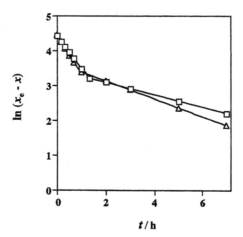

Figure 14 Plots of $\ln(x_e - x)$ vs t for the isomerization of PGA_2 in the presence of GM1–GD1a or GM1–GT1b mixed micelles. Symbols are the same as in Fig. 13.

GM1–GD1a and GM1–GT1b mixed micelles remarkably inhibited the isomerization of PGA_2. The plots of $\ln(x_e-x)$ vs t in the presence of GM1 mixed micelles indicated two linear parts, similar to those found in the presence of GM1 alone micelles. This implies that PGA_2 is also distributed in two regions in the GM1 mixed micellar phase. The rate constants obtained from the slopes of each linear part are summarized in Table 4. The inhibition effect of GM1–GD1a or GM1–GT1b mixed micelles was larger than that of GT1b alone micelles. The inner region from the position of the sialic acid residue of GM1 in GM1–GD1a or GM1–GT1b mixed micelles had a significant inhibition effect on the isomerization of PGA_2: the isomerization rate of PGA_2 decreased to approximately 1/17 and 1/25 in the presence of GM1–GD1a ($X_{GD1a} = 0.1$) and GM1–GT1b ($X_{GT1b} = 0.1$) mixed micelles, respectively. GM1–GD1a and GM1–GT1b mixed micelles have a large aggregation number, similar to that found in GM1 alone micelles, and a larger surface potential $|-\Delta\Psi|$ compared with GM1 alone micelles. The oligosaccharide chains at the surface of the mixed micelles are more complicated and the motion is restricted so that the steric shielding effect of the mixed micelles is larger than that of GM1 alone micelles. Namely, GM1–GD1a and GM1–GT1b mixed micelles exhibit a superior steric shielding effect and an electrostatically repulsive effect from OH^-, thereby the base-catalyzed isomerization of PGA_2 is significantly inhibited.

Taking these results into account, it is expected that ganglioside micelles inhibit the consecutive degradation of PGE_2 as a medicine.

PGs as autacoids are biosynthesized when and where PGs are required. PGE_2 is biosynthesized in the uterine, peripheral blood vessels, stomach and so on [14]. PGD_2 as a sleeping-inducer is biosynthesized in the brain [14]. On the other hand, gangliosides are distributed in the biological membranes. Especially in the brain tissue, 13% GM1, 38% GD1a and 16% GT1b occur [38]. The results shown in this section using ganglioside micelles as a model of the biological membrane suggest that gangliosides inhibit the inactivation of biosynthesized PGs until the physiological function of PGs is fully displayed and that the inhibition effect of gangliosides on the inactivation of PGs is greater in the brain.

D. Effect of Ganglioside G_{M1} Micelles on Autoxidation of Aqueous Solution of Epinephrine Hydrochloride [42]

Epinephrine, 4-[1-hydroxy-2-(methylamino)ethyl]-1,2-benzene-diol, is the principal sympathomimetic hormone produced by the adrenal medulla in most species. Epinephrine hydrochloride (EP) is a medicine found in the Japan Pharmacopoeia. EP undergoes oxidation in the presence of oxygen [43–45]. Adrenochrome (AC), which is an oxidation product of EP, brings about a decrease in the adenine nucleotide content in the heart, and this results in contractile failure of the heart [46]. It is important to suppress the oxidation of EP as a medicine. Antioxidants to stabilize EP have been widely studied [47]. However, micellar shielding and/or electrostatic effects on oxidation of EP have scarcely been studied. In this section, the effect of ganglioside G_{M1} micelles on the autoxidation of EP is described as a model experiment to evaluate the micellar

effect of gangliosides on the stabilization of drugs. Furthermore, the effect of ganglioside G_{M1} on the oxidation of EP is
compared with that of general surfactants.

The oxidation process of EP is shown in Scheme 3.

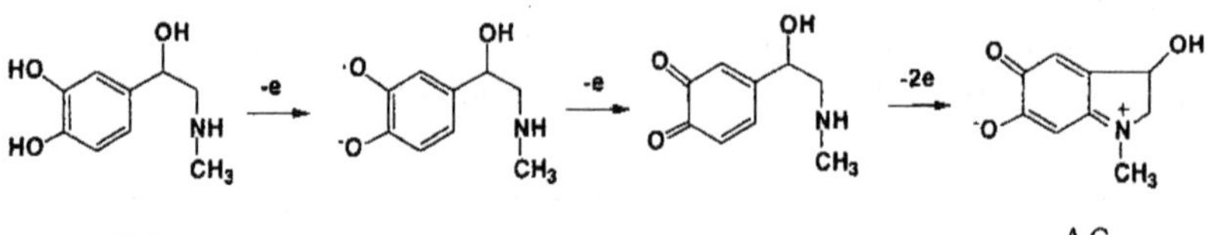

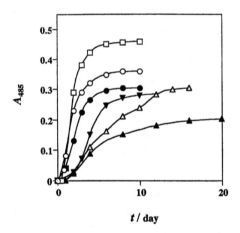

Scheme 3 Structural formulas of EP, AC, and intermediates in the
oxidation of EP.

Figure 15 Time course of autoxidation of EP: Relationship between
absorbance of AC produced from EP and time. ○, 1×10^{-3} mol L^{-1} EP
in the absence of surfactant; ●, 4×10^{-3} mol L^{-1} EP in the absence of
surfactant; □, 4×10^{-3} mol L^{-1} EP in the presence of 0.1 mol L^{-1}
DTAC; △, 4×10^{-3} mol L^{-1} EP in the presence of 0.1 mol L^{-1} SDS; ▼,
4×10^{-3} mol L^{-1} EP in the presence of 6.5×10^{-4} mol L^{-1} GM1; ▲,
4×10^{-3} mol L^{-1} EP in the presence of 1.3×10^{-3} mol L^{-1} GM1.

Figure 15 shows the relationships between the oxidized
EP at pH 5.0 (as the absorbance of AC produced from EP)
and time. The experimental conditions were as follows: GM1,

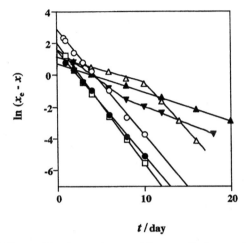

Figure 16 Plots of ln $(x_e - x)$ vs t. The symbols are the same as in Fig. 15.

sodium dodecyl sulfate (SDS) and dodecyltrimethylammmonium chloride (DTAC) were dissolved in pH 5.0 phosphate buffer (the concentrations of surfactants are above the CMCs [40,48,49]); 4 mL portions of EP solutions without and with surfactant were put in 5 mL shaded glass ampoules and immediately sealed, in which 2.2 mL of atmospheric air was contained; the ampoules were preserved at $25 \pm 0.2°C$ in a dark place; each glass ampoule containing the sample was taken out at an appropriate interval, the ampoule was broken, the sample solution was diluted with pH 5.0 buffer solution if necessary, and the concentration of AC was spectrophotometrically determined.

The oxidation of EP is of the first order [45,50]. Thus, the rate constant, k, for the oxidation of EP is defined as Eq. (6), where x_e is the equilibrium value of oxidation and x is the value of oxidized EP during time t. Plots of ln $(x_e - x)$ vs t for EP solutions are shown in Fig. 16, and the values of k are summarized in Table 5.

GM1 micelles inhibited the autoxidation of aqueous solutions of EP. The inhibition effect of 1.3×10^{-3} mol L^{-1} GM1 was larger than that of 0.1 mol L^{-1} SDS. GM1 was stable

Table 5 Rate constants for autoxidation of EP at pH 5.0 and 25°C.

		$k/10^{-6}\,\mathrm{s}^{-1}$	
EP $1 \times 10^{-3}\,\mathrm{mol\,L}^{-1}$		7.60	
EP $4 \times 10^{-3}\,\mathrm{mol\,L}^{-1}$		7.72	
EP $4 \times 10^{-3}\,\mathrm{mol\,L}^{-1}$ + DTAC	$0.1\,\mathrm{mol\,L}^{-1}$	8.80	
EP $4 \times 10^{-3}\,\mathrm{mol\,L}^{-1}$ + SDS	$0.1\,\mathrm{mol\,L}^{-1}$	1.74[a]	7.20[b]
EP $4 \times 10^{-3}\,\mathrm{mol\,L}^{-1}$ + GM1	$6.5 \times 10^{-4}\,\mathrm{mol\,L}^{-1}$	4.63[a]	2.48[b]
EP $4 \times 10^{-3}\,\mathrm{mol\,L}^{-1}$ + GM1	$1.3 \times 10^{-3}\,\mathrm{mol\,L}^{-1}$	2.08	

[a]First linear region.
[b]Second linear region.

at least at pH 5.0. No degradation of GM1 at pH 1.5 and 20°C was found for at least 12 h [40].

The oxidation of EP in the presence of SDS became faster after 11 days. This is considered to be due to the acid-catalyzed hydrolysis of SDS [51] and the consequent disappearance of micelles. The acid-catalyzed hydrolysis of SDS in the micellar state proceeds 50 times faster than that in the free state [51], although the hydrolysis of SDS is significant below pH < 4. The plots of $\ln (x_e - x)$ vs t indicated two linear parts as can be seen in Fig. 15. It is considered that the first and the second linear regions are oxidation reactions of EP in the presence and in the absence of SDS micelles, respectively. The concentration of SDS after 10 days is probably below the CMC. Assuming that SDS micelles disappear when 91% SDS (whose initial concentration is $0.1\,\mathrm{mol\,L}^{-1}$) has been hydrolyzed, the disappearance time is calculated as approximately 9 days using the rate constant [51] for hydrolysis of micellar SDS. The reflection point in the plots of $\ln (x_e - x)$ vs t shown by open triangles in Fig. 16 is roughly consistent with the theoretical time. Thus, the oxidation of EP after 10 days is considered to proceed as in the free EP solution. The value of k, $7.20 \times 10^{-6}\,\mathrm{s}^{-1}$, for the second linear region is nearly equal to that for the free EP.

The rate constant for the oxidation of EP in the presence of DTAC micelles was larger than that of free EP. The accelerated oxidation of EP in the presence of DTAC micelles is interpreted as follows. The deprotonated intermediate [52] is present in the oxidation reaction of EP as shown in Scheme 3. This species with a negative charge is probably located at the surface of positively charged DTAC micelles. EP could not be taken into the inner site of DTAC micelles, thereby the shielding effect was not exhibited.

Summarizing this section, GM1 micelles inhibited the autoxidation of aqueous solutions of EP and the effect of GM1 micelles was more than 70 times larger than that of SDS micelles. GM1 seemed to be more effective at stabilizing drugs than common surfactants. GM1 is of biological origin so that GM1 may be safely utilized to the living body compared with synthetic surfactants.

E. Effect of Mixed Micelles Containing Antioxidant on Autoxidation of Epinephrine Hydrochloride [53]

In section D, the ganglioside G_{M1} (GM1) micellar effect on the autoxidation of epinephrine hydrochloride (EP) was described: the micellar effect of GM1 was superior to that of common surfactants such as sodium dodecyl sulfate; 1.7% EP was, however, oxidized in the micellar phase of GM1 for 30 days at pH 5.0 and 25°C.

In this section, the autoxidation of EP to adrenochrome (AC) in the presence of mixed micelles composed of antioxidants and GM1 is investigated, where one of the antioxidants is a new flavonoid, 3,5,7,4'-tetrahydroxy-2'-methoxyflavone (Fla), isolated from *Anaxagorea luzonensis* A. GRAY and the other one is α-tocopherol (Toc), which is a popular antioxidant. The mole fraction of Fla or Toc in the mixed micelles was 0.28. The final concentration of GM1 was $1.3 \times 10^{-3}\,mol\,L^{-1}$. The measurement was the same as in Section II. D.

The relationships between oxidized EP (as the absorbance of AC produced from EP) and time at pH 5.0 and 25°C in the presence of Fla/GM1 or Toc/GM1 mixed micelles are shown by closed circles and triangles, respectively, in Fig. 17. The autoxidations of EP in the absence of surfactant and in the presence of GM1 micelles without antioxidant (Section II. D)

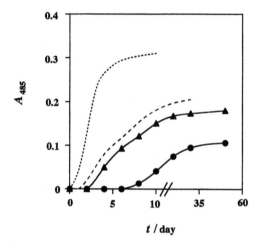

Figure 17 Time course of autoxidation of EP in the presence of Fla/GM1 or Toc/GM1 mixed micelles: Relationship between absorbance of AC produced from EP and time. ●, mixed Fla/GM1 micelles; ▲, mixed Toc/GM1 micelles; dotted line, time course of autoxidation of free EP; broken line, time course of autoxidation of EP in the presence of $1.3 \times 10^{-3}\,\mathrm{mol\,L^{-1}}$ GM1 micelles without antioxidant. Initial concentration of EP: $4 \times 10^{-3}\,\mathrm{mol\,L^{-1}}$.

are also shown by dotted and broken lines, respectively, in Fig. 17.

Mixed Fla/GM1 micelles completely inhibited the autoxidation of EP for 6 days. The autoxidation of EP gradually proceeded after 8 days. Nevertheless, the percentage of oxidized EP was merely 0.87% even on the 50th day. In contrast, mixed Toc/GM1 micelles exhibited no remarkable antioxidant effect on the autoxidation of EP compared with that in the presence of GM1 micelles without antioxidant.

The following condition is necessary to exhibit the antioxidant effect in the micellar phase: antioxidant exists in a similar microenvironment to EP in the micelles. If antioxidant exists in the core of micelles and EP exists at the outer region of the core, the antioxidant effect cannot be exhibited and EP undergoes oxidation. Thus Fla may have a proper hydrophobicity to EP. The autoxidation of EP gradually proceeded after 8 days. This phenomenon is explained as follows. The

3- and 4′-positions of the hydroxy groups of Fla are important to exhibit the greater antioxidant effect [54]. The freedom of hydroxy groups at the 3- and 4′-positions of Fla may be gradually restrained in the mixed micelles by the interaction between the hydroxy groups of Fla, and the oligosaccharide moiety or the hydrophilic part of the ceramide skeleton of GM1.

Mixed Fla/GM1 micelles exhibited a satisfactory potency as a functional micelle species having an antioxidant activity.

REFERENCES

1. K. Ogino, and M. Abe (eds.), *Mixed Surfactant Systems*, Dekker, New York (1993).

2. M. Ueno and H. Asano, In K. Ogino, and M. Abe (eds.), *Mixed Surfactant Systems*, Dekker, New York (1993), pp. 217–233.

3. B. W. Barry, and G. M. T. Gray, *J. Colloid Interface Sci.* **52**: 314 (1975).

4. B. W. Barry, and G. M. T. Gray, *J. Colloid Interface Sci.* **52**: 327 (1975).

5. D. Oakenful, and D. E. Fenwick, *Aust. J. Chem.* **30**: 335 (1977).

6. P. Sauve, and P. Desnuelle, *FEBS Lett.* **122**: 91 (1980).

7. M. Ueno, Y. Kimoto, Y. Ikeda, H. Momose, and R. Zana, *J. Colloid Interface Sci.* **117**: 179 (1987).

8. A. Bandyopadhyay, and S. P. Moulik, *Colloid Polym. Sci.* **266**: 455 (1988).

9. H. Asano, K. Aki, and M. Ueno, *Colloid Polym. Sci.* **267**: 935 (1989).

10. H. B. Kostenbauder, *J. Pharm. Sci.* **54**: 1243 (1965).

11. M. Nakagaki, and S. Yokoyama, *Bull. Chem. Soc. Jpn.* **59**: 19 (1986).

12. M. Nakagaki, and S. Yokoyama, *Bull. Chem. Soc. Jpn.* **59**: 1925 (1986).

13. S. Yokoyama, T. Nakazawa, and M. Abe, *Mater. Technol.* **17**: 283 (1999).

14. N. Ueda, and S. Yamamoto, *Chiryogaku (Therapeutics)* **13**: 730 (1984).

15. S. M. Karim, J. Delvin, and K. Hillier, *Eur. J. Pharmacol.* **4**: 416 (1968).

16. T. J. Roseman, B. Sims, and R. G. Stehle, *Am. J. Hosp. Pharm.* **30**: 236 (1973).

17. D. C. Monkhouse, L. Van Campen, and A. J. Aguian, *J. Pharm. Sci.* **62**: 576 (1973).

18. N. H. Andersons, *J. Lipid Res.* **10**: 320 (1969).

19. G. F. Thompson, J. M. Collins, and L. M. Schmalzried, *J. Pharm. Sci.* **62**: 1738 (1973).

20. H. C. Brummer, *J. Pharm. Pharmacol.* **23**: 804 (1971).

21. B. R. Hajratwala, *Aust. J. Pharm. Sci.* **5**: 39 (1975).

22. S. Yokoyama, K. Hashizaki, and M. Abe, *J. Jpn. Oil Chem. Soc. (J. Oleo Sci.)* **48**: 569 (1999).

23. S. Yokoyama, T. Kimura, M. Nakagaki, O. Hayaishi, and K. Inaba, *Chem. Pharm. Bull.* **34**: 455 (1986).

24. S. Yokoyama, A. Kaneko, and T. Fujie, *Bull. Chem. Soc. Jpn.* **61**: 3451 (1988).

25. S. Yokoyama, J. Obata, T. Fujie, and M. Nakagaki, *Chem. Pharm. Bull.* **41**: 6 (1993).

26. S. Yokoyama, Y. Fujino, T. Kohno, and M. Abe, *J. Jpn. Oil Chem. Soc. (J. Oleo Sci.)* **48**: 1365 (1999).

27. F. Hirayama, M. Kurihara, and K. Uekama, *Chem. Pharm. Bull.* **32**: 4237 (1984).

28. M. Nakagaki, S. Yokoyama, and I. Yamamoto, *Nippon Kagaku Kaishi (The Journal of the Chemical Society of Japan)* 1865 (1982).

29. S. Yokoyama, M. Nakano, T. Takeda, and M. Abe, *Mater. Technol.* **17**: 342 (1999).

30. T. Yamanaka, M. Hayashi, and R. Matsuura, *J. Colloid Interface Sci.* **88**: 458 (1982).

31. T. Yamanaka, T. Tano, O. Kamegawa, D. Excrowa, and R. Cohen, *Langmuir* **10**: 1871 (1994).

32. T. Yamanaka, N. Ogihara, T. Ohhori, H. Hayashi, and T. Muramatsu, *Chem. Phys. Lipids* **90**: 97 (1997).

33. W. Kramp, G. Pieroni, R. N. Pinckard, and D. J. Hanahan, *Chem. Phys. Lipids* **35**: 46 (1984).

34. M. Nakagaki, H. Komatsu, and T. Handa, *Chem. Pharm. Bull.* **34**: 4479 (1986).

35. R. E. Stafford, T. Fanni, and E. A. Dennis, *Biochemistry* **28**: 5113 (1989).

36. R. G. Stehle, and T. O. Oesterling, *J. Pharm. Sci.* **66**: 1590 (1977).

37. S. Yokoyama, T. Takeda, and M. Abe, *Colloid Surf. B: Biointerfaces* **20**: 361 (2001).

38. T. L. Steck, and G. Dawson, *J. Biol. Chem.* **249**: 2135 (1974).

39. R. W. Ledeen, In R. U. Margolis, and R. K. Margolis (eds.), *Complex Carbohydrates of Nervous Tissue,* Plenum Press, New York (1979), pp. 1–23.

40. B. Ulrich-Bott, and H. Wiegandt, *J. Lipid Res.* **25**: 1233 (1984).

41. Y. Barenholz, B. Ceastaro, D. Lichtenberg, E. Freire, T. E. Thompson, and S. Gatt, *Adv. Exp. Med. Biol.* **125**: 105 (1980).

42. S. Yokoyama, T. Takeda, and M. Abe, *Mater. Technol.* **18**: 164 (2000).

43. I. Kruk, *Postepy Fiz. Med.* **13**: 85 (1978).

44. M. Bounias, and I. Kruk, *Comp. Biochem. Physiol. C* **74**: 143 (1983).

45. C. Giulivi, and E. Cadenas, *Free Radical Biol. Med.* **25**: 175 (1998).

46. G. M. L. Taam, S. Takeo, A. Ziegelhoffer, P. K. Singal, R. E. Beamish, and N. S. Dhalla, *Can. J. Cardiol.* **2**: 88 (1986).

47. O. Ortolani, A. Conti, R. Imperatore, P. Sommella, and R. Cuocolo, *Boll. Soc. Ital. Biol. Sper.* **58**: 444 (1982).

48. M. Nakagaki, and S. Yokoyama, *Bull. Chem. Soc. Jpn.* **58**: 753 (1985).

49. M. Nakagaki, *Surface State and Colloid State*, Tokyo Kagaku Dojin, Tokyo (1976), p. 282.

50. J. Dai, L. Bai, Y. Zhang, Y. Zang, D. Jiang, Z. Gu, and X. Zhao, *Wuli Huaxue Xuebau* **7**: 260 (1991).

51. M. Nakagaki, and S. Yokoyama, *J. Pharm. Sci.* **74**: 1047 (1985).

52. T. Kamidate, H. Ichihashi, T. Segawa, and H. Watanabe, *J. Biolumin. Chemilumin.* **10**: 55 (1995).

53. T. Takeda and S. Yokoyama, *Mater. Technol.* **18**: 238 (2000).

54. L. Magnani, E. M. Gaydou, and J. C. Hubaud, *Anal. Chim. Acta* **411**: 209 (2000).

8

A Thermodynamic Study on Competitive or Selective Solubilization of 1:1 Mixed Solubilizate Systems: Solubilization of Different Sterols Mixtures by Bile Salt Micelles in Water

SHIGEMI NAGADOME and
GOHSUKE SUGIHARA
Department of Chemistry, Fukuoka University,
Fukuoka, Japan

I. INTRODUCTION

Since aqueous solubility of sparingly or insoluble substances had been found to remarkably increase with addition of soap as the third component, systematic studies were conducted on this phenomenon using surfactants and led to the establishment of the concept of *solubilization*. This term *solubilization* now has a general definition: "formation of a thermodynamically stable isotropic solution of a substance normally insoluble or slightly soluble in a given solvent with the aid of one or more amphiphilic components." Many solubilization-dependent applications are found in diverse fields of industry, e.g., dissolution of water-insoluble drugs into aqueous solution and their delivery to various parts of the body, preparation of agricultural solutions, production of cosmetics and toiletries, food processing, etc. [1]. Solubilization also plays a very important role in biological processes as found in, for example, digestion and absorption of lipids by bile acids/salts.

On the other hand, extensive studies on mixed surfactant systems have been performed to examine the synergistic effects or molecular interactions resulting from mixing of different surfactants in varied combinations from the practical standpoint of allowing the systems to develop a high performance as well as the scientific standpoint of obtaining detailed information about the interactions between different surfactants. Although there have been many studies on solubilization using surfactant mixtures, few thermodynamic studies on *competitive or selective solubilization of solubilizate mixtures* have so far been reported.

In this chapter, we develop thermodynamic equations for evaluating various thermodynamic parameters concerning solubilization by surfactant micelles of 1 : 1 binary solubilizate mixtures. We then relate the "solubilizing power" defined previously in the literature to the free energy change of solubilization, which is regarded as the free energy change in the transition of a given sparingly soluble solubilizate from a solid state to a dispersed state in micellar solution.

II. THEORY OF SOLUBILIZATION: SOLUBILIZING POWER, EXCESS GIBBS ENERGY AND MISCIBILITY OF DIFFERENT SOLUBILIZATES IN MICELLES

A. Solubilization Equilibrium and Solubilizing Power

1. Single Solubilizate Systems

According to Moroi [1], when n surfactant molecules form a micelle the equilibrium between surfactant S and micelle M is expressed as

$$nS \overset{K_n}{\rightleftharpoons} M \tag{1}$$

where K_n is the equilibrium constant

$$K_n = \frac{[M]}{[S]^n} \tag{2}$$

In the case where a micelle consecutively solubilizes i molecules of solubilizate R, the concentration of solubilizing micelles is given as

$$[MR_i] = K_n \left(\prod_{j=1}^{i} K_j' \right) [S]^n [R]^i \tag{3}$$

Here K_j' denotes the equilibrium constant between solubilizate R plus MR_{i-1} and MR_i. It has been pointed out that if the concentration of solubilizate incorporated into micelles is close to or higher than the micellar concentration, distribution of solubilizate among micelles should be taken into account [1,2]. For such cases where the number of solubilizate molecules per micelle lies around unity, the Poisson distribution can be applied [1]. When a micelle M solubilizes, on average a molecule of R to form MR, the following equation is given

$$M + R \overset{K_1'}{\rightleftharpoons} MR$$
$$[MR] = K_1' \, [M][R] \tag{4}$$

Note that the equlibrium constant K_j' serves as the interaction parameter between the solubilizate and the micelle. If the total concentration of surfactant is denoted by C_t and the total concentration of micelles by $[M_t]$, we have $n[M_t] = C_t - \text{CMC}$ (CMC, the critical micellization concentration). Equation (4) leads to $[R_t] - [R] = [MR] = K_1' \, [M_t] \, [R]$ when the total concentration of solubilizate in the system is $[R_t]$.
  We then have

$$\frac{[R_t] - [R]}{[R]} = K_1' \, [M_t] \tag{5}$$

Here we need to know the value of $[R]$. However, there are many cases in which the solubility in water of the solubilizate concerned is too low to be determined $[R]$ (e.g., the solubility of cholesterol is 10^{-8}–$10^{-7} \, \text{mol} \, \text{dm}^{-3}$).
  When a solubilization experiment is performed on a compound with a very low solubility in water while shaking test tubes containing solids of the solubilizate, the solubilization equibria are expressed as [3]

Slightly soluble solubilizate crystals $\overset{K_d}{\rightleftharpoons}$ Singly dispersed

species in bulk $\overset{K_1'}{\rightleftharpoons}$ Solubilized species in micelles

Equation (5) is useful for solubilization to which the Poisson distribution can be applied. This equation is applicable even when the aggregation number of micelles itself has a polydispersity [1]. Here, the solubility constant K_d is thermodynamically given as

$$K_d = \frac{\text{Activity of singly dispersed solubilizate}}{\text{Activity of solubilizate solid}} \cong \frac{[R]}{1} = [R]$$

(6)

meaning that K_d approximates to the molarity $[R]$. Equation (5) may then be approximated as below, since $[R]$ is negligibly small compared with $[R_t]$

$$[R_t] = K_d K'_1 [M_t] = K_s[M_t]$$

(7)

where the constant $K_s = K_d K'_1$.

This relation originally involves the equilibrium between the states in solid and micellar phases for a given solubilizate [3]. The solubilization of a slightly soluble solid substance corresponds to its transition from the solid state to the solubilized state in micelles. The standard Gibbs energy change between these states is

$$\Delta G^0 = -RT \ln K_s$$

(8)

Previously we defined the solubilizing power S_p as the slope of $[R_t]$ vs C_t curve as follows [3,4]

$$\frac{d[R_t]}{d(C_t - \text{CMC})} = \frac{d[R_t]}{dC_t} \cong \frac{\Delta[R_t]}{\Delta C_t} \equiv S_p$$

(9)

Since we have many cases where a good linearity is found to hold for solubilization of sparingly soluble solubilizates such as cholesterol (Ch) by bile salts except for sodium cholate (NaC) as previously reported [3–5] or shown in the present paper, S_p is approximately related to K_s as

$$S_p \cong \frac{[R_t]}{C_t - \text{CMC}} = \frac{[R_t]}{n[M_t]} = \frac{K_s}{n}$$

(10)

meaning that K_s is n times larger than S_p, i.e., $\ln S_p = \ln K_s - \ln n$. Here it is noted that the aggregation number n is assumed to be constant within a restricted solubilizer

Table 1 The reciprocal of solubilizing power (S_p^{-1}) corresponding to the molecular number required for solubilizing a sterol molecule [4].

			S_p^{-1}		
	Ch	Stig	Chsta	Ch + Stig	Ch + Chsta
NaDC	14.0	24.3	17.3	13.4	19.0
NaC[a]	27.6	60.0	31.4	25.2	35.2

[a] S_p was determined at 100 mM NAC.

concentration range. If the n value is separately determined, K_s and thus the Gibbs energy change in Eq. (8) can be calculated. In the case where n is unknown, we may regard the solubilizing power as the solubility of solubilizate solid per mole of BS in micellar solution without recourse to the strict concept of solubilization by micelles, and then the Gibbs energy change is defined on the basis of the solubility concept as [3]

$$\Delta G_s^\theta = -RT \ln S_p \tag{11}$$

This Gibbs energy change may be used to compare solubilization data which depend on the respective solubilizate species. S_p is usually smaller than unity (the reciprocal S_p ranges of tens to hundreds, see Table 1) so that ΔG_s^θ values are positive [3].

2. Mixed Solubilizate Systems

Next, we consider the cases in which two solid solubilizates, A and B, both with a very low solubility in water, coexist. Such solid solubilizates include cholesterol (Ch), cholestanol (Chsta), and stigmasterol (Stig). A and B are equilibrated among three phases, i.e., solid phase (C), singly dispersed state in bulk solution phase (D) and solubilized state in micelles (M). The chemical potentials are expressed for species i ($i =$ A, B) as:

$$\mu_i^{(C)} = \mu_i^{(D)} = \mu_i^{(M)} \tag{12}$$

where

$$\mu_i^{(C)} = \mu_i^{(C)0} + RT \ln a_i^{(C)} \quad \text{(in solid)}$$

$$\mu_i^{(D)} = \mu_i^{(D)0} + RT \ln a_i^{(D)} \quad \text{(in bulk)}$$

$$\mu_i^{(M)} = \mu_i^{(M)0} + RT \ln a_i^{(M)} \quad \text{(in micelle)}$$

Here, if the activities of A and B in bulk solution, $a_A^{(D)}$ and $a_B^{(D)}$ are approximated by molarities $[R_A]$ and $[R_B]$ because of their very low concentrations, Eqs (7) and (10) yield the relations

$$[R_t] = [R_t^A] + [R_t^B] = (K_d^A K_1'^A + K_d^B K_1'^B)[M_t]$$
$$= (K_s^A + K_s^B)[M_t] = K_s^{(mix)}[M_t] \quad (13)$$

This equation means that A and B are solubilized independently of each other in a given micellar solution. $K_s^{(mix)}$ is related to the solubilizing power of a surfactant for a mixed solubilizate system, $S_p^{(mix)}$, in a way similar to that used before.

$$K_s^{(mix)} = \frac{[R_t]}{[M_t]} \qquad \frac{K_s^{(mix)}}{n} = S_p^{(mix)} \qquad (14)$$

The Gibbs energy change, $\Delta G^{(mix)\theta}$, is then expressed in terms of $S_p^{(mix)}$

$$\Delta G^{(mix)\theta} = -RT \ln S_p^{mix} \qquad (15)$$

It should be noted here that the equilibrium constant, K_M, between the singly dispersed state in bulk (D) and the solubilized state in micelle (M) for binary mixed solubilizate systems is defined as

$$K_M = \frac{a_A^{(M)} a_B^{(M)}}{a_A^{(D)} a_B^{(D)}}$$

and that Eq. (6) and the approximation used in Eq. (13) give the relation

$$K_M \simeq \frac{a_A^{(M)} a_B^{(M)}}{[R_A][R_B]} \simeq \frac{[R_t^A][R_t^B]}{K_d^{(A)} K_d^{(B)}} \tag{16}$$

The Gibbs energy change, ΔG_M^0, is equated to $-RT \ln K_M$, with the transition of the two solubilizates from bulk phase to micellar phase. The Gibbs energy levels are illustrated for the respective states in Scheme 1 [3].

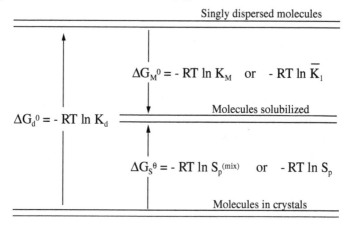

Scheme 1

B. Excess Gibbs Energy and Miscibility of Solubilizates in Micelles

We treat here two solubilizates A and B as the components of a mixed system. For the given mixed solubilizate system, the molar Gibbs free energy change upon solubilization of pure species i ($i = A, B$) is denoted by ΔG_i^θ and the mole fraction of i in the solubilized mixture within the micellar pseudophase by X_i. If mixing occurs in an ideal manner, the molar Gibbs energy change of this ideal mixing, $\Delta G_s^{\theta(mix)}$ (ideal), should satisfy the additivity rule as $\Delta G_s^{\theta(mix)}$ (ideal) $= X_A \Delta G_A^\theta + X_B \Delta G_B^\theta$. The excess Gibbs energy is defined here as the difference between the real free energy change $\Delta G_s^{\theta(\text{mix})}$ determinable from Eq. (15) and the ideal free energy change

$$\Delta G_s^{(\text{ex})} = \Delta G_s^{\theta(\text{mix})} - \{X_A \Delta G_A^\theta + X_B \Delta G_B^\theta\} \tag{17}$$

Only the entropy term contributes to the Gibbs energy change of ideal mixing and the enthalpy of ideal mixing $\Delta H_{\text{mix}}^\theta$ (ideal) is zero, i.e., $\Delta G_s^{\theta(\text{mix})}$ (ideal) is equal to $-T\Delta S_{\text{mix}}^\theta$ (ideal).

Assuming the applicability of regular solution theory to the present system, the real entropy term $-T\Delta S_{\text{mix}}^\theta$ (real) can be put equal to $-T\Delta S_{\text{mix}}^\theta$ (ideal) [6]. In fact, regular solution theory would probably hold for such mixed systems as sterols/stanol because the molecular size and shape are the same for these species. Thus, the excess Gibbs free energy change would be the same as the enthalpy term itself. When the chemical potential of i in micelles is given as $\mu_i = \mu_i^0 + RT \ln \gamma_i X_i$, where μ_i^0 is the standard chemical potential and γ_i, the activity coefficient, then $\Delta G^{(\text{ex})}$ can be expressed as

$$\Delta G^{(\text{ex})} = \Delta H_{\text{mix}}^\theta = RT(X_A \ln \gamma_A + X_B \ln \gamma_B) \tag{18}$$

The activity coefficient γ_i is known to reflect intermolecular interaction and is given below according to regular solution theory

$$\ln \gamma_A = \frac{\omega(1 - X_A)^2}{RT}, \qquad \ln \gamma_B = \frac{\omega(1 - X_B)^2}{RT} \tag{19}$$

where ω is the interaction parameter ascribed to the cohesive forces between unlike molecules [1,6]. From Eq. (19), we have the relation [4]

$$\ln \gamma_A = \frac{(1 - X_A)^2}{X_A^2} \ln \gamma_B \tag{20}$$

Substituting Eq. (20) into Eq. (18), we obtain

$$\frac{\Delta G^{(\text{ex})}}{RT} = \left(\frac{X_B^2}{1 - X_B} + X_B \right) \ln \gamma_B \tag{21}$$

Since all the quantities on the left-hand side of Eq. (21) are known and X_A and X_B on the right-hand side are determinable, the activity coefficients γ_A and γ_B can be calculated. In addition, substitution of Eq. (19) into Eq. (18) allows us to calculate the interaction parameter ω using the equations [3]

$$\omega = \frac{\Delta G^{(ex)}}{X_A X_B} \tag{22}$$

The thermodynamics for competitive solubilization would also be applicable to other mixtures as far as they obey regular solution theory.

III. SOLUBILIZATION OF STEROL MIXTURES BY BILE SALT MICELLES

A. Background

Regarding properties of aqueous bile salt (BS) surfactant systems, Hinze *et al.* [7] have pointed out that compared with bulk solvents, solutions of BS micelles exhibit several unique properties that can be exploited in chemical analysis and separation science applications. These properties include those such as: (a) solubilization, binding and selective compartmentalization of solute molecules; (b) alteration of the rate or pathway of chemical or photophysical processes; (c) shifting of the position of equilibria; (d) alteration of the effective microenvironment (i.e., polarity, viscosity, etc.) sensed by the micellar bound solute; and (e) affecting bulk solution properties (e.g., surface tension, conductance, etc.). Amongst these, the most important property of BS media is their ability to solubilize and bind solute molecules that are insoluble or only marginally soluble in water or other bulk solvents [7–15]. BS micelles solubilize the water-insoluble components in bile (such as cholesterol (Ch) and lecithin) and lipolytic products (monoglycerides and fatty acids) in small intestine [7–9]. It has been known that the amount of solute solubilized by BS generally depends more on solubilizer concentration than by common surfactants with a hydrophobic chain and a simple hydrophilic head group, indicating that the aggregation mode of the former is not so cooperative [2,4]. The micellar structure and the aggregation number depend not only on the experimental conditions such as pH, ionic strength, and temperature but also on the BS species themselves. In fact, the structure of micelles is greatly changeable with concentration, especially around their critical micellization concentration (CMC) even for single BS species [2]. Solubilization of hydrophobic solutes by BS micelles is then determined by the interaction between host and guest molecules or the driving force causing the interaction. Zuman and Fini [5] have extensively discussed the interactions of BSs with organic compounds and listed the collected data for more than 110 compounds solubilized in BS solutions in their paper (Table 1 in Ref. [5]). They also collected the data of the slopes of the linear segments of solubilizate solubility vs BS concentration curves for various solubilizates using a certain number of BS species at concentrations higher than about 20 mM (Table 2 in Ref. [5]) to make it possible to compare the relative solubilizing powers of individual BSs for a single solubilized species and those of a given BS for different solubilized species.

Solute solubility in micellar solution is also a convenient measure of solubilization, given that it is defined as the total

Table 2 The solubilizing powers, S_p, for various sterols and their 1:1 mixtures and the Gibbs energy changes $\left(\Delta G_s^\theta \text{ and } \Delta G_s^{\theta(\text{mix})}\right)$ of solubilization from the respective solid states [4].

	NaDC		NaC	
	S_p	$\Delta G_s^\theta/\text{kJ mol}^{-1}$	S_p	$\Delta G_s^\theta/\text{kJ mol}^{-1}$
Ch	0.0714	6.80	0.0362	8.55
Stig	0.0412	8.05	0.0167	10.55
Chsta	0.0579	7.34	0.0318	8.89
	$S_p^{(\text{mix})}$	$\Delta G_s^{\theta(\text{mix})}/\text{kJ mol}^{-1}$	$S_p^{(\text{mix})}$	$\Delta G_s^{\theta(\text{mix})}/\text{kJ mol}^{-1}$
Ch + Stig	0.0747	6.69	0.0397	8.32
Ch + Chsta	0.0525	7.60	0.0284	9.18

Table 3 The excess Gibbs energy changes, $\Delta G_{\mathrm{s}}^{(\mathrm{ex})}$, produced by mixing of solubilizates at different concentrations of bile salts [4].

		BS/mM					
		25	50	75	100	125	150
Ch + Stig							
$\Delta G_{\mathrm{s}}^{(\mathrm{ex})}$/J mol^{-1}	NaDC	−441	−416	−441	−453	−453	
	NaC		−696	−676	−696	−716	−716
Ch + Chsta							
$\Delta G_{\mathrm{s}}^{(\mathrm{ex})}$/J mol^{-1}	NaDC	252	268	284	317	338	
	NaC		292	292	292	309	312

concentration of the solute species solubilized in micelles and singly dispersed in the bulk. Hinze *et al.* [7] gave the solubility data for selected solutes such as benzene, cyclohexane, cholesterol, bilirubin, azobenzene, naphthalene *n*-butylbenzene and 20-methyl-cholanthren in BS surfactant media (Table 3 in Ref. [7]). Further, the so-called partition coefficient, P, and the equilibrium constant, K_{b}, are useful measures for evaluating the properties of respective species. The former is given by the ratio of solute concentration in the micellar phase to that in the bulk phase while the latter, sometimes called the binding constant, is equal to molar solute concentration in the micellar phase divided by the product of solute concentration in the bulk phase and analytical concentration of micellized surfactant. Partition coefficients (or binding constants) of solutes between water and bile salt surfactant systems [16–31] are summarized in the literature (in Table 4 in Ref. [7]). It is noted here that P and K_{b} relate to each other in the form, $P = K_{\mathrm{b}} \times 55.5$.

We examined the effects of pH, pNa and temperature on micelle formation and solubilization of cholesterol (Ch) in aqueous BS solutions in the earlier works [4]. Thus, the solubilizing power (S_{p}) was defined as the tangential slope of solubilized amount of Ch (W) vs total bile salt concentration (C_{t}) curve, i.e., $S_{\mathrm{p}} = \mathrm{d}W/\mathrm{d}C_{\mathrm{t}}$, and thermodynamic parameters of entropy and enthalpy changes of micelle formation and the degree of counterions binding (H^{+} and Na^{+}) to micelles of

Table 4 The interaction parameter ω/RT values for Ch + Stig and Ch + Chsta mixtures solubilized by NaC and NaDC micelles in water at 37°C [4].

		BS/mM					
		25	50	75	100	125	150
Ch + Stig							
ω/RT	NaDC	−0.89	−0.89	−0.89	−0.89	−0.89	
	NaC		−1.5	−1.5	−1.5	−1.5	−1.5
Ch + Chsta							
ω/RT	NaDC		3.6	2.0	1.2	0.98	
	NaC					2.5	2.1

sodium cholate (NaC) and sodium deoxycholate (NaDC) were calculated. Although we have thereafter determined the S_p value for different BSs [32], we have neither thermodynamically applied the S_p data nor developed thermodynamics of solubilization itself. In fact, no paper has so far dealt with thermodynamic treatment of solubilization by BS micelles except for a few related articles [2,4,5,33].

Ch and phytosterols (plant sterols, abundant in fat-soluble fractions of plants), some of which are investigated as solubilizates in the present work, are absorbed from the upper small intestine in various amounts. The most frequently found phytosterols are campesterol, sitosterol, and stigmasterol, which are found in the normal diet at a total amount of 200–500 mg per day. Chemically resembling Ch, phytosterols can inhibit the absorption of Ch or display hypocholesterolemic properties [34–38]. That is, phytosterol consumption in human subjects under a wide range of medical study conditions has been shown to reduce plasma total and low density lipoprotein (LDL) Ch levels [35–40]. Sugano *et al.* showed first that phytostanols, derived from reduction of the double bonding of phytosterols, can more effectively lower the plasma Ch level in rats than phytosterols [41]. Based on their systematic studies on solubilization of Ch and phytosterols or phytostanols, Ikeda *et al.* revealed that it is an essential condition that phytosterols

Systems Studied

1. Solubilizers

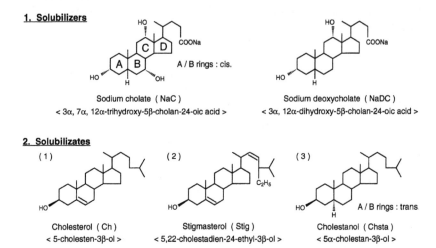

2. Solubilizates

Figure 1 Structural formulas of solubilizers and solubilizates.

inhibit Ch solubilization by BSs in the intestine lumen, or that the suppression of Ch absorption results from the lowering of the relative amount of Ch solubilized by BS micelles when phytosterols or phytostanols coexist with Ch. In other words, the inhibition of Ch absorption takes place at the unstirred water layer (UWL) covering epithelium cells [38a,41,43]. Figure 1 shows the chemical structures of these solubilizates and solubilizers.

In order to examine thermodynamically the effect of phytosterols/phytostanols on the plasma Ch level lowering, this paper describes competitive or selective solubilization of sterol/stanol mixed systems. It should be noted that only one paper has so far reported the thermodynamic treatment of mixed solubilizate systems [3].

B. Experimental Results

First of all, the experimental procedure for studying sterols solubilization by BS micelles is described. BS salts were dissolved in Kolthoff buffer solution ($0.05\,M\ Na_2B_4O_7 + 0.05\,M$

Na_2CO_3) at pH 10.0 and 5 ml each of solutions of the respective BSs at varied concentrations and known amounts of excess solid sterols were placed in test tubes with a glass stopper. The test tubes were hermetically sealed, shaken, and incubated in a thermostated water bath at 37°C for 24 hours. Solubilization was confirmed beforehand to attain an equilibrium within 24 h. A dry N_2 atmosphere was used to avoid oxidation of Ch and stigmasterol during solution preparation. After shaking, each solution was centrifuged (3000 rpm, 10 min) and filtered through a 0.22 μm membrane filter to remove excess (non-solubilized) sterols/stanol crystals.

The concentration of solubilized sterols/stanol was determined by high performance liquid chromatography (HPLC) (HITACHI D-7000). For reverse-phase HPLC studies, the column was filled with octadecylsilane (ODS), 250×4.6 mm, flow rate was set at 1.0 ml/min, and the mobile phase was 100% methanol. The elution process was monitored with a Shodex RI SE-61 differential refractometer and recorded with a HITACHI Integrator D-7500. The calibration curves were obtained with a regression value of 0.999.

Solubilized amounts (W, in mM) of the respective single systems are plotted against BS concentration (C_t), as shown in Fig. 2. The top frame indicates the curves for Ch, Chsta and Stig solubilized by NaC solutions as a function of concentration and the bottom frame, those solubilized by NaDC solutions. As is seen, W decreases in the order: Ch > Chsta > Stig in NaC as well as NaDC solutions. The relations obtained with NaDC solution are linear while the plots for NaC solution show a curvature, suggesting that the state of micelles, especially in terms of aggregation number and structure, changes with concentration, as reported previously [2,4].

In Fig. 3(A) and (B) are shown the relations of W vs C_t for the mixed systems of Ch/Stig in NaC (top) and NaDC (bottom) solutions and for those of Ch/Chsta in NaC (top) and NaDC (bottom) solutions, respectively. The results for the respective single systems (shown by open and closed circles) are included for comparison in all frames. The W values of the respective components in the mixture are shown together with the total W values. Comparison of (A) with (B) reveals that, over the whole

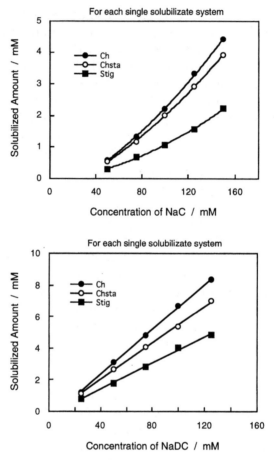

Figure 2 Solubilized amounts in mM of the respective single systems as a function of BS concentration in pH 10 Kolthoff buffer solution at 37°C.

BS concentration range, the total W value of Ch/Stig mixture is higher than that of each single system, whereas the total W value of Ch/Chsta mixture is lower than that of each single system. It is noteworthy that Ch is surprisingly reduced, implying exclusion of Ch by Chsta. At concentrations below 80 mM for NaC and below 40 mM for NaDC, Ch is completely kicked out of micelles by Chasta. Thus, selective solubilization was observed for this combination. It should be noted here that

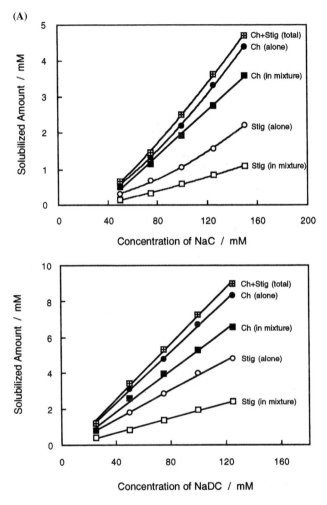

Figure 3 (A) Solubilized amounts of Ch and Stig from their single and mixed systems in micellar solutions of NaC (top) and NaDC (bottom) for total and respective components of the mixed solubilizates. (B) Solubilized amounts of Ch and Chsta from their single and mixed systems in micellar solutions of NaC (top) and NaDC (bottom) for total and respective components of the mixed solubilizates.

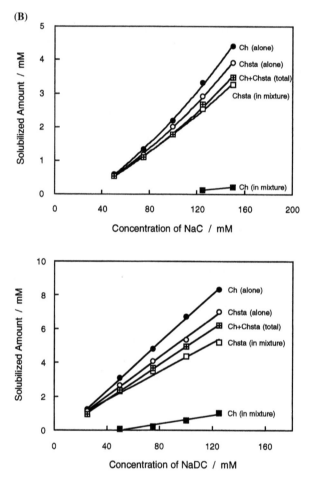

Figure 3   continued.

for the third combination, Stig/Chsta mixed system, the HPLC method we used did not allow us to separately analyze the composition of solubilized mixture, with the exception of the total amount; we should have employed a different column or a different solvent as eluent instead of methanol. No description is then given on solubilization of this mixture.

In order to see the effect of mixing more clearly, W values of the respective singles systems and mixed systems are expressed by histograms in Fig. 4(A) and (B) for Ch/Stig and Ch/Chsta combinations in NaC and NaDC solutions.

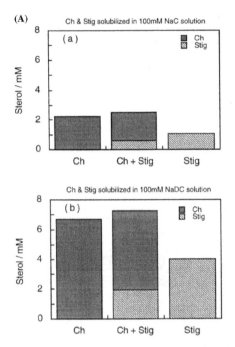

Figure 4 (A) Concentrations of Ch and Stig solubilized by (a) NaC and (b) NaDC micelles from their single and mixed systems in pH 10 Kolthoff buffer solution at 37°C. (B) Concentrations of Ch and Chsta solubilized by (a) NaC and (b) NaDC micelles from their single and mixed systems in pH 10 Kolthoff buffer solution at 37°C.

Mixing results in a decrease in W for each component of Ch/Stig mixture though the total W is slightly increased, suggesting a small decrease in W for Ch caused by the presence of Stig. In contrast, the total W is lower than W for each single system of Ch/Chsta mixture. Attention should be paid again to the fact that most or even all Ch molecules are expelled from micelles. Thus, the histogram (center) clearly demonstrates that BS micelles can selectively solubilize sterol from sterol/ stanol mixtures.

Figure 5 was constructed for easy comparison of selective solubilization. The mole fraction X of each component in the solubilized steroid/stanol mixture is plotted as a function of BS concentration. Stig occupies nearly 20% with no visible BS

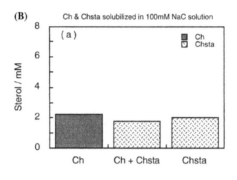

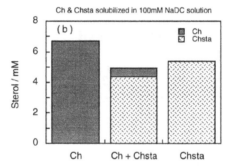

Figure 4   continued.

concentration dependence in the mixed system of Ch and Stig while Chsta is selectively solubilized by BS micelles in the Ch/Chsta system. Ch cannot coexist with Chsta in BS micelles at the physiological concentration, 30 mM, though it is solubilized to some extent at higher BS concentrations. Table 1 lists the reciprocal solubilizing power (S_p^{-1}) of Ch, Stig and Chsta in addition to those for their mixed mixtures solubilized in NaC and NaDC solutions. The S_p^{-1} corresponds to the molecular number of solubilizer required to solubilize a solubilizate molecule.

C. Discussion

In the previous section, the concentrations of sterols and a stanol (hydrogen-saturated sterol) solubilized were determined

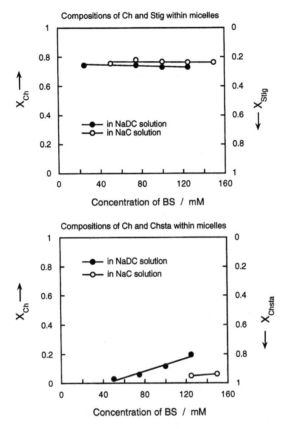

Figure 5 Selectivity as a function of BS concentration.

for the respective single systems and their 1:1 mixtures. Note here that the amounts of solid solubilizates used are present in large excess and their solubilities in water are extremely low (e.g., the solubility of Ch ranges from 10^{-8} to 10^{-7} mol/dm^3 (Table 3 in Ref. [7])). Moreover, the solubilized amount of Ch was lowered in the mixed solubilizate systems. In particular, Ch was found to be almost completely excluded by Chsta for the combination of Ch and Chsta, while Ch/Stig mixture showed a certain level of miscibility in BS micelles. Thermodynamical interpretation of these findings is discussed in this section.

The Gibbs energy changes for solubilizates in the translation from solid into micelles were shown to be determinable from Eqs (11) and (15); the data of S_p, $S_p^{(mix)}$, ΔG_s^θ and $\Delta G_s^{\theta(mix)}$ are tabulated in Table 2. The positive Gibbs energy changes for the respective single systems decrease in the order: Stig > Chsta > Ch. Comparison of the solubilized amounts at any BS concentration among the species sterols and stanol give the increasing order: Stig < Chsta < Ch (see Table 1). The smaller the amount of solubilization, the larger the positive Gibbs energy change, corresponding to the decreasing solubilizing power of a BS for the solubilizate species. On the other hand, the melting points are: Stig (170°C) > Ch (148.5°C) > Chsta (141.5°C). Although we have no conclusive data of their solubility in pure water, the solubility, melting point, and structures of sterol and stanol molecules would be directly related to their solubilization.

The following should be noted in connection with Gibbs energy change. Murata *et al.* [48] determined the aggregation number of NaDC in a buffer solution at pH 7.8 and 30°C as 45, meaning that 2 to 3 Ch molecules are solubilized in a NaDC micelle. Based on these findings, K_s in Eq. (10) is calculated as 3.21 and thus the Gibbs energy change defined by Eq. (8) is estimated to be $\Delta G^0 = -2.94$ kJ/mol. Very recently Matsuoka *et al.* have shown that the aggregation number is 17.5 and S_p is 0.0191 for NaC in pure water at 25°C (private communication, in press in *Biochim. Biophys. Acta*). We obtain $K_s = 0.334$ and $\Delta G^0 = 2.72$ kJ/mol using their data. These results indicate that the sign of ΔG^0 calculated using Eq. (8) depends on solubilizer species; a higher S_p and a larger aggregation number of micelle lead to a more negative value of ΔG^0 (easier spontaneous solubilization) and Ch solubilization by BSs can be classified into two types, those with a negative ΔG^0 and a positive ΔG^0.

As indicated in Section III.B, binary mixed solubilizates were found to exhibit an interesting contrast in terms of selectivity or synergism depending on the combination of the solubilizates involved. Thus, a higher affinity of Chsta to BS molecules results in the exclusion of Ch from the interior of micelles while the combination of Stig with Ch synergistically increases the total solubilized amount of the solubilizates. We then attempt to thermodynamically analyze the experimental findings on the basis of our theory in order to make clear the interaction mode for each combination.

Table 3 lists the excess molar Gibbs energy changes, $\Delta G^{(ex)}$, caused by mixing of solubilizates at different concentrations of BSs for each system. The obtained values of ω/RT and activity coefficient for the respective systems are given in Tables 4 and 5, respectively.

Interestingly, in micellar solutions of both NaC and NaDC, negative values of the excess Gibbs energy change were obtained for Ch/Stig mixture, whereas positive values were obtained for Ch/Chsta mixture. The negative Gibbs energy value suggests an enhanced hydrophobic interaction between Ch and Stig produced by the interaction between their hydrophobic side chains in addition to that between steroid skeletons, giving a possibility for Ch and Stig to form dimers in BS micelles. This would also correspond to the increased solubilizing power or solubilized amount (concentration) of the mixed solubilizates over that of Ch alone in the whole solubilizer concentration range.

In regard to the positive $\Delta G^{(ex)}$ values of Ch/Chsta mixed solubilizate systems, the $\Delta G^{\theta(mix)}$ values calculated using Eq. (15) are so highly positive (because $S_p^{(mix)} << 1$) that Eq. (17) yields no negative value. The positive $\Delta G^{(ex)}$ implies that Ch and Chsta are not spontaneously miscible or a certain repulsive force acts between Ch and Chsta because of poor steric fitness. Values of ω/RT are negative for Ch/Stig mixture in parallel with $\Delta G^{(ex)}$, i.e., -0.8 and -1.5 for NaDC and NaC over the whole range of solubilizer concentration while those for Ch/Chsta mixed systems are positive and concentration-dependent (see Table 4).

The larger absolute values of ω/RT for the Ch/Stig mixture in NaC solution than in NaDC solution may be ascribed to the fact that intermolecular interaction is enforced by a weaker hydrophobicity inside NaC micelles as compared with that inside NaDC micelles. Considering the ω/RT value for Ch/Chsta mixture ranging 3.6–1 and 2.5–2.1 in NaDC and NaC solution, respectively, these very large values (larger than 2) correspond to those for a phase separation, meaning that Ch and Chsta are not miscible with each other [44]. This explains the selective solubilization in the Ch/Chsta mixture or the exclusion of Ch from BS micelles by Chsta. It is also possible that the higher ω/RT values for Ch/Chsta mixture than for the Ch/Stig mixture may arise from larger errors involved in the determination of the Ch concentration which is extremely low due to its exclusion by Chsta.

In fact, in the lower concentration range of BS, smaller Ch amounts determined for the Ch/Chsta mixture involve higher errors and the calculated ω value widely varies even if the mole fraction of Ch fluctuates within a small range as suggested by Eq. (22). The activity coefficient of Chsta in the Ch/Chsta mixture was 1.10–1.14 in both BS solutions. In contrast, an extraordinarily large value was obtained for Ch especially in the lower BS concentration range. Because of this, the activity coefficient of Ch is indicated as "not determinable (n.d.)" in Table 5.

The solubilized state or structure of the solubilizates in BS micelles would be as follows. Miyata *et al.* [45] have studied in detail the molecular assemblies of cholic acid derivatives and use of their inclusion spaces: multiple hydrogen groups are essential to the formation of the assemblies. Slight change in host and guest molecules may frequently lead to great change in the assemblies due to exchange of the corresponding hydrogen bonding networks. In

Table 5 Activity coefficients of cholesterol in the sterol mixtures solubilized at various concentrations of NaDC and NaC in water at 37°C [4].

		BS/mM					
		25	50	75	100	125	150
Ch + Stig							
γ_{Ch}	NaDC	0.941	0.950	0.941	0.937	0.937	
	NaC		0.922	0.929	0.922	0.916	0.916
Ch + Chsta							
γ_{Ch}	NaDC		n.d.	n.d.	n.d.	n.d.	
	NaC					n.d.	n.d.

the crystal structure of cholic acid (CA) and deoxycholic acid (DCA), an inclusion space is offered among belly (α-side of steroid skeleton)-to-belly type dimeric units. This may be true to some extent for structure formation of not only single species micelles but also sterol or stanol solubilizing micelles, even though the hydrogen of the carboxyl group at carbon number 24 (C-24) does not join the hydrogen bonding networks because of its electrolytic dissociation at pH 10. However, when BS micelles are formed apart from crystals in water, most hydroxyl groups (3α, 7α, 12α for NaC and 3α, 12α for NaDC) and C-24 carboxyl group may contact with surrounding water molecules and form hydrogen bonding with water. That is to say, not the belly-to-belly type dimer but the back-to-back type dimer must be a more stable unit. The back-to-back dimer formation as an assembling unit supporting the Small model [46] has been revealed by measurements of membrane potential [2] and gel filtration chromatography [47]. In addition, stepwise increase of aggregation number as 2, 3, 4, 8 and 16, is clearly indicated in micelle formation of NaC [2]. Although the aggregation number of NaDC depends greatly on given conditions such as pH, ionic strength, temperature, etc. [48,49], it mostly ranges around 15 to a few tens.

It should be noted here that Chun and Jou have studied molecular mechanics simulation of CA micelles [50], demonstrating graphic representation of sequential CA micelle (from dimer to dodecamer) formation by monomeric units. The layer structure of CA micelle is highly dissimilar, forming channels of irregular size and shape in a somewhat helical structure. Their simulated model is not the same as the model proposed by Small [46] and supported by Sugioka *et al.* [2] and Funasaki *et al.* [47]. Anyway, according to Chun and Jou in view of the dimensions of channels, guest molecules such as phospholipid, cholesterol, and phenanethrene are likely to be incorporated into the micelle through more than one channel, forming inclusion complexes such as gallstones [11].

The data on the reciprocal S_p that corresponds to the number of BS molecules required to solubilize one solubilizate molecule in Table 1 would allow us to assume that one solubilizate molecule on average is solubilized by one BS micelle, though the number of solubilizate molecules solubilized per BS micelle ranges from zero to three. Even if zero to three solubilizate molecules are solubilized by one BS micelle, the better affinity to BS micelle of one solubilizate than that of another must be recognized by both the solubilizate and the micelle when a restricted space inside the micelle is allowed for a solubilizate molecule to occupy; this may lead to the selection between sterols and stanols.

When Ch is compared with Chsta, Ch is somewhat superior to Chsta in solubilized amount because the double bond of Ch, which is slightly hydrophilic as compared with the B-ring of Chsta, may lead Ch to a favorable positioning or orientation toward the solubilizing site of BS micelles (see Fig. 1). However, when the total figure of B–C–D rings with a side chain is taken into consideration, the difference at carbon 5, double-bonded for Ch and hydrogen-saturated for Chsta, is likely to divide sterol/stanol mixtures into types of more and less easily penetrating toward the given hydrophobic space in micelles. In the case of the Ch/Chsta mixture, the steric structure of B–C–D rings of Chsta seems to avoid steric hindrance more easily than that of Ch when entering into the hydrophobic part inside the micelle. Although while A/B rings take the *trans* form in Chsta and the *cis* form in BSs, more suitable spaces needed for hydrophobic interaction may be given to Chsta. This may allow Chsta molecules to occupy more firmly and more stably the solubilizing sites of micelles than Ch, and consequently, Ch molecules cannot find any more room for solubilization. Sugano *et al.* [51] and Ikeda *et al.* [52] have shown an interesting comparison between sitosterol and sitostanol in regard to their plasma Ch concentration lowering effect using four types of feeds containing (a) Ch only, (b) Ch + sitosterol, (c) Ch + sitostanol and (d) no Ch added. They found the Ch + sitostanol system to be most effective in rats and rabbits. This also indicates that stanol can reduce the micellar solubilization of Ch more greatly than sterol.

Regarding the effect of side chain length on solubilization, Stig, which has the same steroid skeleton as Ch, shows the lowest solubilized amount among the single species. This suggests that the length, bulkiness and flexibility (presence or absence of double bond) of side chain play a role in sterol/ stanol solubilizing-micelle formation and that the side chain attached to the steroid skeleton is more deeply inserted into the hydrophobic space and interacts directly with the inside of micelles. In this sense, the bulky and non-flexible side chain of Stig can fit within the hydrophobic space of micelle interior to somewhat less extent than that of Ch, interacting more weakly with it. When Ch and Stig coexist, however, their side chains seem to interact hydrophobically in a rather favorable way for their penetration into the hydrophobic space, thus the total solubilized amount of the Ch/Stig mixed system is greater than that of the Ch/Chsta mixed system.

The present study is summarized as follows.

1. Thermodynamic evaluation. Solubilization power (S_p) could be related to the Gibbs energy change of solubilization

$$\Delta G_s^\theta = -RT \ln S_p \quad \text{or} \quad \Delta G_s^{\theta(mix)} = -RT \ln S_p(mix)$$

For mixed systems, the miscibility was evaluated in terms of excess Gibbs energy (enthalpy term) of mixing, $\Delta G_s^{(ex)}$, activity coefficient γ_i and interaction parameter ω. As expected, ω values were negative for Stig/Ch and positive for Chsta/Ch mixed solubilizate systems.

2. From a 1 : 1 mixture of Ch with Chsta, Chsta was selectively solubilized (Ch is selectively excluded) by BS micelles.

3. The composition in the solubilized 1 : 1 mixed system of Stig and Chsta could not be determined by the present method, because the HPLC spectrum of Stig overlapped with that of Chsta. However, the solubilized amount of the mixed system was found greater than that of the respective single species.

REFERENCES

1. Y. Moroi, Micelles. Plenum Press, New York and London (1992), Chapter 9.

2. H. Sugioka, and Y. Moroi, *Biochim. Biophys. Acta* **1394**: 99 (1998).

3. S. Nagadome, Y. Okazaki, S. Lee, S. Sasaki, and G. Sugihara, *Langmuir* **17**: 4405 (2001).

4. G. Sugihara, K. Yamakawa, Y. Murata, and M. Tanaka, *J. Phys. Chem.* **86**: 2748 (1982).

5. P. Zuman, and A. Fini, In Hinze, W. L., (ed.), *Organized Assemblies in Chemical Analysis: Bile Acid/Salt Surfactant Systems*, JAI Press, Stamford, CT (2000), Vol. 2. Chap. 3.

6. M. Senoo, *Entropy.* Kyoritsu Shuppan, Tokyo (1993), pp. 114–122.

7. W. J. Hinze, W. Hu, F. H. Quina, and I. U. Mohammadzai, In W. L. Hinze (ed.), *Organized Assemblies in Chemical Analysis: Bile Acid/Salt Surfactant System.* JAI Press, Stanford, CT (2000), Chap. 1.

8. M. C. Carey, In H. Danielsson, and J. Sjovall (eds), *Steroid and Bile Acids.* Elsevier, New York (1985).

9. R. A. Kroc, R. L. Kroc, G. D. Wheedon, and W. Garey, *Hepatology* **4**: 1S–252S (1984).

10. H. Danielsson (ed.), *Sterols and Bile Acids; New Comprehensive Biochemistry.* Elsevier Science, New York (1986), Vol. 12.

11. H. Igimi, and M. C. Carey, *J. Lipid Res.* **22**: 254 (1982).

12. E. Pramauro, and E. Pelizzetli, In S. G. Weber (ed.), *Surfactants in Analytical Chemistry, Vol XXXI of Wilson and Wilson's Comprehensive Analytical Chemistry Series.* Elsevier, New York (1996).

13. U. Wosiewits, and S. Schroebler, *Experientia* **35**: 717 (1979).

14. M. L. E. McBain, and E. Hutchinson, *Solubilization and Related Phenomena.* Academic Press, New York (1995).

15. A. Bandyopadhyay, and S. Y. Moulik, *J. Phys. Chem.* **95**: 4529 (1991).

16. M. A. Schwarz, R. H. H. Neubert, and H. H. Ruttinger, *J. Chromatogr.* **745**: 135 (1996).

17. H. Kawamura, M. Manabe, T. Tokunoh, H. Saiki, and S. Tokunaga, *J. Soln. Chem.* **20**: 817 (1991).

18. X. Tan, and S. Lindenbaum, *Int. J. Pharmaceutics* **74**: 127 (1991).

19. C. T. Yim, X. X. Zhu, and G. R. Brown, *J. Phys. Chem.* **103**: 597 (1999).

20. F. M. Menger, and M. McCreey, *J. Amer. Chem. Soc.* **258**: 6362 (1974).

21. S. Shinkai, and T. Kunitake, *Bull. Chem. Soc. Jpn.* **44**: 3086 (1971).

22. J. D. McIntyre, F. Schroeder, and W. D. Bohnke, *Biophys. Chem.* **38**: 143 (1990).

23. E. Mukidjam, S. Barnes, and G. A. Elgavish, *J. Amer. Chem. Soc.* **108**: 7082 (1986).

24. S. M. Meyerhoffer, T. J. Wenzel, and L. B. McGrown, *J. Phys. Chem.* **96**: 1961 (1992).

25. C. L. Lin, U. K. Jian, W. L. Higuchi, and N. A. Mazer, *J. Colloid Interface Sci.* **162**: 437 (1994).

26. M. C. Carey, J. C. Montet, M. C. Philips, M. J. Armstrong, and N. A. Mazer, *Biochemistry*, **20**: 3637 (1981).

27. N. A. Mazer, M. C. Carey, and G. B. Benedek, In K. L. Mittal and E. J. Fendler (eds), *Solution Behavior of Surfactants*, Vol. 1, Plenum Press: New York (1982), p. 595.

28. T. Saitoh, T. Fukuda, H. Tani, T. Kamidate, and H. Watanabe, *Anal. Sci.* **12**: 569 (1996).

29. K. D. Black, S. Kothari, P. A. Sharp, J. W. Quesnel, V. A. Escobar, W. E. Kurthin, and M. M. J. Bushey, *Liq. Chromatogr.* **23**: 113 (2000).

30. S. K. Wiedmer, M. L. Riekkola, M. Nyden, and O. Sodenam, *Anal. Chem.* **69**: 1577 (1997).

31. C. H. Spink, and S. Colgan, *J. Phys. Chem.* **87**: 888 (1983).

32. S. Nagadome, H. Miyoshi, G. Sugihara, Y. Ikawa, and H. Igimi, *J. Jpn. Oil Chem. Soc.* **39**: 542 (1990).

33. G. Sugihara, D. S. Shigematsu, S. Nagadome, S. Lee, Y. Sasaki, and H. Igimi, *Langmuir* **16**: 1825 (2000).

34. J. L. Weihrauch, and J. M. Gardner, *J. Am. Diet Assoc.* **73**: 39 (1978).

35. I. Heinemann, G. Axtmann, and K. Von Bergmann, *Europ. J. Clinical Invest.* **23**: 827 (1993).

36. P. J. H. Jones, D. E. MacDougall, F. Ntanios, and C. A. Vanstone, *Can. J. Physiol. Pharmacol.* **75**: 217 (1997).

37. G. V. Vahouny, and Kritchevsky, In G. Spiller (ed.), *Nutritional Pharmacology*, Alan R. Liss, New York (1981), p. 32.

38. (a) I. Ikeda, K. Tanaka, M. Sugano, G. V. Vahouny, and L. L. Gallo, *J. Lipid Res.* **29**: 1573 (1988). (b) I. Ikeda, K. Tanaka, M. Sugano, G. V. Vahouny, and L. L. Gallo, *J. Lipid Res.* **29**: 1583 (1988).

39. N. Goto, H. Mori, Y. Katsuragi, T. Toi, T. Yasukawa, and H. Shimasaki, *Nippon Yukagaku Kaishi (J. Jpn. Oil Chem. Soc.)* **48**: 235 (1999).

40. H. Hibino, *J. Jpn. Oil Chem. Soc.* **46**: 1127 (1997).

41. M. Sugano, F. Kamo, I. Ikeda, and H. Morioka, *Atherosclerosis* **24**: 301 (1976).

42. I. Ikeda, and M. Sugano, *Biochim. Biophys. Acta* **732**: 651 (1983).

43. K. Tanaka, I. Ikeda, and M. Sugano, *Biosci. Biotech. Biochim.* **57**: 2059 (1993).

44. K. Shinoda, *Yoeki to Yokaido (Solution and Sobulility)*, Maruzen, Tokyo (1974), pp. 99–101.

45. (a) M. Miyata, K. Sada, and Y. Miyake, In W. L. Hinze (ed.), *Organized Assemblies in Chemical Analysis: Bile Acid/Salt Surfactant Systems*. JAI Press, Stamford, CT (2000), Chap. 7, p. 205. (b) M. Miyata, and K. Sada, *J. Jpn. Oil Chem. Soc.* **49**: 447 (2000).

46. D. M. Small, In E. D. Goddard (ed.), *Molecular Association in Biological and Related Systems*. ACS Advances in Chemistry Series 84, American Chemical Society, Washington D.C. (1968), Chap. 4.

47. N. Funasaki, R. Ueshiba, S. Hada, and S. Neya, *J. Phys. Chem.* **98**: 1541 (1994).

48. Y. Murata, G. Sugihara, N. Nishikido, and M. Tanaka, In K. L. Mittal and E. Fendler (eds), *Surfactants in Solution.* Plenum Press, New York (1984), Vol. 2, p. 455.

49. J. P. Kratohvil, *Hepatology* **4**: 855 (1984).

50. (a) P. W. Chun, In W. L. Hinze (ed.), *Organized Molecular Assemblies in Chemical Analysis: Bile Acid/Salt Surfactant Systems.* JAI Press: Stamford, CT (2000), Chap. 6, p. 187. (b) W. S. Jou, P. W. Chun, *J. Mol. Graphics* **9**: 237, 240 (1991).

51. M. Sugano, H. Morioka, and I. Ikeda, *J. Nutr.* **107**: 2011 (1977).

52. I. Ikeda, A. Kawasaki, K. Samejima, and M. Sugano, *J. Nutr. Sci. Vitaminol.* **27**: 243 (1981).

9

Phase Behavior and Microstructure in Aqueous Mixtures of Cationic and Anionic Surfactants

ERIC W. KALER

Department of Chemical Engineering,
University of Delaware, Newark, Delaware, U.S.A.

KATHI L. HERRINGTON

W. L. Gore and Associates, Newark,
Delaware, U.S.A.

DANIEL J. IAMPIETRO

Merck and Company, Rahway,
New Jersey, U.S.A.

BRET A. COLDREN

Advanced Encapsulation, Santa Barbara,
California, U.S.A.

HEE-TAE JUNG

Department of Chemical and Biomolecular
Engineering, Korea Advanced Institute of
Science and Technology, Daejon, Korea

JOSEPH A. ZASADZINSKI

Department of Chemical Engineering and
Materials, University of California,
Santa Barbara, California, U.S.A.

1. SYNOPSIS

Mixtures of anionic and cationic surfactants in water display interesting phase behavior and a range of microstructures, including small micelles, rod-like micelles, lamellar phases and vesicles. This chapter reviews the properties of these mixtures with a focus on the experimental and theoretical aspects of vesicle formation and stability.

I. INTRODUCTION

Electrostatic forces are typically the dominant interaction in colloidal systems, and application of surfactants in solution or at surfaces calls for early consideration of the desired sign of the surface charge. It is no surprise then that oppositely charged surfactants in solution mix in a highly non-ideal way, and in fact are able to form spontaneously structures (in particular, unilamellar vesicles) that are uncommon in other surfactant mixtures. The range of self-assembled microstructures in mixtures of cationic and anionic surfactants also includes small spherical micelles, cylindrical or worm-like micelles, and other bilayer lamellar or L_3 phases. The state of equilibrium of all of these microstructures has not yet been fully established, but some vesicle phases have been observed to be stable for well over a decade. Although the folklore in surfactant formulations suggests that oppositely charged surfactants should never be mixed because of the potential for precipitation of the insoluble surfactant ion pair, precipitation is generally found only near equimolar compositions or in samples below their Krafft temperature, and even this precipitation can sometimes be blocked.

Of particular interest in this review are mixtures of oppositely charged surfactants from which the companion co-ion simple salt has not been removed. This is in contrast to true 'catanionic' surfactants, as first described by Jokela and Wennerstrom [1], in which solutions contain only water and the two oppositely-charged surfactant ions. These catanionic surfactants are also called "ion-pair amphiphiles" (IPAs), and they are most easily formed by combining an organic acid and an organic hydroxide, so the common ancillary product is water rather than a salt. IPA properties have been reviewed recently by Tondre and Caillet [2]. A spectacular example of the assembly of IPAs is the observation by Zemb *et al.* [3] of self-assembled structures about 1 μm in dimension having a regular hollow icosahedral form. The icosahedral particles are stabilized by pores located at their vertices, and they dissolve when salt is added.

Historically the addition of oppositely charged moieties has been used most frequently to control the growth of rod-like or worm-like micelles. This is typically done by adding an anionic hydrotrope (such as salicylate or tosylate salt) to a solution of cationic surfactant [4]. The hydrotrope partitions into the surfactant headgroup region and results in dramatic micellar growth. Similarly, a cationic hydrotrope can drive the growth of anionic micelles [5]. As the molecular weight (or length) of the hydrophobic portion of the hydrotrope is increased, the hydrotrope begins to behave more like a second surfactant and can alter the self-assembly in solution, leading to the formation of vesicles. For example, when alkyl sulfates ($C_nSO_4^-$–Na^+) with n greater than six are added to alkyl ammonium halide surfactants, vesicles can form on both the cationic and anionic-rich side of the phase diagram [6]. Such vesicles form spontaneously and without the input of shear, and their structural and thermodynamic properties are discussed below. Both the static and dynamic properties of these mixtures have recently been reviewed [7].

Vesicles are, of course, not newly discovered. Nearly forty years ago, Bangham *et al.* showed that phospholipids dispersed in water formed closed, multibilayer aggregates called liposomes, capable of separating an internal compartment from the bulk solution [8]. As bilayers are relatively impermeable to many ions and nonelectrolytes, it became possible to create small domains of different composition and to explore some of the properties that nature has designed into cells and organelle membranes. Unilamellar vesicles are distinguished from multilayer liposomes by their single-bilayer closed shell structure that encapsulates a single aqueous compartment. While vesicles often form spontaneously *in vivo*, they have only rarely been observed to form *in vitro* without the input of considerable mechanical energy or elaborate chemical treatments. Hence, a variety of methods have been developed to create vesicles with sizes ranging from about 20 nm to more than 20 μm [9].

The stability of such vesicles is limited because the bulk lamellar phase, which at high water fractions can exist as a colloidal dispersion of multilamellar liposomes, is the equilibrium form of aggregation under typical conditions. Hence, unilamellar vesicles formed from lamellar phases are metastable and eventually will revert to multilamellar liposomes. This reversion is invariably accompanied by a release of the vesicle contents and failure of the vesicle carriers. For multilamellar liposomes to be stable at high water fractions, the bilayers must have a net attractive interaction. The stability of mechanically formed vesicles against aggregation can be enhanced in many of the same ways that other colloidal systems are stabilized. These include the incorporation of bulky, polymer-like head groups on some fraction of the lipids [10–12], or the addition of charged lipids or surfactants to the bilayers to enhance electrostatic repulsion [13–15]. Both additives decrease the net bilayer attraction and help vesicles form spontaneously or with little energy input [16]. Once formed, unilamellar vesicles can also be made mechanically stable by polymerization of the head or tail groups [17,18]. These methods limit the kinetics of reversion, but leave the question of equilibrium stability unanswered.

There have been surprisingly few demonstrations of equilibrium vesicle phases in the literature, in comparison to the immense body of literature on vesicles and liposomes in general. By equilibrium, we refer to the following three criteria:

1. Unilamellar vesicles are formed spontaneously upon dispersing dry surfactant into water without mechanical or chemical perturbation;
2. Vesicles do not aggregate with time; and

3. Any physical or chemical process to which the vesicles are exposed will result in spontaneous reformation of unilamellar vesicles on reversing the process.

Surprisingly, nearly all of the several reports of spontaneous vesicle formation in the literature have involved surfactant mixtures. Single component bilayers usually form multilamellar phases, although there are reported exceptions [19]. Nonetheless, the question of equilibrium is a difficult one to address in these mixtures because the state of aggregation changes in some cases only on the time scale of years, and may thus reflect chemical degradation of the surfactant (e.g., hydrolysis) rather than a physical progression to a more thermodynamically stable structure.

The fundamental questions are why vesicles form in mixtures of oppositely charged surfactants, and why, once formed, do they remain stable? The formation and stability of any surfactant aggregate depends on whether that aggregate represents the global minimum in free energy for a given composition, and there are different approaches to describing the thermodynamics of these mixtures. As a way to begin, it is useful to start not with a molecular-level description, but instead with the more mechanical description of the properties of the bilayer first given by Helfrich [20]. In this case the elastic free energy of a bilayer is written in terms of its local state of curvature, described by the two principle curvatures c_1 and c_2. For spherical vesicles, $c_1 = c_2 = 1/R$, in which R is the vesicle radius. To terms second order in curvature, the free energy of a bilayer, per unit area, is

$$E/A = \tfrac{1}{2}\kappa(c_1 + c_2 - 2c_0)^2 + \bar{\kappa}c_1 c_2 \qquad (1)$$

in which κ is the bending modulus, $\bar{\kappa}$ is the saddle splay or Gaussian modulus, and the spontaneous curvature of the bilayer is c_0. By definition, $\kappa > 0$ but $\bar{\kappa} > 0$ for surfaces that prefer hyperbolic shapes (saddle shaped surfaces in which the centers of curvature are on opposite sides of the surface, $c_1 c_2 < 0$) and $\bar{\kappa} < 0$ for surfaces that prefer elliptical shapes (spheres, ellipsoids, etc., in which the centers of curvature are on the same side of the surface, $c_1c_2 > 0$) [21]. For a chemically and physically symmetric bilayer, the spontaneous curvature $c_0 = 0$. Hence, a single component bilayer cannot have a nonzero spontaneous curvature. It is necessary for the bilayer or its local environment to be chemically asymmetric for spontaneous curvature to exist [22].

Thermal fluctuations of the bilayers lead to a repulsive interaction as the bilayers come into contact and the are damped [23]. For bilayers separated by a distance, d, the so-called undulation interaction energy is [23]:

$$E_{\text{fluct}} = \frac{3\pi^2}{128} \frac{(k_B T)^2}{\kappa d^2} \tag{2}$$

The repulsive undulation interaction can overwhelm the van der Waals attraction between bilayers (which is also proportional to d^{-2}) when κ is small, leading to a net repulsive interaction between bilayers and hence, stable unilamellar vesicles, especially when combined with electrostatic repulsion in charged systems [24].

Within this Helfrich framework, vesicles may be made stable by either an entropic or an enthalpic mechanism. Entropically stabilized vesicles have a low bending constant ($\kappa \sim k_B T$, where k_B is Boltzmann's constant). The bending energy is therefore low and the population of vesicles is stabilized both by the entropy of mixing and the undulation interaction. The resulting size distribution is broad. Enthalpically stabilized vesicles, on the other hand, require a non-zero spontaneous curvature and a larger value of the bending constant ($\kappa > k_B T$). In this case the vesicles are narrowly distributed around a preferred size set by the spontaneous curvature. These theories are discussed in more detail below.

Work in our laboratories has demonstrated the existence of spontaneous, apparently equilibrium vesicles with both narrow [25–27] and broad size distributions [6,25,28–32] in aqueous mixtures of a wide range of surfactants. The most common cationic surfactant used has been an alkytrimethylammonium bromide or tosylate, e.g., cetyltrimethylammonium bromide (CTAB) or tosylate (CTAT) as well as cetylpryidinium chloride (CPCl), while the common anionic surfactants are the sodium alkylsulfates, e.g., sodium octyl (SOS) or decyl sulfate (SDS), or dodecylbenzene sulfonate (SDBS), which may have either a branched or comb structure [6,33]. In addition, vesicles form with CTAB when the anionic surfactant is a salt of perfluorinated carboxylic acid such as sodium salts of perfluorohexanote (FC_5) or perfluorooctanoate (FC_7) [25,34]. Unilamellar monodisperse equilibrium vesicles also form spontaneously when cholinergics (for example, choline chloride) are added to aqueous solutions of sodium bis[2-ethylhexyl] sulphosuccinate (AOT) [35]. In this case, the cationic "surfactant" is actually a hydrotrope and is water-soluble while the anionic surfactant alone forms bilayers in water. Thus, vesicle formation in this case is due to a change in curvature of the AOT bilayer induced by addition of the cholinergic compound.

A substantial number of other researchers have also explored vesicle formation in other mixtures of oppositely charged surfactants, and the recent review of Gradzielski [7] is especially comprehensive. Vesicles also form when a single-tailed anionic surfactant (SDS) is added to a double-tailed cationic surfactant (DDAB) [36], and these vesicles also form interesting phases and complexes in the presence of DNA and other polyelectrolytes [37]. Certain bacterial surfactants known as siderophores can undergo micelle to vesicle transitions on complexation of multivalent ions [38], which may be important for sequestering sufficient iron in marine environments. A range of zwitterionic/ionic pairs have been studied, as have surfactant pairs when one of the ionic species is produced by a chemical reaction [39]. The kinetics of formation of vesicles in these mixtures has been probed by light and small angle scattering as well as by cryo-TEM [40–42]. A variety of unstable rod and disk-like structures are found as the aggregate morphology progresses from simple micelles to vesicles.

This chapter is organized first around experimental observations of the surface chemistry of dilute mixtures, phase diagrams, and microstructure characterization. This section is followed by a review of the relevant theories and a discussion of the origin of stability of vesicles in cationic/anionic mixtures.

II. EXPERIMENTAL OBSERVATIONS

A. Surface Tensions and Nonideal Mixing

The first striking observation about cationic–anionic mixtures is their high degree of nonideality. This is clearly shown by the strong variation in critical aggregation concentration given by surface tension measurements [33]. For example, mixtures of CTAT and SDBS display critical aggregation concentrations (*cacs*) orders of magnitude lower than the individual Critical Micellization Concentration (CMC) values (Fig. 1). Mixtures of CTAT and SDBS also produce a lower surface tension than is observed for either surfactant alone. Visual and light scattering observation of samples with intermediate concentration a few times the *cac* are revealing. The samples appear somewhat blue in color, suggesting the presence of colloidal structures larger than micelles. This is borne out by light scattering measurements of dimensions of *ca.* 100 nm, and, as described below, unilamellar vesicles are the first aggregates to form in these highly dilute solutions.

Similar results hold for mixtures of SOS and CTAB. The CMC of pure SOS is 3 wt% (120 mM) and that of pure CTAB is 0.03 wt% (0.88 mM). The critical aggregation concentrations for mixtures of CTAB and SOS are significantly lower and range from 0.001 wt% to 0.002 wt% (3–7×10^{-5} M) [31], and mixtures of CTAB and SOS also produce a lower surface tension than either pure surfactant. Precipitate appears in samples in the vicinity of the *cac* after equilibration for several days, thus the *cac* for mixtures of CTAB and SOS corresponds to the solubility of the equimolar precipitate. The value of the solubility product $[= (a_{\text{CTA}+})(a_{\text{SOS}-})]$ for this salt, calculated from the dependence of the *cac* on bulk mixing ratio, is 9×10^{-10} mol^2/l^2, and similarly low values are expected for other surfactant ion pairs.

Surface Tension of CTAT/SDBS Solutions

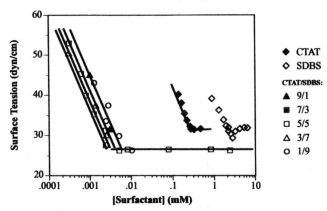

CAC from Surface Tension Measurements

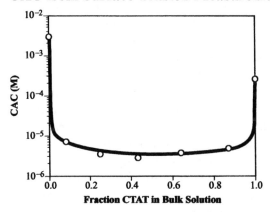

Figure 1 Surface tensions of SDBS, CTAT, and their mixtures at 25°C (top). The mixtures are highly nonideal, as shown by the variation of the critical aggregation concentration with CTAT fraction (bottom).

B. Phase Behavior

Aqueous mixtures of cationic (R^+X^-) and anionic surfactants (R^-X^+) are actually five-component systems according to the Gibbs phase rule: R^+X^-, R^-X^+, R^+R^-, X^+X^-, and water, and are subjected to an electroneutrality constraint. Therefore, the pseudoternary phase diagram for R^+X^-, R^-X^+, and water at constant temperature represents only a portion of the phase prism. The full prism is needed to represent compositions in multiphase regions when the surfactant ion and associated counterion separate into different phases, as is the case when precipitate forms. Nonetheless, when precipitate is not present the pseudo-ternary map is a useful guide, although the phase alternation rule [43] need not apply.

A canonical phase diagram for anionic and cationic mixtures is that of CTAT and SDBS (Fig. 2), where in this case the SDBS hydrophobe is of the "soft" type and has rake-like branching. (Hard-type SDBS has approximately a dodecane tail bonded to the aromatic ring at a single carbon.) The phase boundaries in Fig. 2 and all other phase diagrams of catanionic mixtures can only be established after visual observations remain unchanged over an extended period of time. Most compositions equilibrate within one to two weeks, but samples that contain vesicles or a viscous phase require longer equilibration times. Vesicle phases are identified first by their characteristic isotropic blue appearance and then by determining the mean diameter using quasielastic light scattering (QLS) to verify that aggregate sizes are in the range typical of vesicles. Cryo-transmission and/or freeze-fracture electron microscopy (cryo-TEM) can be used to confirm the unilamellar vesicle structure [26,32], as can small-angle neutron scattering methods [26,30]. The long equilibration time is necessary for the samples containing vesicles in order to distinguish between single-phase vesicle regions and two-phase vesicle/lamellar regions, since small amounts of lamellar structure develop slowly compared to other phases. Visual observations and QLS measurements over a time period of nine months or longer (in one case over 10 years) confirm the stability of the vesicle phase relative to a lamellar phase in one-phase regions.

At higher surfactant concentrations, Fig. 2 shows that the vesicle phases are in equilibrium with one of two lamellar phases: (1) CTAT-rich vesicle solutions (V+) are in equilibrium with a CTAT-rich lamellar phase ($L_\alpha+$) of lower density; and (2) SDBS-rich vesicle solutions (V−) coexist with an SDBS-rich lamellar phase ($L_\alpha-$) which undergoes an inversion in density

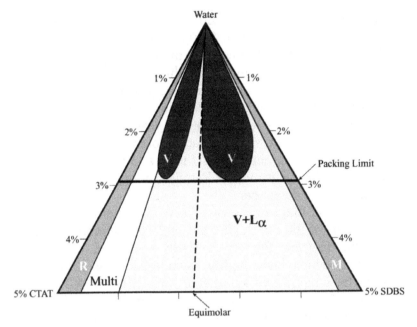

Figure 2 Ternary phase map for CTAT/SDBS/water at 25°C. One phase regions are marked by M, R, or V. The V region on the CTAT side of the diagram contains CTAT-rich vesicles that are positively charged, while the V region on the SDBS side of the diagram contains SDBS-rich negatively charged vesicles. Two-phase regions show vesicle and lamellar phase coexistence, and the "multi" region contains several viscous phases that are slow to separate. R corresponds to CTAT-rich rodlike micelles, and M to small globular SDBS-rich micelles. A precipitate forms on the equimolar line. Compositions are on a weight percent basis. The "packing limit" indicates the concentration at which vesicles of this size become close-packed (see Section IV.D). Modified from [33].

at $\sim 35/65$ CTAT/SDBS. Isotropic one-phase regions border each binary surfactant–water boundary. The rod-like micelle region on the CTAT side extends into the ternary diagram and is separated from the vesicle lobe by several multiphase regions. On the other side of the diagram, SDBS forms spherical micelles in water at the water concentrations shown in the phase diagram. Within the micellar region, the viscosity is low and constant and there is no increase in scattered light intensity as CTAT is added. Thus it is unlikely that adding CTAT causes significant micellar growth in SDBS-rich micellar samples. A two-phase region, in which L_α^- is in equilibrium with an isotropic solution presumably containing vesicles, separates the one-phase micellar region from the vesicle lobe. The general qualitative features of two vesicle lobes, which may differ significantly in size, an equimolar precipitation line, lamellar phases at higher concentration and micellar phases on the edges of the diagram are found in most cationic–anionic surfactant phase diagrams.

The phase diagrams evolve in a systematic way as the chain lengths of the two surfactant molecules change. Figure 3 shows the diagram for CTAT/SDBS compared to that for CTAB/SOS, which have highly asymmetric tails [31] and to DTAB/SDS, which have symmetric tails [6]. The CTAB/SOS diagram displays a very large vesicle lobe on the SOS-rich (shorter hydrophobic chain) side of the diagram, while there is only a small composition range wherein CTAB-rich vesicles form. The DTAB/SDS phase diagram is dominated by a large composition range wherein precipitate forms. As the ionic head groups are the same in the two surfactant pairs, this strongly suggests an important thermodynamic role in vesicle stability for the chain packing configurations within the hydrophobic core of the bilayer.

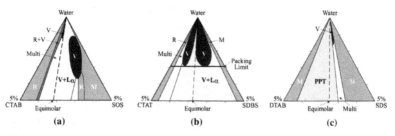

Figure 3 Ternary phase maps of three cationic-anionic mixtures in water at 25°C. CTAB/SOS (left) shows a large vesicle lobe on the SOS-rich side of the diagram along with other phases identified in Fig. 2 (and see Fig. 4). The DTAB/SDS mixture (right) shows a wide range of precipitate formation.

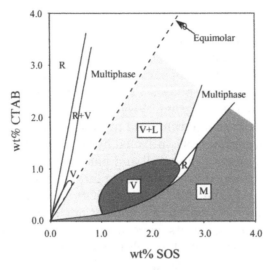

Figure 4 Phase behavior of CTAB/SOS/H$_2$O with no added salt plotted on a rectangular diagram. Dotted lines represent equimolar CTAB/SOS composition, at 61.4% CTAB. One-phase vesicle lobes (V) exist at dilute CTAB-rich and SOS-rich compositions. Samples here appear bluish and are isotropic. One-phase rod-like (R) and spherical (M) micelles form near the CTAB and SOS axes, respectively. Rod-like and spherical micellar phases are both clear, yet scatter more light than pure water. Rod-like micellar samples are viscous and viscoelastic. At intermediate mixing ratios, much of the phase behavior is dominated by vesicles in equilibrium with a lamellar phase (L), which appears as birefringent clouds above the vesicles. The CTAB-rich R and V phases are separated by a narrow two-phase region of rods and vesicles in equilibrium. SOS-rich micelles transform abruptly to vesicles at most concentrations, though around 3.0 wt%, an intervening region of rod-like growth occurs. Unresolved multiphase regions are at concentrations above those of the vesicle lobes. Modified from [29].

The implications of the CTAB/SOS diagram are seen more clearly when plotted in rectangular coordinates (Fig. 4) [29]. When samples are initially prepared, vesicles appear to form over a wider range of compositions. However, as the samples age, the range of compositions that yields stable vesicles shrinks considerably, particularly at the lamellar phase boundary. Small amounts of the lamellar phase form in two-phase samples close to the vesicle phase, and become visible only after aging for days or weeks. Boundaries are assigned only when the sample appearance does not change with time. As expected, micellar phases exist on the binary surfactant–water axes of the diagram. Vesicles form in the water-rich corner of the phase diagram, and are found in both CTAB-rich and SOS-rich samples as confirmed by QLS and cryo-TEM. The CTAB-rich vesicle lobe is small and narrow in extent, while the SOS-rich vesicle lobe is considerably larger. Note that the lobe extends to nearly the CMC value of SOS, so that compositions where vesicles form all have compositions such that the SOS concentration is below its pure component CMC. As SOS is added to CTAB-rich micellar solutions, there is strong rod-like micellar growth as indicated by increased viscosity and viscoelasticity. Samples become increasingly more viscoelastic at high surfactant concentrations. SOS-rich micelles are spherical at low amounts of added CTAB, while at higher ratios of CTAB to SOS and higher concentrations, rod-like micelles form.

The micelle-to-vesicle phase transition is of considerable interest. For CTAB-rich samples and SOS-rich samples at higher surfactant concentration, there is an intervening two-phase region of rod-like micelles and vesicles. Samples separate into two phases: one phase scatters more light than a micellar solution and is viscous, while the other phase is not viscous. Depending on the composition, the appearance of the second phase ranges from clear and colorless to bluish and somewhat turbid. At higher surfactant concentrations, samples contain more than two phases and may contain vesicles, rod-like micelles, and liquid crystalline microstructures. SOS-rich samples exhibit different behavior for low surfactant concentrations (see Fig. 4). In these samples there is limited micellar growth with added CTAB or increased dilution, and the colorless micellar solutions progressively scatter more light than micellar solutions of pure SOS as the micellar phase boundary is approached. Samples become noticeably turbid over a very narrow increment of concentration, and the phase boundary between micelles and vesicles is set at the point at which samples appeared turbid. In no case is an intermediate two-phase region observed, so this is not a first-order phase transition.

A crystalline precipitate, presumably the equimolar salt $CTA^+ : OS^-$, forms in equimolar mixtures as well as in dilute (<0.1–0.2 wt%) solutions at all mixing ratios. Dilute samples in the vesicle phase close to the precipitate phase boundary often contain turbid wisps that are so easily dispersed by mixing that attempts to characterize them with optical microscopy are unsuccessful. Because of the proximity of the precipitate phase boundary, the wisps may be a fine precipitate, or a dilute dispersion of multilamellar vesicles (MLVs).

Electrostatic attractions between the cationic and anionic surfactants are responsible for this interesting phase behavior, so it is no surprise that addition of excess electrolyte markedly alters the phase behavior of catanionic solutions [29]. This has been examined in some detail for mixtures of CTAB/SOS with added NaBr (Fig. 5). The most conspicuous effect is the destabilization of the large one-phase vesicle lobe with increasing amounts of added NaBr. The vesicle region shrinks in all dimensions. Vesicles still form in the water-rich corner of the lobe, but small turbid clouds form over the vesicle phase after the addition of salt. These clouds are either lamellar or multilamellar vesicle (MLV) phases, or a fine dispersion of precipitate. Adding NaBr also drives the two-phase lamellar/ vesicle region on the CTAB-rich side of the vesicle lobe to more SOS-rich compositions. The lamellar phase becomes increasingly favorable as salt is added, as observed visually by the increasing predominance of a turbid white birefringent phase over the vesicle phase. This is consistent with a decreased electrostatic interaction between the vesicle bilayers leading to a more attractive interaction [44]. The vesicle/micelle phase boundary, however, shifts to compositions richer in CTAB with increasing NaBr. Thus, overall, the most SOS-rich section of the larger vesicle lobe becomes unstable with respect to the neighboring micelle phase, while the most CTAB-rich section of the lobe becomes unstable with respect to the lamellar phase. It is striking that the addition of salt in some cases drives the vesicles to form small micelles rather than larger MLVs or

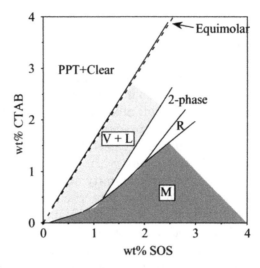

Figure 5 Phase behavior of CTAB/SOS/H$_2$O with 4.0 wt% NaBr plotted on a rectangular diagram. The entire CTAB-rich side of the phase diagram is dominated by a clear isotropic liquid in equilibrium with precipitate. Samples here are no longer viscous or viscoelastic. The one-phase micellar region M extends farther toward more CTAB-rich mixing ratios. The multiphase region is at some compositions a clear streaming birefringent phase over an isotropic bluish phase and at other compositions is a clear isotropic phase that scatters more light than water over a clear phase. Modified from [29].

lamellar phases, as would be the prediction based on either colloidal stability arguments (aggregation) or by appeal to analogy to charged phospholipids, in which salt addition drives fusion.

The differences in behavior of mixtures of homologous surfactants with identical headgroups but different hydrocarbon tails further points to the important role of the packing of surfactant tails in the bilayer. This can be explored more directly by examining mixtures of a hydrogenated cationic surfactant with a fluorinated anionic surfactant [25,26]. The antipathy of the tails, and even their potential demixing, is balanced by the attractions between the head groups. Figure 6 shows the phase map for the mixture of CTAB/FC$_5$/water in the water-rich corner (<8 wt%) at 25°C. The CMC of each

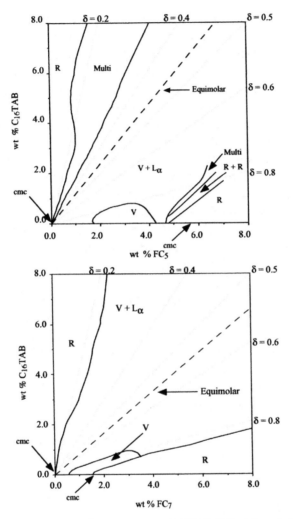

Figure 6 Phase maps for mixtures of CTAB/FC$_5$ (a) and CTAB/FC$_7$ (b) measured at 25°C. The gray lines in each graph correspond to constant weight mixing ratios (δ) as indicated. The CMC for each surfactant is shown by the arrows on the binary axis. At intermediate mixing ratios in each system, the phase behavior is dominated by a two phase region consisting of vesicles in equilibrium with a denser lamellar phase (V + L$_\alpha$). Rodlike micelles (R) exist on the CTAB-water binary axis for both FC$_5$ and FC$_7$ mixtures. The most important feature is the existence of a single vesicle phase (V) on the fluorinated surfactant-rich side of the phase map.

surfactant is indicated by an arrow on the appropriate binary axis and the gray lines represent constant values of the surfactant mixing ratio, $\delta(= \text{wt.} FC_5/(\text{wt.} FC_5 + \text{wt.} CTAB))$). The majority of the phase space at intermediate values of δ, including equimolar mixtures, is composed of a two-phase region denoted by $(V + L_\alpha)$ containing a vesicle phase in equilibrium with a denser, birefringent lamellar phase. When CTAB is present in large excess, a viscoelastic solution of long rodlike micelles is formed. The phase behavior on the FC_5-rich side on the phase map is more interesting. Along the FC_5–water binary axis $(\delta = 1)$, at concentrations below the CMC, the surfactant presumably exists as monomers. Over the majority of this range, the addition of small amounts of CTAB leads to the formation of larger aggregate including vesicles between 2 and 4 wt%. The vesicle region extends to about $\delta = 0.8$. Vesicles do not form in FC_5–water binary mixtures, but quasielastic light scattering measurements and cryo-TEM images indicate that large structures form with the addition of only a small amount of CTAB $(\delta \sim 0.995)$. Thus on this scale, the vesicle lobe appears to almost touch the binary axis. For δ below about 0.80, the samples also contain a more dense birefringent phase, with the phase boundary between the vesicle lobe (V) and the two-phase region $(V + L_\alpha)$ set by the formation of this dense phase. Mixtures of CTAB and FC_7 show a similar phase diagram, and neither of these mixtures shows a substantial change in phase behavior with temperature, suggesting that there is no demixing of the tails.

C. Microstructure Characterization

Once phase behavior has been established the microstructure can be explored with a variety of techniques. Especially useful methods are quasielastic light scattering, which provides a rapid size estimate and can be used to track the evolution of structure with time; small angle neutron scattering, which can provide information about aggregate shape, structure and, under ideal conditions, size distribution; and cryo-TEM, which can provide unambiguous size, shape, and size distribution information.

As an example of the evolution of apparent vesicle radius determined by QLS as a function of time, consider the properties of CTAB/FC$_5$ and FC$_7$ mixtures. Figure 7(a) shows the measured hydrodynamic radius plotted as a function of the sample age for CTAB/FC$_5$ concentrations 3 wt% and $\delta = 0.80$, 0.90, and 0.99. At the two higher δ values, vesicle growth is observed over the first 50 days with a leveling off to an equilibrium size seen at longer times. For the lowest δ, there is a more dramatic change in size initially and the measured size continues to increase even after 110 days. In these samples a lamellar phase is eventually observed visually. The situation in the FC$_7$ case is different, with vesicles at 2 wt% concentration reaching a constant size more quickly at δ values of 0.8 and 0.85.

Cryo-TEM provides the simplest and most model-free measure of microstructure, although artifacts due to sample preparation are possible [45]. Figure 8 is a cryo-micrograph of a 2 wt% CTAB/SOS mixture (7 : 3 wt : wt). This sample shows unilamellar vesicles with radii varying from about 20 to >100 nm. Note that the bilayers appear flexible and the vesicles are not all spherical. Such shapes may be the equilibrium conformations of vesicles either because of the low bending rigidity of mixed bilayers or may reflect deformations due to shear during sample preparation. The radii of the vesicles imaged are in good agreement with those measured with QLS, although the cryo-TEM average radius is typically less than the QLS average. It is extremely important to note that QLS sizes measure a higher moment of the size distribution than the number average measured by cryo-TEM, so the QLS size is very sensitive to the presence of negligible (below 1 part in 1000) numbers of larger aggregates [32], and this may explain some size growth reported by others [46]. Such large aggregates can result from even a slight degree of chemical or physical degradation of the surfactants.

Small-angle neutron scattering (SANS) is also a useful method for determining average vesicle dimensions and, to some extent, size distributions [47] and is very complementary to cryo-TEM. SANS scattering spectra (intensity I as a function of the scattering vector magnitude q) can be analyzed either

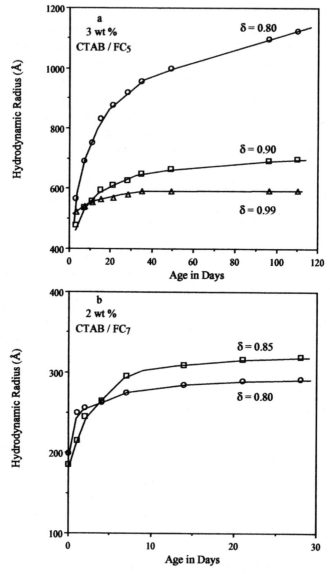

Figure 7 Quasielastic light scattering results for $CTAB/FC_m$ vesicles. (a) $CTAB/FC_5$ vesicles at a surfactant concentration of 3 wt% and δ values of 0.99, 0.90, and 0.80. (b) $CTAB/FC_7$ vesicles measured at a surfactant concentration of 2 wt% and δ values of 0.80 and 0.85. The lines are shown to guide the eye. All measurements were performed at 25°C.

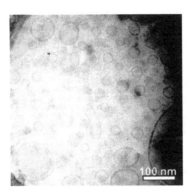

Figure 8 Cryo-TEM image of $7:3$ (wt:wt) CTAB:SOS vesicle phase at $2\,\mathrm{wt\%}$ total surfactant in water. The broad size distribution and formation of floppy and odd shaped vesicles is consistent with a bending constant, $K = 0.2k_{\mathrm{B}}T$, and a spontaneous curvature, $R_0 = 30\,\mathrm{nm}$ [27].

by fitting a model function describing the presumed structure to the data, or by using an indirect fourier transform (IFT) method to extract model-free estimates of microstructure dimensions. Both methods have been used to analyze the spectra from samples known (by cryo-TEM) to contain vesicles. Generally spectra are fitted using a polydisperse core-shell model [48] where the vesicles are assumed to have a poly-disperse core with constant shell thickness (t). For a poly-disperse system of unilamellar non-interacting vesicles, the scattered intensity as a function of the scattering vector is given by

$$I(q) = \frac{\mathrm{d}\sum}{\mathrm{d}(q)} = n \int_0^\infty G(r_{\mathrm{c}})P^2(qr_{\mathrm{c}})\,\mathrm{d}r_{\mathrm{c}} \qquad (3)$$

where n is the number density of vesicles, $P(qr_{\mathrm{c}})$ is the form factor of a single particle (e.g., vesicles) consisting of a core and an outer shell, and $G(r_{\mathrm{c}})$ is the normalized probability of finding a particle with a core radius between r_{c} and $r_{\mathrm{c}} + \mathrm{d}r_{\mathrm{c}}$. $G(r_{\mathrm{c}})$ is modeled as a Schulz distribution, so

$$G(r_{\mathrm{c}}) = \frac{r_{\mathrm{c}}^Z}{\Gamma(Z+1)}\left(\frac{Z+1}{\bar{r}_{\mathrm{c}}}\right)^{Z+1}\exp\left(\frac{-r_{\mathrm{c}}}{\bar{r}_{\mathrm{c}}}(Z+1)\right) \qquad (4)$$

where $\bar{r}_c$ is the mean core radius and Z is related to the polydispersity of the particles (p) and the variance (σ^2) of the core radius by

$$p^2 = \frac{1}{Z+1} = \frac{\sigma^2}{\bar{r}_c^2} \tag{5}$$

The form factor is

$$P(x) = \frac{4\pi}{x^3}(\rho_b - \rho_c)\left\{J_1\left(x + \frac{t}{r_c}x\right) - J_1(x)\right\} \tag{6}$$

where x is the dimensionless variable qr_c, $J_1(x) = \sin(x) - (x)\cos(x)$, and ρ_b and ρ_c are the scattering length densities (SLDs) of the bilayer and the core (taken as the solvent), respectively. The SLDs were calculated by adding the scattering amplitudes of each group or atom in a molecule and dividing the total by the corresponding molecular volume. The SLD of the bilayers is calculated assuming the bilayer is made of an equimolar composition of the oppositely charged components. This assumption has little influence on the results. Interestingly, use of one deuterated surfactant allows the SLD of the bilayer to be changed, and combining this information with mass balances allows the composition of the vesicle to be determined [49]. The vesicle composition is not, of course, generally equal to the bulk composition.

Typical scattering spectra for polydisperse vesicles are relatively featureless and show only a broad q^{-2} decay of intensity, which is consistent with the presence of a two-dimensional bilayer (Fig. 9a). More monodisperse populations (Fig. 9b) yield spectra with undulations that reflect the oscillations of the form factor. In this case, the vesicles are about 20–22 nm in number average radius with *ca.* 20% polydispersity. Further analysis shows the bilayers to be rather thin (2–3 nm), and IFT methods allow the cross-sectional SLD profile to be determined [34].

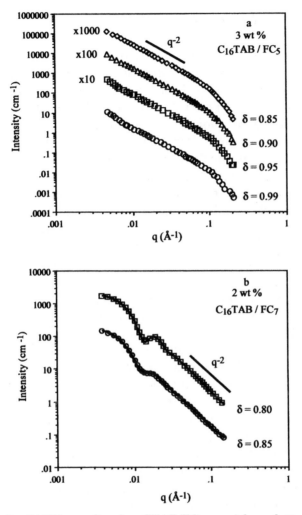

Figure 9 SANS results for CTAB/FC$_m$ vesicles plotted as the scattering intensity in absolute units versus the scattering vector q. (a) CTAB/FC$_5$ vesicles measured at a surfactant concentration of 3 wt% and δ values ranging from 0.85 to 0.99. The top three curves have been offset by the indicated scale factors for clarity. Each scattering curve shows the characteristic q^{-2} decays as indicated by the line in the graph. (b) CTAB/FC$_7$ vesicles measured at a surfactant concentration of 2 wt% and δ values of 0.80 and 0.85. The top curve is offset by a factor of 10 for clarity. The q^{-2} region is observed at higher q values and minima from the form factor are evident around $q \sim 0.01$–0.02.

III. THEORETICAL APPROACHES

Much work has focused on development of theoretical
models that can account for the behavior of surfactant solutions. These models range from the simple but useful geometric
packing model of Tanford [50] and Israelachvili [44] to
more quantitative molecular thermodynamic models. The
level of detail required from a model varies widely with the
circumstance of its application. A simple conceptual model is
useful when attempting to organize results or guide experi-
mentation, and both the geometric packing and the curvature
elasticity models fulfill this need. These models require
the input of parameters that characterize the surfactant
properties, and given these parameters, both models can be
used to rationalize observations and to suggest an experimental
plan.

Often, solution properties such as micellar composition or
monomer concentrations for mixtures of surfactants are
required for a given surfactant application. In this case, a
model based on classical thermodynamics in which aggregates
are treated as a single phase in equilibrium with an aqueous
monomer phase may be suitable. In this approach, deviations
from ideal mixing in the aggregate phase are accounted for by
activity coefficients for each surfactant in the mixed aggregate
[51,52]. Activity coefficients can be calculated using a suitable
theory for nonideal mixing (such as regular solution theory)
with interaction parameters determined from the experimental values of the CMC of each surfactant and the *cac*
measured for at least one mixing ratio. Finally, there are
powerful predictive models based on a detailed statistical
thermodynamic analysis of aggregate formation [53–58].
These molecular thermodynamic models are capable of predicting the equilibrium state of aggregation, aggregate sizes
and composition, and equilibria between multiple phases,
given data about the pure surfactants that is readily available
in the literature.

The surfactant inventory in monomeric and aggregated
(micellar or vesicular) form is needed in many applications of
surfactant mixtures. The inventory may be calculated in

several ways, with each method having a different degree of complexity and accuracy. In a first approach, micellar size distributions can be computed by considering each aggregate to be a distinct chemical species. From thermodynamics, the chemical potential of a surfactant molecule has the same value regardless of the size of the aggregate it is in. Combining the condition of chemical equilibrium with a mass balance on surfactant yields the aggregate size distribution. This approach is the basis of the mass action model, which has been widely applied in the analysis of surfactant properties to predict the micellar size distribution, monomer concentration, electrical conductivity of micellar solutions, and degree of counterion binding to micellar aggregates (see, e.g., Kamrath and Franses [59]).

In a second model, the effect of finite aggregate size on solution properties is neglected. Instead, the collection of aggregates is viewed as a separate phase with uniform properties. This approach, known as the pseudo-phase separation model, can account for many properties, such as the micellar composition and monomer concentration in mixtures of surfactants [60]. Since this model ignores the finite size of aggregates, the monomer concentration above the CMC is predicted to be constant, which is at odds with experimental results and the predictions of more realistic theories, both of which show a moderate decrease in monomer concentration above the CMC. This decrease is a consequence of the mass action effects for a collection of micelles. Also, the pseudo-phase separation approach yields no information about the optimal aggregate size or distribution of sizes.

A third class of models is based on a molecular-level thermodynamic analysis, and provides monomer and micellar compositions as well as detailed information on aggregate structure and stability. This last approach requires significantly more computational effort than the other two, but yields more information, ideally without requiring data for mixtures of the two surfactants.

Some theoretical work has been directed at modeling the important properties of mixtures of anionic and cationic surfactants [1,51,61,62]. Regular solution theory, combined with the pseudo-phase separation model, has been used to predict the monomer concentrations, micellar compositions, and critical aggregation concentrations (*cac*) for anionic and cationic surfactant mixtures [51]. In the pseudo-phase separation approach, the contribution to the solution entropy by individual aggregates is assumed small and the aggregate phase is treated as a bulk phase. Although realistic predictions are made for the dependence of the *cac* on the solution composition, the theory is strictly applicable only to molecules of similar size and functionality [63], which is clearly not the case for mixtures of oppositely-charged surfactants (also see Hoffmann and Pössnecker [52]). Further, some of the reported interaction parameters actually characterize precipitate/monomer equilibria rather than micelle/monomer equilibria [61].

The phase equilibria between monomer, micelles, and precipitate in mixtures of SDS and dodecylpyridinium chloride has been reported and interpreted in terms of a model that combines regular solution theory to describe the micellar phase and a solubility product to characterize the precipitate phase [61]. The thermodynamic machinery needed to combine an expression for Gibbs free energy for micellar aggregates with the mass action law has been developed for the calculation of micellar properties, size distributions, and phase separation [58]. This model was used to make qualitative predictions of trends for micellar solutions of anionic and cationic surfactants. Application of Puvvada and Blankschtein's model [62] to mixtures requires the input of interaction parameters for each of the binary components. Finally, a statistical thermodynamic cell model has been used to predict the phase equilibrium between various phases for mixtures of amphiphilic molecules [64]. Of these approaches, the cell model structure provides the most careful way to account for the main contributions to the aggregate free energy. In the cell model, the nonidealities due to electrostatic interactions are calculated using the nonlinearized Poisson–Boltzmann equation and the effect of excluded volume is accounted for by the dependence of cell volume on surfactant concentration and aggregate size.

A. Thermodynamics of Aggregation

The surfactant solution consists of monomeric surfactant, surfactant ions contained in aggregates of a given geometry with aggregation number N, water, and counterions. Each aggregate is treated as a distinct species with chemical potential μ_N, and at equilibrium, the Gibbs free energy of the system is minimized. For a mixture of two surfactants, A and B, dispersed in water, the Gibbs free energy is

$$G = n_w \mu_w^0 + n_{A,mon} \mu_{A,mon}^0 + n_{B,mon} \mu_{B,mon}^0 + \sum_{N>1,X_A} n_{N,X_A} \mu_N^0$$

$$+ k_B T \left\{ n_w \ln x_w + n_{A,mon} \ln x_{A,mon} + n_{B,mon} \ln x_{B,mon} \right.$$

$$+ \sum_{N>1,X_A} n_{N,X_A} \ln x_{N,X_A} \Bigg\}$$

$$+ N X_A \lambda_A + N X_B \lambda_B$$

$$(7)$$

where X_A is the fraction of component A in the mixed aggregate, n_i is the number of molecules of species i in the aqueous solution: $n_{i,mon}$ is the number of surfactant ions i ($i = A$ or B) in monomeric form, n_w is the number of water molecules, and n_{N,X_A} is the number of surfactant aggregates with aggregation number N (an "N-mer") with composition X_A, μ_i^0 is the standard chemical potential of species i (i refers to water, monomeric surfactant A or B, or N-mer of composition X_A), k_B is the Boltzmann constant, T is the temperature, and the mole fraction of species i, x_i, is defined as

$$x_i = \frac{n_i}{n_w + n_{A,mon} + n_{B,mon} + \sum_{N>1,X_A} n_{N,X_A}} \qquad (8)$$

The summation ($N > 1, X_A$) includes all possible combinations of aggregates of aggregation number N and composition X_A, excluding monomeric surfactant since this is already accounted for by the terms $n_{A,\text{mon}}$ and $n_{B,\text{mon}}$. The Lagrangian multipliers, λ_A and λ_B, are chosen to satisfy the mass balance constraints for components A and B

$$\sum_{N,X_A} NX_A n_{N,X_A}\lambda_A - n_{A,\text{total}} = 0$$

$$\sum_{N,X_A} N(1-X_A)n_{N,X_A}\lambda_B - n_{B,\text{total}} = 0 \tag{9}$$

In this equation, $n_{i,\text{total}}$ is the total number of molecules of species i in the sample. Minimization of Eq. (7) relative to n_{N,X_A} yields the size distribution for mixed aggregates [65]

$$x_{N,X_A} = x_{A,\text{mon}}^{N_A} x_{B,\text{mon}}^{N_B}$$

$$\times \exp\left[-\left(\mu_{N,X_A}^0 - N_A \mu_{A,\text{mon}}^0 - N_B \mu_{B,\text{mon}}^0\right)\Big/ k_B T\right] \tag{10}$$

where N_A and N_B are the number of surfactant ions of type A or B contained within the mixed aggregate having aggregation number N and composition X_A. Note that for monomers, $N = 1$.

The aggregate size distribution is calculated by simultaneously solving Eqs (7)–(9) for the monomer mole fractions, $x_{A,\text{mon}}$ and $x_{B,\text{mon}}$. Typically, the mole fraction of monomeric surfactant is very small and the exponent of the difference in chemical potentials is large, so that solution of these equations may present numerical problems [66]. A good initial guess for solution of the equations is the mole fraction of the pure surfactant at the CMC, which is often available in the literature.

With the help of the size distribution defined in Eq. (10), various useful quantities can be calculated, including the average aggregation number (neglecting the contribution from monomeric surfactant)

$$\langle N \rangle = \sum_{N>1,X_A} Nx_{N,X_A} \Big/ \sum_{N>1,X_A} x_{N,X_A} \tag{11}$$

the mixing ratio, $Y = n_{A,\text{total}}/(n_{A,\text{total}} + n_{B,\text{total}})$, in the bulk solution

$$Y = \frac{x_{A,\text{mon}} + \sum\limits_{N>1, X_A} N X_A x_{N, X_A}}{x_{A,\text{mon}} + x_{B,\text{mon}} + \sum\limits_{N>1, X_A} N x_{N, X_A}} \tag{12}$$

the average mole fraction of component A in aggregate form

$$\langle X_A \rangle = \sum_{N, X_A} N_A x_{N, X_A} / \sum_{N, X_A} N x_{N, X_A} \tag{13}$$

and the monomer mixing ratio

$$Y_{\text{mon}} = x_{A,\text{mon}}/(x_{A,\text{mon}} + x_{B,\text{mon}}) \tag{14}$$

The equations developed in this section are very general and may be used to predict size distributions of aggregates of various geometries, given a suitable function for the variation of the chemical potential of the aggregate as a function of size and composition. Thus, size distributions of micelles or vesicles may be calculated. In addition, this approach can be used to predict the distribution of surfactant between coexisting microstructures such as small micelles and large vesicles.

B. Curvature Elastic Energy

1. Overview

Models based on the concept of the curvature elastic energy of a surfactant film are well suited to describe the properties of aggregates composed of monolayers (microemulsions) or bilayers (lamellar bilayers, vesicles, sponge phases) [20,23, 67–70]. In this approach, the bilayer is modeled as a two-dimensional elastic film. Trends in morphology are then predicted as a function of the elastic constants and the spontaneous curvature of the film. This model is of limited use in the treatment of very small aggregates, such as micelles, because the quadratic curvature expansion breaks down at high curvature. To apply this model quantitatively, the variation of the elastic constants and spontaneous curvature with composition or surfactant properties is required.

Helfrich wrote the curvature elastic energy of a bilayer per unit area, E/A, in a quadratic approximation in curvatures as given in Eq. (1). For spherical deformations, $c_1 = c_2 = 1/R$, where R is the radius at the surface of inextension or "neutral" surface. For a symmetric bilayer, the area per molecule is invariant with curvature at the midplane, and so the neutral surface is the midplane. The energy to bend a bilayer away from the spontaneous curvature is proportional to κ and the cost of making a saddle splay deformation is proportional to $\bar{\kappa}$.

Application of the Gauss–Bonnet theorem shows that the integral of the Gaussian curvature, $c_1 c_2$, over the bilayer surface is independent of the shape and size of a closed surface and depends only on the number of handles, n_h, or components, n_c, of the surface

$$\int c_1 c_2 d^2 S = 4\pi(n_c - n_h) \tag{15}$$

Handles are defined as pores or passages from one side of the surface to the other and each closed surface is one component. A vesicle has no handles and one component per vesicle. Hence, shape variations of simple closed vesicles produce no change in the Gaussian curvature energy. However, for spherical vesicles, the saddle splay ("Gaussian") contribution to the bending energy is $4\pi \bar{\kappa}$ *per vesicle*, while for infinite lamellar bilayers it is zero [71]. Hence, the Gaussian curvature can play a significant role in determining vesicle size distributions [26,71].

At a molecular level, the tendency of a surfactant molecule to bend towards or away from the aqueous region is determined by the chemical make-up of each surfactant [22,71]. For example, a *mono*layer composed of single-tailed ionic surfactants with large head groups and small tail volumes will tend to bend away from the water, while a monolayer of a double-tailed zwitterionic surfactant such as lecithin will tend to curve more towards the aqueous region. By convention, positive curvature is away from the aqueous region. If a bilayer is made up of two chemically and physically identical monolayers, the bilayer has no net spontaneous curvature, $c_0 = 0$ in Eq. (1). As the bilayer is bent, the molecules in one monolayer may be approaching their preferred curvature, but the molecules in the other monolayer are far from theirs. The bending energy is given by Eq. (1) with $c_0 = 0$. It is important to note that, in addition to allowing for a nonzero spontaneous curvature, a multicomponent bilayer consisting of monolayers with different compositions may have bending constants, κ and $\bar{\kappa}$, considerably lower than for a uniform bilayer [71–73].

2. Implications of Curvature Elastic Energy for Vesicle Formation

Vesicles form spontaneously typically only in mixtures of two or more surfactants in water. Considering curvature elasticity alone, energy is required to bend a symmetric bilayer away from the planar configuration [Eq. (1)]. Thus, for single component (i.e., symmetric) bilayers with large bending rigidity, vesicles are formed only with the input of chemical or mechanical energy. However, this need not be the case for bilayers containing two or more components. Two factors are at work here. First, the elastic constants are strongly affected by bilayer composition, as observed experimentally [24,74] and as supported by theoretical calculations [72,73]. In mixtures, the bending rigidity can be reduced an order of magnitude, thus reducing the bending penalty that opposes the formation of an ensemble of vesicles. This mechanism gives rise to vesicles stabilized by the entropy of mixing. Alternately, a bilayer composed of two surfactants could have an asymmetric composition, thus producing a bilayer with a spontaneous curvature. This could occur if the two components in the bilayer have different spontaneous curvatures, so that when mixed the surfactants assemble into monolayers of equal and opposite curvature, resulting in an effective bilayer spontaneous curvature. This situation can lead to an enthalpic stabilization.

Consider first the case of a vesicle with a small bending constant ($\kappa, \bar{\kappa} \leq k_B T$) in the absence of a spontaneous curvature. Unilamellar vesicles can be stabilized if the bending energy per vesicle relative to the infinite lamellar phase [$8\pi\kappa + 4\pi\bar{\kappa}$, Eq. (1)] is offset by the increased translational entropy of the larger number of independent vesicles. If there is a nonzero spontaneous curvature, the effective bending energy is zero for a vesicle population with a mean radius of $\sim 1/c_0$ and entropy dictates that vesicles are the stable phase. A small value of κ also promotes unilamellar vesicles as the repulsive undulation force (Eq. 2) reduces attraction between bilayers, leading to stable unilamellar vesicles, especially when combined with electrostatic repulsion in charged systems [24]. Theory and experiment have shown that surfactant mixing can lead to sufficiently low values of κ for entropic stabilization to be effective [24,72,73].

Undulations of the bilayer may also decrease the effective bending constant at long length scales. Helfrich and others have developed an expression relating the effective bending rigidity of a membrane to the length scale of observation, $\eta (= R$ for vesicles) and a molecular distance, δ ($\sim$bilayer thickness) [68,69,75].

$$\kappa = \kappa_0\left[1 - \alpha\frac{k_B T}{4\pi\kappa_0}\ln(\eta/\delta')\right] \tag{16}$$

The numerical value of α is probably between 1 and 3 [68,75]. Similarly, the effective value of the Gaussian modulus is

$$\bar{\kappa} = \bar{\kappa}_0\left[1 + \bar{\alpha}\frac{k_B T}{4\pi\bar{\kappa}_0}\ln(\eta/\delta')\right] \tag{17}$$

where the prefactor $\bar{\alpha}$ is estimated to be zero [69] or 10/3 [75]. The form of the bending constants, renormalized for thermal fluctuations, represents a logarithmic decay in bilayer rigidity with distance. The effect of these thermal fluctuations is to increase the apparent membrane rigidity at short distance (high curvature or small vesicles) and to decrease the effective rigidity at long distances (large vesicles or free bilayers). When both entropy and undulations are accounted for, there are various predictions of how the vesicle size distribution should vary with surfactant concentration [24,76–78], but these have not been verified experimentally except for a single system [24].

Safran and coworkers have studied in detail enthalpic stabilization that results if the bilayer has a spontaneous curvature and if the bending rigidity is of the appropriate magnitude [22,79]. They recast Eq. (1) in terms of the interior and exterior spontaneous curvatures, c_I and c_E

$$E/A = \tfrac{1}{2}K\big[(c + c_E)^2 + (c - c_I)^2\big] \tag{18}$$

again with the convention that the curvature of the outer monolayer is positive. For symmetric bilayers, $c_I = c_E$ and the free energy is minimized for $c = 0$ (flat bilayers). However, for surfactant mixtures, nonideal mixing of surfactant molecules in the bilayer could allow the interior and exterior monolayers to have equal and opposite curvatures: $c_I = -c_E = c$. For this to happen, the effective head group size of the mixed surfactants on the inside monolayer must differ from that of molecules in the outside monolayer. This can be achieved either with a mixture of amphiphiles with widely different areas per head group, or in a mixture in which surfactant complexes form such that the complex has a small area per head group. By placing more of the smaller head group component (or complex) in the inside monolayer (and more of the larger head group or uncomplexed component in the exterior monolayer) of the vesicle, the spontaneous curvatures could be adjusted to suit a particular composition of the vesicle [22,79].

Vesicles with such a curvature would then be stable with respect to flat, symmetric bilayers, especially in the limit that the bending modulus of the bilayer is large compared to $k_B T$ [22,79]. The net result is a composition-dependent spontaneous curvature for the bilayer that determines the radius and size distribution of the vesicle population. In the event that vesicle formation is promoted by a spontaneous bilayer curvature, the size distribution is predicted to be a Gaussian peaked at a size near $1/c_o$ with a relative standard deviation that is inversely proportional to the sum of the Helfrich bending constants, $K = \kappa + (\bar{\kappa}/2)$. The model does not treat bilayer interactions that undoubtedly become important at higher surfactant concentrations, but it does account for several features of the experimental phase diagram [79].

In the spontaneous curvature model, interactions between surfactants are critical in stabilization of a phase of vesicles. In particular, attractive interactions are necessary to alter the balance between the area required for head groups relative to that needed for the tail groups. Applied to mixtures of oppositely charged surfactants, the interaction between the oppositely charged head groups can produce a surfactant pair that occupies a smaller area per molecule at the interface than the two individual surfactants when separated. Within this model, spontaneous vesicle formation with a well-defined size distribution is predicted for mixtures of surfactants with equal length tails as well as in mixtures with asymmetric tail groups.

IV. EXPERIMENTAL MEASUREMENTS OF BENDING CONSTANTS

A. Size Distributions

To distinguish between these two mechanisms of vesicle stability, the bending elasticity and bilayer spontaneous curvature has been measured for a number of different systems of spontaneous, unilamellar vesicles [26,27,32] by analysis of the vesicle size distribution using cryo-microscopy and freeze-fracture replication. The size distribution is calculated as follows.

The equilibrium curvature, calculated from the minimization of the curvature energy given by Eq. (1), is

$$c_{eq} = c_0 \frac{2\kappa}{2\kappa + \bar{\kappa}} \tag{19}$$

The stability criterion for the bilayer bending energy is $2\kappa + \bar{\kappa} > 0$. When $\kappa > -\bar{\kappa}/2$, this theory predicts that the stable state of the bilayer, even for the case of $c_0 = 0$, is a spherical deformation, with higher order terms in the free energy expansion required to limit the vesicle size from becoming vanishingly small. For $\bar{\kappa} > 0$, infinite films are unstable towards formation of a surface with many handles and in this case cubic or sponge-like phases are predicted to be the stable bilayer configuration. For sufficiently dilute systems with finite bilayer fragments, vesicles are still possible [27] as systems with only negative Gaussian curvature cannot close on themselves, which results in bilayers with energetically unfavorable edges in contact with water.

The equilibrium size distribution of a population of vesicles is determined by a subtle competition between the entropy of mixing and the curvature elasticity of the bilayers. For the case of spherical vesicles, $R_1 = R_2 = R$, and Eq. (1) can be simplified [22,79] to

$$E/A = 2K\left(\frac{1}{R} - \frac{1}{R_0}\right)^2 \quad 2K = 2\kappa + \bar{\kappa}R_0 = \frac{2\kappa + \bar{\kappa}}{2\kappa}r_0 \qquad (20)$$

R_0 is the radius of the minimum energy vesicle ($=1/c_{\mathrm{eq}}$), and K is an effective bending constant [22,79].

The distribution of surfactant between vesicles of aggregation number M, corresponding to the minimum energy radius, R_0 ($M = 8\pi R_0{}^2/A_0$, in which A_0 is the mean molecular area), relative to vesicles of aggregation number N and radius R, is dictated by a balance between the entropy of vesicle mixing and the curvature energy [80], and can be written in terms of the law of mass action [cf. Eq. (10)]

$$\frac{X_N}{N} = \left\{\frac{X_M}{M}\exp\left[\frac{M(\mu_M^0 - \mu_N^0)}{k_{\mathrm{B}}T}\right]\right\}^{N/M} \qquad (21)$$

X_M, μ_M^0 and X_N, μ_N^0 are the mole fraction of surfactant and the standard chemical potential per molecule in vesicles of aggregation numbers M and N, respectively. Equation (21) assumes ideal mixing of the vesicles (not the molecules within the bilayers, which will be assumed to be nonideal allowing for a spontaneous curvature) and is valid for dilute vesicle dispersions in which the Debye length is small in comparison to the inter-vesicle distance. In practice, this is always the case for catanionic vesicles as the surfactant co-ions result in at least

10–100 millimolar electrolyte concentrations. The chemical potential difference is due to the change in curvature energy per molecule for surfactant distributed between a vesicle of radius R and aggregation number N and the minimum energy vesicle of radius R_0 and aggregation number M

$$\left(\mu_N^0 - \mu_M^0\right) = \frac{4\pi R^2 (E/A)}{N} = \frac{8\pi K [1 - (R/R_0)]^2}{N} \tag{22}$$

Inserting Eq. (22) into Eq. (21) and substituting $M = 8\pi R_0^2/A_0$ and $N = 8\pi R^2/A_0$, in which A_0 is the area per surfactant molecule, gives a two parameter vesicle size distribution as a function of R_0 and K [80,81]

$$C_N = \left\{ C_M \exp\left[\frac{-8\pi K}{k_B T} \left(1 - \frac{R_0}{R} \right)^2 \right] \right\}^{R^2/R_0^2} \tag{23}$$

$C_M\ (= X_M/M)$, and C_N are the molar or number fractions of vesicles of size M and N, respectively. A consequence of Eq. (23) is that vesicles stabilized by thermal fluctuations $(K \sim k_B T)$ have a much broader size distribution than vesicles stabilized by the spontaneous curvature $(K \gg k_B T)$. This is the opposite of vesicle size distribution models that do not include a spontaneous curvature [24,44]; larger bending constants predict more polydisperse vesicles of larger size.

B. Experimental Results

Cryo-TEM images were used to determine the radius of vesicles [32]. Histograms of the vesicle size distributions were built up by measuring the size of ∼3000 spherical vesicles per concentration ratio taken from many different samples over several weeks. The measured distribution was fit to Eq. (23) (solid line) to determine R_0 and K. Figure 10(a) shows a typical image of cetyltrimethylammonium bromide (CTAB)/sodium octyl sulfate(SOS)/water (0.3/0.7/99% by wt) vesicles, while Fig. 10(b) shows the measured size distribution of vesicles and excellent agreement with the equilibrium distribution. The best fit to Eq. (23) gives $K = 0.7 \pm 0.2\, k_B T$ and $R_0 = 37\,\text{nm}$,

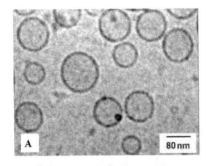 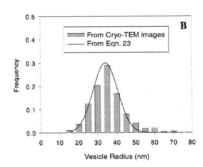

Figure 10 (A) Cryo-TEM image of 0.3/0.7/99% by wt CTAB/SOS vesicles in water and the histogram (B) of their size distributions. The solid line is a fit to the size distribution predicted by a mass-action model [i.e., Eq. (23)] using a spontaneous curvature, determined to be 37 nm, and an elastic constant $K = 0.7k_{\mathrm{B}}T$ as fitting parameters.

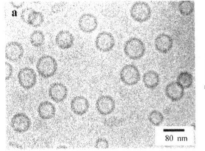 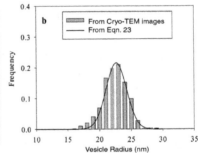

Figure 11 (a) Cryo-TEM image of fluorinated CTAB/FC7 vesicles show a much more narrow size distribution and smaller mean size (b), but that is also fit by Eq. (23). The bending constant, $K = 6k_{\mathrm{B}}T$, is about an order of magnitude greater than the CTAB/SOS vesicles. The spontaneous radius of curvature is 23 nm.

indicating these vesicles have a low bending constant and thus are entropically stabilized. Even though the $K \sim k_{\mathrm{B}}T$, single parameter size distributions with no spontaneous curvature could not fit the experimental size distribution [24,82].

Figure 11(a) shows that the vesicles in a 2 wt% 2/8 CTAB/perflurooctanoate (FC_7) dispersion are smaller, and much more monodisperse (Fig. 11b) than the CTAB/SOS dispersion (Fig. 10b). Fitting this distribution to Eq. (23) gives $R_0 = 23$ nm and $K = 6 \pm 2k_{\mathrm{B}}T$, indicating that the hydrocarbon–fluorocarbon (CTAB/FC$_7$) bilayers are much stiffer than the hydrocarbon–hydrocarbon bilayers (CTAB/SOS). Replacing the FC$_7$ fluorocarbon surfactant with the shorter chain sodium perfluorohexanoate (FC$_5$) lowers K to $0.5k_{\mathrm{B}}T$ while increasing R_0 to 56 nm [26,27]. For the CTAB/SOS and CTAB/perflurohexanoate (FC$_5$) systems, thermal undulations due to the small value of K stabilize the hydrogenated vesicles against formation of multilamellar liposomes, even in the absence of electrostatics. CTAB/SOS vesicles are stable even with 1.4 wt% added salt, at which point a phase transition to micelles occurs. The large bending constant and narrow size distribution suggest that CTAB/FC$_7$ vesicles are stabilized by the energy costs of deviations from the spontaneous curvature.

C. Theoretical Estimation of Elastic Constants

Factors such as the chemical identity of the surfactants, the addition of cosurfactants, and the value of the ionic strength strongly influence the elastic properties of surfactant films. Theoretical treatments of both electrostatic [83–87] and chain contributions [73,88] to the overall bending moduli are available. To summarize the results of theoretical predictions, the contribution to the bending moduli from the chain region is found to be the dominant factor, with bending moduli ranging from $>10k_{\mathrm{B}}T$ for pure surfactant bilayers to several $k_{\mathrm{B}}T$ in mixed bilayers. The electrostatic contribution to the bending modulus is found to be of order $k_{\mathrm{B}}T$, and is predicted to exceed $k_{\mathrm{B}}T$ only for highly charged membranes with low salt concentrations [83–86,89,90]. However, in mixtures of short- and long-chained surfactants, and in microemulsions, the chain contribution to the bending moduli is reduced considerably, and in these cases, the electrostatic contribution can be of similar magnitude and should not be neglected.

Szleifer and coworkers have calculated the elastic moduli as a function of surfactant chain length, area per molecule, and composition for mixed surfactant bilayers [73,88]. The model is based on a mean field treatment of the organization of the surfactant chains within an aggregate of a given geometry.

The model contains no adjustable parameters and the only assumption made is that the chains pack at uniform density within the aggregate, as is seen experimentally [91,92]. Calculations based on chain conformation free energy show that the bending constants increase strongly with increasing chain length or decreasing area per molecule. The bending modulus k scales with chain length as $n_c^{3.2}$ and with area per molecule as $a^{-7.5}$ [73,88]. The bending constants are predicted to decrease sharply in mixtures of short- and long-chain surfactants. For example, in mixtures of chains with 16 (C16) and 8 (C8) carbons, the bending constant is reduced from $\sim 40 k_B T$ for pure C16 bilayers to as low as $\sim k_B T$ in mixtures [72,88].

D. Close-packed Vesicle Dispersions and Vesicle "Nesting"

The hollow bilayer shells of vesicles exclude a substantial volume, and so vesicles can overpack at relatively low surfactant concentrations. Overpacking occurs roughly when the volume fraction of solution enclosed by vesicles, Φ

$$\Phi = \frac{4\pi}{3V_{total}} \sum_{N, X_A} R_N^3 n_N \tag{24}$$

is of order 60%. Clearly, when the distribution shifts towards large vesicles, the surfactant mole fraction at which vesicles become close packed, x^*, will drop considerably. This concentration is relevant experimentally, and will occur at surfactant concentrations of a few percent for larger vesicles. It probably marks the high concentration end of the vesicle "lobes" in the phase diagram, as shown in Fig. 2.

At surfactant concentrations above x^*, the added surfactant can be accommodated either by shifting the vesicle size distribution to lower radii, by forming multilamellar vesicles, or by forming a lamellar phase. Simons and Cates give a thermodynamic analysis of the stability of the various phases in the region of the overlap concentration c^* [76]. Their model combines curvature elasticity with entropy of mixing and focuses on symmetric bilayers with zero spontaneous curvature. They find that as the overlap concentration is approached, the large vesicle sizes are eliminated from the size distribution. This is accompanied by a decrease in polydispersity and an accompanying decrease in the entropic contribution. At the same time, there is an energetic penalty for forcing the surfactant to assemble into more highly curved aggregates. Consequently, the entropic term that favors formation of unilamellar vesicles decreases, and the lamellar phase, which has a lower bending energy than an ensemble of vesicles, becomes more favorable. This corresponds to a first order phase transition from vesicles to a lamellar liquid crystalline phase [76]. Another possibility is that above the overlap concentration, the vesicles can "nest" within each other, forming multilamellar vesicles. In this case, surfactant can more efficiently pack into the available volume, and the favorable entropic contribution arising from a polydisperse distribution of vesicles continues to stabilize the vesicle phase relative to the lamellar phase. Simons and Cates found that MLVs can be stable at concentrations intermediate to the unilamellar vesicle phase and the lamellar phase [76].

There is an additional interesting possibility for vesicle nesting that arises for dispersions stabilized by spontaneous curvature. In this case it is possible for vesicles to form with a discrete number of bilayers, depending on the magnitude and sign of the bilayer interactions. This was observed experimentally when 1 wt% NaBr was added to screen any residual short-range electrostatic interactions between the bilayers, the result was the spontaneous formation of a population of primarily two-layered vesicles (Fig. 12). In Fig. 12, two layer vesicles are distinguished from one layer vesicles by the darker rim on the inside edge of the apparent vesicle membrane (arrows). This dark rim is due to the greater projection of the electron beam through both the interior and exterior vesicle bilayers. From examining many images, about 90% of the vesicles with added salt have two bilayers, while the rest have one bilayer. There were essentially no vesicles with three layers or more. The vesicles in 1% NaBr sample also had a greater tendency to adhere to each other and the carbon coated electron microscope

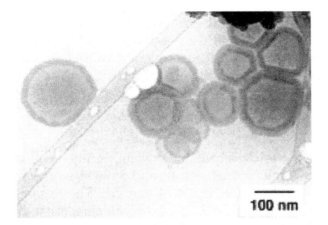

100 nm

Figure 12 Cryo-TEM image of CTAB : FC$_7$ (2 wt% total surfactant, CTAB : FC$_7$ ratio of 2/8 by weight) in 1 wt% NaBr. Two layer vesicles are distinguished from one layer vesicles by the darker rim on the inside edge of the vesicle membrane. This dark inside rim is due to the increased projection of the electron beam through both the interior and exterior vesicle bilayers; the single bilayer vesicles have membranes with a uniform intensity and do not show the interior dark rim. From examining many images, about 90% of the vesicles with added salt have two bilayers, with the rest appearing to have one bilayer. There were essentially no vesicles with three layers or more.

grid and flatten, consistent with the enhanced attraction between the vesicle bilayers [93].

The distribution between one layer and two-layer vesicles can be derived using the mass action model for vesicles with a spontaneous bilayer curvature. The analysis shows that vesicles stabilized by spontaneous curvature can have a narrow distribution of the number of bilayers when the attractive interactions just balance the curvature energy. In the absence of a spontaneous bilayer curvature, each additional layer added to a vesicle has a decreasing curvature energy, but the attractive interaction energy grows with the net bilayer area in contact, and a polydisperse population of multilamellar liposomes result. Hence, typical phospholipid vesicles, with $1/R_0 = 0$, are unstable relative to multilamellar liposomes. The combination of a narrow size distribution, a large bending elastic constant, and the formation of two-bilayer vesicles shows that the CTAB/FC$_7$ vesicles are stabilized by spontaneous curvature. Interesting other shapes (rods and disks) can also be analyzed quantitatively [27].

V. EQUILIBRIUM?

Unilamellar vesicles form in a wide variety of mixtures of cationic and anionic surfactants, and do so spontaneously; that is, without substantial input of energy. Whether or not the vesicles are the equilibrium state of aggregation has been explored by several authors [46,94,95], all of whom conclude for various reasons that the vesicles are unstable with respect to a lamellar phase. A corollary argument is that the vesicles that are observed are the consequence of shear forces, which might be vanishingly small, that disrupt the stacked bilayers of a lamellar phase. Experimental conformation is hampered by the extremely slow evolution of these structures with time, which in turn must reflect a combination of small thermodynamic driving forces and slow mass transfer processes.

The question is clearer from a theoretical point of view. Consider first a patch of bilayer with zero spontaneous curvature in water in the limit of low concentrations. The patch of bilayer has hydrophobic edges exposed to water. This "edge energy" is recovered by closing the bilayer into a vesicle, with the energy penalty given by Eq. (1). For large enough values of edge energy and small enough values of the bending constant, the vesicle is stable. The situation is more interesting when the bilayer has a preferred curvature, as can happen for a mixed bilayer. Several theoretical approaches all yield the result that the vesicle is the thermodynamically preferred structure.

Thus, at least at low concentrations, bilayers with either a low bending constant or with a non-zero spontaneous curvature form vesicles are absolutely the preferred thermodynamic state compared to a stacked bilayer for the reasons laid out above. The experimental data show that vesicles are stabilized by one of two distinct mechanisms depending on the value of the bending constant. Helfrich undulations ensure that the interbilayer potential is always repulsive when the bending constant, K, is of order k_BT. When $K \gg k_BT$, unilamellar vesicles are stabilized by the spontaneous curvature that picks out a particular vesicle radius; other radii are energetically disfavored. Measurements of the bilayer elastic constant and the spontaneous curvature, R_0, for three different systems of vesicles by an analysis of the vesicle size distribution determined by cryo-transmission electron microscopy show cases for both entropically and enthalpically stabilized vesicles.

An absolute resolution of this question is probably impossible, but there are several relevant observations at hand. In particular, Hoffmann and co-workers have produced both lamellar phases and vesicles by chemical synthesis to produce one of the surfactant species in situ, thereby avoiding shear. They see the formation of lamellar phases that are stable in time but that form vesicles upon a single inversion of the sample tube. These vesicles are also stable with time, so the experiment simply does not allow resolution of the question of which is the stable phase. Further, the lamellar phases are observed by freeze-fracture electron microscopy, which requires exposing a thin layer of solution to a solid surface, and solid surfaces are well known to nucleate lamellar sheets from vesicles [96]. Thus the question of the metastability of the lamellar phase is open. For the reasons given above, creating a lamellar phase by specific chemical or physical treatments that is unstable to small perturbations does not show that the lamellar phase is the preferred state of organization.

Almgren [46] recently summarized many observations around swollen lamellar phases, and notes that they can be easily dispersed to form vesicles. This observation was also offered by Laughlin as the explanation for the observed vesicles [94]. Almgren also reports QLS measurements of mixtures of CTAB and SOS at a given mixing ratio that show an increase in mean radius with time, which agrees with our results at that mixing ratio [31]. Other ratios yield vesicles whose average size does not change with time. Careful, long term studies of catanionic mixtures at certain well-defined compositions have shown uniform phases of unilamellar vesicles for well over a decade [28]. The size distributions of these vesicles remain constant over time and are well described by equilibrium theories of self-assembly [26,27,32]. Of course, not every mixture of anionic and cationic surfactants will form equilibrium vesicles and care must be taken to understand the phase diagrams and carefully delimit the concentration ranges that do form such vesicles. That various experimental observations are often at odds highlight the experimental challenges in resolving the issue of equilibrium vesicles.

REFERENCES

1. P. Jokela, B. Jönsson, and H. Wennerström. *Prog. Colloid and Polym. Sci.* **70**: 17–22 (1985).

2. C. Tondre, and C. Caillet, *Adv. Colloid and Interface Sci.* **93**: 115–134 (2001).

3. T. Zemb, M. Dubois, B. Deme, and T. Gulik-Krzywicki, *Science* **283**: 816–820 (1999).

4. F. Kern, R. Zana, and S. J. Candau, *Langmuir* **7**: 1344–1351 (1991).

5. P. A. Hassan, S. R. Raghavan, and E. W. Kaler, *Langmuir* **18**: 2543–2548 (2002).

6. K. L. Herrington, E. W. Kaler, D. D. Miller, J. A. Zasadzinski, and S. Chiruvolu, *J. Phys. Chem.* **97**: 13792–13802 (1993).

7. M. Gradzielski, M. Bergmeier, H. Hoffmann, M. Muller, and I. Grillo, *J. Phys. Chem. B* **104**: 11594–11597 (2000).

8. A. D. Bangham, M. M. Standish, and J. C. Watkins, *J. Molec. Biol.* **13**: 238–252 (1965).

9. D. D. Lasic, *Liposomes: From Physics to Applications.* Elsevier, Amsterdam (1993).

10. L. Rydhag, P. Stenius, and L. Odberg, *J. Colloid and Interface Sci.* **86**: 274–276 (1982).

11. R. Joannic, L. Auvray, and D. D. Lasic, *Phys. Rev. Lett.* **78**: 3402–3405 (1997).

12. M. Rovira-Bru, D. H. Thompson, and I. Szleifer, *Biophys. J.* **83**: 2419–2439 (2002).

13. H. Hauser, *Proc. Natl Acad. Sci.*, USA **86**: 5351–5355 (1989).

14. M. P. Nieh, T. A. Harroun, V. A. Raghunathan, C. J. Glinka, and J. Katsaras, *Phys. Rev. Lett.* **91** (2003).

15. M. P. Nieh, C. J. Glinka, and J. Katsaras, *Biophys. J.* **84**: 134A–134A (2003).

16. D. D. Lasic, *J. Colloid Polym. Sci.*, **140**: 302–304 (1990).

17. D. F. O'Brien, B. Armitage, A. Benedicto, D. E. Bennett, H. G. Lamparski, Y. S. Lee, W. Srisiri, and T. M. Sisson, *Accounts of Chem. Res.* **31**: 861–868 (1998).

18. J. T. Lei, T. M. Sisson, H. G. Lamparski, and D. F. O'Brien, *Macromolecules*, **32**: 73–78 (1999).

19. L. Cantu, M. Corti, E. Del Favero, and A. Raudino, *J. Phys. II (France)* **4**: 1585–1604 (1994).

20. W. Helfrich, *J. Phys. (Paris)* **47**: 321–329 (1973).

21. D. Hilbert, and S. Cohn-Vossen, *Geometry and the Imagination*, Chelsea, New York (1983).

22. S. A. Safran, P. Pincus, and D. Andelman, *Science* **248**: 354–355 (1990).

23. W. Helfrich, *Z. Natur.* **33A**: 305–315 (1978).

24. P. Hervé, D. Roux, A. M. Bellocq, F. Nallet, and T. Gulik-Krzywicki, *J. Phys. II (France)* **3**: 1255–1270 (1993).

25. D. Iampietro, and E. W. Kaler, *Langmuir* **15**: 8590–8601 (1999).

26. H. T. Jung, B. Coldren, J. A. Zasadzinski, D. J. Iampietro, and E. W. Kaler, *Proc. Natl Acad. Sci.* **98**: 1353–1357 (2001).

27. H. T. Jung, Y. S. Lee, E. W. Kaler, B. Coldren, and J. A. Zasadzinski, *Proc. Natl Acad. Sci* **99**: 15318–15322 (2002).

28. E. W. Kaler, A. K. Murthy, B. E. Rodriguez, and J. A. N. Zasadzinski, *Science* **245**: 1371–1374 (1989).

29. L. L. Brasher, K. L. Herrington, and E. W. Kaler, *Langmuir* **11**: 4267–4277 (1995).

30. L. L. Brasher, *Phase Behavior and Microstructure of Surfactant Mixtures*. PhD dissertation, University of Delaware of Delaware, Newark, DE (1996).

31. M. T. Yatcilla, K. L. Herrington, L. L. Brasher, E. W. Kaler, S. Chruvolu, and J. A. Zasadzinski, *J. Phys. Chem.* **100**: 5874–5879 (1996).

32. B. A. Coldren, R. van Zanten, M. J. Mackel, J. A. Zasadzinski, H. T. Jung, *Langmuir* **19**: 5632–5639 (2003).

33. E. W. Kaler, K. L. Herrington, A. K. Murthy, and J. A. N. Zasadzinski, *J. Phys. Chem.* **96**: 6698–6707 (1992).

34. D. J. Iampietro, L. L. Brasher, E. W. Kaler, A. Stradner, and O. Glatter, *J. Phys. Chem. B* **102**: 3105–3113 (1998).

35. A. K. Murthy, E. W. Kaler, and J. A. N. Zasadzinski, *J. Colloid and Interface Sci.* **145**: 598–600 (1991).

36. E. Marques, A. Khan, M. D. G. Miguel, and B. Lindman, *J. Phys. Chem.* **97**: 4729–4736 (1993).

37. M. G. Miguel, A. Pais, R. S. Dias, C. Leal, M. Rosa, and B. Lindman, *Colloids and Surfaces A–Physicochemical and Engineering Aspects* **228**: 43–55 (2003).

38. J. S. Martinez, G. P. Zhang, P. D. Holt, H. T. Jung, C. J. Carrano, M. G. Haygood, A. Butler. *Science* **287**: 1245–1247 (2000).

39. K. Horbaschek, H. Hoffmann, and J. Hao, *J Phys. Chem. B* **104**: 2781–2784 (2000).

40. A. J. O'Connor, T. A. Hatton, and A. Bose, *Langmuir* **13**: 6931–6940 (1997).

41. A. Shioi, and T. A. Hatton, *Langmuir* **18**: 7341–7348 (2002).

42. Y. Xia, I. Goldmints, P. W. Johnson, T. A. Hatton, and A. Bose, *Langmuir* **18**: 3822–3828 (2002).

43. R. G. Laughlin, *The Aqueous Phase Behavior of Surfactants*. Academic Press, New York (1994).

44. J. Israelachvili, *Intermolecular and Surface Forces: With Applications to Colloidal and Biological Systems*, Academic Press (1992).

45. S. Chiruvolu, E. Naranjo, and J. A. Zasadzinski, In C. A. Herb, and R. K. Prud'homme (eds.), *Microstructure of Complex Fluids by Electron Microscopy*. American Chemical Society (1994).

46. M. Almgren, *Australian J. Chem.* **56**: 959–970 (2003).

47. E. W. Kaler, In H. Brumberger (ed.) *Small-Angle Scattering From Complex Fluids*. Kluwer Academic Publishers (1995).

48. P. Bartlett, and R. H. Ottewill, *J. Chem. Phys.* **96**: 3306–3318 (1992).

49. L. L. Brasher, and E. W. Kaler, *Langmuir* **12**: 6270–6276 (1996).

50. C. Tanford, *The Hydrophobic Effect: Formation of Micelles and Biological Membranes*. John Wiley & Sons, New York (1973).

51. P. M. Holland, and D. N. Rubingh, *J. Phys. Chem.* **87**: 1984–1990 (1983).

52. H. Hoffmann, and G. Pössnecker, *Langmuir* **10**: 381–389 (1994).

53. G. Gunnarsson, B. Jönsson, and H. Wennerström, *J. Phys. Chem.* **84**: 3114–3121 (1980).

54. B. Jönsson, P. Jokela, B. Lindman, and A. Sadaghiani, *Langmuir* **7**: 889–895 (1991).

55. R. Nagarajan, and E. Ruckenstein, *J. Colloid and Interface Sci.* **71**: 580–604 (1979).

56. R. Nagarajan, and E. Ruckenstein, *Langmuir* **7**: 2933–2969 (1991).

57. M. Bergstrom, *Langmuir* **12**: 2454–2463 (1996).

58. P. Puvvada, and D. Blankschtein, *J. Phys. Chem.* **96**: 5579–5592 (1992).

59. R. F. Kamrath, and E. I. Franses, *J. Phys. Chem.* **88**: 1642–1648 (1984).

60. J. N. Israelachvili, and H. Wennerström, *J. Phys. Chem.* **96**: 520–531 (1992).

61. K. L. Stellner, J. C. Amante, J. F. Scamehorn, J. H. Harwell, *J. Colloid and Interface Sci.* **123**: 186–200 (1988).

62. S. Puvvada, and D. Blankschtein, *J. Phys. Chem.* **96**: 5567–5579 (1992).

63. S. Sandler, *Chemical and Engineering Thermodynamics*. John Wiley and Sons, New York (1989).

64. B. Jönsson, and H. Wennerström, *J. Colloid and Interface Sci.* **80**: 482–496 (1981).

65. R. Nagarajan, *Langmuir* **1**: 331–341 (1985).

66. M. M. Stecker, and G. B. Benedek, *J. Phys. Chem.* **88**: 6519–6544 (1984).

67. W. Helfrich, *Z Natur.* **28c**: 693–703 (1973).

68. W. Helfrich, *J Phys. (France)* **46**: 1263–1268 (1985).

69. W. Helfrich, *J Phys. (France)* **47**: 321–329 (1986).

70. S. A. Safran, *Statistical Thermodynamics of Surfaces, Interfaces and Membranes*, Addison-Wesley, Reading, Mass. (1994).

71. S. A. Safran, *Adv. Phys.* **48**: 395–448 (1999).

72. I. Szleifer, A. Ben-Shaul, and W. M. Gelbart, *J. Phys. Chem.* **94**: 5081–5089 (1990).

73. I. Szleifer, D. Kramer, A. Ben-Shaul, W. M. Gelbart, and S. A. Safran, *J. Chem. Phys.* **92**: 6800–6817 (1990).

74. C. R. Safinya, D. Roux, G. S. Smith, S. K. Sinha, P. Dimon, N. A. Clark, and A. M. Bellocq, *Phys. Rev. Lett.* **57**: 2718–2721 (1986).

75. L. Peliti, and S. Leibler, *Phys. Rev. Lett.* **54**: 1690–1693 (1985).

76. B. D. Simons, and M. E. Cates, *J Phys. II (France)* **2**: 1439–1451 (1992).

77. D. C. Morse, and S. T. Milner, *Europhys. Lett.* **26**: 565–570 (1994).

78. D. C. Morse, and S. T. Milner, *Phys. Rev. E* **52**: 5918–5985 (1995).

79. S. A. Safran, P. A. Pincus, D. Andelman, and F. C. MacKintosh, *Phys. Rev. A* **43**: 1071–1078 (1991).

80. J. N. Israelachvili, D. J. Mitchell, and B. W. Ninham, *J Chem. Soc. Faraday Transactions II* **72**: 1526–1568 (1976).

81. N. D. Denkov, H. Yoshimura, T. Kouyama, J. Walz, and K. Nagayama, *Biophys. J.* **74**: 1409–1420 (1998).

82. B. A. Coldren, *Phase Behavior, Microstructure and Measured Elasticity of Catanionic Surfactant Bilayers.* PhD dissertation, University of California of Delaware, Newark, DE (2002).

83. M. Winterhalter, and W. Helfrich, *J. Phys. Chem.* **92**: 6865–6867 (1988).

84. M. Winterhalter, and W. Helfrich *J. Phys. Chem.* **96**: 327–330 (1992).

85. H. N. W. Lekkerkerker, *Physica A* **159**: 319–328 (1989).

86. D. J. Mitchell, and B. W. Ninham, *Langmuir*, **5**: 1121–1123 (1989).

87. A. Fogden, and B. W. Ninham, *Adv. Colloid and Interface Sci.* **83**: 85–110 (1999).

88. I. Szleifer, D. Kramer, A. Ben-Shaul, D. Roux, and W. Gelbart, *Phys. Rev. Lett.* **60**: 1966–1969 (1988).

89. A. Fogden, I. Carlsson, and J. Daicic, *Phys. Rev. E* **57**: 5694–5706 (1998).

90. A. Fogden, D. J. Mitchell, and B. W. Ninham, *Langmuir* **6**: 159–162 (1990).

91. D. W. R. Gruen, *J. Phys. Chem.* **89**: 146–153 (1985).

92. D. W. R. Gruen, *J. Phys. Chem.* **89**: 153–163 (1985).

93. S. Bailey, M. Longo, S. Chiruvolu, and J. A. Zasadzinski, *Langmuir* **6**: 1326–1329 (1990).

94. R. G. Laughlin, *Colloids and Surfaces A* **128**: 27–38 (1997).

95. J. C. Hao, H. Hoffmann, and K. Horbaschek, *J. Phys. Chem. B* **104**: 10144–10153 (2000).

96. J. T. Groves, N. Ulman, S. G. Boxer, *Science* **275**: 651–653 (1997).

10

Phase Behavior and Microstructure of Liquid Crystals in Mixed Surfactant Systems

CARLOS RODRIGUEZ and
HIRONOBU KUNIEDA
Graduate School of Engineering,
Yokohama National University,
Yokohama, Japan

SYNOPSIS

The phase behavior and microstructure of several mixed nonionic and anionic surfactants systems in the liquid crystal state are addressed in this chapter. The relationships and parameters for microstructural analysis of liquid crystals by Small Angle X-ray Scattering (SAXS) are summarized, and then used to explain certain changes in self-organization. Comparison of theory and experiments is presented for nonionic surfactant mixtures. The solubilization of oils in mixed systems is also discussed. Systems closely related to applications such as sucrose and polyglycerol fatty acid esters as well as novel surfactants, such as poly(dimethylsiloxane) surfactants, are also included.

1. INTRODUCTION

Commercial products containing surfactants are usually mixtures, either because this is inherent to the process of synthesis or because a synergistic effect is desired. Many of the applications of surfactants derive from their unique capability to form self-organized structures in solution. At concentrations above the critical micellar concentration (CMC), those structures can be found in a liquid crystalline state [1–7], namely, they show long-range order but the liquid-like disorder remains in the short-range order, unlike solid crystals. The enthalpy of transition to the liquid phase is then usually small. Since the properties of these liquid crystals are strongly dependent on the concentration of the solvent, they are referred to as lyotropic liquid crystals. Although some nematic phases have been found in surfactant systems [8], lyotropic liquid crystals are mainly of the smectic type, namely, they show both positional and orientational order.

Liquid crystal phases in surfactant systems show a wide variety of shapes and structures. Some of these phases are listed below.

1. *Lamellar phases* ($L_\alpha, L_\beta, P_\beta$): these phases are built up of bilayers of surfactant molecules alternating with solvent layers. There are several types of lamellar phases [9–12]. The L_α phase consists of one-dimensional lamellar lattice with hydrocarbon chains in liquid-like (disordered) state and oriented perpendicular to the lamellar plane (Fig. 1a). The L_β

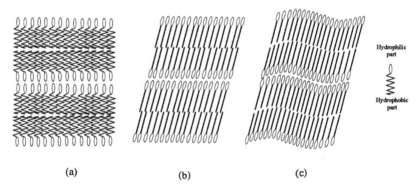

(a)   (b)   (c)

Figure 1   Lamellar phases (a) L_α (b) L_β (c) P_β.

phase is built up with a one-dimensional lamellar lattice with hydrocarbon chains in solid-like (ordered) state and oriented perpendicularly to the lamellar plane, sometimes interdigitated (Fig. 1b). Finally, the P_β phase consists of two-dimensional monoclinic lattice with the bilayer distorted by a periodic ripple (Fig. 1c).

Small angle X-ray scattering (SAXS) measurements allow us to estimate, from geometrical considerations, some structural parameters of the L_α phase. The half-thickness of the apolar domain consisting of oil and the lipophilic chain r_L and the effective cross sectional area per surfactant molecule for the lamellar phase, a_{sL} are given by

$$r_L = \frac{d_1(\phi_L + \phi_0)}{2} \tag{1}$$

$$a_{sL} = \frac{v_L}{r_L N_A}\left(\frac{\phi_L + \phi_0}{\phi_L}\right) \tag{2}$$

where v_L is the molar volume of the lipophilic part of surfactant, ϕ_0 and ϕ_L are the volume fractions of oil and the lipophilic part of surfactant, respectively, N_A is Avogrado's number and d_1 is the interlayer spacing corresponding to the first diffraction peak.

2. *Hexagonal phases* (H_1, H_2): these phases are built of long cylindrical micelles with a lipophilic core arranged in a hexagonal pattern [13]. In the normal hexagonal phase (H_1), the hydrophobic chains form the core of the cylinders while the hydrophilic chains face the polar solvent (Fig. 2a). On the other

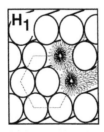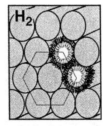

Figure 2 Hexagonal phases: (a) normal hexagonal (b) inverse hexagonal.

hand, in the inverse hexagonal phase (H_2), the core of the cylinders consists of the hydrophilic chains with the hydrophobic chains directed towards a nonpolar solvent (Fig. 2b).

The radius of the lipophilic part in the cylinders, r_{H_1}, and the effective cross-sectional area per surfactant molecule for the hexagonal H_1 phase, a_{SH_1}, are given by

$$r_{H_1} = d_{10}\left[\frac{2(\phi_L + \phi_o)}{\sqrt{3}\pi}\right]^{1/2} \tag{3}$$

$$a_{sH_1} = \frac{2v_L}{r_{H_1}N_A}\left(\frac{\phi_L + \phi_o}{\phi_L}\right) \tag{4}$$

where d_{10} is the interlayer spacing corresponding to the first diffraction peak.

Similar equations can be derived for the H_2 phase:

$$r_{H_2} = d_{10}\left[\frac{2(1 - \phi_L - \phi_o)}{\sqrt{3}\pi}\right]^{1/2} \tag{5}$$

$$a_{sH_2} = \frac{2v_L}{r_{H_2}N_A}\left(\frac{1 - \phi_L - \phi_o}{\phi_L}\right) \tag{6}$$

3. *Bicontinuous cubic phases*: in these phases surfactant molecules form a porous connected structure in three dimensions (Fig. 3). In this structure, the curvature of aggregates is at all points positive and negative (as in a saddle), giving the net curvature equal to zero [14]. The bicontinuous cubic phase consists of multiply connected bilayers

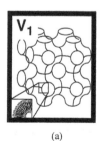 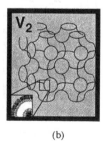

(a)   (b)

Figure 3 Bicontinuous cubic phases: (a) normal bicontinuous (b) reverse bicontinuous.

separating two distinguishable and continuous domains of the same solvent. In the V_1 phase (Fig. 3a), a polar film consisting of the hydrophilic groups and the polar solvent divides two apolar domains. On the other hand, the V_2 phase (Fig. 3b) can be described as a nonpolar film consisting of the hydrophobic groups dividing two polar domains.

The midplane of the layer consisting of water and hydrophilic chains in the bicontinuous cubic phase (oil-in-water) forms a periodic minimal surface [15]. The water/oil interface is assumed to consist of two parallel surfaces at a distance L from the midplane [16]. The total interfacial area per unit volume (A/V) is expressed by [14]

$$\frac{A}{V} = \frac{\phi_s N_A}{V_L} a_s = \frac{2c}{a}\left(1 + \frac{2\pi\chi^u}{c}\left(\frac{L}{a}\right)^2\right) \tag{7}$$

where χ^u is the Euler characteristic per unit cell of the minimal surface, c is the dimensionless area [16,17] and a is the lattice parameter. Furthermore, the volume fraction of polar domain consisting of water and hydrophilic chain ($\phi_H + \phi_W$) is given by the equation [16]

$$\phi_H + \phi_W = 1 - (\phi_L + \phi_o) = \frac{2cL}{a}\left(1 + \frac{2\pi\chi^u}{3c}\left(\frac{L}{a}\right)^2\right) \tag{8}$$

By solving Eqs (7) and (8) simultaneously, a_s can be obtained [18].

4. *Discontinuous micellar cubic phases*: these phases are formed of discrete aggregates packed in a three-dimensional array [19]. The normal discontinuous cubic phase (I_1) (Fig. 4a) is built up with discontinuous hydrocarbon regions embedded in continuous aqueous medium (for example, normal micellar aggregates in water). The inverse discontinuous cubic phase (I_2) (Fig. 4b) consists of discontinuous water regions embedded in a continuous lipid medium (for example, reversed micellar aggregates in a hydrocarbon). Although only spherical micelles are shown in the schematics for the discontinuous cubic phases (Figs 4a and 4b), the possibility of the presence of short cylindrical micelles in these phases cannot be ruled out. Due to

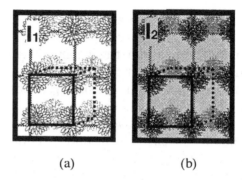

(a)   (b)

Figure 4   Discontinuous cubic phases: (a) normal (b) reverse.

the spherical symmetry of aggregates, discontinuous cubic phases are optically isotropic. Additionally, their viscosity is very high, similar to that of gels.

For spherical, normal micelles packed in a cubic lattice, the following equations can be derived for the radius of the hydrophobic part of micelles, r_{I_1}, and the effective cross-sectional area per surfactant molecule, a_{sI_1} in the I_1 phase [20]

$$V_{Lo} = \frac{(\phi_L + \phi_0)p^3}{n_m} \tag{9}$$

$$r_{I_1} = \left[\frac{V_{Lo}}{4\pi/3}\right]^{1/3} \tag{10}$$

$$a_{sI_1} = \frac{3v_L}{r_{I_1}N_A}\left(\frac{\phi_L + \phi_0}{\phi_L}\right) \tag{11}$$

where p is the unit cell parameter, n_m is the number of micelles per unit cell, and V_{Lo} is the volume of the hydrophobic part of micelles.

For spherical, inverse micelles packed in a cubic lattice, the radius of the hydrophilic part of micelles, r_{I_2}, and the effective cross-sectional area per surfactant molecule, a_{sI_2} in the I_2 phase are calculated from [21]

$$V_w = \frac{(1 - \phi_L - \phi_o)p^3}{n_m} \qquad (12)$$

$$r_{I_2} = \left[\frac{V_w}{4\pi/3}\right]^{1/3} \qquad (13)$$

$$a_{sI_2} = \frac{3v_L}{r_{I_2}N_A}\left(\frac{1 - \phi_L - \phi_o}{\phi_L}\right) \qquad (14)$$

V_w is the volume of the hydrophilic part of micelles.

Geometrical factors are considered to play a very important role in the mechanism of phase transitions between liquid crystal phases. The concept of critical packing parameter [22] is useful to relate the properties of the surfactant molecule to the shape of aggregates taking into account the concept is valid (quantitatively) for room temperature, relatively low concentrations, absence of added salt, or organic solute. This parameter is defined as

$$CP = \frac{v}{l_c a_s} = 1 - \frac{1}{2}\left(\frac{1}{R_1} + \frac{1}{R_2}\right) + \frac{l_c^2}{3R_1 R_2} \qquad (15)$$

where R_1 and R_2 are the local radii of curvature, a_s is the interfacial area per surfactant molecule, v is the volume of lipophilic fluid chains, and l_c is their effective length. Then, $CP < 1/3$ for spheres, $CP < 1/2$ for rods, $CP < 1$ for layers, and $CP > 1$ for inverted structures. Since l_c cannot be larger than the fully extended length (all-*trans* length) of the surfactant molecule, for a given shape, the lowest value of a_s is determined by the values of CP. In a mixture of two surfactants with different values of CP and water the packing parameter of the mixture (CP_{mix}) and the curvature of the mixed micelles depend on the mixing ratio, since the values of v, a_s and l_c are average values [23].

Studies of systems of mixed surfactants and water in the concentrated region are scarce [24–27]. The purpose of this chapter is to review new experimental findings regarding this subject.

II. PHASE BEHAVIOR AND MICROSTRUCTURE IN WATER/MIXED NONIONIC SURFACTANT SYSTEMS

A. Polyoxyethylene Alkyl Ethers

The use of poly(oxyethylene)-type surfactant is advantageous because the hydrophile–lipophile balance can be freely varied by changing the hydrophilic poly(oxyethylene) chain [28]. On the other hand, as the lipophilic chain of surfactant gets longer, the cohesive energy increases, and rich phase behavior is expected [28,29]. Moreover, branched EO-chain surfactants have been synthesized, so it is possible to study the effect of molecular geometry without changing the hydrophilic–lipophilic volume ratio [23].

Mixtures of two nonionic surfactants with similar head group usually show an almost ideal behavior, namely, the values of the interaction parameter β are small [30,31]. Here, β represents the type and extent of the interaction between the two different surfactant molecules forming the mixed micelle in aqueous solution; the larger the value of β, the stronger the interaction between the two surfactants [32]. In the case of polyoxyethylene alkyl ether surfactants, the β is small and negative (weak attraction). However, as the difference in EO chain length increases, the values of β tend to increase. Nonideal behavior has been found [32] in mixtures of fluorinated and hydrogenated nonionics in aqueous mixed systems, although every one-phase system (isotropic or liquid crystal) is constituted by only one type of mixed aggregate. Segregation between fluorinated and hydrogenated chains seemed to increase with addition of water.

The pseudobinary phase diagrams of mixtures of different poly(oxyethylene) dodecyl ethers ($C_{12}EO_n$) with the same average number of six oxyethylene groups per surfactant molecule, are shown in Fig. 5, together with the phase diagram of the homogeneous surfactant. The average EO chain length n

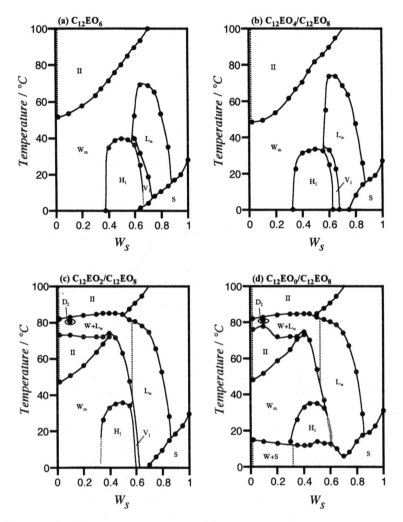

Figure 5 The binary or pseudo binary phase diagrams of aqueous system of $C_{12}EO_6$ (a), $C_{12}EO_4$–$C_{12}EO_8$ (b), $C_{12}EO_2$–$C_{12}EO_8$ (c), and $C_{12}EO_0$–$C_{12}EO_8$ (d). W_s is the surfactant concentration in the system. W, W_m, H_1, V_1, L_α, D_2, S and II indicate water phase, aqueous micellar phase, hexagonal liquid crystalline phase, bicontinuous cubic phase, lamellar liquid crystalline phase, sponge phase, solid phase, and two isotropic equilibrium phases [33].

is calculated by $n = n_1 X_1 + n_2 X_2$, where n_1 and n_2 are the EO chain lengths of surfactants 1 and 2 and X_1 and X_2 are their mole fractions. As surfactant concentration is increased, different phases appear in the sequence $W_m \rightarrow H_1 \rightarrow V_1 \rightarrow L_\alpha$ (W_m is a normal micellar solution). The effect of mixing surfactants on the phase behavior is more pronounced as the difference in the EO chains of mixed surfactants is larger. For the $C_{12}EO_2$–$C_{12}EO_8$–water system, the L_α region shifts to lower concentrations and intrudes into the two-phase region. This might be attributed to an increase in the rigidity of the surfactant layer. The isolated isotropic phase D_2 (or L_3 phase) appears inside the $W + L_\alpha$ region. In general, this phase forms at temperatures above those for the $W + L_\alpha$ region in homogeneous surfactants. In the case of the $C_{12}EO_0$–$C_{12}EO_8$ system, the V_1 phase disappears and the solid-present region expands. In general, the behavior of the lamellar L_α phase is sensitive to the difference in EO-chain length between the mixed surfactants, but the cloud point and the hexagonal H_1 phase are almost unaffected.

Change in the effective cross-sectional area per surfactant molecule a_s is considered to have a strong influence on the interfacial behavior. Figure 6 shows the values of a_s for the lamellar L_α phase as a function of the number of EO groups, in single and mixed $C_{12}EO_n$ systems. Even if the average EO chain is the same, a_s tends to decrease with mixing of surfactants, and this tendency becomes stronger as the difference in EO number between the surfactants becomes larger. This indicates that the surfactant molecules are more tightly packed at the interface in mixed systems.

For the single surfactant system, a_s can be written as [34]

$$a_s = \sqrt{\frac{k}{\gamma} n} + a_H \qquad (16)$$

The first term on the right side of Eq. (16) corresponds to the repulsion force between EO chains, k is the repulsion constant of one EO unit [34], n is the number of EO units, γ is the bare surface tension between water and the hydrophobic part of the surfactant, and a_H is the excluded area of the hydrocarbon chain ($0.2\,\mathrm{nm}^2$). If the mixing of surfactants is ideal, the average $\bar{a}_s$ can be expressed as

$$\bar{a}_s = a^0_{S,1} X_1 + a^0_{S,2} X_2 \tag{17}$$

where $a^0_{S,1}$ and $a^0_{S,2}$ are the effective cross-sectional areas of surfactant 1 and 2 in the single surfactant system. The value of $\bar{a}_s$ can also be derived from thermodynamic considerations [35]

$$\bar{a}_s = X_1 \sqrt{\frac{k}{\gamma} n_1} + X_2 \sqrt{\frac{k}{\gamma} \left\{ n_1 + (n_2 - n_1) \frac{X_2}{b + (1 - b)X_2} \right\}} + a_H \tag{18}$$

where $b = a^0_{S,1}/a^0_{S,2}$ and n_1, n_2 are the number of EO units of surfactant 1 and 2.

Figure 6 shows the cross-sectional areas calculated using Eq. (18). The values of a_s at points P1 and P2 indicate the hypothetical area of the L_α phase for $C_{12}EO_7$ and $C_{12}E_8$ calculated using Eq. (16) (the L_α phase is not formed in these systems at the given surfactant concentrations). The predicted areas are slightly smaller than the experimental ones. In the $C_{12}E_0$–$C_{12}EO_8$ system, the deviation from the experimental data is larger than that in other systems. In deriving Eq. (18) the interaction between bilayers, which might increase a_s, is neglected. Additionally, the repulsion of the end hydroxyl group in $C_{12}E_0$ is not considered in the calculations.

B. Polyglycerol Fatty Acid Esters

Polyglycerol fatty acid esters are widely used in formulations for foods, cosmetics, and toiletries. Usually, commercial polyglycerol fatty acid esters are a mixture of compounds, since the polyglycerols have a wide distribution of chain lengths and degree of esterification. Figure 7 shows the phase behavior in water of polyglycerol polydodecanoates with a narrow distribution (10G*0.7L) and a wide distribution (10G0.7L) of polyglycerol chains. The average number of glycerol groups per polyglycerol chain is 10.65 for 10G*0.7L and 10.9 for 10G0.7L

(a)

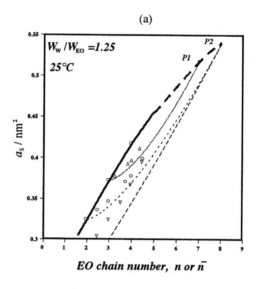

EO chain number, n or n̄

(b)

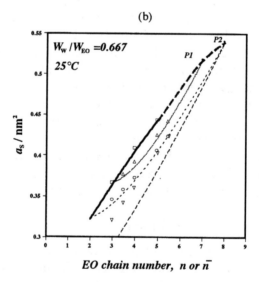

EO chain number, n or n̄

Figure 6 Comparison between experimental and theoretical a_s in the L_α phase in the mixed surfactant system. The weight ratio of water to EO chain is 1.25 (a) and 0.667 (b). ($\square$, —) single homogenous $C_{12}EO_n$. ($\triangle$, - - -) $C_{12}EO_3$–$C_{12}EO_7$. ($\bigcirc$, – – –) $C_{12}EO_2$–$C_{12}EO_8$. (∇, ——), $C_{12}EO_0$–$C_{12}EO_8$. The a_s at points P1 and P2 for $C_{12}EO_7$ and $C_{12}EO_8$ are calculated by Eq. (16). Each theoretical curve for the mixed surfactant is calculated by Eq. (18) [35].

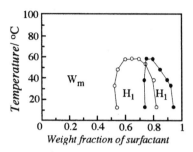

Figure 7 Phase diagram of 10G*0.7L /water (●) and 10G0.7L/ water (○) systems. W_m: micellar solution; H_1: hexagonal liquid crystal [36].

(10G0.7L contains a larger proportion of long polyglycerol chains than 10G*0.7L). The average number of alkanoate groups per surfactant molecule is 2.23 for 10G*0.7L and 2.64 for 10G0.7L. Both surfactants form aqueous micellar solution and hexagonal H_1 liquid crystal. However, the liquid crystal region shifts to lower concentrations when the distribution of chain gets wider. It is well known that hydration of head groups increases the effective volume fraction of aggregates, facilitating the structuring of aggregates below the critical packing concentration. Consequently, 10G0.7L is likely to form liquid crystals at lower surfactant concentrations than 10G*0.7L due to the larger proportion of hydrated, long hydrophilic chains in 10G0.7L.

For the hexagonal H_1 phase, the following equation holds

$$d^2 = \frac{\sqrt{3}}{2} r_s^2 \left(\frac{\rho_s}{\rho_w}\right)\left(\frac{1}{W_s}\right) + \frac{\sqrt{3}}{2} r_s^2 \left(\frac{\rho_w - \rho_s}{\rho_w}\right) \tag{19}$$

where d is the interlayer spacing, r_s is the radius of cylindrical micelles, ρ_s and ρ_w are the densities of surfactant and water, respectively, and W_s is the weight fraction of surfactant. The plot of d^2 vs $1/W_s$ gives a straight line (Fig. 8), which indicates that the radius of micelles does not change very much with surfactant concentration. Then, from the slope of the plot, the value of r_s is estimated to be 2.1 nm for 10G*0.7L and 2.33 nm for 10G0.7L.

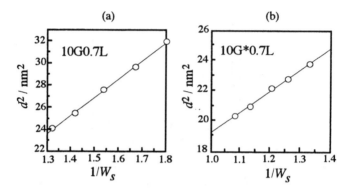

Figure 8 d^2 as a function of $1/W_s$ in 10G0.7L/water (a) and 10G*0.7L /water (b) systems [36]. Lines are best fits to experimental data.

The radius of the lipophilic part of the surfactant r_{sL} and the average effective cross-sectional area per hydrophobic chain, a_s, can be calculated using the equations

$$r_{sL} = r_s\sqrt{V_L/V_s} \tag{20}$$

$$a_s = 2V_L/r_sN_Am \tag{21}$$

where V_L and V_s are the molar volumes of the lipophilic part and of the surfactant, respectively, N_A is Avogrado constant, and m is the average number of fatty acid molecules bound to one polyglycerol chain. The results are summarized in Table 1.

Since the length of the lipophilic chain at its full extension is 1.54 nm, the lipophilic chain of 10G0.7L is almost fully stretched, whereas that of 10G*0.7L is somewhat contracted. On the other hand, a_s is smaller for the surfactant with a wide distribution of hydrophilic groups. Although the difference in the average number of polyglycerol units per fatty acid chain may contribute, the main factor seems to be the reduction in the repulsion between hydrophilic groups, the extent of which increases as the surfactant layer becomes more heterogeneous, i.e., the chain distribution becomes wider.

Table 1 Parameters for hexagonal H_1 liquid crystals in polyglycerol fatty acid ester systems [36].

Parameter	10G*0.7L	10G0.7L
r_s (nm)	2.10	2.33
r_{sL} (nm)	1.34	1.57
a_s (nm^2)	0.46	0.41

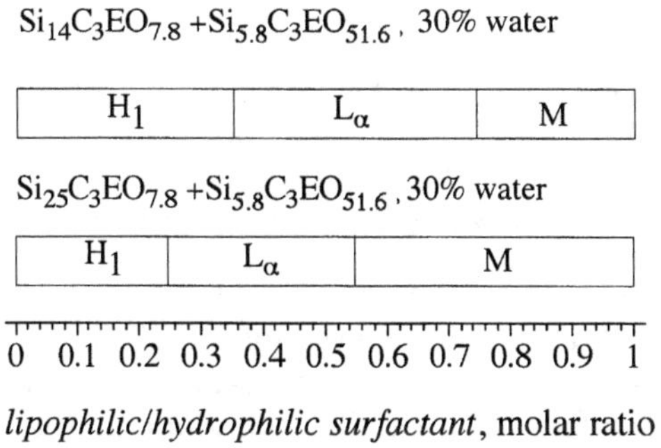

$Si_{14}C_3EO_{7.8} + Si_{5.8}C_3EO_{51.6}$, 30% water

H_1	L_α	M

$Si_{25}C_3EO_{7.8} + Si_{5.8}C_3EO_{51.6}$, 30% water

H_1	L_α	M

```
0   0.1  0.2  0.3  0.4  0.5  0.6  0.7  0.8  0.9  1
```

lipophilic/hydrophilic surfactant, molar ratio

Figure 9 Schematic phase behavior of $Si_{5.8}C_3EO_{51.6}/Si_{14}C_3EO_{7.8}/$ water and $Si_{5.8}C_3EO_{51.6}/Si_{25}C_3EO_{7.8}/$water systems at 25°C. The water weight fraction is kept at 0.3. M is a multiphase region [38].

C. Silicone Surfactant Systems

Siloxane surfactants have unique surface-active properties and they find applications in the production of polyurethane foams, textiles, cosmetics, agricultural adjuvants, and paints. Siloxane surfactants are commonly used in combination with other surfactants and polymers; therefore, it is important to understand the behavior of these mixtures [37].

The phase behavior of mixtures of poly(oxyethylene) poly(dimethylsiloxane) surfactants $(Me_3SiO-(Me_2SiO)_{m-2} -Me_2SiCH_2CH_2CH_2-O-(CH_2CH_2O)_nH$, abbreviated as $Si_mC_3EO_n)$ in water is shown in Fig. 9. When the hydrophilic surfactant $Si_{5.8}C_3EO_{51.6}$ is mixed with a lipophilic one with a similar molecular weight, $Si_{25}C_3EO_{7.8}$, a phase transition from H_1 to lamellar (L_α) phase takes place. This phase transition can be attributed to a change in the hydrophile-lipophile balance of

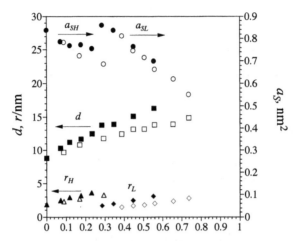

Figure 10 SAXS data for $Si_{5.8}C_3EO_{51.6}/Si_{14}C_3EO_{7.8}$/water and $Si_{5.8}C_3EO_{51.6}/Si_{25}C_3EO_{7.8}$/water systems at 25°C. ($\square$ $\blacksquare$) interlayer spacing, d; ($\bigcirc$ $\bullet$) effective cross-sectional area per surfactant molecule, a_s; ($\triangle$ $\blacktriangle$) radius of lipophilic cylinders in the hexagonal H_1 phase, r_H; ($\diamond$ $\blacklozenge$) lipophilic thickness in the lamellar L_α phase, r_L. Open symbols: $Si_{5.8}C_3EO_{51.6}/Si_{14}C_3EO_{7.8}$/water; filled symbols: $Si_{5.8}C_3EO_{51.6}/Si_{25}C_3EO_{7.8}$/water [38].

the surfactant mixture. Further addition of lipophilic surfactant leads to phase separation. Although both are polyoxyethylene-type silicone surfactants, the lipophilic surfactant cannot participate in formation of aqueous micelles with the hydrophilic surfactant due to the difference in lipophilic chain length.

When the number of polydimethylsiloxane groups of the lipophilic surfactant increases from 14 to 25 while keeping the EO number constant, the H_1–L_α transition and the phase separation shift to lower lipophilic surfactant molar ratios. This indicates that the lipophilic chain length affects the curvature of the surfactant layer and that the surfactant molecules are more incompatible with each other in aggregate formation.

As shown in Fig. 10, a_s decreases with increasing $Si_{14}C_3EO_{7.8}$ ratio within the H_1 and L_α phases. This means that the surfactant curvature decreases with addition of a lipophilic surfactant having almost the same molecular size but a different hydrophile-lipophile balance number. In this process, the hydrophobic part is elongated in the aggregates. Interestingly, the transition H_1–L_α takes place when the radius of cylindrical micelles (r_H) almost reaches as the value for the fully extended length of the $Si_{14}C_3$-chain, $l_{max} = 3.9$ nm [39].

When $Si_{14}C_3EO_{7.8}$ is replaced by $Si_{25}C_3EO_{7.8}$, the Si chain length of the lipophilic surfactant is increased keeping the EO chain length constant. Again, a_s decreases while r_H and r_L increases with increasing $Si_{25}C_3EO_{7.8}$ ratio within the H_1 and L_α phases. However, r_H and r_L show practically no change when compared with the $Si_{14}C_3EO_{7.8}$ system, even if the lipophilic chain is almost twice as long. This means that the Si chain is coiled in the aggregates. As a matter of fact, the actual Si chain length is less than one half of the fully extended length of the $Si_{25}C_3$-chain, $l_{max} = 6.7$ nm [39]. Figure 10 shows that a_s increases with increasing lipophilic chain length even if EO-chain length is fixed. According to a polymer theory [40], a long polymer chain has a short-bulky shape compared with a short polymer chain, because the entropy loss is very large when a long chain is in its extended form. Namely, it is energetically unfavorable for the long lipophilic chain to pack in a cylindrical micelle because the chain has to be elongated. Hence, at a certain point, the phase transition takes place to expand a_s and to decrease the radius. The bulky conformation of the Si_{25} chain causes stronger steric repulsions and induces a larger a_s than the Si_{14} chain.

III. SOLUBILIZATION OF OIL IN MIXED LIQUID CRYSTALS AND RELATED SYSTEMS

A. Nonionic Surfactant Systems

There has been increasing public interest in environmentally-friendly surfactants. Sucrose alkanoates derived from sugars and natural fatty acids are a good example of biocompatible surfactants. The study of their phase behavior is therefore important, particularly in the presence of water and oil.

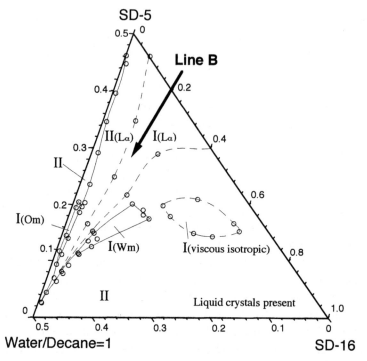

Figure 11 Partial phase diagram of water/SD-16/SD-5/decane system at 30°C (water to decane weight ration is unity). The weight ratio of mixed surfactant is varied from 0 to 0.5. Line B indicates a constant mixing ratio of SD-16/SD-5 = 2/8, along the line toward water–oil apex SAXS measurements were performed. SD-5 is a lipophilic sucrose ester surfactant (30.3% monododecanoate, 39.3% didodecanoate and 30.4% tridodecanoate). SD-16 is a hydrophilic sucrose ester surfactant (83.6% monododecanoate, 15.2% didodecanoate and 1.2% tridodecanoate) [41].

Addition of a cosurfactant usually induces the desired changes in the phase behavior.

Sucrose alkanoates possess a bulky hydrophilic group. The hydrophilic–lipophilic balance can be modified by changing the length of the fatty acid chain. Short-chain, hydrophobic surfactants with a low monomeric solubility in both water and oil are suitable as cosurfactants in mixed systems. Figure 11 shows the phase behavior of a mixture of lipophilic (SD-5, HLB = 5)

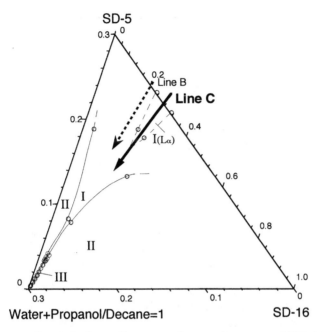

Figure 12 Partial phase diagram of water/1-propanol/SD-16/SD-5/ decane system at 30°C (water + propanol/decane weight ration is unity). I and III indicate one- and three-phase microemulsion regions respectively. The weight ratio of mixed surfactant is varied from 0 to 0.3. Lines B and C indicate constant mixing ratio of SD-16/SD-5 = 2/8 and 27/73, respectively, along the lines toward the water–oil apex SAXS measurements were performed [41].

and hydrophilic commercial (SD-16) sucrose alkanoates in water and oil. The lamellar L_α liquid crystal phase appears in a dilute region along line B (SD-16/SD-5 = 2/8). No three-phase region is observed. Since the CMCs of both surfactants are extremely low, most surfactant molecules should be located at the interface. Their solubility in oil is also low, and hence, the composition of the bilayer is considered to be almost constant, and the rigidity of the bilayer can be preserved. This rigidity inhibits the formation of a middle-phase microemulsion.

Middle chain alcohols are usually used to destabilize mesophases. This property is shown in Fig. 12. 1-Propanol

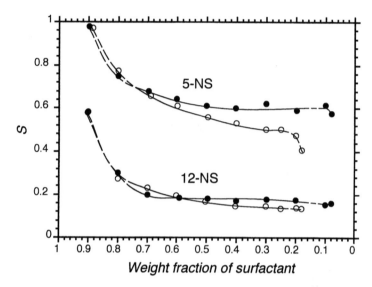

Figure 13 Apparent order parameter, S, of 5-NS (hydrophilic probe) and 12-NS (hydrophobic probe) along the line B in water/SD-16/SD-5/decane system (●) and line C in water/1-propanol/SD-16/SD-5/decane system (○) [41].

penetrates the surfactant layer and increases the flexibility of the bilayer, and consequently, the middle-phase microemulsion appears.

Decreasing in rigidity can be inferred from the values of the order parameter as measured by electron-spin resonance (ESR) [42]. Figure 13 shows that the decrease in the order parameter of the hydrophilic spin probe with dilution is more pronounced when propanol is added. On the other hand, the curves for the hydrophobic spin probe almost overlap each other. These results indicate that propanol molecules are situated mainly in the bilayer and reduce the attractive forces between surfactant molecules.

SAXS results for the previous systems are shown in Fig. 14. The value of interlayer spacing d increases linearly with the dilution ratio in the concentrated region, although there is a deviation from this tendency in the dilute region. As a matter of fact, if all surfactant molecules are situated in the bilayer and

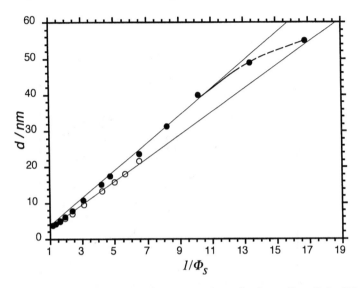

Figure 14 Change in interlayer spacing, d, along line B in Fig. 11 (●) and C in Fig. 12 (○). d is plotted against the reciprocal of the volume fraction of the surfactants (SD-16 + SD-5) [41].

the cross-sectional area per surfactant molecule remains constant, a linear relationship between d and the volume fraction of the surfactant, ϕ_s is expected.

B. Ionic-Nonionic Surfactant Systems

It is known [43] that lamellar L_α phases are stabilized by repulsive interactions mainly of electrostatic (ionic surfactants) or steric (nonionic surfactants) origin. If the lamellar L_α phase is composed of a mixture of nonionic and ionic surfactants, the electrostatic and steric interactions combine. Figures 15 and 16 show the phase diagrams of pseudo-ternary mixed ionic (sodium dodecyl sulfate, SDS)–nonionic ($C_{12}EO_3$) surfactant system. The single-phase microemulsion region extends from the $C_{12}EO_3$ apex to the brine/decane corner. Along the micro-emulsion region, lamellar L_α liquid crystals are present over a wide range of composition.

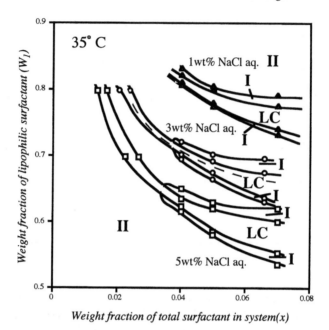

Figure 15   Phase diagrams of 1 wt% NaCl aq./SDS/$C_{12}EO_3$/decane
($\triangle$), 3 wt% NaCl aq./SDS/$C_{12}EO_3$/decane ($\bigcirc$), and 5 wt% NaCl aq./
SDS/$C_{12}EO_3$/decane ($\square$) systems in a dilute region at 35°C. I and II
indicate isotropic one-phase and two-phase regions, respectively. LC
means the region in which lamellar L_α liquid crystal is present [44].

If we assume that the bilayer composition is unchanged
and the solubility of the non-ionic surfactant in the oil part
of the L_α phase S_1 is constant, then the composition of the
bilayer would be unchanged along line A of Fig. 16. Under this
assumption, the following relations hold

$$d = 2d_s/\varphi_s' \tag{22}$$

$$\varphi_s' = \varphi_s - \frac{S_1}{1 - S_1}\frac{\rho_0}{\rho_1}\varphi_0 \tag{23}$$

where d is the interlayer spacing, d_s is the effective length of
the surfactant in the bilayer, φ_s is the volume fraction of mixed
surfactant in the system, φ_s' is the volume fraction of mixed

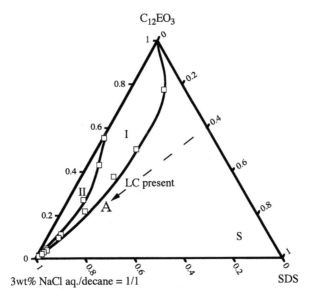

Figure 16 Phase diagram of $3\,\text{wt}\%$ NaCl aq./SDS/$C_{12}EO_3$/decane system at $35°C$. The brine/decane weight ratio is unity. The line A corresponds to the broken curve in Fig. 15 [44].

surfactant in the bilayer, φ_0 is the volume fraction of oil in the system, and ρ_1 and ρ_0 are the densities of nonionic surfactant and decane, respectively.

A straight line is obtained for the plot of d vs $1/\varphi'_s$ (Fig. 17) which is in agreement with Eq. (22). The average value of a_s calculated using Eq. (2) is practically constant at $0.37\,\text{nm}^2$, a value similar to the value measured for a lamellar L_α phase with a composition 50% $C_{12}EO_3$ in water [44]. However, SDS alone forms a hexagonal H_1 phase at the same composition, with a value of a_s much larger than the value indicated above. Since the charge density in the mixed surfactant layer decreases, the lateral attractive interactions increase over those of the ionic surfactant alone. Additionally, $C_{12}EO_3$ has a long lipophilic chain compared with ordinary cosurfactant (i.e. hexanol); therefore, the cohesive energy is strong and the lamellar L_α phase is stable even in a very dilute region.

The role of short EO chain alkyl ethoxylates as cosurfactants was pointed out earlier [45]. In $C_{12}EO_2$/SDS/water

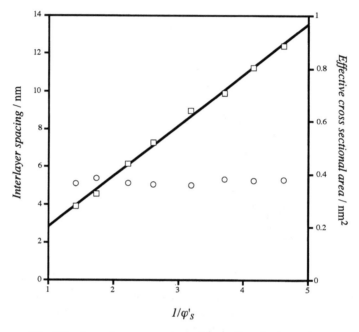

Figure 17 The interlayer spacing of lamellar L_α liquid crystal ($\square$) and effective cross sectional area per one surfactant molecule in the bilayer ($\bigcirc$) are plotted against $1/\phi'_s$ on the line A in Fig. 16. φ'_s is the volume fraction of the mixed surfactant bilayer in the system [44].

systems, addition of long hydrocarbon oil destabilizes the cubic phases to form W/O microemulsions. Furthermore, this addition promotes the formation of inverse structures (inverse hexagonal H_2 phase).

C. Cationic–Cationic Surfactant Systems

The phase behavior and microstructures of double-tail cationic surfactants have been extensively investigated [46]. Figure 18 shows the effect of adding oil (m-xylene) on the phase behavior of dodecyl trimethyl ammonium bromide (DTAB)/didodecyl dimethyl ammonium bromide (DDAB)/water systems. When a trace of oil is added to W_m and L_α regions, their compositions shift appreciably to the DTAB-rich side. With further addition of m-xylene, the W_m region moves back to the DDAB-rich side

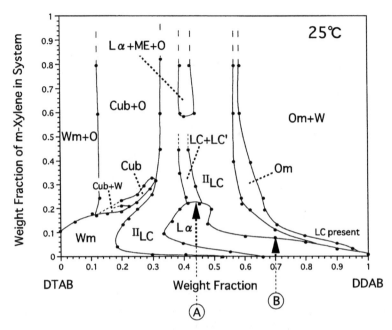

Figure 18 Partial phase diagram of a water/DTAB/DDAB/m-xylene system at 25°C. Volume fraction of lipophilic tail of a DTAB-DDAB mixture in aqueous surfactant solution is fixed at 0.2. W_m, W, O_m, O, Cub, L_α, II_{LC}, LC + LC', ME represent normal-type micellar solution, water containing monomeric surfactant, reverse-type micellar solution, oil containing monomeric surfactant, cubic liquid crystal, lamellar liquid crystal, two phase equilibrium of a birefringent liquid crystal and solvent (water or oil), two phase equilibrium of two kinds of birefringent liquid crystals, microemulsion, respectively. SAXS measurement was performed on lines A and B, whose DTAB/DDAB weight ratios are 56/44 and 3/7, respectively [47].

and the L_α region extends upward. The inverse micellar solution phase O_m region also shows a similar behavior.

The hydrophile–lipophile property of mixed surfactants is balanced in the L_α region, because its curvature is zero. With increasing size of oil molecule, the L_α region shifts to the DDAB-rich side as is shown in Fig. 19. In other words, a more lipophilic surfactant is needed when long hydrocarbon chain oil is added.

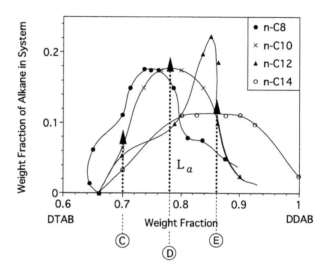

Figure 19 The L_α region for the system with n-octane (filled circle), n-decane (cross), n-dodecane (filled triangle), and n-tetradecane (open circle). SAXS measurements in the n-decane system were performed on the lines C, D and E, whose DTAB/DDAB weight ratios are 3/7, 22/78 and 14/86, respectively [47].

It is known that solubilization of oil occurs mainly by two mechanisms: penetration in the surfactant palisade layer and swelling of aggregates [48]. In the case of complete penetration, the value of the length of the lipophilic part of the surfactant r_L does not change with addition of oil. Then, the change in interlayer spacing d for the L_α phase is given by

$$d = \frac{2r_L^0}{\phi_L + \phi_0} \tag{24}$$

where r_L^0 is the value of r_L in the absence of oil. On the other hand, in the case of complete swelling, oil molecules do not mix with the lipophilic tails of surfactant in the bilayer, and a_s does not change. Then, the change in d is given by

$$d = \frac{2v_L}{a_s^0 \phi_L} \tag{25}$$

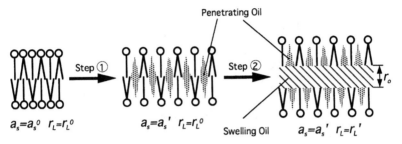

Figure 20 Derivation scheme of the penetration parameter, P_e, Step 1 and 2 correspond to the complete penetration and the complete swelling, respectively [47].

where a_s^0 is the a_s value for an oil-free L_α phase.

The degree of penetration of oil (P_e) can be evaluated if solubilization proceeds in two steps (Fig. 20). In the first step, oil molecules are solubilized penetrating into the palisade layer, and a_s' can be estimated as

$$a_s' = \frac{v_L}{r_L^0} \frac{(\phi_L + P_e \phi_0)}{\phi_L} \tag{26}$$

In the second step, the rest of the oil is solubilized only by swelling of the bilayer, and the value of a_s' does not change. Then, combination of Eq. (20) and Eq. (21) yields

$$P_e = \frac{2r_L^0}{\phi_0 d} - \frac{\phi_L}{\phi_0} \tag{27}$$

The P_e monotonically decreases with oil concentration, as shown in Fig. 21. Enhanced penetration is also observed at high DTAB ratios and m-xylene molecules penetrate much more than n-decane molecules.

D. Anionic–Cationic Surfactant Systems

The phase behavior of mixtures of anionic and cationic surfactants has been investigated in detail [49–53]. These systems display a rich polymorphism, and those phases that are not present in binary systems (single surfactant–water) appear.

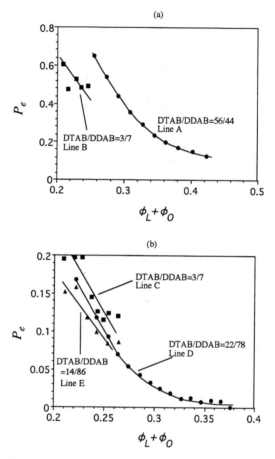

Figure 21 Variation of the penetration parameter, P_e, as a function of m-xylene (a) or n-decane (b) contents. Lines A–E correspond to Figs 18 and 19 [47].

The electrostatic interaction between oppositely charged groups, which decreases the interfacial area, and the steric interaction between hydrophobic tails, which increases the effective hydrophobic volume, both lead to an increase in the packing parameter CP [Eq. (15)] so that aggregates with low curvature are favored. In the system didodecyl dimethyl ammonium bromide (DDAB)–sodium dodecyl sulfate (SDS)–water [50] two lamellar phases are formed as well as a cubic phase of reversed structure (I_2). Equimolar mixtures of DDAB and sodium di-(2-ethylhexyl) sulfosuccinate (AOT) in water apparently form a pseudo four-tailed zwitterionic surfactant which results in a smaller effective head group and a larger hydrophobic moiety than those for individual surfactants [51]. This leads to formation of aggregates with reverse curvature. Addition of small amounts of DDAB to the L_α phase in the AOT/water system causes partial neutralization of the lamellar surface and new phases, as the bicontinuous cubic phase, are preferred. Interestingly, in the ternary system, two different reverse hexagonal (H_2) phases coexist.

The ternary phase diagrams for water–dodecane–SDS, water–dodecane–SDS–DTAB and water–dodecane–SDS–DDAB systems at 25°C are shown in Fig. 22. For the single surfactant SDS, only a micellar solution W_m and a hexagonal H_1 phase appear along the SDS–water axis. The oil solubilization in both W_m and H_1 phases is limited to 3% and further addition of oil to the H_1 phase leads to the formation of the discontinuous cubic phase I_1. The maximum solubilization capacity of the I_1 phase is around 8%.

Upon addition of a small amount of DTAB to the SDS system ($Y = $ cationic/(anionic + cationic) $= 0.2$), the W_m, H_1 and I_1 phases shift towards the dodecane apex, namely, the solubilization in the aggregates increases.

The I_1 can also be formed from the lamellar L_α phase in SDS–DDAB–dodecane–water systems at $Y = 0.2$ and solubilization in the I_1 phase is larger than that of SDS–DTAB. A vesicle-present region appears in the vicinity of the water apex. Compared with the single surfactant SDS, mixing of anionic and cationic surfactants results in formation of the I_1 phase at higher oil contents. The cubic I_1 phases with the lowest water content are also those that contain the largest amount of oil and have the highest oil/surfactant ratios, since the surfactant layer curvatures are also low at low water volume fractions.

A certain amount of oil is required to make the curvature more positive in anionic–cationic surfactant systems, which induces the formation of the I_1 phase (Fig. 23), and this amount

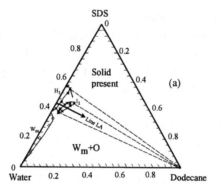

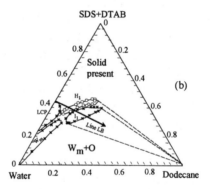

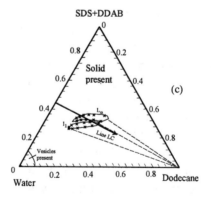

Figure 22 Phase diagrams for water–dodecane–SDS, $Y=0$ (a), water–dodecane–SDS–DTAB, $Y=0.2$ (b) and water–dodecane–SDS–DDAB, $Y=0.2$ (c) systems at 25°C. Y represents the cationic surfactant weight fraction in the surfactant mixture. Compositions are given in weight fractions [53].

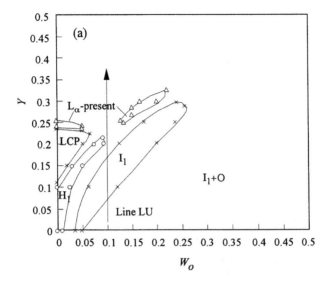

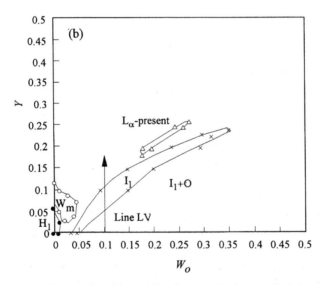

Figure 23 Phase diagrams if the cationic surfactant weight fraction Y as a function of the oil weight fraction W_o for the water-dodecane-SDS-DTAB (a) and water-dodecane-SDS-DDAB (b) systems at 25°C. The surfactant to water ratio is fixed at 40/60 [53]. LCP is a region in which liquid crystals are present.

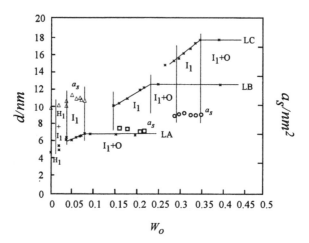

Figure 24 Interlayer spacing and corresponding cross-sectional area of micellar cubic (I_1) phases as a function of oil weight fraction at surfactant to water ratio equal to 45/55 along lines LA, LC and LB of Fig. 22 [53].

increases with Y, namely, as the curvature of the original oil-free system becomes more negative.

Figure 24 shows that the interlayer spacing d in anionic–cationic surfactant systems increases upon addition of oil and then becomes constant when the saturation limit is reached and phase separation occurs. This indicates that the micellar I_1 cubic phase is an isotropic single liquid crystal and that micelles grow by incorporating oil in the micellar core. On the other hand, a_s shows almost no change as oil is added, indicating little penetration of oil into the surfactant palisade layer.

Figure 25 shows that a_s decreases upon addition of cationic surfactant due to the electrostatic interactions between the hydrophilic heads of the cationic and anionic surfactants. Mixing of anionic and cationic surfactants and addition of oil work in the same direction in stabilizing the cubic phases and reduce the disparity between high preferred curvature and low mean curvature required by constraints in volume fractions.

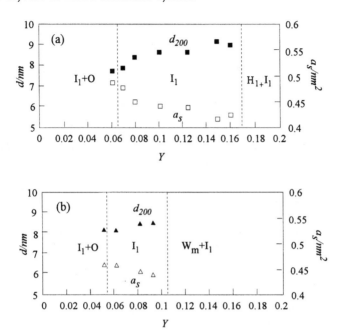

Figure 25 Interlayer spacing and corresponding cross-sectional area of micellar cubic (I_1) phases measured along lines LU and LV of Fig. 23 in SDS–DTAB (a) and SDS–DDAB (b) systems as a function of the cationic surfactant weight fraction at surfactant to water ratio of 40/60 and $W_o = 0.10$ [53].

REFERENCES

1. R. G. Laughlin, *The Aqueous Phase Behavior of Surfactants.* Academic Press, London (1994), pp. 200–237.

2. P. J. Collings, *Liquid Crystals.* Princeton Science Library, Princeton (1990), pp. 147–161.

3. G. H. Brown, and J. J. Wolken, *Liquid Crystals and Biological Structures.* Academic Press, New York (1979), pp. 22–44.

4. D. F. Evans, and H. Wennerström, *The Colloidal Domain.* Wiley-VCH, New York (1999), pp. 494–495.

5. C. Fairhurst, S. Fuller, J. Gray, M. C. Holmes, and G. J. T. Tiddy, In D. Demus, J. Goodby, G. W. Gray, H.-W. Spiers, and V. Vill

(eds.), *Handbook of Liquid Crystals*. Wiley-VCH, New York (1998), Vol. 3, Chapter 7, pp. 341–386.

6. P. Ekwall, In G. H. Brown (ed.), *Advances in Liquid Crystals*. Academic Press, New York (1975), pp. 1–141.

7. K. Fontell, and L. Mandell, *Colloid Polym. Sci.* **271**: 974–991 (1993).

8. H. Hoffmann, S. Hofmann, and J. C. Illner, *Prog. Colloid Polym. Sci.* **97**: 103–109 (1994).

9. M. Janiak, D. Small, and G. Shipley, *J. Biol. Chem.* **254**: 6068–6078 (1979).

10. J. Ulmius, H. Wennerström, G. Lindblom, and G. Arvidson, *Biochemistry* **16**: 5742–5745 (1977).

11. M. Ruocco, and G. Shipley, *Biochimica et Biophysica Acta* **684**: 59–66 (1982).

12. B. Mädler, H. Binder, and G. Klose, *J. Colloid Interface Sci.* **202**: 124–138 (1998).

13. V. Luzzati, In D. Chapman (ed.) *Biological Membranes*. Academic Press, New York (1968).

14. S. Anderson, S. T. Hyde, K. Larsson, and S. Lidin, *Chem. Rev.* **88**: 221–242 (1988).

15. V. Luzzati, R. Vargas, P. Mariani, A. Gulik, and H. Delacroix, *J. Mol. Biol.* **229**: 540 (1993).

16. D. Anderson, H. Wennerström, and U. Olsson, *J. Phys. Chem.* **93**: 4243–4253 (1989).

17. P. Alexandridis, U. Olsson, and B. Lindman, *J. Phys. Chem.* **100**: 280–288 (1996).

18. H. Kunieda, K. Shigeta, and M. Suzuki, *Langmuir*, **15**: 3118–3122 (1999).

19. K. Fontell, *Colloid Polym. Sci.* **268**: 264–285 (1990).

20. C. Rodriguez, K. Shigeta, and H. Kunieda, *J. Colloid Interface Sci.* **223**: 197–204 (2000).

21. M. H. Uddin, K. Watanabe, C. Rodriguez, A. Lopez-Quintela, T. Kato, H. Furukawa, A. Harashima, and H. Kunieda, *Langmuir* **17**: 5169–5175 (2001).

22. J. Israelachvili, *Intermolecular and Surface Forces.* London, Academic Press, (1991).

23. K. Kratzat, C. Stubenrauch, H. Finkelmann, *Colloid Polym. Sci.* **273**: 257–262 (1995).

24. D. Zhou, P. Alexandridis, and A. Khan, *J. Colloid Interface Sci.* **183**: 339–350 (1996).

25. K. Tamori, K. Esumi, and K. Meguro, *J. Colloid Interface Sci.* **142**: 236–243 (1991).

26. R. Boonbrahm, and A. Saupe, *Mol. Cryst. Liq. Cryst.* **109**: 225 (1984).

27. T. Wärnheim, B. Bergenstahl, U. Henriksson, A. C. Malmvik, and P. Nilsson, *J. Colloid Interface Sci.* **118**: 223–242 (1987).

28. H. Kunieda, K. Shigeta, K. Ozawa, and M. Suzuki, *J. Phys. Chem. B* **101**: 7952–7957 (1997).

29. K.-L. Huang, K. Shigeta, and H. Kunieda, *Prog. Colloid Polym. Sci.* **110**: 171–174 (1998).

30. J. C. Ravey, A. Gherbi, and M. J. Stebe, *Prog. Colloid Polym. Sci.* **79**: 272–278 (1989).

31. D. Rubingh, In K. L. Mittal (ed.) *Solution Chemistry of Surfactants.* Plenum, New York (1979), pp. 337–354.

32. M. J. Rosen, *Surfactants and Interfacial Phenomena,* 2nd Edn. John Wiley, New York (1989).

33. H. Kunieda, H. Kabir, K. Aramaki, and K. Shigeta, *J. Mol. Liq.* **90**: 157–166 (2001).

34. H. Kunieda, G. Umizu, Y. Yamaguchi, and M. Suzuki, *Yukagaku* **47**: 879–888 (1998).

35. H. Kunieda, G. Umizu, and Y. Yamaguchi, *J. Colloid Interface Sci.* **218**: 88–96 (1999).

36. M. Ishitobi, and H. Kunieda, *Colloid Polym. Sci.* **278**: 899–904 (1999).

37. R. M. Hill, In P. M. Holland, and D. N. Rubingh (eds.) *Mixed Surfactant Systems.* ACS Symposium Series. Maple Press, York (1992), pp. 278–291.

38. C. Rodríguez, M. H. Uddin, H. Furukawa, A. Harashima, and H. Kunieda, *Prog. Colloid Polym. Sci.* **118**: 53–56 (2001).

39. H. Kunieda, M. H. Uddin, M. Horii, H. Furukawa, and A. Harashima, *J. Phys. Chem. B* **105**: 5419–5426 (2001).

40. W. Kuhn, and F. J. Grun, *Polymer Sci.* **1**: 183 (1946).

41. N. Nakamura, Y. Yamaguchi, B. Hakansson, U. Olsson, T. Tagawa, and H. Kunieda, *J. Disp. Sci. Technol.* **20**: 535–557 (1999).

42. S. Schreier, C. F. Polnaszek, and I. C. P. Smith, *Biochimica et Biophysica Acta* **515**: 375–436 (1978).

43. R. Schomacker, and R. Strey, *J. Phys. Chem.* **98**: 3908–3912 (1994).

44. H. Kunieda, K. Ozawa, K. Aramaki, A. Nakano, C. Solans, *Langmuir* **14**: 260–263 (1998).

45. H. Sagitani, and S. E. Friberg, *Coll. Polym. Sci.* **261**: 862–867 (1983).

46. S. J. Chen, D. F. Evans, B. W. Ninham, D. J. Mitchell, F. D. Blum, and S. Pickup, *J. Phys. Chem.* **90**: 842 (1986).

47. K. Aramaki, and H. Kunieda, *Colloid Polym. Sci.* **277**: 34–40 (1999).

48. H. Kunieda, K. Ozawa, and K.-L. Huang, *J. Phys. Chem. B.* **102**: 831–838 (1998).

49. A. Khan, and E. Marques, In D. Robb (ed.) *Specialist Surfactants*. Blackie A&P, London (1997), pp. 44–51.

50. E. Marques, and A. Khan, *J. Phys. Chem.* **97**: 4729–4736 (1993).

51. A. Khan, O. Regev, A. Dimitrescu, and A. Caria, *Prog. Colloid Polym. Sci.* **97**: 146–150 (1994).

52. X. Li, and H. Kunieda, *J. Colloid Interface Sci.* **231**: 143–151 (2000).

53. X. Li, and H. Kunieda, *Langmuir* **16**: 10092–10100 (2001).

11

Sponge Structures of Amphiphiles in Solution

KAZUHIRO FUKADA

Department of Biochemistry and Food Science,
Faculty of Agriculture, Kagawa University,
Kagawa, Japan

KAZUO TAJIMA

Department of Chemistry, and Hightech Research
Center, Faculty of Engineering, Kanagawa
University, Kanagawa, Japan

SYNOPSIS

Formation of sponge structure was described for amphiphiles in solution. Experimental evidence for sponge-like structure was showed as a L_3 phase on a nonionic surfactant solution, and moreover effects of ionic additives to L_3 phase were described together with some theoretical interpretations for L_3 phase. Sponge state may be essentially recognized in the phase diagram for the surfactant–water system. On the contrary, sponge structure of phospholipid NaDMPG was observed as a network structure of bilayer-assembly with and without additive salt. However, sponge-like structure for the phospholipid was absolutely different from that of surfactant solution. Sponge-like structure appeared only at an intermediary state in a region of a phase transition from vesicles to lamellar or vice versa, but no thermodynamically stable state. An intermediary state of phospholipid bilayer-assembly was indicated on the conversion into a new gel state with very slow time aging.

I. INTRODUCTION

Amphiphilic molecules such as surfactants and lipids can self-associate into a variety of molecular aggregates in solution, which can transform from one into another when the solution conditions are changed. While the geometry of aggregates depends on the molecular packing constraints or the inter-molecular forces within the aggregates, the equilibrium phase state of the whole system is determined by the strength of inter-aggregate forces. When the concentration of amphiphilic molecules is high, the inter-aggregate interactions increase and the tendency of amphiphile to organize the system finds its fullest expression in the lyotropic liquid crystalline phases. Examples are hexagonal, cubic, and lamellar phases. If the concentration is not high enough to bring about a liquid crystalline phase, but more than high enough to self-association of amphiphilic molecules, an isotropic fluid phase containing a finite concentration of micelles or vesicles can be formed. In the case of surfactants which can form a highly swollen lamellar phase owing to the long-range repulsion between the bilayers, there is now strong evidence that in the region of intermediate concentrations sufficient to produce a structured or complex fluid but not so much as to produce a liquid crystalline phase, the whole system consists of *sponge-like disordered three-dimensional network of bilayers* as the equilibrium phase state. This phase is called sponge or L_3 phase, and which is of great interest to us.

The first part of the present article is concerned with the thermodynamically stable sponge-like structured complex fluids, i.e., L_3 phase of surfactant–water systems. In the literature one can find examples of L_3 phase for the systems of pentaoxyethylene dodecylether ($C_{12}E_5$)–water [1,2], cetylpyridinium halide–hexanol–brine [3–5], sodium dodecyl sulfate (SDS)–pentanol–brine [6], SDS–pentanol–dodecane–brine [6], didodecyldimethylammonium acetate–water [7], Aerosol OT–brine [8], and fluorinated nonionic surfactant–water [9]. Surfactant systems exhibiting flexible membranes, where the flexibility is induced by the presence of co-surfactant (long chain alcohol) in some cases, high-amplitude thermal undulation of the membranes lead long-range repulsive interactions between membranes, which also play the major role for the sponge phase formation. Brief explanation on the theoretical basis for the L_3 phase formation will be made in this section.

In the second part, on the other hand, we will look at the sponge-like structures found for charged phospholipids in aqueous medium. Contrary to the L_3 phase of surfactant systems, however, the sponge-like network structure of phospholipids has been observed only in a restricted condition or as a transitional intermediate [10]. For molecular assemblies of phospholipids, it is often quite difficult to identify experimentally the true equilibrium state, and so, slight differences in the preparation method and/or the experimental conditions can lead different structures for phospholipid aggregates in many cases. It is therefore important to clarify the meaning of thermodynamically equilibrium state of phospholipid–water systems. Discussion will be made concerning the true equilibrium conditions of such systems.

II. L₃ (SPONGE) PHASE IN WATER–SURFACTANT SYSTEMS

A. What is L₃ Phase?

In the systems of water and surfactants, the surfactant molecules can self-assemble into a variety of structures depending on the HLB-balance or structural constraints of the molecules. These molecular assemblies can organize themselves on a large scale, with either long-range order (liquid crystalline phases) or only short-range correlation (liquid isotropic phases). Concerning the liquid isotropic phases, the most stable structural unit in many cases is a bilayer, which is locally flat, but tends to wander entropically at large distances. The bilayer sheets formed by surfactant molecules play a role similar to the monolayers in water–oil–surfactant systems. Thus, one can expect in the water–surfactant system an analogue of the bicontinuous microemulsion in the water–oil–surfactant system. This analogue is denoted by L_3, or sponge phase.

The L_3 phase is rather viscous, shows flow birefringence, and scatters light strongly, showing the presence of extended surfactant aggregates. When the surfactant concentration is extremely low, flow birefringence may not be noticed by visual observation even though the sample shows some internal turbidity [7]. Figure 1 shows a schematic illustration of the randomly connected bilayer network proposed for the structure of L_3 phase. The sponge-like random surface of bilayer divides space into two interpenetrating solvent labyrinths. In the phase diagram, the L_3 domain is generally very narrow, and is found in the vicinity of a highly swollen lamellar phase stabilized by an undulation force. In some cases the L_3 phase is in equilibrium with both the dilute micelle solution (L_1) and the swollen lamellar phase (L_α). A typical example is shown for water–$C_{12}E_5$ system in Fig. 2 [1,2,11]. With such systems, L_3 simply phase separates at high dilution and expels a very dilute micelle solution (L_1). In phase diagrams for other systems, however, the L_3 domain is continuously connected to the domain of the dilute micelle solution [12]. So, it is possible in

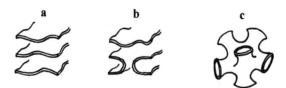

Figure 1 Structures of lamellar and L_3 phases. (**a**) Three adjacent flexible bilayers in a swollen lamellar L_α phase. (**b**) Formation of one passage or "handle" in the L_α phase. (**c**) Multiconnected membrane in the L_3 sponge structure.

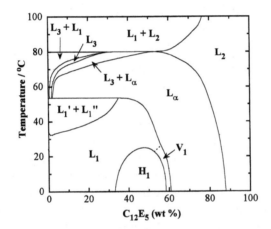

Figure 2 Phase diagram of binary water-pentaoxyethylene dodecylether ($C_{12}E_5$) system, redrawn based on [1,2]. L_1, L_2, and L_3 denote isotropic solutions. H_1 is a hexagonal liquid crystal, V_1 is a cubic liquid crystal, and L_α is a lamellar liquid crystal phase, respectively.

this case to move from the domain of multiconnected membrane structure in L_3 phase towards the domain of dipersed disconnected micelles (or vesicles) with no phase separation along the dilution line.

It is worthy to note that L_3 phase can also be found in water–oil–surfactant systems [6,11]. Usually, oil can be incorporated into the water–surfactant L_3 phases. When the oil content is small, the interconnected bilayer simply swells and the sponge-like structure of the phase is unaffected. When a large amount of oil is added, an oil-rich microemulsion phase separates. In some cases a continuous path is seen in the phase diagram connecting the L_3 domain to the classical bicontinuous microemulsion. A topological transformation must occur at some point in this case, since the bilayer in L_3 phase separates two identical water sub-volumes, whereas a monolayer in the microemulsion separates interwoven oil and water domains. One may expect some critical behavior to accompany this topological change.

B. Experimental Evidence for Sponge-like Structure of L_3 Phase

Experimental facts leading to a geometrical description of the sponge-like structure of interconnected bilayers for L_3 phase (Fig. 1c) are based on several measurement techniques: light scattering properties; transport properties measured using conductivity and self-diffusion; neutron scattering; and freeze-fracture electron microscopy [13].

In L_3 phase, typical neutron scattering profiles over a wide range of wave vector (q) exhibit a low-q broad maximum at q_c followed by a long decreasing tail at higher-q [3,4]. The quantitative analysis of the high-q decay showed that on a local scale the structure of L_3 consists of bilayers of the same thickness of bilayers in neighboring L_α phase with random orientations. On the other hand, the position q_c of the broad maximum was of the same order of magnitude as this of the Bragg wave vector for bilayer spacing in L_α phase at the same dilution. Therefore q_c can be interpreted as a signature of the average distance d between the bilayers in the L_3: $d = 2\pi/q_c$. For a random distribution of connected bilayers, one can expect $d = \gamma\delta/\phi$ (δ and ϕ are the thickness and volume fraction of bilayers, respectively) with γ, a number larger than 1 (the lattice model leads to $\gamma = 1.5$). The neutron scattering peak position in L_3 phase indeed showed a dilution behavior of d with $\gamma = 1.4$–1.6. These data indicate no long-range order in L_3 phase, but a well-defined characteristic length d related to the average inter-bilayer spacing. For the L_3 phase of quasi-ternary brine-1-hexanol-cetylpyridinium chloride system, the neutron scattering patterns described above had been confirmed [3,4], and the image of the L_3 phase of this system was directly obtained by freeze fracture electron microscopy [5], which clearly showed an apparently bicontinuous structure of two aqueous sub-volumes separated by a random bilayer network.

C. Effects of Ionic Additives to L_3 phase of Nonionic Surfactant Systems

Effects of additives on the phase behavior and properties of nonionic bilayers such as the bending modulus have been investigated by several researchers. It has been suggested that addition of ionic surfactants to nonionic bilayers suppress the undulation of bilayers, indicating an increase in the bending modulus [14–19].

Figure 3a shows the effect of addition of lauric acid, $C_{12}H_{25}COOH$, on the phase behaviors of $C_{12}E_5$–water system

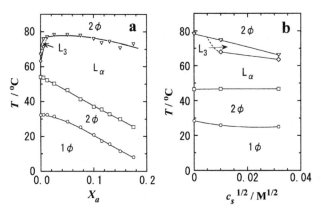

Figure 3 Phase diagrams of water–$C_{12}E_5$–lauric acid system [19]. L_3, L_α, 1ϕ, and 2ϕ denote a sponge phase, a lamellar liquid crystal phase, a micelle solution, and coexistence of two isotropic solutions, respectively. (**a**) Influence of lauric acid on phase behavior where the mole fraction of lauric acid in the total mixed solute (X_a) is varied from 0 to 0.175 while the concentration of the total mixed solute (c_{tot}) is kept at 1.4 wt%. (**b**) Effect of sodium chloride on phase behaviors for $X_a = 0.05$ and $c_{tot} = 1.4$ wt%. The abscissa is the root of the molarity of sodium chloride.

keeping the total concentration of the mixed solute ($C_{12}H_{25}$ COOH + $C_{12}E_5$), $c_{tot} = 1.4$ wt% [19]. The solubility of lauric acid in water is extremely low and almost all molecules are incorporated in $C_{12}E_5$ bilayers. The mole fraction of lauric acid in the total mixed solute is denoted by X_a. As X_a increases, the L_α L_3 transition temperature first increases rapidly, and then the L_3 phase disappears when X_a becomes *ca.* 0.012. After the L_3 phase disappears, the high-temperature boundary of the L_α phase takes a maximum and then gradually decreases. The space distance between bilayers in the L_α phase at 58°C was determined by light scattering measurements, and a slight decrease of the spacing with increasing X_a was confirmed (data not shown here). According to Strey *et al.* [2], the space distance of thermally undulating bilayers, d, is expressed as

$$d = \frac{\delta}{\phi}\left(1 + \frac{k_B T}{4\pi}\ln\frac{c}{\phi}\left(\frac{\kappa}{k_B T}\right)^{0.5}\right) \cong \frac{\delta}{\phi}\left(1 - \frac{k_B T}{4\pi\kappa}\ln\phi\right) \qquad (1)$$

where ϕ, δ, and κ are the volume fraction, thickness, and bending modulus of bilayer, respectively, c the numerical constant of the order of unity, k_B the Boltzmann constant, and T the absolute temperature. It should be noted that the second relation is derived under the approximations $c \cong 1$ and $\kappa \cong k_B T$. From Eq. (1), the decrease in d with increasing X_a can be interpreted from the fact that the dissociation of lauric acid incorporated in $C_{12}E_5$ bilayers increases the bending modulus, κ. Further, one may say that this effect of electric charge leads to the disappearance of L_3 phase when X_a becomes above 0.012.

To see the effects of screening of electrostatic repulsion by the addition of salt, the phase behavior of $C_{12}H_{25}COOH + C_{12}E_5$ in NaCl aqueous solution was investigated [19]. Figure 3b shows changes in the phase behavior at $c_{tot} = 1.4$ wt% and $X_a = 0.05$ as the concentration of NaCl is increased. The low-temperature boundary of the L_α phase is not affected by the addition of NaCl in the concentration range studied, but the high-temperature boundary decreases with increasing salt concentration. Also, one can see that L_3 phase reappears when the concentration of NaCl is above $10^{-3}\,\mathrm{mol\,dm^{-3}}$. It would be worthy to note here that for the samples with $c_{tot} = 1.4\,\mathrm{wt\%}$ and $X_a = 0.05$ at 50°C in the L_α region, the light scattering diffraction peak becomes very broad with increase in NaCl concentration. This observation suggests that the screening of electrostatic repulsion increases the fluctuations of the bilayer space distance remarkably as a result of the decreased bending modulus of bilayers. This screening effect must be the cause of the reappearance of L_3 phase in the higher NaCl concentration region around 70°C.

These experimental results mentioned above support the idea that the bending properties of bilayers are the foundation for the formation and stability of sponge-like structure [20].

D. Theoretical Interpretations of L₃ Phase

In this subsection, we will outline the basis of theoretical approaches for L_3 phase formation. For the interested readers, the review articles should be referred to get more detailed information about the theoretical treatments [13,20,21].

For the systems of surfactant bilayers in aqueous medium where no strong Coulombic interactions exist, the theoretical description on the stability of L_3 phase can be made based on the continuum elasticity of two-dimensional flexible films embedded in three-dimensional space. The starting point for such analysis is the harmonic bending free energy of a curved surface, H, described as [22]

$$H = \int \left(\frac{\kappa}{2}(C - C_0)^2 + \bar{\kappa}K \right) dA \qquad (2)$$

in which C is the mean curvature $[C = (1/R_1) + (1/R_2)$, where R_1 and R_2 are the two principal radii of curvature of the surface], C_0 the spontaneous curvature of the surface, $K = 1/CR_1R_2$ (R_1R_2) the Gaussian curvature, and dA the area element. The parameters κ and $\bar{\kappa}$ are the mean and Gaussian bending modulus, respectively. The integral over Gaussian curvature gives the following relation by the Gauss–Bonnet theorem

$$\int K dA = 4\pi(1 - g) \tag{3}$$

where g is the number of "handles" (see Fig. 1b) of the surface under analysis. Consequently, $\bar{\kappa}$ is included in the calculation of H as the chemical potential for the formation of handles, whereas κ alone determines the topology-preserving deformations of the bilayer. If the local conditions of the bilayer are changed so as to make $\bar{\kappa}$ less negative within the range $-2\kappa < \bar{\kappa} < 0$, structures with many handles, such as L_3 phase, will be relatively favored in free energy compared to a phase of vesicles or lamellae.

To understand the effect of thermal fluctuations on the bending free energy, one has to compare κ to k_BT. When $\kappa \sim k_BT$, thermal fluctuations will become important. At finite temperature, thermal fluctuations at small scales affect the change in the bending free energy when imposing a deformation of length scale ξ. As a consequence of the length scale dependent renormalizations of bending rigidities [23], $\kappa(\xi)$ decreases with ξ and finally vanishes at a characteristic length called the persistence length, $\xi_\kappa = a \exp(4\pi\kappa/3k_BT)$, where a is the molecular size cutoff. For lengths larger than ξ_κ, a bilayer is expected to thermally crumple.

The first interpretations of L_3 phase were focused on its disorder character: it was described as a random self-avoiding surface [24]. For simplicity, a single bending constant approximation was considered based on a scale dependent $\kappa(\xi)$ but with $\bar{\kappa}$ being set to zero. L_3 was thus found to have a lower free energy than L_α at low surfactant concentrations. In this *scenario*, L_3 is stabilized mainly by entropic effects, which melt the lamellar phase when its layer space distance becomes too large, compared to the persistence length, ξ_κ.

An alternative view was proposed which emphasized the difference in topology between L_α and L_3, hence using $\bar{\kappa}$ as the leading control parameter [4]. This scenario explains the stability of L_3 phase found at rather high concentrations by adding just the right amount of alcohol, which is incorporated into the bilayer as a co-surfactant. In this situations, the Gaussian bending modulus, $\bar{\kappa}$, is tuned to be very close to zero: $0 \gtrsim \bar{\kappa} \gg -\kappa$, and this allows handles of small curvature to form at almost no cost in free energy. In this case, L_3 structure is far from a random surface, resembling rather a disordered analogue of a cubic minimal surface phase [11].

III. FORMATION OF SPONGE-LIKE STRUCTURE FOR PHOSPHOLIPIDS

A. Three-dimensional Structure of Phospholipid Bilayers

We have a few papers reporting that phospholipid bilayers take a network formation of a structure resembling the sponge structure observed in the surfactant solutions [25,26]. Only a few phospholipids were confirmed to form a three-dimensional structure of bilayer-assembly in the form of a sponge-like structure [25–27]. In practice, no sponge-like structure can be formed in the bilayer-assembly of dimyristoylphosphatidylcholine (DMPC)–water system due to the thermal instability, but the only possibility for formation of three-dimensional (or network) structure is a bilayer-assembly for sodium dimyristoylphosphatidylglycerol (NaDMPG) with or without sodium chloride. Schneider *et al.* have suggested the possibility of network formation for NaDMPG dispersion in a unique temperature region as schematically illustrated in Fig. 4 [25]. They showed a sponge-like structure of NaDMPG taken by the freeze fracture method (Fig. 5), and mentioned that the sponge-like structure was formed in a range of 25–28°C in a dispersion with dilute salt solution. They also found that the formation of sponge-like structure was limited to a region of an appropriate concentration of ionic strength, and no sponge-like structure can be formed at high concentrations of coexisting salt. Figure 6(a) shows the possible region of salt concentration for the network formation with NaDMPG, and Fig. 6(b) shows the expected temperature region of network formation for another kind of phospholipid glycerol.

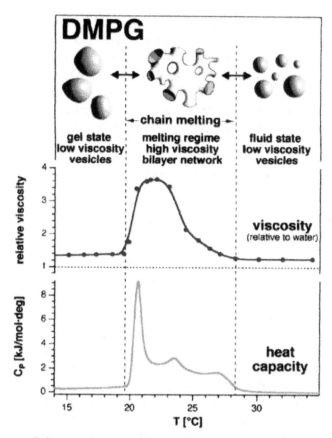

Figure 4 Schematic summary of the chain of events deduced from calorimetry, viscosity measurements, and electron microscopy. Structural changes represent changes in the three-dimensional membrane arrangement. These changes correlate with viscosity increases and distinct heat capacity maxima. Viscosities are given relative to that of water at the same temperature.

We have known many papers concerning the bilayer properties of NaDMPG dispersed in salt solution [28,29]. In particular, Parsegiane *et al.* reported a lot of sophisticated theoretical studies and elaborated experimental data on the effect of additive salts on the hydration repulsion between the hydrated biomolecular membrane arrangements [30,31]. However, they failed to either show or describe the formation

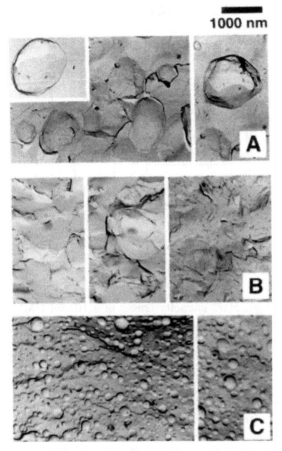

Figure 5 Freeze fracture electron micrographs of a 45×10^{-3} mol dm^{-3} DMPG dispersion. (A) Large vesicles at 9°C (representative features of different areas are given). The membrane sheets are overlaid by short periodic ripples. (B) Membrane sheets at 24.5°C arranged randomly toward the fractured surface and continuous membrane segments (representative features of different areas are given). The membrane sheets are overlaid by large periodic ripples. In some segments these ripples can be seen in a cross section. (C) Perfectly spherical vesicles of various sizes at 50°C (representative features of different areas are given). The vesicles are on average significantly smaller than those in A.

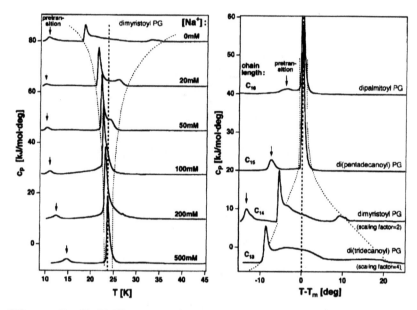

Figure 6 (*Left*) Heat capacity traces of a 10 mM DMPG dispersion under various ionic strength conditions (in a 2 mM Hepes, 1 mM EDTA, pH 7.5 buffer; the top trace was measured in distilled water). At low ionic strength the C_p traces show a complex behavior as in Fig. 4. With increasing NaCl concentration the chain of events occurs over a narrower temperature interval. At 500 mM Na^+ one single, a highly cooperative heat capacity peak is found. Below the main transition the pretransition can be found (indicated by an arrow). (*Right*) Heat capacity traces of 120 mM dispersions of phosphatidylglycerols with various chain lengths (in distilled water, pH ~ 7.5). PG (C_{13} chains) and dimyristoyl PG (C_{14} chains) show a complex behavior as in Fig. 4. With increasing chain length the chain of events occurs over a narrower temperature interval. Dipalmytoyl PG (C_{16} chains) displays only one very cooperative heat capacity peak.

or existence of sponge-like structure in any bilayer phase transitions. On the other hand, Gershfeld *et al.* found a new transition temperature, which they called a critical temperature, T^*, in their equilibrium spread pressure measurements of phospholipids as a function of temperature [32], but they could give no clear interpretation for the physical properties of T^*. According to their many studies [33–35], T^* may be related to a

Table 1 T^* for various compositions of NaDLPG–NaDMPG mixed bilayer.

X_M	0	0.25	0.50	0.75	1.0
$T^*/°C$	19.2	23.7	29.0	30.5	31.5

X_M, Mole fraction of NaDMPG in mixed bilayer.

bilayer phase transition, corresponding to the formation temperature of sponge-like structure for NaDMPG bilayer, though they did not give any direct evidence for the presence of sponge-like structure in the dispersion. Recently, they reported the possible formation of a sponge-like structure for NaDMPG and NaDLPG–NaDMPG bilayers [36,37]. Assuming the formation of a three-dimensional structure of bilayer at T^*, they reported the values of T^* for various phosphlipid bilayers as shown in Table 1.

However, no direct evidence for the formation of sponge-like structure was given in Gershfeld's papers. The T^* could easily be determined for NaDMPG bilayer, but it was unclear for DMPC bilayer in the DSC measurements. The T^* for DMPC bilayers may be determined from the ESR spectra measured as a function of temperature [38]. Also, we showed that T^* could be determined more easily from the temperature dependence of d spacing for DMPC bilayers [39].

The observations on T^* of DMPC bilayers could be conformed to the temperature dependence of surface chemical phenomena at the oil–water interface. Generally, when hexadecane was emulsified with a DMPC dispersion, the emulsion prepared was absolutely different in the surface chemical properties and stability below and above T^* [40]. The droplet surface was coated with a bilayer-assembly of DMPC below T^*, whereas it was covered with a monolayer of DMPC molecules above T^*. The physical properties of bilayer-assembly should then be said to be quite different below and above T^*. We found the same tendency for the ionic phospholipids of NaDMPG and NaDLPG in emulsification of hexadecane using them [41].

Possible formation of sponge-like structure for phospholipids would be expected as an intermediary phase when the

Vesicle ⟺ Network Structure ⟺ Lamellar Structure

Figure 7 Schematic representation for formation of sponge-like structure as an intermediate state in bilayer transition from vesicles into lamellar structures.

bilayer transition occurs from a vesicle structure phase to a lamella phase as schematically shown in Fig. 7. However, we have no comprehensive explanation for the fact that DMPC bilayer forms no sponge-like structure, while DMPG bilayer is capable of forming sponge-like structure only when the concentration of NaCl as additive is limited to a unique region. Also, we have no evidence to imply that thermodynamical stability is one of the essential requisites for sponge-like structure to appear in phospholipid bilayer as that observed in the surfactant bilayer.

Then, the formation of sponge-like structure and vesicles in the phospholipid bilayer would not always be comparable to the appearance of a similar structure in the surfactant solutions [1–9], because the formation of sponge-like structure or vesicles may not be one of the essential properties of the phospholipid bilayer in contrast to the surfactant solution as described in the above section. We believe that sponge structure appears as a temporary one in the transition process in a thermodynamic sense as shown in a later section. We would like to propose here the following processes that allow a sponge-like structure to appear for the phospholipid bilayer in the transition of bilayer-assembly from vesicles to lamellae or vice versa.

First, we describe how the formation of sponge-like structure can be considered for the phospholipid bilayer-assembly as compared with that for morphological liquid crystal of surfactant solution. (1) Two types of thermal undulation for surfactant bilayers would be taken account of. One is symmetric vibration and the other is asymmetric vibration. Symmetric undulation would occur in the dilute bilayer solution, but when the concentration of surfactant is increased, the undulating motion should transform into the asymmetric motion. (2) Asymmetric undulating motion induces the formation of a node and loop structure by two bilayers. The node is possible to conduct each of the two bilayers. The concentration of surfactant at the contact part in the nodes must increase. Then, surfactant in the nodes should dissolve in water, and therefore the undulating motion leads to hole formation, when the nodes of bilayers are again widened by the vibration motion. (3) The formation mechanism of sponge-like structure is likely the same as that of the coalescence of liquid droplets in emulsion or the rupture of foam films.

B. Thermal Properties and Sponge-like Structure of NaDMPG

Recently, Tajima *et al.* have reported the effects of temperature, incubation and aging on NaDMPG and DMPC bilayer-assemblies. Figure 8 shows the DSC charts of dispersions of DMPC, NH_4DMPG and NaDMPG incubated overnight at 20°C [42]. Lamellar liquid crystals of DMPC and NH_4DMPG gave reproducible DSC charts in repeated measurements, whereas NaDMPG and NaDLPG, though the data are not shown here, indicated quite different DSC charts in repeated measurements after heating once up to their T^* [43]. An explanation for such abnormal thermal behavior of NaDMPG and NaDLPG bilayer-assemblies was given in terms of the molecular theory. Namely, the abnormal thermal properties of NaDMPG bilayer-assemblies produced by irreversible conformational changes in the terminal glycerol moiety in NaDMPG molecules are shown in Fig. 9 when the bilayer-assembly was heated once over T^*.

In general, we can say that when the temperature of the dispersion was varied, the entire bilayer-assembly itself could not immediately establish a thermodynamically stable structure, even though the respective molecules in the

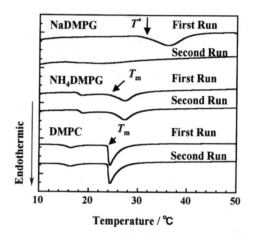

Figure 8 Repeated DSC measurements for phospholipid dispersions. The thermal properties of NH₄DMPG dispersion was quite resemble those of DMPC dispersion, while NaDMPG dispersion was absolutely different from DMPC dispersion. The second run for NaDMPG dispersion gave no clear transition peaks.

bilayer-assembly could promptly be rearranged to take a thermally stable conformation accordingly. As shown in Fig. 10, therefore, the physical states of bilayer-assembly gradually shifted to a more stable arrangement, when the states were continuously measured for a long period of time [44]. In other words, temperature change or a progress of hydration phenomena caused a gradual change in the bilayer-assembly dispersed in water toward a thermally stable state, so that it is very difficult to judge if the bilayer is in a thermodynamically stable state or not. For instance, it took about 14 or 20 days for the bilayer to reach a stable state.

Figure 11 shows the XRD patterns for NaDMPG dispersions incubated at 35°C. Any XRD peaks for NaDMPG bilayers could not be observed during a period of about 7 days. After about 10-day aging, we could observe the XRD pattern clear enough to recognize a bilayer-structure for NaDMPG. This abnormal phenomenon may be explained as showing that when the dispersion was heated once over the T^* (31.5°C), the ordered arrangements of the polar head moieties in the bilayer

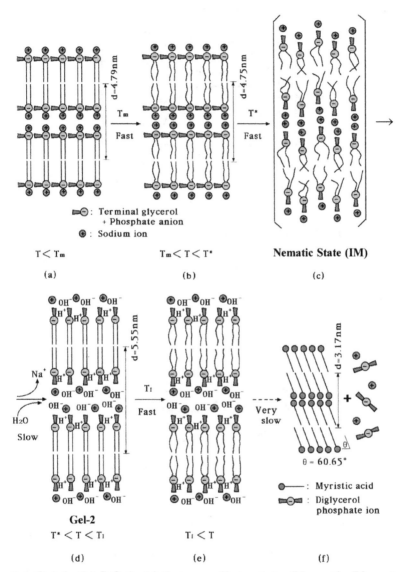

(a) Gel-1, (b) LC-1, (c) Intermediate state (Nematic Phase),
(d) New Phase Gel-2, (e) LC-2, (f) Myristic acid.

Figure 9 Schematic representation for various kinds of thermal transitions in NaDMPG bilayer-assembly, in consideration of the effect of aging for bilayer-assembly from (a), (b), and (c) states to (d) gel state.

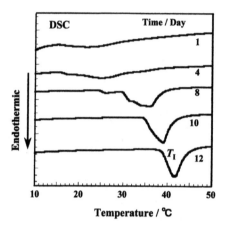

Figure 10 Time dependence of T_m transition for NaDMPG. T_I, transition temperature for a new gel phase of Gel 2. The new gel phase Gel 2 appeared about 12 days after dispersion in water at 35°C.

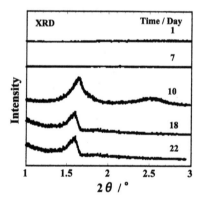

Figure 11 Time dependence of XRD patterns for NaDMPG dispersion. The bilayer was transformed from a nematic to smectic structure after being aged for about 12 days at 35°C.

would slip into the direction vertical to the bilayer surface to take a structure similar to the nematic one. After being aged for about 10 days, the disordered surface of the polar moieties was rearranged to make a renewed NaDMPG-bilayer with a smectic structure with a dissociation of water.

The sponge-like structure of NaDMPG, which was found by Schneider *et al.* [25], was confirmed undoubtedly by the freeze fracture method. However, we need to know whether or not the sponge-like structure is able to exist as a thermodynamically stable phase in the phase diagram. Because the phase transition in the bilayer-assembly of phospholipid is extremely slow as shown in Fig. 9, we have to pay attention to the possible occurrence of sponge-like structure as an intermediary state when the bilayer transforms from a phase into another. In particular, since the polar moiety in NaDMPG consists of an ionic part and a terminal glycerol part, the former part quickly put the ionic repulsive force into effect between molecules, and the latter part takes a very slow hydration. Since the phase transition proceeds very slowly, the reproducible physical states of bilayer-assembly are very difficult to obtain and are sensible to the experimental conditions.

C. Possibility for Formation of Sponge-like Structure as an Intermediary Phase in Transition from Spontaneous Vesicle Phase to Lamella Phase

No phase diagram for the NaDMPG–water system has yet been established due to variations in the aging effect and significant thermal history in the bilayer-assembly. Recently, Tajima *et al.* have reported the detailed phase behavior of a dispersion of NaDMPG (5.0 wt%) as a function of salt concentration [45,46]. Figure 12 shows variations in the physical state for NaDMPG bilayer with the concentration of NaCl. The physical state of NaDMPG was clearly dependent on changes in the aging temperature and the concentration of added salt. For instance, when the dispersion was aged overnight at 20.0°C, a large endothermic peak was found to appear at 31.5°C (T^*) by DSC, while no peak was observed at the corresponding temperature when the dispersion was incubated at 40°C overnight and then cooled below 0°C for DSC measurements [43].

We would like to describe the bilayer behavior in more detail as a function of salt concentration (C_s) following the results in Fig. 11. At the incubation temperature of 40°C,

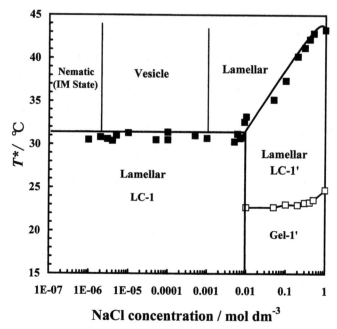

Figure 12   Effect of added salt on the phase transitions of NaDMPG bilayer. ■, T^* transition; □, T_m transition.

since the NaDMPG bilayer with salt at $C_s < 2 \times 10^{-6}$ mol dm^{-3} took a nematic state, no peak was found in the DSC and XRD measurements. In the salt concentration region of $2 \times 10^{-6} < C_s < 1 \times 10^{-3}$ mol dm^{-3}, no transition peak was observed by DSC, whereas a very sharp XRD peak caused by the bilayer structure of NaDMPG was found. These findings therefore suggest spontaneous vesicle formation. Figure 13a is a typical TEM micrograph taken by the freeze fracture method of vesicles (LULV) of NaDMPG [46]. At salt concentrations higher than $C_s > 1 \times 10^{-3}$ mol dm^{-3}, DSC gave two peaks at T_m and T^*, while XRD indicated a clear bilayer structure. Then, we can say that the vesicle structure was destroyed to form a lamella structure by the addition of salt at concentrations higher than 1×10^{-3} mol dm^{-3} as shown in Fig. 13b.

As shown in Fig. 2 for the surfactant solution [5], a sponge-like structure has been confirmed to appear in the transition process from cubic phase to lamella phase in the

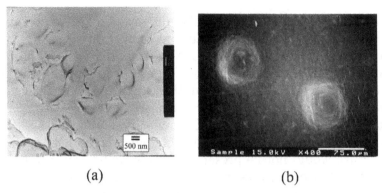

(a)   (b)

Figure 13 (a) A typical TEM photo taken by the freeze fracture method for NaDMPG dispersion in NaCl solution at 5×10^{-4} mol dm^{-3}. The photo indicates the formation of vesicles at salt concentrations of 2×10^{-6} mol dm$^{-3} < C_s < 1 \times 10^{-3}$ mol dm^{-3}. (b) SEM image for the NaDMPG dispersion in coexistence with large multilamellar vesicles at salt concentration at higher than 1×10^{-3} mol dm^{-3}.

phase diagram. Similarly, as to sponge-like structure formation in phospholipid bilayer, we believe that the structure is possible to appear as an intermediary state when spontaneously formed vesicles transform into lamellae caused by variations in salt concentration as shown in Fig. 12. Another possibility in Fig. 12 is that if vesicles are formed from lamellae by simply heating NaDMPG dispersion aged at 20°C with salt less than 1×10^{-3} mol dm^{-3}, sponge-like structure could be present as an intermediary state. This possibility has never been checked by reliable experiments except of Schneider's report [25]. The abnormal events reported near the T^* transition temperature for NaDMPG bilayer by Gershfeld et al. [34] might be based on the temporal formation of a sponge-like three-phase structure in the bilayer-assembly.

D. Differences in Stability between Sponge-like Structures of Phospholipid and Surfactant

As described in the previous section, sponge-like structure formation would be possible as an intermediary state in the transition from vesicle to lamella and vice versa. The apparent similarity between phospholipid and surfactant in sponge-like structure formation may be understood in terms of bilayer liquid crystal structure. However, there are important differences in the stability between these cases. With surfactant bilayers, sponge-like structure is reasonably recognized in the phase diagram as a thermodynamically stable phase [34]. In contrast, in the case of phospholipid bilayers, sponge-like structure would not be present as a thermodynamically stable phase, because the appearance of sponge-like structure reported by Schneider [25] is based on the unsuitable treatments on the bilayer-assembly in NaDMPG dispersion.

The uncertainties in Schneider's study [25] can be summarized as follows: first, in order to promote the swelling process in the preparation of NaDMPG dispersion, NaDMPG powder was added to salt solution or water and the dispersion was heated and cooled repeatedly many times. Such heating and cooling process should not been carried out because the process produces very serious results as shown in Fig. 8. When the NaDMPG dispersion was heated once over T^*, the DSC measurements gave no reproducible DSC curve for NaDMPG bilayer due to an incredibly grave thermal history, quite different from that for DMPC bilayer, even though the dispersion was incubated at a constant temperature for a while.

Secondly, the aging of NaDMPG bilayer was not taken into consideration. We emphasized that the physical state of NaDMPG dispersion depends on the incubation temperature as well as the aging time in the pretreatment as shown in Fig. 10. The presence of sponge-like structure in Fig. 4 is therefore understandable as indicating that the physical state of the dispersion changes with aging time. In other words, the sponge-like structure of NaDMPG bilayer must change into other physical states with aging, similar to the fact that the DMPC vesicle structure is spontaneously broken.

Finally, they prepared the NaDMPG dispersion using a buffer solution to prevent the lipid from being degraded. However, the physical state of NaDMPG bilayer could not be determined, because the bilayer is quite sensitive to added ions or solution pH, as our results showed. The time dependent

Table 2 Thermal transition temperatures for various kinds of phospholipid bilayers with water.

Phospholipid	Temperature /$°C$		
	T_m	T^*	T_I
NaDLPG	4.5	19.0	20.5
NH$_4$DLPG	6.5	19.5	22.5
NaDMPG	23.5	31.7	40.0
NH$_4$DMPG	23.5	31.7	42.0
DMPC	23.5	29.0	40.0
DPPC	40.5	45.0	55.0

phenomena occurred in the bilayer-assembly seem to be unavoidable for the present due to the requirement of thermodynamic stability, with or without a buffer solution. Unfortunately, trace amounts of coexisting salts produce variations in the bilayer structure of NaDMPG as shown in Fig. 9. It is very important therefore that if one wishes to determine the physical state of bilayer-assembly, the dispersion should be prepared using a bufferless solution.

We have confirmed through solid NMR and DSC measurements [46] that all states observed in Fig. 12 converted into a new state, a hydrated gel (Gel 2), after the system was aged for about 20–30 days below or above T^*. This Gel 2 state changes reversibly to a new liquid crystal (LC2) at the T_1 temperature, like a gel–liquid crystal state for DMPC bilayer-assembly. Table 2 shows the T_I temperatures with some other transition temperatures for various lipids.

The Gel 2 state of NaDMPG which comes out in the dispersion after a few weeks is just in a thermodynamic equilibrium state. We therefore believe that all phases appeared before reaching the Gel 2 state after the crystal powder was dispersed into water are undoubtedly at intermediary pseudo-states and are not true and thermodynamically stable phases. Consequently, we do not hesitate to say that the sponge-like structure and vesicles formed with only phospholipid are thermodynamically unstable and they should spontaneously break with elapsed time. Such characteristic of phospholipid bilayers evidently distinguishes them from surfactant bilayers. In conclusion, we believe that the sponge-like structure or vesicles formed with phospholipids cannot exist as a thermodynamically stable state if some kinds of proteins or lipids are added into the phospholipid bilayer-assembly.

REFERENCES

1. F. Harusawa, S. Nakamura, and T. Mitsui, *Colloid Polymer Sci.* **252**: 613–619 (1974).

2. R. Stray, R. Schomäcker, D. Roux, F. Nallet, and U. Olsson, *J. Chem. Soc. Faraday Trans.* **86**: 2253–2261 (1990).

3. G. Porte, J. Marignan, P. Bassereau, and R. May, *J. Phys. France* **49**: 511–519 (1988).

4. G. porte, J. Appell, P. Bassereau, and J. Marignan, *J. Phys. France* **50**: 1335–1347 (1989).

5. R. Strey, W. Jahn, G. Porte, and P. Bassereau, *Langmuir* **6**: 1635–1639 (1990).

6. D. Gazeau, A. M. Bellocq, D. Roux, and T. Zemb, *Europhys. Lett* **9**: 447–452 (1989).

7. E. Z. Radlińska, T. N. Zemb, J.-P. Dalbiez, and B. W. Ninham, *Langmuir* **9**: 2844–2850 (1993).

8. I. Carlsson, and H. Wennerström, *Langmuir* **15**: 1966–1972 (1999).

9. J. C. Ravey, M. J. Stébé, S. Sauvage, and C. Elmoujahid, *Colloids Surfaces A* **99**: 221–231 (1995).

10. M. F. Schneider, D. Marsh, W. Jahn, B. Kloesgen, and T. Heimburg, *Proc. Nat. Acad. Sci. USA* **96**: 14312–14317 (1999).

11. D. Anderson, H. Wennerström, and U. Olsson, *J. Phys. Chem.* **93**: 4243–4253 (1989).

12. M. Filali, G. Porte, J. Appell, and P. Pfeuty, *J. Phys. II* **4**: 349–365 (1994).

13. D. Roux, C. Coulon, and M. E. Cates, *J. Phys. Chem.* **96**: 4174–4187 (1992).

14. M. Jonströmer, and R. Strey, *J. Phys. Chem.* **96**: 5993–6000 (1992).

15. R. Schomäcker, and R. Strey, *J. Phys. Chem.* **98**: 3908–3912 (1994).

16. C. B. Douglas, and E. W. Kaler, *Langmuir* **7**: 1097–1102 (1991).

17. C. B. Douglas, and E. W. Kaler, *Langmuir* **10**: 1075–1083 (1994).

18. C. B. Douglas, and E. W. Kaler, *J. Chem. Soc. Faraday Trans* **90**: 471–477 (1994).

19. N. Awata, K. Minewaki, K. Fukada, M. Fujii, and T. Kato, *Colloids Surfaces A* **183–185**: 449–455 (2001).

20. G. Porte, *Curr Opinion Colloid Interface Sci.* **1**: 345–349 (1996).

21. D. C. Morse, *Curr. Opinion Colloid Interface Sci.* **2**: 365–372 (1997).

22. W. Helfrich, *Z. Naturforsch.* **28c**: 693–703 (1973).

23. P. G. deGenns, and C. Taupin, *J. Phys. Chem.* **86**: 2294–2304 (1982).

24. M. E. Cates, D. Roux, D. Andelman, S. T. Milner, and S. A. Safran, *Europhys. Lett* **5**: 733–739 (1988).

25. M. F. Schneider, D. Marsh, W. Jahn, B. Kloesgen, and T. Heimburg, *Proc. Natl. Acad. Sci.* **96**: 14312–14317 (1999).

26. R. G. Laughlin, Micelles Microemulsions, and Monolayers; Science and Technology, Dekker, New York, pp. 73–99.

27. Y. S. Tarahovsky, A. L. Arsenault, R. C. MacDonald, and T. J. McIntosh, *Biophys. J.* **79**: 3193–3200 (2000).

28. K. A. Riske, H.-G. Doebereiner, L. Freund, and M. Teresa, *J. Phys. Chem., B* **106**: 239–246 (2002).

29. P. Garidel, G. Forster, W. Richter, B. H. Kunst, G. Rapp, and A. Blume, *Phys. Chem. Chem. Phys.* **2**: 4537–4544 (2000).

30. M. E. Loosley-Millman, R. P. Rand, and V. A. Parsegian, *Biophys. J.* **40**: 221–232 (1982).

31. R. P. Rand, and V. A. Parsegian, *Biochim. Biophys. Acta* **988**: 351–376 (1989).

32. N. L. Gersheld, and K. Tajima, *Nature* **279**: 489–500 (1979).

33. K. Tajima, and N. L. Gershfeld, *Biophys. J.* **47**: 203–209 (1985).

34. N. L. Gershfeld, and L. Ginsberg, *J. Membrane Bio.* **156**: 279–286 (1997).

35. N. L. Gershfeld, K. Tajima, C. P. Mudd, and R. L. Berger, *Biophys. J.* **65**: 1174–1179 (1993).

36. M. Koshinuma, K. Tajima, A. Nakamura, and N. L. Gershfeld, *Langmuir* **15**: 3430–3436 (1999).

37. K. Tajima, M. Koshinuma, A. Nakamura, and N. L. Gershfeld, *Langmuir* **16**: 2576–2580 (2000).

38. K. Tajima, Y. Imai, T. Horiuchi, M. Koshinuma, and A. Nakamura, *Langmuir* **12**: 6651–6658 (1996).

39. K. Tajima, Y. Imai, A. Nakamura, and M. Koshinuma, *Chem. Lett.* **1995**, 527 (1995).

40. K. Tajima, Y. Imai, A. Nakamura, and M. Koshinuma, *J. Jpn Oil Chem. Soc.* **46**: 283–291 (1997).

41. K. Tajima, Y. Imai, A. Nakamura, and M. Koshinuma, *Colloids and Surfaces*, **155**: 311–322 (1999).

42. K. Tajima, T. Tsusui, Y. Imai, A. Nakamura, and M. Koshinuma, *Chem. Lett.* **2002**: 50–51 (2002).

43. K. Tajima, Y. Imai, A. Nakamura, and M. Koshinuma, *Adv. Colloid and Interface Science* **88**: 79–97 (2000).

44. K. Tajima, Y. Imai, A. Nakamura, and M. Koshinuma, *J. Oleo Sci.* **50**: 475–484 (2001).

45. K. Tajima, Y. Imai, A. Nakamura, and M. Koshinuma, Phase behavior of bilayer-assembly for dimyristoylphosphatidylglycerol sodium salt dispersed in NaCl aqueous solution: spontaneous vesicle formation before transformation to a stable phase by gelation (in press).

46. K. Tajima, Y. Imai, and T. Tsusui, *J. Oleo Sci.* **51**: 285–296 (2002).

12

Surfactant Gels from Small Unilamellar Vesicles

M. MÜLLER and H. HOFFMANN
University of Bayreuth, Physical Chemistry I,
Bayreuth, Germany

SYNOPSIS

While vesicles are usually associated with phospholipids [1] or double chain surfactants it is well known now that vesicles can also be formed from single chain surfactants. Actually, a whole variety of methods are known by which vesicles can be prepared. In general, vesicles are found in systems where one expects to find L_α-phases because vesicles can always be produced by shear from L_α-phases [2,3]. Vesicles are therefore observed in aqueous mixtures of cat-anionic surfactants at a certain mixing ratio, in mixtures of surfactants and cosurfactants and in ionic surfactant solutions in combination with strongly hydrophobic counterions [4–6]. Furthermore, mixtures of Ca-salts of ionic surfactants and zwitterionic surfactants can also reveal vesicles under special conditions [7]. The size of the vesicles in these systems usually depends on the way the phases have been prepared and therefore on the history of the phases. While shear stress was involved in the preparation of many of these vesicle phases it is also true that vesicles are formed under conditions where no shear is involved. There actually is evidence that vesicles can be thermodynamically stable species. That means that the properties of these thermodynamically stable vesicle systems do not depend on the way they are prepared; the same phase composition will always lead to the same microscopic and macroscopic properties (e.g., size distribution of the vesicles, rheological behavior) if the system is given enough time for reaching equilibrium. So, on discussing surfactant vesicles we have to be careful in making general statements. What might be valid in one particular situation might not be true in another situation. The vesicle phases which can be prepared with the given methods have usually a broad size distribution including small unilamellar vesicles together with big multilamellar vesicles.

1. INTRODUCTION

The phases under such conditions can be viscoelastic if the concentration is high enough for a dense packing of the vesicles. Actually, under these conditions the system is really a L_α-phase in the single-phase region in which the bilayers have been transformed into vesicles.

However, recently it was observed that under the same conditions under which one finds the viscoelastic vesicle phases one also could find very stiff isotropic phases. These isotropic phases consist of densely packed monodisperse unilamellar vesicles [8]. As far as we know today, these phases are thermodynamically stable. In this article we would like to summarize the available results on these systems. We would also like to point out that vesicle phases might have been discovered before the term "vesicle" was used and before the micellar structure of vesicles was proven by electron microscopy. A number of ternary phase diagrams that have been determined by Ekwall *et al.* [9] contain liquid crystalline (LC) isotropic phases around areas where vesicles normally can be expected. The micellar structures that were discussed to exist in these would be called vesicles today.

II. RESULTS

A. The Phase Diagrams

Stiff vesicle phases have been found in the sodium-oleate/octanol/water system. The phase diagram of this system is given in Fig. 1. It contains a region that consists of a completely transparent, stiff, and optically isotropic phase that had an unknown structure when discovered and for this reason was called a "gel" in the phase diagram. In other studies on ternary

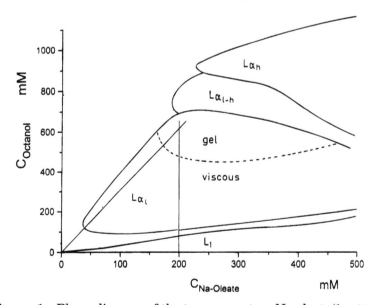

Figure 1   Phase diagram of the ternary system Na oleate/1-octanol/water at 25°C. L_1, micellar phase $L_{\alpha l}$, vesicle phase; $L_{\alpha l-h}$, two phase region of vesicles and lamellae; $L_{\alpha h}$, lamellar phase.

phase diagrams with cosurfactants no phase with such properties had been found. Usually, viscous birefringent L_α-phases had been observed in the composition region of other ternary systems. For constant cosurfactant/surfactant ratio and increasing concentration the gel phase is placed between an isotropic viscous phase and a birefringent $L_{\alpha h}$-phase. The isotropic viscous phase may actually be a two-phase region in which no macroscopic phase separation occurs. On further dilution one observes however a two-phase region where an isotropic L_1- and a birefringent L_α-phase separate from each other.

Further studies showed that phases with similar properties could be found in other similar systems like sodium isostearate. Sodium linolate, in contrast, did not show any transparent stiff gel phase in the respective region of the phase diagram. In Fig. 2 different phases are indicated that are observed in sodium oleate, sodium isostearate, and sodium linolate when the concentration of cosurfactant is increased. The investigations show that the formation of the "gel"-phase

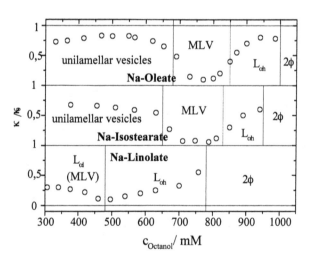

Figure 2 Cut through the phase diagrams of the systems Na oleate, Na isostearate and Na linolate respectively with octanol as cosurfactant. The surfactant concentration is kept constant. The conductivity vs increasing octanol concentration.

is rather subtle. While it occurs in the sodium-oleate/octanol/
water and sodium-isostearate/octanol/water systems, it does
not occur in the system sodium-linolate/octanol/water system.
Otherwise, the two systems behave in very similar ways.
Figure 2 also contains the results of conductivity measurements. These data show that the conductivities in the "gel"-
phase drop with respect to the L_1-phase, but the drop is much
less pronounced than in systems where classic L_α- or multilamellar vesicles phases (MLV) had been observed. This is an
indication that the vesicles in the gel-phase are unilamellar and
the number of ionic groups and counterions is larger on the
outer monolayer than on the inner monolayer.

B. Preparations of the Gel-Phases

The gel-phases can be prepared in two ways that are quite
different from each other but lead to the same final properties.
In the first way a 200 mM solution of the ionic surfactant is
prepared, which is a low viscous micellar solution. In the next
step an appropriate amount of cosurfactant is added under
stirring the micellar solution. Within seconds we obtain a
turbid but homogenous phase that then becomes transparent
after a few minutes but is still of low viscosity. The gelation of
this phase can be followed by rheological measurements. At
room temperature the whole process takes about 1 h. This
shows that the micellar reorganization in the transparent
isotropic phase is a rather slow process. The development of the
final structures with time in the stagnant phase can easily be
followed by various physicochemical methods, in particular by
small angle neutron scattering (SANS)-measurements and by
oscillating rheological measurements. The results of these
measurements show that on mixing the compounds rod-like
micelles are formed at first and they are then transformed into
small unilamellar vesicles that become monodisperse with time.

The gel phase can also be prepared in a way in which the
micellar structures are not exposed to shear. In this process
the micellar solution and a previously prepared emulsion of
the cosurfactants, both phases having an appropriate concentration, are mixed quickly to reach a homogenously dispersed

state of the emulsion in the micellar solution. This turbid phase can then be left to its own. As in the first method, the phase becomes transparent and develops into the final transparent stiff gel-phase. Obviously in the last method, the cosurfactants molecules have to leave the emulsion droplets and have to diffuse to the micellar structures. The whole process can thus be divided into several stages. The fact that the properties of the gel phase are independent of the route on which the phase is prepared is evidence for the phase being an equilibrium phase. Since the phase consists of unilamellar vesicles, this experiment demonstrates that vesicles do indeed form spontaneously without shear stress forces when a right composition in a micellar phase is given.

C. Rheological Properties

A rheogram of the gel phase is given in Fig. 3. One typical feature is the shear modulus that is practically independent of

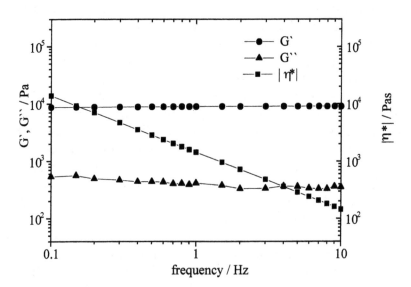

Figure 3 Rheogram of a monodisperse densely packed vesicle gel (182 mM Na-Isostearate/567 mM Octanol). Frequency dependence of the storage modulus (G'), the loss modulus (G'') and the complex viscosity (η^*).

the frequency over the whole applied frequency range of more than three orders of magnitude. The loss modulus is more than an order of magnitude lower and decreases with increasing frequency. The storage modulus is about 10^4 Pa and two orders of magnitudes higher than the values that are attained in vesicle phases when these vesicles are multilamellar. In Fig. 4 a rheogram of a typical vesicle phase with comparable surfactant and cosurfactant concentrations is shown. The rheogram looks very much the same but the G' and G'' parameters are very much lower. This comparison shows clearly that the unilamellar vesicles form only under very subtle conditions, which depend on the chemistry of the system. Besides the mixing ratio, additional parameters determine the formation of unilamellar vesicles. In the following chapter several parameters of the systems will be varied in order to characterize the stability criteria for the vesicle gel phase.

If a stiff gel-phase with a constant cosurfactant/surfactant ratio is diluted below a certain concentration one obtains a clear viscous solution. The solution exhibits shear thinning

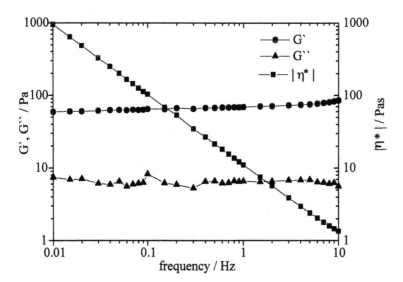

Figure 4 Rheogram of vesicle system of multilamellar, polydisperse and densely packed vesicles. The system consists of 182 mM Na-linolate and 567 mM octanol.

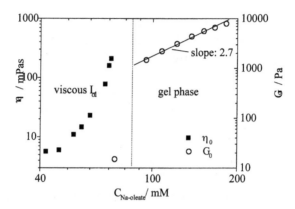

Figure 5 Shear modulus and zero-shear viscosity for samples with a constant molar ratio of octanol and Na oleate of $1:3.1$ as a function of the surfactant concentration.

with a finite zero-shear viscosity. A rheogram of such a phase is given in Fig. 5. It is still possible to determine a frequency-independent shear modulus for this diluted sample at high frequencies. The modulus is falling with the concentration with a slope of about 3 as can be seen in Fig. 6. This is an interesting result for the theoretical discussion.

D.   The Influence of Excess Salt on the Vesicle Phase

From previous investigations it is known that the rheological properties of MLV-phase depend very much on the salt concentration added to shield the ionic charge of the bilayers. Similar measurements on the gel phases reveal a more refined situation shown in Fig. 7.

Within a salt concentration range up to 10 mM there is hardly any effect on the shear modulus. In contrast to the modisperse small unilamellar vesicles (SUV) gel, the modulus of the polydisperse MLV gel is sensitive even to very small amounts of added salt. Other measurements, in particular SANS measurements, show that in the region where the modulus decreases the system has lost its single-phase character and

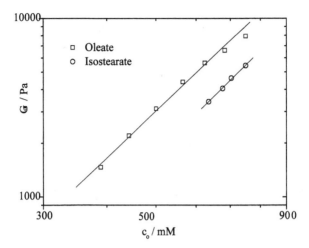

Figure 6 Concentration depending shear modulus of monodisperse SUV gels. The total molar concentration of Na oleate (Na isostearate) plus octanol was increased while the molar ratio of Na oleate : octanol was kept constant at 1 : 3.1.

has become a two-phase system where domains of MLV phases are dispersed in a matrix of SUVS. These results indicate that the shear modulus in the vesicle gel phase is not determined by the electrostatic interaction but by the number density of the vesicles. However, when more than 10 mM salt is added the bilayer becomes more and more flexible and MLV-vesicle and typical L_α-phases can be formed.

E. The Influence of Charge Density

In an ionic surfactant system more than 50% of the counterions will be condensed at the micellar interface in the Helmholtz fixed double layer. The counterions do not contribute to the ionic strength and the electrostatic interaction. The fraction of ionic surfactant with condensed counterions may therefore be replaced with uncharged surfactants without changing the electrostatic interaction between the micellar structures. In order to test this hypothesis samples were prepared in which part of ionic surfactants were replaced by

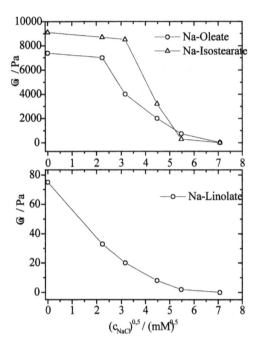

Figure 7 Differences between the influence of salt on the stiffness of SUV and a polydisperse MLV gel. Dependence of the shear modulus on additional salt (NaCl) for gels with 182 mM surfactant and 567 mM octanol.

alkyldimethylaminoxide. The results show that about up to 50% of the ionic surfactants could be replaced by the zwitterionic surfactant without loss of the gel-phase. The shear modulus of the gel-phases obtained is plotted in Fig. 8 against the mole fraction of the zwitterionic surfactant. The modulus remains practically constant up to a mole fraction of 50% and then breaks down at higher contents. At a mole fraction of 90% alkyldimethylaminoxide the shear modulus reaches that of a typical multilamellar vesicle system.

F. Chain Length Dependence of Cosurfactant

It is known that n-alcohols can be used as cosurfactants up to a rather long chain length. Even oleylalcohol can form L_α-phases

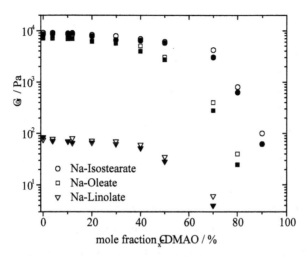

Figure 8 Influence of the charge density on the gel stiffness. The carboxy surfactant was replaced by zwitterionic surfactant $C_{14}DMAO$ (open symbols) and $C_{16}DMAO$ (solid symbols) at a constant octanol concentration ($c_{surfactant} = 182$ mM, $c_{octanol} = 567$ mM).

when combined with other surfactants. It was therefore of interest to find out whether all the n-alcohols that are generally used as surfactants can form vesicle gels.

The results showed that it is only possible to prepare the vesicle-phases with intermediate chain length alcohols between hexanol and octanol. For longer n-alcohols the normal MLV-systems were obtained. This is shown in a plot of the shear modulus against the chain length of the alcohol (Fig. 9). It is very surprising that a micellar phase that can be prepared from mixtures of Na–oleate and octanol cannot be formed with mixtures of Na–oleate and nonanol. This result clearly shows that a subtle balance of interaction is necessary for the formation of stiff vesicle phases.

While stiff gels were all observed with a surfactant concentration of about 200 mM and cosurfactant concentrations of around 600 mM it is noteworthy that Ekwall reported an isotropic gel-like phase in the system potassium oleate/decanol/water. This phase is located in a very small region with the composition in wt% 14.35, 23.17% and 62.47 wt%. This phase

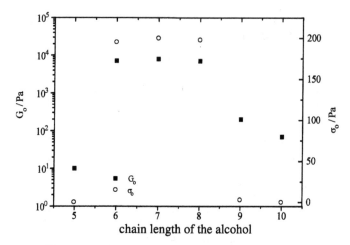

Figure 9 Shear modulus and yield stress value as a function of chain length of the alcohol ($c_{surfactant} = 182$ mM, $c_{octanol} = 567$ mM).

contains more than twice as much surfactant and cosurfactant as the gel phase reported in this work. It is noteworthy, however, that the mole ratio of cosurfactant to surfactant is about the same as that for the vesicle phases in this work. Potassium oleate in Ekwall's isotropic gel phase can be replaced by sodium oleate while the characteristic properties of the phase remain unaffected. At the time of its discovery, the term "vesicles" had not yet been coined. The description that was given by Ekwall for the micellar structures that exist in the phase—he describes these structures as spheres made up from bilayers—is however consistent with vesicles. These vesicles are extremely small and have a radius of about 50 Å. The macroscopic properties of this reported phase were very much like the properties of gel phases of the novel systems reported in this work.

G. The Melting of the Gel-Phases

The stiff phases become viscous phases at a well-defined temperature. This process does not occur over a large temperature region but occurs within a small temperature range and the melting is associated with an enthalpy that can be

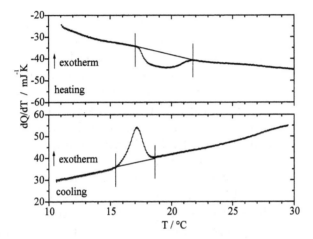

Figure 10 Melting and crystallization of a monodisperse SUV gel (14.35 wt% Na oleate, 23.17 wt% decanol and 62.47% water. Heating and cooling rate: 0.1 K/min.

seen in a DSC-signal given in Fig. 10. These results are an indication that the stiff gel is actually in a LC state and not in a micellar liquid state. It is likely that the micellar structures in the gel-phase have a long-range order. Typically melting and crystallization peaks had only been observed for the highly concentrated Na oleate/decanol/water gel. Neither melting peaks in differential scanning colorimetry (DSC) measurements up to 100°C nor softening or fluidization of the gel phase has been observed for the lower concentrated ($c_{surfactant} \approx 200$ mM, $c_{cosurfactant} \approx 600$ mM) system.

As with other crystallization processes, melting and crystallization do not occur exactly at the same temperature because the system can be undercooled before crystallization sets in.

H. Cryo-TEM and FF-TEM

In retrospect, it is interesting to note that it was expected that the stiff gel phase would consist of densely packed small unilamellar vesicles before any scattering data and TEM micrographs were available. The properties of the gel-phase remind us very much of the properties of cubic phases. The so-called ringing-gel phases—that are formed from globular micelles or from globular microemulsion droplets—have an inner core of solubilized hydrocarbon. The phases are optically isotropic and appear in the phase diagrams where the formed structures of surfactants and cosurfactants are bilayers and the volume fraction is in the order of 0.1 to 0.15. The stiffness of the phase, which is given by the shear modulus, reflects the number density of the aggregates.

Several possible aqueous surfactant/cosurfactant phases are known to exhibit a typically high viscosity or a gel-like appearance. Those are cubic phases, formed from globular micelles or from globular microemulsion droplets, hexagonal phases or L_α-phases. After careful consideration, one can rule out these structures as a possible structure. For a cubic phase a higher volume fraction is necessary to ensure contact of the globular aggregates, and hexagonal or L_α-phases would be birefringent and would have a lower stiffness. The only aggregate that fulfills these requirements would be a phase built from closely packed SUVs.

It was thus no longer surprising that when micrographs were made from the phase and both cryo- and FF-TEM micrographs clearly showed vesicles. Figure 11 shows a typical FF-TEM micrograph. At first glance, it is clear that the vesicles are densely packed, they are unilamellar, and they are rather monodisperse. A subtler question to be answered is whether the phase is a kind of glassy state with only a short-range order or whether the vesicles really show long range orders. In principle, both states would be possible and have been observed for colloidal systems from globular particles. In fact, theoretically there could be two glassy gel states: a gel state due to repulsive interaction and one due to long range orders. The attractive state can be ruled out in the present situation. Based on the high electrostatic interaction indicated in the micrograph, it is very likely that the phase has a long-range order. In some region of the micrograph the long-range order extends over distances of at least 20 times the diameter of the vesicles.

Because of their stiffness, it is practically impossible to prepare cryo-TEM micrographs of the gel-phases. Preparation

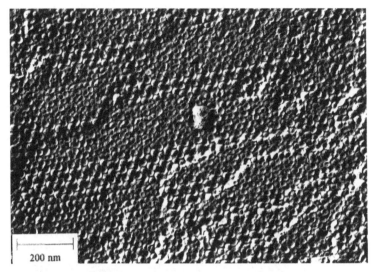

Figure 11 FF-TEM of a monodisperse SUV gel. Sample consists of 182 mM Na isostearate and 567 mM octanol in a water/glycerol mixture that contains 20 wt% of glycerol.

of such micrographs requires one to produce thin films on a perforated polymer membrane. Cryo-TEM micrographs could be prepared from viscous samples, which were beyond the phase boundary of the stiff gel-phase. An example of such a micrograph is Fig. 12. The sample had the same mole ratio of 3.1 of cosurfactant to surfactant but a total concentration of only 163 mM. Most of the micellar structures are present as unilamellar vesicles. A few vesicles with two shells are shown. They have an interlamellar distance of about 20 nm. The unilamellar vesicles have some polydispersity. They have diameters between 15 and 40 nm. There is some fractionation within the grid structure: larger vesicles are usually found towards the periphery of the film where it has the largest thickness while smaller vesicles are found in the middle part between the thin films. If the samples are further diluted, as is the case in Fig. 13, vesicle polydispersity is increased. We find diameters between 10 and 100 nm.

In summary, the cryo-TEM micrographs show that vesicles are already present in the phase region called $L_{\alpha 1}$. At

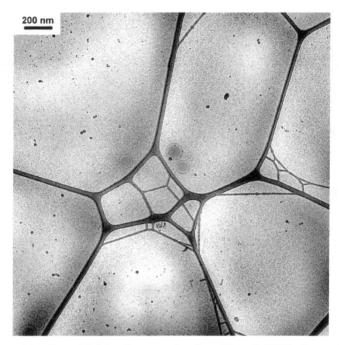

Figure 12 Cryo-TEM micrograph of a diluted SUV phase. Molar ratio Na oleate: octanol = 1 : 3.1, total molar concentration 163 mM.

the phase boundary the vesicles have some polydispersity and even contain vesicles with several shells. With increasing concentration the polydispersity becomes smaller and in the gel-phase the vesicles become really monodisperse and as a consequence of the monodispersity the vesicles can now build up a long-range order in the system.

The results thus strongly indicate that the size of the vesicles and its distribution is not only determined by the packing parameter of surfactants and cosurfactants but at the same time by the intervesicular interaction. It is mainly this interaction that is responsible for the monodispersity of the system.

I. Scattering Data

Scattering data of ionically charged colloidal systems usually show a strong correlation peak due to the structure factor of

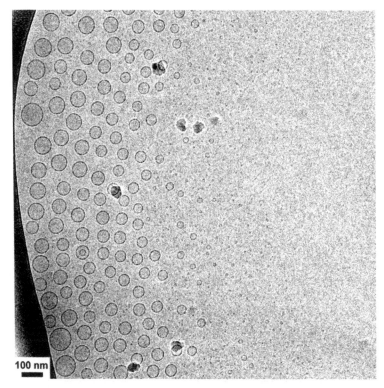

Figure 13 Cryo-TEM micrograph of a diluted SUV phase. Molar ratio Na oleate: octanol = 1 : 3.1, total molar concentration 82 mM.

the particles. In densely packed SUV systems, this peak is overlaid by the minimum of the form factor of the vesicles so that the broad peak splits into two halves (for details of the scattering functions for vesicles see Fig. 14 and [8]). Because of this special interplay between form factor and structure factor, the scattering functions of densely packed vesicle gels revealed then to be fairly uncommon.

Some scattering curves of the vesicle phases are given in Fig. 14. The characteristic features of the scattering functions shift with the total concentration but remain approximately the same. This is already a qualitative indication that the particle size of the vesicles changes with the total concentration. The evaluated results are given in Fig. 15. Within the gel-phase,

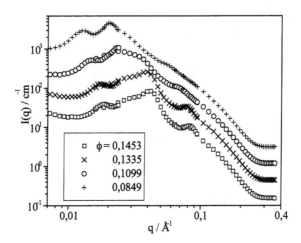

Figure 14 SANS curves for samples located in the isotropic gel phase of the system Na oleate/octanol/water. All samples have a constant molar ratio of Na oleate : octanol of 1 : 3.1 but differ in the volume fraction of the dispersed material. The absolute units are valid for the most concentrated sample; each subsequent curve is multiplied by a factor 3 for better lucidity.

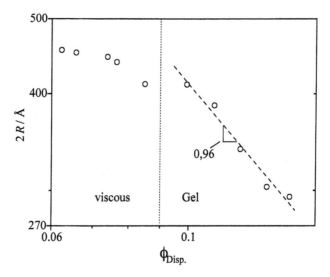

Figure 15 Vesicle diameter as a function of the volume of dispersed material of the SUV at a constant molar ratio of Na oleate: octanol $= 1 : 3.1$.

the diameters decrease linearly with the total volume fraction of surfactant and cosurfactant at constant ratio.

III. TIME RESOLVED GELATION PROCESS

A. Rheology

As gelation takes place in a time range typically of several minutes, the process can be monitored by changes of the characteristic parameters. Mixing a surfactant solution slightly with a certain amount of cosurfactant produces a low viscous turbid dispersion of cosurfactant droplets. Such dispersions clarify without further agitation accompanied by an increase of viscosity. Figure 16 illustrates the evolution of the viscosity of two samples 30 s after mixing surfactant solution and

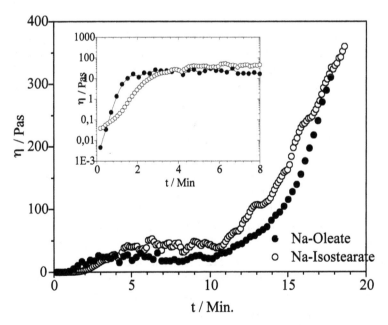

Figure 16 Evolution of the viscosity during gelification of two samples (182 mM Na oleate and Na isostearate respectively with 567 mM octanol). Shear rate $= 0.22\,\text{s}^{-1}$, $T = 25^\circ\text{C}$.

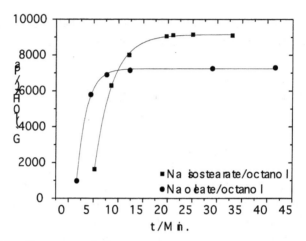

Figure 17 Storage modulus at 10 Hz as function of time for samples containing 182 mM Na oleate and Na isostearate respectively with 567 mM octanol.

cosurfactant. Increase of viscosity takes place in several time scales. In the first two (Na oleate) and four (Na isostearate) minutes, respectively, the viscosity increases about three orders of magnitude till a plateau value is reached. A second increase of the viscosity occurs about 10 min after the first data points were taken. Although oleate and isostearate differ in their initial increase of viscosity, the viscosity values of the samples diverge at about 15–18 min. Of course no final viscosity plateau can be reached with such experimental set up since permanent shearing prevents the build-up of long range three-dimensional structures which is a precondition for any gel structure. However, further gelation can be monitored by less destructive oscillating measurements. Time resolved measurements of the storage modulus at 10 Hz and a maximum deformation of 0.1% reveal dependencies such as illustrated in Fig. 17. Surprisingly, the modulus has already reached its final magnitude at the end of the viscosity plateau or shortly after. Obviously, the number density of particles, which is responsible for the modulus, does not change dramatically after the plateau is passed through.

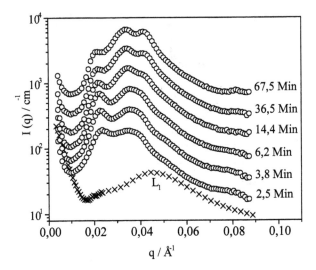

Figure 18 Time depending SANS curves taken at various times after mixing a sample containing 182 mM Na oleate and 567 mM octanol. The nth curve has been multiplied with 2^{n-1} beginning with $n = 1$ for $t = 2.5$ min due to better illustration. The curve denoted by L_1 corresponds to a sample containing 182 mM Na-Oleat/68 mM Oktanol, that means near the phase border micellar (L_1) solution–vesicle solution ($L_{\alpha l}$).

B. SANS

Additionally, the gelation process can be monitored by means of SANS. The experiments were carried out by mixing a Na oleate solution in D_2O with octanol in the quartz cell and detecting SANS intensities with an inevitable time delay of 2 min (Fig. 18). The first scattering pattern already differs significantly from that of an equilibrated L_1 phase with a lower content of alcohol. The broad peak of the micellar solution is immediately shifted towards smaller q-values corresponding to a higher mean inter-particle spacing. A detailed analysis of the evolution of the innermost minimum at $q \approx 0.01 \, \text{Å}^{-1}$ revealed a time range of about 30 min in which the scattering intensity is falling till a constant value is reached. This is coincident with the time the gel needs to clear up totally. The evolution of the

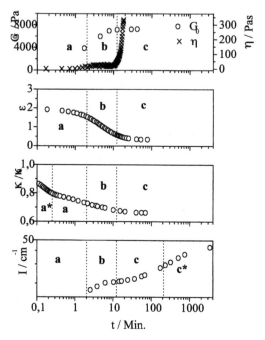

Figure 19 Comparison of different characteristic parameters changing during the gelification process. Shear modulus G_0, viscosity η, extinction ε, normalized electric conductivity κ/κ_0 and SANS intensity at $q = 0.04\,\text{Å}^{-1}$ for a sample with $182\,\text{mM}$ Na oleate and $567\,\text{mM}$ octanol. Ranges: (**a**) transition of rodlike micelles to lamellae, (**b**) transition of lamellae to unilamellar vesicles, (**c**) long time ordering processes of the vesicles.

bump can be interpreted by the increased monodispersity of the vesicles and the splitting of the structure factor into two distinct peaks. The increase of the $0.04\,\text{Å}^{-1}$ peak reaches an intermediate plateau after $10\,\text{min}$ and rises again after $30\,\text{min}$. Figure 19 compares the characteristic parameters that are changing during gelation on the same time scale.

C. Cryo TEM

The scattering model proposed above is confirmed by cryo-TEM images. The sample in Fig. 20 was blotted and fixed $1.5\,\text{min}$ after mixing. The bigger ($\approx 0.5\,\mu\text{m}$) dark regions are caused by

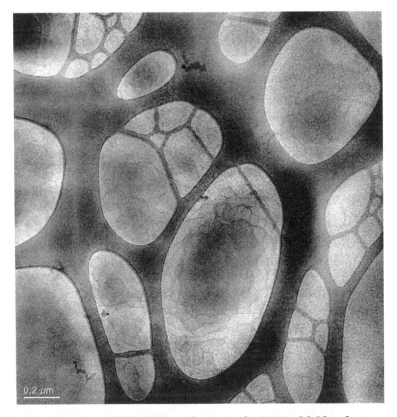

Figure 20 Cryo-TEM image of a sample (182 mM Na oleate and 567 mM octanol) blotted and fixed 1.5 min after mixing. The dark spots (arrows) are caused by octanol droplets. Lamellae fragments are generated at the surface of the droplets that are partly connected to each other.

octanol droplets on whose surface lamella fragments are generated that are partly connected to each other. The mean distance of the fragments can be estimated from the picture as 30 nm, which correspond to the position of the innermost scattering peak ($q = 0.01 \, \text{Å}^{-1}$) of the SANS data.

D. Structural Transformations

If the above observations are summarized, we can derive the following structural transformations that take place during the spontaneous gelation process of the vesicle gel. Generally, the process can be divided up into four different time scales.

1. $t = 0$–2 min: Immediately after slightly mixing, a portion of the cosurfactant has been incorporated into surfactant micelles so that the micelles are saturated with cosurfactant molecules. The maximum solubilization concentration is reached very fast as the total amount of cosurfactant is much higher than the maximum amount of solubilized cosurfactant. Further addition cosurfactant generates bilayer structures with a nearly zero average curvature. The formation of lamellae fragments and first vesicles raise the viscosity up to a factor 1000 higher than that of the primary L_1-phase.

2. $t = 2$–12 min: During the next 10 min the processes do overlap. The dispersed cosurfactant droplets shrink as the cosurfactant material is transformed into lamellae. This can take place through molecular diffusion of the cosurfactant from the droplet to the saturated micelles or by generating fractal structures at the surface of the droplet. Simultaneously, lamella fragments form isolated spherical bodies, which turn into unilamellar vesicles. Besides, monodispersity of the vesicles increases.

3. $t > 12$ min: When the volume fraction of the vesicles and their monodisperstity is sufficiently high, viscosity diverges. Also, the still existing octanol droplets disappear as evidenced by the constancy of the extinction.

4. $t > 100$ min: A long time ordering process can be monitored by means of SANS. A pronounced structure peak develops.

IV.   THEORETICAL CONSIDERATIONS AND A SIMPLE MODEL FOR SUVs

All the experimental evidence that has been presented shows unambiguously that the gel-phase consists of densely packed unilamellar vesicles. The question that can now be asked is whether the size of the vesicles can be calculated with the help of a single model and if it is possible to predict the macroscopic properties of this phase, in particular the rheological properties. This is indeed possible as the following simple model shows.

If densely packed spheres are considered, the number density N of particles is proportional to R^{-3} with R being the radius of the spheres. For a vesicle system that keeps a densely package and constant packing fraction when it is diluted, the radius increases according to $R \sim \phi^{-1}$ with ϕ being the volume fraction of dispersed material. This inverse proportionality is demonstrated in Fig. 21 for the Na oleate/octanol sample.

Then, the number density of particles has to increase with ϕ^3. Actually, this ϕ^3 dependency is found for the shear modulus (Fig. 6). This implies a linear proportionality of the modulus with the number density of particles. So, the elasticity of the vesicle system can be described as a network of connected springs. Of course, this comparison is valid only for very small deformations of the vesicle gel. As can be seen in Fig. 22 the complex and the storage modulus break down when the maximum oscillating deformation exceeds 2%. Beyond this

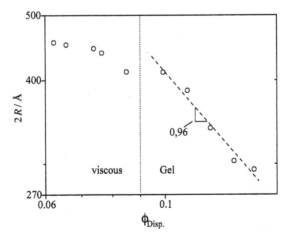

Figure 21 Vesicle diameter ($2R$) as function of the total volume fraction of Na oleate and octanol at a constant molar ratio of Na oleate : octanol of 1 : 3.1.

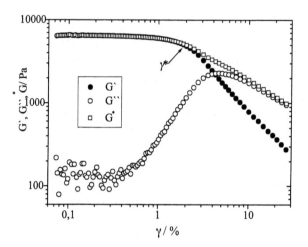

Figure 22 Storage modulus (G'), loss modulus (G'') and complex modulus (G^*) of monodisperse SUV gel at increasing maximum oscillating (1 Hz) deformation. The sample contains 164 mM Na isostearate and 510 mM octanol.

linear viscoelastic region, shear force and shear stress respectively are not proportional to the resulting deformation. In addition to a maximum deformation, the stiff gel phase can be characterized by a maximum stress, which is identical with the yield stress value.

Hence, the examined gel of monodisperse unilamellar vesicles is an ideal model system to correlate the microscopic structure of the gel with its rheological (macroscopic) behavior. Ordinary vesicle phases commonly show polydispersity if they are densely packed. Monodisperse vesicles have only been described in nondensely packed systems before. We have been able, for the first time, to prove the suggested relationship between number density of vesicles and the shear modulus G_0 experimentally.

V. SUMMARY

Stiff hydrogels are described that are found in ternary phase diagrams of sodium salts of soaps (oleate or stearate) and cosurfactants (n-alcohols) in water at a molar ratio of about 1 : 3. The volume ratio of surfactant and cosurfactant in the gel phase ranges from 0.1 to 0.15 depending somewhat on the cosurfactant/surfactant ratio.

The gels are transparent, optically isotropic, and highly elastic. They have been characterized by rheological measurements, and the storage moduli are around 10^4 Pa. The properties of the gels are reminiscent of the properties of the "ringing gels" that have already been known for some time, and that consist of densely packed microemulsion droplets or globular micelles but contain a larger volume fraction of surfactant and hydrocarbon ($\phi \simeq 0.5$).

It is shown by different physicochemical methods (FF-TEM, cryo-TEM, SANS) that the presently described gels are composed of densely packed small unilamellar vesicles (SUV). The size of the vesicles decreases linearly with the volume fraction of surfactant and cosurfactant. On the basis of the preparation of the samples the vesicles in the gel phases seem to be thermodynamically stable and they form spontaneously. There is evidence from FF-TEM and from the SANS-data that the samples show a long-range order and that the "gels" are actually cubic phases derived from SUVs.

It is shown in addition that the composition of the gel phases can be varied considerably without loosing the "gel" state. The ionic carboxylates can be substituted by uncharged alkyldimethylaminoxides up to 50%. The cosurfactants from hexanol to octanol can be used for the preparation of the phases. The "gel" phases also tolerate the presence of some excess salt.

REFERENCES

1. L. Saunders, J. Perrin, and D. Gammack, *J. Pharm. Pharmacol.* **14**: 567 (1962).

2. M. Bergmeier, H. Hoffmann, and C. Thunig, *J. Phys. Chem. B* **101**: 5767 (1997).

3. H. Hoffmann, M. Bergmeier, M. Gradzielski, and C. Thunig, *Prog. Colloid. Polymer Sci.* **109**: 13 (1998).

4. H. Hoffmann, *Adv. Mater.* **6**: 116 (1994).

5. E. W. Kaler, A. K. Murthy, B. E. Rodriguez, and J. A. Zasadzinski, *Science*, **245**: 1371 (1989).

6. H. Hoffmann, C. Thunig, P. Schmiedel, and U. Munkert, *Langmuir* **10**: 3972 (1994).

7. H. Hoffmann, D. Gräbner, U. Hornfeck, and G. Platz, *J. Phys. Chem. B* **103**: 611–614 (1999).

8. M. Gradzielski, M. Müller, M. Bergmeier, H. Hoffmann, and E. Hoinkis, *J. Phys. Chem. B* **103**: 1416 (1999).

9. K. Fontell, L. Mandell, and P. Ekwall, *Acta Chem. Scand.* **22**: 11 (1968).

13

Microscopic Phase Separation in Mixed Micellar Solutions

SHIGEYOSHI MIYAGISHI,
TSUYOSHI ASAKAWA, and AKIO OHTA

Kanazawa University,
Kanazawa, Japan

SYNOPSIS

This chapter deals with the microscopic micelle behavior in binary surfactant solutions containing hydrocarbon and fluorocarbon surfactant revealed by fluorescent probe methods together with macroscopic phenomena. The fluorescent probe methods were very useful to elucidate de-mixing micellization (transition from one kind of micelle to two kinds of micelles) and de-micellization (transition from two kinds of micelles to one kind of micelle), and microscopic (nano-scale) phase separation in a micelle.

I. INTRODUCTION

This chapter will describe the micellar behavior of surfactants from the microscopic point of view, regarding binary surfactant solutions containing hydrocarbon and fluorocarbon surfactant, and relations between microscopic events in a micelle and macroscopic phenomena in a micellar solution, together with microscopic phase separation in micellar solutions.

Surfactant molecules form self-assemblies in aqueous solutions such as micelles and vesicles having the specific functional properties that never appear in monomeric surfactant solution. As a result, the self-assembly provides an environment for control of molecular orientation, concentration effect of reactants, reaction control, and so on. The shape and size of a self-assembly depend on surfactant structure and concentration, temperature, additive concentration, etc. Alteration in structure of such organized assembly often brings about changes in the microscopic properties such as micro-viscosity and micro-polarity. Distribution of reactants between the bulk and micelle phases is also influenced, and the reaction rate in the assembly changes at the transition point from spherical to rod-like micelles or upon phase transition of vesicles due to an increase in micro-viscosity produced by the transition. Such micellar properties depend not only on the surfactant head group but also on the hydrophobic tail forming the micellar core. The micellar core consists essentially of hydrocarbon chains and/or fluorocarbon chains. Hydrocarbon chains are flexible and give micelles with a liquid-like core. Since fluorocarbon chains are rigid and bulky, fluorocarbon surfactants often make small and loose micelles open to water penetration because of non-flexibility of the chains. Consequently, the hydrophobic groups in the fluorocarbon micelles tend to contact more with aqueous solution than the hydrocarbon micelles. Chothia [1] predicted that fluorocarbon exhibits a higher hydrophobicity than the hydrocarbon, taking into account that the former has a wider contact area with water due to the bulkiness of a fluorine atom. The hydrophobicity of fluorocarbon was estimated to be 1.7 fold as large as that of hydrocarbon based on the fact that a CF_2 group has 1.7 fold greater surface area compared with a CH_2 group, and then this estimation was confirmed experimentally [2,3]. Fluorocarbons have a larger moral volume and density than hydrocarbons [4]. Low interfacial tension of fluorinated surfactant solution results from the extremely weak cohesive forces of fluorocarbon [4]. In addition, the weak cohesive forces produces nonideal miscibility for mixtures of fluorocarbons and hydrocarbons. Such unique properties of the fluorocarbon chain has prompted us to undertake a systematic investigation on fluorocarbon surfactants and their applications to industries.

Since Mukerjee and Yang [5] suggested in 1976 a possibility that two kinds of micelles coexist in mixed solutions of sodium perfluoro-octanoate and sodium dodecylsulfate, the behavior of aqueous mixed solutions of hydrocarbon and fluorocarbon surfactants has been investigated both experimentally and theoretically by many scientists, and the results were summarized by Funasaki [6]. Some significant findings are as follows:

1. Two kinds of mixed micelles coexist in solution under a certain specified condition.
2. Chain length of alkyl chain and temperature for mutual solubility are factors as important in mixed micelles of hydrocarbon and fluorocarbon surfactants as in liquid mixtures containing hydrocarbons and fluorocarbons.
3. The upper critical solution temperature in mixed micellar solutions increases with alkyl chain length.
4. The composition of each mixed micelle is constant under the condition at which two kinds of micelles coexist, although the concentration of each kind of micelle varies with the total composition of the surfactant mixture.

Most of these researches, however, were carried out from the macroscopic or thermodynamic point of view. That is, light scattering technique, surface tension, conductivity, gel filtration, ultrafiltration, etc. have been used for estimation of the macroscopic properties such as critical micelle concentration (CMC), micellar size, micellar shape, micellar composition, and thermodynamic parameters. On the other hand, microscopic information on micellar solutions can be obtained by means of spectroscopic techniques such as fluorescence, nuclear magnetic resonance (NMR), and electron spin resonance (ESR) [7]. Several fluorescent probe methods have often been utilized owing to their ease, even since Infelta *et al.* [8] pioneered the use of fluorescent probes to study such organized assemblies as micelles and polymers. We will find in this chapter that the fluorescent probe methods are very useful to monitor microscopic events in mixed micelle systems and microscopic phase separation (strictly speaking, phase separation in a nano-scale region). Then, by combining the microscopically determined results with the macroscopic point of view, a more detail picture will be delineated about the binary surfactant solutions. Some examples are shown in the following paragraphs, where microscopic events and distributions of substrates among (or in) the mixed micelles of hydrocarbon and fluorocarbon surfactants are also discussed.

II. FLUORINATED SURFACTANTS

Since the interaction between fluorocarbon and hydrocarbon chains is a very important factor determining the miscibility of hydrocarbon and fluorocarbon surfactants in mixed micelles, it is necessary to shed light on the difference in the interaction between the micelle phase and bulk phase to elucidate the characters of mixed micelles. That is, exact estimation of the contribution of a hydrophobic group to micelle formation energy is significant. However, most researchers have used mixtures of a hydrocarbon surfactant and a fluorocarbon surfactant having different hydrophilic head groups, because no surfactant mixtures with the same head group were available. Then, in the thermodynamic equation for mixed micellization, the term of the interaction between hydrocarbon and fluorocarbon surfactants always contained the effect coming from the difference between their head groups, in addition to the interaction between their hydrophobic groups. In order to eliminate any effect of the difference in head group and to simplify the problem, Miyagishi and Asakawa and co-workers [9,10] prepared two series of binary surfactant systems having the same hydrophilic group. The first series are both hydrocarbon and fluorocarbon surfactants, each of which has alanine as the anionic head group. The second surfactant series, which are cationic, are alkyl pyridinium chlorides.

A. Sodium N-Acyl-Alaninates [9]

N-Lauroyl-alanine (LauAla) and N-myristoyl-alanine (MyrAla) were prepared by acylation of alanine with the corresponding acyl-chlorides. N-perfluoro-nonanoylalanine (PFNAla) was synthesized by the reaction of methylperfluoro-nonanoate with alanine. Surface tensions of solutions of these surfactants were measured and compared with those of the corresponding hydrocarbon surfactants. The values of CMC, surface tension at the CMC, and molecular cross section are listed in Table 1. The fluorinated surfactant molecules packed together very closely at the air–water interface and gave an extremely low surface tension at the CMC. The surface tension (15 mN/m)

Table 1 Surface activities and CMCs of sodium N-acylalaninates in 0.1 M NaOH solution at 30°C.

	γ at CMC/$mN\,m^{-1}$	A/$\mathring{A}^2$ molecule^{-1}	CMC/ $mol\,dm^{-3}$
$C_8F_{17}CONHCH(CH_3)COONa$			
Na PFN-l-Ala	15.5	27	0.49
Na PFN-dl-Ala	15.5	27	0.52
$C_{11}H_{23}CONHCH(CH_3)COONa$			
Na Lau-l-Ala	40.0	52	3.6
Na Lau-dl-Ala	39.9	51	3.8
$C_{13}H_{27}CONHCH(CH_3)COONa$			
Na Myr-l-Ala	37.5	35	0.32
Na Myr-dl-Ala	37.5	35	0.34

Most of values were estimated from data of Ref. 9.

was similar to the value (11.5 mN/m) of perfluorohexane [6] , as if the surface were fully covered by the fluorinated surfactants. On the other hand, the hydrocarbon surfactants, Na LauAla and Na MyrAla, lowered the surface tension of water to 37–40 mN/m. The lowering was smaller than that by the fluorinated surfactants. The number of methylene groups in a perfluoro-nonanoyl group is 8 and that in a myristoyl group is 13. The CMC of PFNAla (0.49 mmol dm^{-3}) was similar to that of MyrAla (0.32 mmol dm^{-3}), whereas the CMC of sodium octylsulfate (140 mmol dm^{-3}) was much larger than that of sodium tridecylsulfate (4.3 mmol dm^{-3}) [11,12]. Namely, a fluorinated surfactant is very surface active and is highly liable to form micelles because of the remarkably low cohesive energy of its perfluorocarbon chain.

   Amino acid surfactant has stereoisomers, the optically active isomer of which has a smaller CMC than the corresponding racemic isomer [13,14]. The difference in CMC results from the difference in conformation of the amino acid part on the micellar surface. Similar CMC differences are also found for the surfactants listed in Table 1, irrespective of their hydrophobic group.

B. Alkyl Pyridinium Chlorides [10]

Alkyl pyridinium halide with a hydrocarbon chain is easily available but its homologue with a fluorocarbon chain was absent until Asakawa *et al.* [10] prepared it. Cationic fluorinated alkylpyridinium iodides were synthesized from 1H,1H, 2H,2H-perfluoroalkyl iodides and pyridine and the iodides obtained were ion-exchanged to give the chlorides. Some properties of the cationic surfactants having a pyridinium head group are listed in Table 2, where the fluorinated surfactants exhibit a higher surface activity and associate to micelles at a lower concentration than the corresponding hydrocarbon surfactants. However, less remarkable differences were found for the cationic surfactants in Table 2 than those observed for the anionic surfactants in Table 1. Two methylene groups, which were introduced between the perfluorocarbon group and the pyridinium head group in the fluorinated

Table 2 Surface activities and CMCs of alkylpyridinium chlorides.

	γ_{cmc} mN m^{-1}	A Å^2 molecule^{-1}	CMC/mol dm^{-3}			
			a	b	c	d
$C_6F_{13}CH_2CH_2NC_5H_5Cl$ HFOPC	27.5	61.4	18.0	21.7	21.1	20.5
$C_8F_{17}CH_2CH_2NC_5H_5Cl$ HFDePC	26.1	54.2	2.8	2.7	2.5	2.5
$C_{10}F_{21}CH_2CH_2NC_5H_5Cl$ HFDPC	25.6	39.8	0.32	0.33	0.30	0.32
$C_{12}H_{25}NC_5H_5Cl$ DPC	44.6	65.2	14.8	15.0	14.5	16.0
$C_{14}H_{29}NC_5H_5Cl$ TPC	42.2	69.6	4.1	4.1	4.2	4.0
$C_{16}H_{33}NC_5H_5Cl$ CPC	42.5	61.3	1.1	1.0	0.74	0.90

[a]surface tension. [b]conductivity. [c]Electro motive force. [d]Fluorescence of pyrene.
Source: Ref. 66.

cationic surfactant in order to make its synthesis easier, might somewhat weaken the surface activity. However, the groups hardly perturbed the hydrophobic interaction in micellization because the groups adjacent to the head group of the surfactant were not important in micellization.

Table 2 also indicates that the experimental methods useful for the hydrocarbon surfactants are also applicable to the fluorinated surfactants. In addition, the four methods for determining CMC (surface tension, conductivity, electromotive force, and fluorescence quenching) gave similar CMC values. The fluorinated surfactants had very low CMCs, for example, the CMC of 1H,1H,2H,2H-perfluoro-dodecylpyridinium chloride (HFDPC) was one fiftieth that of dodecylpyridinium chloride (DPC). When the number of methylene groups in the perfluoro-alkyl group increased by 4 (that is, from 1H,1H,2H, 2H-perfluoro-octylpyridinium chloride (HFOPC) to HFDPC), the change in cmc was 1.8 in logarithmic units, while the cmc of DPC was larger in a logarithmic scale than that of CPC by a factor of 1.2. The change in the former case was 1.5 fold that in the latter. This value of 1.5 corresponds to a relative measure of hydrophobicity for fluorocarbons and coincides with the result reported by Shinoda *et al.* [15].

It will become evident later that the perfluoro-alkyl pyridinium chlorides work as an excellent quencher toward pyrene and are very useful in research on microscopic phase separation.

III. FLUORESCENT PROBES FOR FLUORINATED SURFACTANTS

Light scattering methods, which have generally been utilized to determine the micellar size, size distribution, micellar shape, CMC of hydrocarbon surfactants, are not suitable for fluorinated surfactants due to the similarity in refractive index between water and fluorocarbon (1.33 and 1.35 for water and fluorocarbon, respectively), except for those fluorinated surfactants having a head group of high or low refractive index.

Many fluorescent probes have been utilized to estimate the micro-polarity, micro-viscosity, CMC, aggregation number, etc. of surfactant micelles. Turro *et al.* and Zana reviewed the results of fluorescence experiments in micellar systems of hydrocarbon surfactants in 1980 [16] and 1987 [6], respectively. On the other hand, there had been no application of fluorescent probe methods to the fluorinated surfactants until Miyagishi's group [17] reported that some fluorescent probes successfully work in fluorinated surfactant solutions. However, it should be noted that not all fluorescent probes and quenchers are applicable to fluorinated surfactants, because hydrocarbon molecules are often insoluble in micelles of fluorinated surfactants. Some examples will be described below.

A. Micro-polarity

1-Anilinonaphthalene-8-sulfonate, one of the classical probes for micro-polarity, is useful to estimate the micro-polarity [18,19] and CMCs of surfactants [20]. However, this probe is very soluble in water and thus is partitioned into both micelle and water phases. Then, we should use this probe with great care in the determination of micro-polarity, especially for fluorinated surfactants. Many water-soluble probes are not solubilized in fluorinated surfactant micelles or they adsorb only on the micele surface [17].

Some hydrophobic probes such as pyrene and octafluoronaphthalene (OFN) [17,21] are solubilized in the so-called palisade layer of micelles and provide information about the palisade layer. It is well known [22] that the ratio of the intensity of the first vibration band to that of the third one of pyrene fluorescence spectrum (I_1/I_3) can be a measure of micropolarity. The value of this ratio is 1.85, 1.2, and 0.65 in water, ethanol, and hexane, respectively. The ratios in micellar solutions were compared between hydrocarbon surfactants (lithium dodecylsulfate (LiDS) and CPC) and fluorocarbon surfactants (lithium 1H,1H,2H,2H-perfluorodecyl sulfate (LiHFDeS) and HFDePC) as shown in Fig. 1 [23]. They were similar to that in water below CMC and then decreased rapidly with increasing concentration above CMC. Although the ratio decreased to 1–1.3 in LiDS and CPC micellar solutions, the decrease was small for the fluorinated surfactants, i.e., the ratio was 1.6 in LiHFDeS and HFDePC micellar solutions and was similar to that for aqueous environment rather than for methanol. The high polarity perceived by pyrene strongly

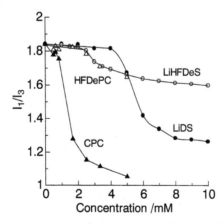

Figure 1 Fluorescence intensity ratio I_1/I_3 as a function of surfactant concentration. Pyrene concentration was fixed at 1.0×10^{-7} mol/dm^3. (From Ref. 27.)

suggests that pyrene lies close to the bulk phase side of palisade layer in a fluorinated micelle. The location of pyrene solubilized in the fluorinated micelles may result not only from its large size but also from its low miscibility with fluorocarbon. According to a packing model of spherical micelle [24], the aggregation number and surface area per surfactant molecule is estimated to be 60 and $60\,\text{Å}^2$ for a LiDS micelle. This surface area, which is much larger than the cross-sectional area of a sulfate group $27\,\text{Å}^2$, indicates that 55% of the micelle surface remains in contact with water [24]. On the other hand, a fluorinated micelle is small and is open enough for water to penetrate due to short and rigid fluorocarbon chain [25,26]. The high polarity in LiHFDeS and HFDePC micelles can then be ascribed to their micellar structures.

The above results suggest that new probes of a high miscibility with fluorocarbons are necessary for fluorinated surfactants. Since fluorinated aromatic compounds such as OFN has fluorescence quantum yield higher than that for its nonsubstituted analog and is miscible with fluorocarbons, it is one of the probes that are suitable for fluorinated micelles. Figure 2 shows that the fluorescence intensity of OFN is stronger in LiHFDeS micelles than in LiDS micelles [27]. This

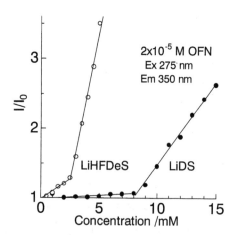

Figure 2 Dependence of OFN fluorescence intensity on surfactant concentration. (From Ref. 27.)

fact strongly suggests that fluorinated fluorescent probes are more appropriate for fluorocarbon surfactants because of their high miscibility in fluorinated micelles. Decafluorobiphenyl (FBIP) is one such probe [28]. The fluorescence intensity of FBIP increases with decreasing polarity of the solvent. In surfactant solutions, the fluorescence intensity starts to increase upon micelle formation and continues to increase with increasing surfactant concentration. Micelle growth is known to result in a decrease in micro-polarity and an increase in micro-viscosity [29]. The decrease of micro-polarity in micellar solutions can be detected by FBIP as a probe. The relative fluorescence intensity of FBIP increased in LiDS solutions with addition of LiCl and reached a constant value as a result of micelle growth, while the fluorescence intensity passed a maximum value and then decreased in both LiFOS and LiPFN solutions [28]. The decrease in the fluorescence intensity corresponds to an increase in polarity and its decrease after passing the maximum value implies that salt addition induces a micelle shape transition. This transition in micelle shape was confirmed by viscosity data [27]. Abrupt increases in viscosity were observed at 1.2, 0.6, and 1.2 M LiCl for LiDS, LiFOS, and LiPFN solutions, respectively. The intrinsic viscosities of LiDS, LiFOS, and LiPFN were 4.2, 2.5, and 2.7 cm^3/g, respectively at low LiCl concentrations, while the corresponding values at high LiCl concentrations were 5.0, 12.0, and 16.3 cm^3/g, respectively. These results evidently show that addition of salt induces the micelle growth from a spherical to a larger size one in LiFOS and LiPFN solutions and that there are non-spherical LiDS are present in LiCl solutions. Bededouch and Chen reported the formation of a prolate ellipsoidal micelle of LiDS in LiCl solutions from small angle neutron scattering experiments [30]. For fluorocarbon micelles, the shape transition from a spherical micelle to a prolate ellipsoidal micelle accompanies a large expansion of micellar surface per surfactant molecule in contrast to hydrocarbon micelles [28]. According to Nagarajan and Ruckenstein [31], the micellar surface area per surfactant molecule can be calculated by using Eq. (1) for a spherical micelle with a hydrophobic core of radius r and using Eq. (3) for a prolate ellipsoidal micelle with a length of major axis b and minor axis L (extended length of surfactant tail), respectively.

$$A/g = 4\pi(r+\delta)^2/g \tag{1}$$

$$r = (3gv/4\pi)^{1/3} \tag{2}$$

$$A/g = 2\pi(L+\delta)^2[1 + (\sin^{-1} E)/\{E(1-E^2)^{1/2}\}]/g \tag{3}$$

$$E = [1 - \{(L+\delta)/(b+\delta)\}^2]^{1/2} \tag{4}$$

$$gv = 4\pi L^2 b/3 \tag{5}$$

Here, A is the micellar surface area, g the aggregation number, v the volume of hydrophobic core per surfactant, and δ the distance of separation between micelle core and head group ($\delta = 3.6$ for LiDS, $\delta = 2.3$ for LiFOS [32]). Molecular volume and chain length are given by Eqs (6) and (7) for a hydrocarbon [33], and by Eqs (8) and (9) for a fluorocarbon [34], respectively.

$$v = 27.4 + 26.9n_c(\text{Å}^3) \tag{6}$$

$$L = 1.50 + 1.265n_c(\text{Å}) \tag{7}$$

$$v = 88.7 + 38.6n_c(\text{Å}^3) \tag{8}$$

$$L = 1.25 + 1.25n_c(\text{Å}) \tag{9}$$

where n_c is the number of chain carbons. We can know from comparison of Eq. (6) with Eq. (8) that a fluorocarbon has a large volume. The surface area per LiFOS molecule was estimated to be 75 and 105 for spherical and prolate ellipsoidal micelles with $g = 100$, while the surface area per LiDS was about 70 for both micelles with the same aggregation number. The expansion in micellar surface area makes water penetration easier and then this results in a higher micro-polarity. The micellar shape transition of a fluorinated micelle caused by salt addition, therefore, results in the abnormal fluorescence behavior of probe (increase in fluorescence intensity of FBIP) in LiFOS and LiPFN systems.

The above results indicate that water molecules are easily penetrable into fluorinated surfactant micelles and the micellar core behaves as if it were a globular aggregate composed of loosely packed rigid rods. Generally, fluorinated surfactant forms smaller micelles than hydrocarbon surfactant because its hydrophobic group is bulky, rigid, and short. A small micelle has a large cross-sectional area per surfactant molecule and then becomes more open to water penetration. This is the second reason for a high polarity index of fluorinated surfactants. The third reason is that aromatic hydrocarbons (for example, pyrene) are solubilized on the hydrophilic side of the palisade layer because of low miscibility with perfluorocarbons.

B. Solubilization

Aromatic fluorocarbons are expected to be solubilized more in fluorinated surfactant solutions than aromatic hydrocarbon. Asakawa *et al.* [28] found that fluorocarbon surfactants solubilized a larger amount of FBIP than hydrocarbon surfactants, as shown in Fig. 3. Table 3 shows a comparison of the

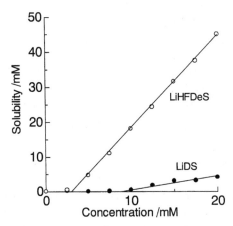

Figure 3 Solubility of FBIP in LiHFDeS and LiDS solutions. (From Ref. 28.)

Table 3 Comparison in solubilization ability among anionic surfactants.

Surfactants	n		
	Pyrene[a]	OFN[a]	FBIP[b]
$C_{12}H_{25}$ SO_4Li LiDS	0.73	0.26	0.19
C_8F_{17} C_2H_4 SO_4Li LiHFDeS	–	–	0.82
C_8F_{17} SO_4Li LiFOS	0.03	0.13	0.17
$C_9F_{19}COOLi$ LiPFN	0.05	0.19	0.17
$C_8F_{17}COONH_2(C_2H_5)_2$ DEAPFN	0.05	4.61	–

n = number of solubilized molecules per micelle.
[a]Calculated from data of Ref. 21.
[b]Calculated from data of Ref. 28.

solubilization ability between hydrocarbon and fluorocarbon surfactants. Hydrocarbon surfactants solubilized pyrene more than FBIP and OFN, while fluorocarbon surfactants tended to solubilize more FBIP and OFN. LiDS was similar in solubilization power toward the aromatic fluorocarbons to LiFOS and LiPFN. It should be noted that the solubilized amount of FBIP in LiHFDeS micelles (0.82) was comparable to the amount of pyrene per LiDS micelle (0.73) but the amount of FBIP was 0.17 for the other fluorocarbon surfactants. On the other hand, a much larger amount of OFN was solubilized in DEAPFN than in the other surfactants [27]. LiHFDeS and DEAPFN evidently exhibit a specific solubilization ability toward FBIP and OFN, respectively.

Addition of salt generally promotes micelle growth, thereby raising solubilization ability. Solubilization of FBIP and OFN into LiDS micelles followed this trend. However, an unusual dependence of solubilization on salt concentration was observed with fluorinated surfactant solutions [28]. The solubility of FBIP in a 20 mM LiFOS micellar solution first increased and then decreased after passing through a maximum with increasing salt concentration. Similar behavior was also found for LiPFN solutions. These phenomena can be explained by the same mechanism as that for abnormal fluorescence behavior in the foregoing section (Section III. **A**). That is, addition of salt to fluorinated surfactant solutions produces a micellar shape transition from spherical to ellipsoidal micelles in which fluorocarbon chains are loosely packed; as a consequence, the micelle becomes more open to water penetration and solubilizes a lesser amount of hydrophobic substances than the spherical micelles, in contrast to hydrocarbon surfactant micelles.

C. Aggregation Behavior

Micellar aggregation numbers can also be estimated from the geometrical procedure proposed by Israelachivili and Ninham [35]. The method was used for estimation of aggregation numbers for HFDPC and DPC having a dodecyl group. The aggregation number (n_{agg}) and cross-sectional area per surfactant molecule ($A_{micelle}$) calculated using the method were 34.1 and 97.4 $Å^2$ for HFDPC and 54.9 and 63.2 $Å^2$ for DPC, respectively. Difference in the value between HFDPC and DPC results from the difference in the volume of alkyl chain, as shown in Eqs (6) and (8). The large $A_{micelle}$ of HFDPC makes it possible for water to penetrate easily into the micelle core and this is supported by the high polarity probed by pyrene (see Fig. 1).

Experimental methods applicable to determination of aggregation number are limited for fluorinated surfactants because the determination by conventional light scattering techniques is extremely difficult as described before. An alternative method, however, is provided by small angle neutron scattering and this method has mainly been used before [30,36,37]. Fluorescence probe methods were found to be applicable to fluorinated surfactants [38,39]. Cyclic voltammetry is also useful for examining the diffusion coefficients of micelles and micelle sizes in fluorinated surfactant solutions [40].

Table 4 Aggregation numbers by pyrene fluorescence quenching method in 20 mM surfactant solutions.

LiDS	LiHFDeS	LiFOS	LiPFN	DEAPFN	SPFO[a]
55	23	23	25	47	20

SPFO: $C_7F_{15}COONa$.
[a]50 mM solution.
Source: Ref. 39.

Aggregation numbers determined by a fluorescence quenching method are given in Table 4. However, it should be noted for exact estimation of aggregation number that both fluorescent probe and quencher are perfectly solubilized in micelles and its quenching occurs only in a micelle. Pyrene and CPC are a typical set of fluorescence probe and quencher. This set satisfies the above condition in many hydrocarbon surfactant solutions [41]. The set was tried for determination of aggregation numbers of fluorinated surfactant micelles. The fluorinated surfactants given in Table 4 formed relatively small micelles, as expected. On the other hand, Hoffmann *et al.* [42] found vesicles in diethylammonium perfluoro-octanoate solutions and Fontell and Lindman [43] reported the presence of a liquid crystal phase in many perfluoro-octanoate systems with different counterions. Asakawa *et al.* [40] revealed by using cyclic voltammetry that an addition of salt induced micelle growth in anionic fluorocarbon surfactant solutions as well as in hydrocarbon surfactants. Recently, cationic fluoro-carbon surfactants were reported to self-assemble into various aggregates such as micelles, thread-like micelles, vesicles, and other lamellar aggregates, depending on salt concentration [44]. Although fluorinated surfactants have been believed to form relatively small micelles compared with hydrocarbon surfactants, fluorinated surfactants are likely to behave in a roughly similar way to hydrocarbon sur-factants, forming micelles of various structures. Recently, the aggregation behavior of fluorinated surfactants was reviewed [45,46].

IV. NON-IDEAL MIXED MICELLE FORMATION AND TWO KINDS OF MICELLES

Liquid mixtures of hydrocarbon and fluorocarbon give the upper critical solution temperatures [4], below which de-mixing into two phases occurs and each phase is partially miscible. Mutual miscibility in each phase increased with increasing temperature. According to phase rule, the composition of each phase is constant at temperatures below the upper critical solution temperature regardless of the total composition and only the volume of each phase can vary. Similar phenomena were observed with mixed micellar solutions of hydrocarbon and fluorocarbon surfactants [47–49]. A low cohesive energy of fluorocarbons and partial miscibility with hydrocarbons cause interesting phenomena in adsorption, micelle formation and growth, solubilization, and so on. Some topics will be discussed in this paragraph.

A. Selective Adsorption at the Air–Water Interface

A fluorinated surfactant is more compatible with a water surface than a hydrocarbon surfactant and predominantly adsorbs on the solution surface from their mixture [9,15,48,50,51]. Figure 4 shows the surface tensions at CMCs in PFNAla–LauAla system

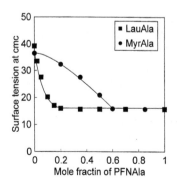

Figure 4 Surface tensions at mixed CMCs in PFNAla–LauAla and PFNAla–MyrAla systems. (From Ref. 9.)

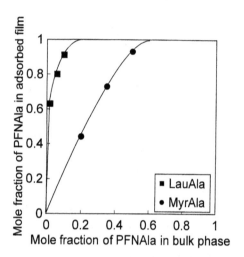

Figure 5   Compositions of adsorbed films in PFNAla – LauAla and PFNAla–MyrAla systems. (From Ref. 9.)

[9]. At mole fractions of PFNAla above 0.15, the surface tension was the same as that of pure PFNAla solution ($15.5\,\mathrm{mN\,m^{-1}}$). It was as if PFNAla molecules cover the whole surface of the solution. Surface compositions of PFNAla estimated thermodynamically are plotted as a function of its bulk composition in Fig. 5. This figure indicates that PFNAla is extremely rich at the surface of the mixed surfactant solution and adsorption of LauAla is negligible at high mole fractions of PFNAla. The higher the surfactant concentration was, the richer in PFNAla the surface composition became. PFNAla predominantly adsorbed also in the PFNAla–MyrAla system, but its predominance was less than in the PFNAla–LauAla system. Perfluoroalkane is more hydrophobic than the corresponding hydrocarbon by a factor of 1.5 [15], that is, a C_8F_{17} group exhibits a similar hydrophobicity to that of a $C_{12}H_{25}$ group. In the PFNAla–MyrAla system, MyrAla is then expected to be predominant in the surface adsorbed film because PFNAla and MyrAla have a C_8F_{17} group and a $C_{13}H_{27}$ group, respectively, and the former has a higher CMC than the latter (Table 1). Contrary to the expectation, however, the experimental result showed the opposite trend in surface adsorption.

Per-fluorination of an alkyl group thus enhanced surface activity more than hydrophobicity.

B. Azeotropic Phenomena in Micellar Solutions

When a surfactant solution contains sufficient salt, one can measure the concentration of monomeric surfactant molecule in equilibrium with micelles by ultrafiltration [52,53]. In ultrafiltration of a single surfactant solution, the surfactant concentration in the filtrate is identical with that in the filtrand below its CMC, and the concentration becomes constant above the CMC and equals the CMC. That is, the concentration in the filtrate corresponds to the monomeric surfactant concentration. For binary solutions of hydrocarbon and fluorocarbon surfactants, however, two types of curves were found for the plots of filtrate concentration–total surfactant concentration [54]. The first type is where the plot of total monomer concentration has two inflection points (first and second CMCs) and the filtrate concentration increases even above the second inflection point with increasing total concentration. The sodium perfluoro-octanoate (SPFO)–SDS system is one example of this type. This type is also typical for binary hydrocarbon surfactant solutions forming miscible mixed micelles [52,53]. In the second type, the total monomer concentration curve also has two inflection points, however, the total monomer concentration (the filtrate concentration) remains constant after the total concentration reaches the second inflection point. Lithium perfluoro-octanesufonate (LiFOS)–LiDS and LiFOS–LiTS systems belong to the second type. For an equimolar LiFOS–LiTS system, the concentration of monomeric LiFOS was nearly equal to the total concentration of LiFOS up to about 5 mM and became constant above the second CMC (Fig. 6). On the other hand, the concentration of monomeric LiTS remained constant beyond the first CMC and the total concentration of LiTS at the first CMC was equal to the CMC of pure LiTS. These results imply that LiTS rich micelles form at the first CMC and micellar phase de-mixing (coexistence of two kinds of mixed micelles) occurs at the second CMC. The monomer composition in this system is

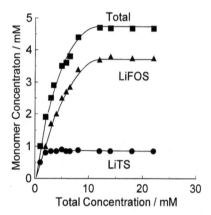

Figure 6 Dependence of monomeric surfactant concentration on total surfactant concentration for the equimolar LiFOS–LiTS system. (From Ref. 54.)

plotted in Fig. 7. Each curve represents the relation between the monomer composition and the total concentration at a fixed mole fraction of the surfactant mixture. Below the first CMC, the monomer composition is identical with the given mole fraction of LiFOS in the solution. Above the CMC, the monomer composition approaches a certain mole fraction ($X_s = 0.84$) at high total concentrations. When the given mole fraction is 0.84, the monomer composition remains constant even if the total concentration is beyond the CMC. This phenomenon is known as 'azeotropy.' The value, $X_s = 0.84$, is the mole fraction under azeotropic conditions. Once the azeotropic composition (X_{AZ}) and concentration (C_{AZ}) are known, one can calculate the second CMC using the following relation obtained from material balances

$$\text{CMC2} = \frac{X_F - X_{AZ}}{X_F - \alpha} C_{AZ} \qquad (10)$$

$$\text{CMC2} = \frac{X_{AZ} - X_H}{\alpha - X_H} C_{AZ} \qquad (11)$$

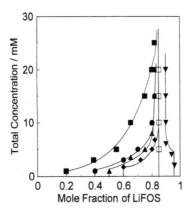

Figure 7 Plots of total surfactant concentration against monomeric surfactant composition at fixed overall compositions for LiFOS–LiTS system. (From Ref. 54.)

where X_F and X_H are the mole fractions of fluorocarbon surfactant in fluorocarbon rich micelles and hydrocarbon rich micelles, respectively. The larger value of CMC2 in the two equations is the second CMC. The azeotropic composition and concentration can be obtained from the ultrafiltration experiment [54]. Experimental values for the second CMCs can be determined by ultrafiltration [54], NMR [17], gel filtration [49], and probe methods [17,23], and some experimental values for the second CMCs in LiFOS–LiDS system are plotted in Fig. 8 and they are in agreement with the predicted values.

Conductivity measurement has frequently be utilized for determination of the first and second CMCs, although the second break point on the plot of conductivity against surfactant concentration corresponds only to the concentration that gives the maximum monomer concentration but is not always the second CMC [55]. The slope of the plot above the first CMC varies with mixing ratio in binary surfactant system, so that the monomer concentration and micellar composition vary with increasing surfactant concentration. The monomer concentration is constant and independent of the total surfactant concentration above the second CMC and as a result a break point appears on the plot. The break point

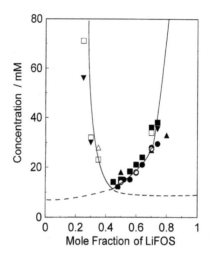

Figure 8 Second CMCs for LiFOS–LiDS system: (●) second CMC by ANS fluorescence; (▲) pyrene I_3/I_1; (■) auramine fluorescence; (○) pyrene solubility; (△) ^{1}H NMR; (□) ^{13}C NMR. The dotted line and solid line are the mixed CMC and the calculated second CMC. (From Ref. 17.)

therefore corresponds to the second CMC when the monomer concentration is maximum at the break point.

C. Synergistic Solubilization

Treiner *et al.* [56] found synergistic solubilization of 1-pentanol in mixed solutions of fluorocarbon and hydrocarbon surfactants, and theoretical researches [56–59] have progressed the theoretical treatments of solubilization in mixed micelles. Here, two examples are given illustrating the enhancement of solubilization power due to micellar growth in mixed solution of fluorocarbon and hydrocarbon surfactants.

First, let us discuss the solubilization of OFN as a function of the mole fraction of DEAPFN (X_F) in DEAPFN–DEATS mixed solutions [21]. Solubilization capacity toward OFN increased very slightly up to $X_F = 0.4$ and remarkably above $X_F = 0.5$, and then decreased after passing through a maximum around at $X_F = 0.8$. Similarly, the micelle aggregation number of DEAPFN–DEATS mixed micelles remained at about 100 up to $X_F = 0.3$, started to increase above $X_F = 0.5$, and passed through a maximum at $X_F = 0.8$. That is, the solubilization capacity of the mixed micelle toward OFN was strongly dependent on micellar size. In contrast, solubilization capacity toward pyrene exhibited a monotonous decrease with increasing mole fraction of DEAPFN. As shown in Table 3, fluorocarbon surfactants have a very low solubilization capacity toward pyrene compared with hydrocarbon surfactants. As a result, the solubilization capacity of the mixed micelle would decrease with increasing mole fraction of the fluorocarbon surfactant that is lipophobic against pyrene. The lipophobic nature of fluorocarbon surfactants would be a predominant factor for solubilization of pyrene when compared with the preferred solubilization effect due to micelle growth.

LiDS and LiFOS had very similar solubilization capacity toward FBIP, as seen in Table 3. Their mixtures had an almost constant solubilization capacity at all mixing ratios in the absence of salt [28]. In the mixed systems, addition of salt increased the solubilized amount of FBIP and induced positive synergistic solubilization. The synergistic behavior was enhanced with increasing salt concentration. A similar synergistic effect was observed in the LiHFDeS–LiDS mixed system. Salt-induced micelle growth explained these phenomena [28].

D. Models of Nonideal Mixed Micelle Solution

Among many thermodynamic models proposed for mixed micelle formation, the phase separation model and the mass action model have been used in most cases. The Rubingh model [60], which is a combination of the phase separation model and regular solution theory, has frequently been used to calculate the interaction parameter (β) in mixed micelle systems. The value of β is obtained by fitting the experimental values of CMCs. In the mixed systems of fluorinated and hydrocarbon surfactants, the value of β is a very important criterion to judge whether micellar phase separation does occur or not. According to this model, the mixed surfactant system with $\beta > 2$ gives two kinds of mixed micelles in aqueous solutions. However, the

Rubingh model was too simple to estimate the exact value of β and then a pseudo-phase model taking into account an electrostatic interaction term was proposed by Shinoda and Nomura [61]. The pseudo-phase separation model could successfully explain the experimental results of CMCs in many mixed surfactant systems. Unfortunately, some experimental values of CMCs often tended to deviate from the CMC curve predicted by the model for the binary surfactant mixture where one of them had remarkably higher CMC compared with the other. It must be kept in mind that the value of β is not always the same as that defined originally in the regular solution theory, in which the entropy of mixing is ideal and nonideality results only from the heat of mixing. In most cases, it is reasonable to consider that the requirement of ideal mixing entropy is not met in nonideal mixed micelle formation, especially for mixtures of hydrocarbon and fluorocarbon surfactants. In conclusion, the interaction parameters obtained should be carefully interpreted, although the model is very useful.

On the other hand, a group contribution method had been used for the prediction of activity coefficients in the non-electrolyte liquid mixtures in the field of chemical engineering for estimation of thermodynamic quantities [62–65]. Asakawa, *et al.* [2] proposed a model for estimation of CMC by combining the method with the pseudo-phase model. This model (it is called group method), enabled to predict mixed CMCs without experimental values of mixed CMCs. On the contrary, none of the models mentioned above permitted an estimate of the CMCs without knowing experimental values of mixed CMCs, even for those mixed surfactant systems where only alkyl chain length varied systematically. However, the prediction was easily achieved using the group method.

In the group method, the monomeric concentration of each component is given by the following equations

$$C_{1m}(C_{1m} + C_{2m} + C_{ion})^{kg} = C_1(C_1 + C_{ion})^{kg_1}x_1F_1 \qquad (12)$$

$$C_{2m}(C_{1m} + C_{2m} + C_{ion})^{kg} = C_2(C_2 + C_{ion})^{kg_2}x_2F_2 \qquad (13)$$

where C_m is the monomer concentration of surfactant in the mixed system, C_1 the CMC in the single component system, C_{ion} the concentration of added counterion, and x the micellar composition. kg is the effective electrostatic parameter (sometimes called micelle counterion binding parameter) and F is the activity coefficient of surfactant in the micelle. The subscripts 1 and 2 refer to surfactants 1 and 2.

$$kg = \alpha_1 kg_1 + (1 - \alpha_1) kg_2 \tag{14}$$

$$\alpha = C_{1m}/(C_{1m} + C_{2m}) \tag{15}$$

$$\ln F_i = \ln F_i^C + n F_i^R \tag{16}$$

where $\ln F_i^C$ is the combinatorial term and $\ln F_i^R$ the residual component. The former is essentially based on differences in size and shape of the constituent groups of a surfactant, while the latter corresponds to the interaction energy between the groups of two surfactants. Basic equations [Eqs (12) and (13)] are the same as those in the pseudo-phase model [61] and activity coefficient of each component Eq. (16) is calculated using the group contribution method.

For the LiFOS–LiDS system [2], the CMCs estimated using the group method exhibited an excellent agreement with the experimental results compared with those based on the pseudo-phase model. That is, the standard deviations in the group method and the pseudo-phase model were 0.123 mM and 0.236 mM, respectively, over the range of 15–40°C. This method and the pseudo-phase model both predicted formation of two types of mixed micelles. In addition, the group method indicated that fluorocarbon rich micelles sparingly solubilized hydrocarbon surfactant while hydrocarbon rich micelles solubilized fluorocarbon species to some extent. Several other research groups reported this kind of asymmetrical mutual solubility behavior in mixed micelle systems [2,48,51].

In the systems of LiFOS–Li alkylsulfates [2], coexistence of two kinds of mixed micelles was predicted and the immiscible region widened with increasing carbon number of the hydrocarbon surfactant. A similar trend was found for cationic mixed

Table 5 Immiscible regions predicted by group method at 25°C.

	HFOPC	HFNPC	HFDePC X_H-X_F	HFDPC X_H-X_F
DePC	Miscible	Miscible	0.41–0.45	0.08–0.79
DPC	Miscible	Miscible	0.41–0.55	0.09–0.83
TPC	Miscible	Miscible	0.40–0.65	0.11–0.87
CPC	Miscible	Miscible	0.40–0.72	0.12–0.90

DePC: decylpyridinium chloride.
HFNPC: 1*H*,1*H*,2*H*,2*H*-perfluorononylpyridinium chloride.
Source: Ref. 10.

Table 6 Temperature effect on micellar miscibility.

Temperature °C	HFOPC–DPC	HFNPC–DPC	HFDePC–DPC X_H-X_F	HFDePC–CPC X_H-X_F
25	Miscible	Miscible	0.41–0.55	0.40–0.72
30	Miscible	Miscible	Miscible	0.40–0.70
35	Miscible	Miscible	Miscible	0.42–0.68
40	Miscible	Miscible	Miscible	0.47–0.64
45	Miscible	Miscible	Miscible	0.555

Source: Ref. 10.

surfactant systems [10]. That is, the mutual solubilities of hydrocarbon and fluorocarbon chains in the micellar phases were estimated by use of the group method for the mixed surfactant systems having a pyridinium group as their hydrophilic groups. The immiscible region expanded when the alkyl chain in either surfactant molecule became longer (Table 5). The region narrowed with increasing temperature (Table 6). Some examples of the phase separation in cationic surfactant mixtures [66] are shown in Fig. 9, together with CMC data. Comparison of the region of phase separation among the three methods (pseudo-phase model, Motomura's method [67] and the group method) is given in Table 7. The experimentally determined phase separation regions were asymmetrical at about a mole fraction of 0.5 (Fig. 9 and Table 7). The regions calculated using the group method and

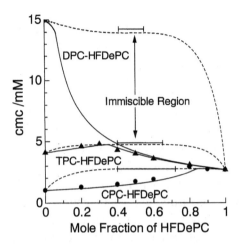

Figure 9 CMCs of mixtures of alkyl pyridinium chlorides. The plotted points are experimental values. The solid lines and dotted lines are the CMC curves and the micellar composition curves predicted by the group method. (Experimental data were cited from Ref. 66.)

Table 7 Comparison in pseudo-phase separation region among several methods for LiFOS–LiDS system in the presence of 10 mM LiCl.

	X_H	X_F	X_{AZ}	C_{AZ}/mM
Pseudo-phase model ($\beta = 2.2$)	0.24	0.76	0.43	8.24
Motomura's method	0.23	0.78	0.43	8.30
Group method	0.25	0.90	0.46	8.08
Experimental results[a]	0.24	0.78	0.43	8.30

[a]*Source:* Ref. 49.

Motomura's method were also asymmetrical, while the region determined using the pseudo-phase model was always symmetrical.

The group method requires many group interaction parameters determined in advance from vapor–liquid equilibrium and liquid–liquid equilibrium data, and CMC data in other mixed surfactant systems [2]. This requirement is a disadvantage. However, once the group parameters are given, activity coefficients are very easily calculated. This method is thus useful in predicting mixed CMCs in the systems showing significant deviation from an ideal mixture. In addition, it is possible to estimate which groups contribute to the nonideal behavior in the micellar phase.

The molecular theory of mixed micelle developed by Nagarajan [68] was also applicable to mixed solutions of hydrocarbon and fluorocarbon surfactants [32]. According to the theory, the average mole fraction and aggregation number were 0.13 and 71 for hydrocarbon rich micelles and 0.87 and 84 for fluorocarbon rich micelles, respectively, in the LiFOS–LiDS system [32]. The calculation indicated that mixing of surfactants and addition of salt promoted micellar growth.

V. MICROSCOPIC PHASE SEPARATION

Since Mukerjee and Yang [5] predicted the coexistence of two kinds of micelles, much attention has been paid to micellar miscibility in mixed surfactant solutions and various research techniques have been applied to understand this phenomenon. It has become consequently evident that separation of a pseudo-phase micelle into two kinds of mixed micelles can occur in solutions containing a fluorinated surfactant and a hydrocarbon surfactant. If two kinds of micelles coexist, the mixed micellar solution will be a system of three phases (one aqueous bulk and two micellar phases) and three components (water and two surfactants). Under the condition of constant temperature and pressure, the number of degrees of freedom is zero. Then, the composition in each phase becomes constant. That is, when two kinds of micelles (two phases) coexist with the monomeric surfactant solution (bulk phase), the composition of each phase should remain unchanged independently of increasing surfactant concentration. It was confirmed experimentally (gel filtration [49], ultrafiltration [54], surface tension [9], etc.) that this condition is kept so long as two kinds of mixed micelles are in equilibrium with the monomer surfactant solution. Gel filtration of the LiFOS–LiTS mixed surfactant solution gave two peaks corresponding to a hydrocarbon

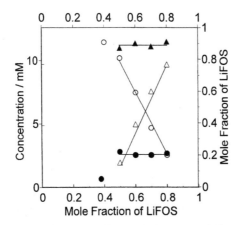

Figure 10 Concentrations and compositions of the mixed micelles in the 20 mM LiFOS–LiTS system: ($\bigcirc$, $\bullet$) LiTS rich micelle concentration and composition, ($\triangle$, $\blacktriangle$) LiFOS rich micelle concentration and composition. (From Ref. 49.)

surfactant rich micelle and a fluorocarbon rich micelle, respectively [49]. The composition in each micelle was constant, whereas the concentration of each micelle varied linearly with the total composition in the region where two kinds of micelles coexisted, as expected (see Fig. 10).

In addition to the macroscopic point of view, however, the pseudo-phase separation to two kinds of micelles should be verified directly from a microscopic viewpoint. The fluorescence probe method is one of the informative ways to study the microscopic aspects of intra-micellar events. This paragraph shows that the fluorescence quenching behavior of the fluorinated quencher is very useful to verify de-mixing micellization (transition from one kind of micelle to two kinds of micelles) and de-micellization (transition from two kinds of micelles to one kind of micelle), and microscopic phase separation in a micelle.

A. Distribution of Probe and Quencher Among Mixed Micelles

The pair of pyrene (fluorescence probe) and CPC (fluorescence quencher) has been very often utilized to determine the micellar aggregation number as described before. Under the condition that the quencher has a Poisson distribution among micelles, the aggregation number can be determined using the equation [69,70].

$$\ln\left(\frac{I_0}{I}\right) = \frac{N_{\text{agg}}[Q]}{C-\text{CMC}} \tag{17}$$

where I and I_0 are the fluorescence intensities in the presence of quencher and in its absence, N_{agg}, $[Q]$, and C are the micellar aggregation number, and quencher and surfactant concentrations, respectively. Table 4 shows some examples of aggregation numbers obtained using the equation. The Poisson distribution, however, does not hold for some combinations of quencher and surfactant [71]. It is then necessary to choose the most suitable quencher for a given micelle.

Cetyl pyridinium chloride (CPC), a typical quencher, has a pyridinium ring that contributes to quenching [72], but neither its alkyl group nor Cl ion quench pyrene emission. It is therefore, expected that pyridinium salts having a fluorocarbon chain are also effective in quenching pyrene emission. Recently, a perfluoro-alkyl pyridinium salt (HFDePC) was synthesized [10] and was confirmed to quench the fluorescence of pyrene in a micelle as CPC [23] does. The fluorinated quencher, HFDePC, forms micelles in its aqueous solution (see Section II. **B**) and can solubilize pyrene [23]. The second-order quenching rate constant for pyrene in CPC micelle is very similar to that of HFDePC micelle (6.3×10^9 and $5.1 \times 10^9 \,\text{mol}^{-1}\text{dm}^3\text{s}^{-1}$, respectively) [73]. However, the fluorocarbon is not always miscible highly with the hydrocarbon as described before. As a result, the apparent quenching ability of HFDePC often becomes smaller than that of CPC because of the difference in solubilization ability [71].

In order to determine the exact aggregation number using Eq. (17), all of quencher molecules are required to be solubilized and to obey the Poisson distribution. When quencher molecules distribute among the micellar and bulk water phases, Eq. (17) gives a smaller aggregation number. This case was found when HFDePC was used as a quencher in

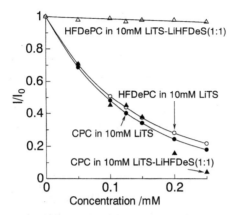

Figure 11 Fluorescence quenching ratios of pyrene in 10 mM total surfactant concentration as a function of quencher concentration. Pyrene concentration was fixed at 1.0×10^{-5} M. (From Ref. 74.)

hexadecyltrimethylammonium chloride (CTAC) solutions [23]. This is due to lower miscibility between fluorocarbon and hydrocarbon chains. However, the quencher was completely solubilized in micelles with sufficient addition of salt as well as CPC. In the presence of 1 M NaCl, there was little difference in quenching ability between CPC and HFDePC [23].

Next, we show an interesting example that the quenching behavior of pyrene emission is remarkably different from each other for CPC and HFDePC. A cationic quencher is easily solubilized in anionic micelles through electrostatic interaction. The quenching behavior of pyrene in LiTS and LiTS–LiHFDeS systems [74] is illustrated in Fig. 11, where the surfactant concentration is kept at 10 mM and the micelle is allowed to solubilize 1–2 quencher molecules. The values of I_0 and I in the figure are the pyrene fluorescence intensities in the absence and presence of quencher, and the minimum and maximum values of I/I_0 are 0 and 1, meaning the perfect quenching and non-quenching, respectively. In the LiTS system, HFDePC effectively quenched the fluorescence of pyrene, similarly to CPC. On the other hand, the quenching behavior of HFDePC was remarkably different from that of CPC in the LiTS–LiHFDeS mixed system. That is, CPC quenched the fluorescence of pyrene both in pure LiTS micelles and in mixed micelles, while HFDePC hardly quenched the emission in the mixed micelles and the depression of pyrene emission was very little, even under the condition that a micelle contained two HFDePC molecules. Since the relative fluorescence intensity (I/I_0) corresponds to the probability of finding a micelle without a quencher molecule among micelles [41], the large value of I/I_0 means that there are few quencher molecules in the micelles solubilizing pyrene. When LiHFDeS-rich and LiTS-rich micelles coexist in aqueous solutions, most of HFDePC molecules must be partitioned into the LiHFDeS-rich micelles because of the mutual phobicity of hydrocarbon and fluoro-carbon chains. Pyrene is solubilized in an overwhelmingly larger amount in a hydrocarbon micelle than in a fluorocarbon micelle (see Table 3). In micellar solutions containing pyrene and CPC, intra-micellar quenching is dominant and inter-micellar transfer of CPC not only via the bulk phase, but also via micellar collision, is negligible in the lifetime of exited pyrene [75,76]. The collision between pyrene (in LiTS-rich micelles) and HFDePC (in LiHFDeS-rich micelles) would therefore be inhibited within the lifetime of exited pyrene. In the case of CPC, the quencher is localized in LiTS-rich micelles together with pyrene and this results in effective quenching. On the other hand, HFDePC is solubilized in LiHFDeS-rich micelles without pyrene and LiTS-rich micelles solubilizing pyrene but no quencher can quench the pyrene emission. Almgren *et al.* [73] found from time-resolved fluorescence quenching experiments that no quenching by HFDePC was observed in 10 mM equimolar LiDS–LiPFN solution containing 0.1 M LiCl. Similar depression in quenching was also found in LiFOS–LiDS and lithium perfluoro-nonanoate (LiPFN)–LiDS systems [71], in which the coexistence of two kinds of mixed micelles had been reported [32]. Figure 11 strongly suggests that two kinds of mixed micelles coexist in the LiTS–LiHFDeS mixed system. This conclusion is in accordance with the macroscopic view obtained from the results of surface tension and conductivity measurements [74]. These results mean that the pyrene and the fluorinated quencher distribute in a nonuniform manner among two kinds of mixed micelles, respectively.

Let us examine in more detail the relation between depression of quenching and coexistence of two kinds of mixed micelles under the condition that pyrene and quencher concentrations are fixed. The quenching behavior of HFDePC was compared with that of CPC at various fixed concentrations of LiTS (high enough compared with the CMCs) in the LiTS–LiHFDeS mixed system [74]. Although HFDePC quenched effectively pyrene fluorescence in pure LiTS micelles, the quenching was sharply depressed with addition of LiHFDeS into 10 mM LiTS until the LiHFDeS concentration reached around 1.5 mM and then the depression became almost perfect at higher LiHFDeS concentrations. The total surfactant concentration above which almost perfect depression was found agreed with the second CMC determined by the surface tension method [74]. LiHFDeS rich micelles begin to appear at the second CMC and both LiTS rich and LiHFDeS rich micelles coexist above this concentration. Below the second CMC, only LiTS micelles exist and solubilize the added LiHFDeS molecules. When the added amount of LiHFDeS exceeds the solubilized limit, the second type of micelles rich in LiHFDeS can appear. Under this condition, the collision probability between the separately solubilized pyrene and HFDePC will be lowered depending on the immiscibility of hydrocarbon and fluorocarbon species. If perfect de-mixing of micelle systems might occur (that is, there are either pure LiTS or pure LiHFDeS micelles between the first and second CMCs, and both pure LiTS and LiHFDeS micelles coexist above the second CMC), the quenching would not vary at every concentration below the cmc of LiHFDeS (2.9 mM). Based on this assumption, there would be only pure LiTS micelles containing no LiHFDeS in the region of the total mole fraction rich in LiTS and also the ratio of pyrene/HFDePC would be kept constant in the micelles. Then, the depression of quenching at low LiHFDeS concentrations indicates solubilization of LiHFDeS into LiTS micelles, since only one kind of mixed micelle exists below the second CMC. In the mixed systems, the fluorinated quencher, HFDePC, will interact more strongly with LiHFDeS than

LiTS because of the difference in lipophobicity between hydrocarbon and fluorocarbon chains, while pyrene will be located nearer LiTS. (Note the fact that the value of I_1/I_3 of pyrene over the range of the mole fraction of LiHFDeS from $\alpha = 0$ to 0.5 is similar to that in a LiTS pure micelle [71]). The perfect depression of quenching means no contact of HFDePC with pyrene and it corresponds therefore to the separated solubilization of pyrene and HFDePC into different micelles, respectively. This inclined solubilization might decrease the collision probability between pyrene and HFDePC and then might result in less quenching of pyrene fluorescence. Once the second type of micelles rich in LiHFDeS appears with further addition of LiHFDeS, pyrene, and HFDePC are separately solubilized in LiTS rich micelles and LiHFDeS rich micelles, respectively, depending on their miscibility with the surfactants. On the contrary, when CPC was used instead of HFDePC, pyrene fluorescence was effectively quenched and the fluorescence intensity of pyrene scarcely increased due to the increase in micelle concentration.

Not only surfactant concentration, but also composition in mixed surfactant solutions significantly affects the quenching behavior of pyrene fluorescence [74,77]. For the LiTS–LiHFDeS system, the ratio I/I_0 is plotted in the presence of 0.25 mM HFDePC as a function of the total surfactant concentration in Fig. 12. HFDePC did not signifcantly quench the fluorescence of pyrene in water, reflecting a low collision probability at their low concentrations. That is, there was no specific interaction between pyrene and HFDePC. However, a small addition of LiTS produced large depression of the ratio I/I_0 and the depression was largest at the CMC. The result indicates that LiTS forms a complex with HFDePC through the coulombic interaction, and that the complex interacts with pyrene and is most concentrated at the CMC. Above the CMC, the ratio increases slightly with increasing LiTS concentration. The increase is ascribable to the decrease in the numbers of both pyrene and HFDePC molecules in a micelle due to the increase in micelle concentration. In contrast to the pure LiTS system, the value of I/I_0 increased remarkably with increasing total surfactant concentration after its initial decrease in

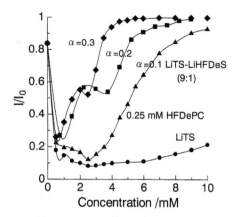

Figure 12 Fluorescence quenching ratios vs total surfactant concentration with constant mole fraction for LiHFDeS–LiTS systems. Pyrene and HFDePC concentrations were fixed at 1×10^{-7} M and 2.5×10^{-4} M, respectively. (From Ref. 74.)

LiHFDeS–LiTS mixed systems. In higher concentration regions at certain mole fractions of LiHFDeS ($\alpha = 0.2$ and 0.3), the values eventually approached 1.0 (the value of 1.0 means no-quenching). We can conclude from this result that pyrene and the quencher are solubilized in different micelles, respectively. Because such a large depression of quenching cannot be explained only by the increase in micelle concentration, the result in the pure LiTS system was taken into account. The total surfactant concentration at which I/I_0 became almost constant was in fairly good agreement with the second CMC determined independently by the surface tension method [74]. At total surfactant concentrations above the second CMC, the micelle composition is constant [49]. This fact explains the constant I/I_0 above the second CMC. Over the range of total surfactant concentrations between the first and second CMCs, the micellar composition and concentration continuously vary with increasing total surfactant concentration. Partition of quencher molecules between the mixed micelle and bulk phase, distributions of the quencher and pyrene among micelles, and their collision probability must therefore vary with surfactant concentration. The behavior of I/I_0 in the region of Fig. 12 reflects this situation.

Since addition of salt causes a decrease in the monomeric surfactant concentration and sometimes promotes micellar growth, the ratio I/I_0 is expected to vary with salt concentration. However, the ratio hardly varied with addition of LiCl up to 0.1 M in a 10 mM LiTS–LiHFDeS mixed solution and it remained at almost 1 at the mole fraction of LiHFDeS $\alpha = 0.2$ in spite of sufficient addition of LiCl. As pointed out before, the value of "1" for I/I_0 means that fluorescent probe cannot meet quencher molecules within its fluorescence lifetime. We can therefore conclude that the presence of salt does not change the situation that there are two kinds of micelles (LiHFDeS rich micelles and LiTS rich micelles) in which HFDePC and pyrene are solubilized separately.

B. Demicellization

Demicellization is defined as the phenomenon that a solution containing two kinds of mixed micelles changes to a solution having only one kind of mixed micelles at a specific concentration with increasing surfactant concentration. The specific concentration is the "critical demicellization concentration (CDC)." The concept of demicellization was proposed by Mysels in 1978 [78], and was confirmed experimentally by Funasaki and Hada [48] and Lake [79]. At the CDC, one of two kinds of mixed micelles disappears because it is solubilized into the other one. Funasaki and Hada [48] studied the phenomenon in Neos Ftergent (NF)–sodium tridecyl sulfate system by the surface tension method, and Lake [79] estimated the CDC for sodium decanoate–SPFO system by kinetic dialysis. Asakawa and Miyagishi [71] determined the CDC for the SPFO–SDS system by pyrene fluorescence quenching. The fluorescence quenching method is very appropriate to detect microscopic events in micelles. We will hereafter discuss the phenomenon of demicellization from a microscopic point of view.

The ratios I_1/I_3 and I/I_0 are plotted in Fig. 13 against surfactant concentration for an equimolar SPFO–SDS mixed

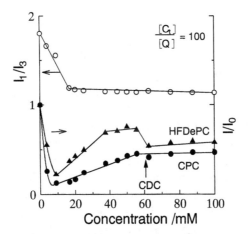

Figure 13 Fluorescence intensity ratios of pyrene in SPFO–SDS equimolar mixture as a function of total surfactant concentration. Concentration of quencher was kept to be 1/100 of total surfactant concentration. (Data of I/I_0 were cited from Ref. 71.)

system containing pyrene and HFDePC (or CPC) as the fluorescence probe and quencher, the concentration of which was kept at 1% of the total surfactant concentration. The value of I_1/I_3 that reflects the polarity of the environment surrounding pyrene in mixed micelles decreased with increasing surfactant concentration and became constant ($I_1/I_3 = 1.2$) above the mixed CMC. The constant value was nearly the same as that in pure LiDS micelles (see Fig. 1). This result suggests that pyrene is selectively solubilized in SDS rich micelles regardless of surfactant concentration. Quenching was very effective for I/I_0 at the mixed CMC due to the concentration effect of quencher in micelles. The ratio I/I_0 increased with increasing surfactant concentration because of the dilution effect of quencher solubilized into an increasing number of micelles. CPC quenched more effectively than HFDePC in this system as well as in other systems (compare with Fig. 11). A significant difference was observed between CPC and HFDePC in a concentration range above 40 mM. When HFDePC was used as quencher, the ratio I/I_0 gave a maximum around 40–55 mM. Since an increase in the ratio (depression in quenching) corresponds to a decrease in the chance that a HFDePC molecule comes close to a pyrene molecule, the depression in quenching indicates the coexistence of SPFO rich and SDS rich micelles in this concentration range. HFDePC tends to partition into fluorinated micelles due to a low affinity of its fluorocarbon chain for the hydrocarbon environment when the two kinds of micelles coexist. Consequently, the fluorescence probe (pyrene) and the quencher (HFDePC) are solubilized separately in SDS rich micelles and SPFO rich micelles, and thus the chance of collision between the probe and the quencher is remarkably diminished and this inhabitation is responsible for the significant depression of quenching.

As seen in Fig. 13, the ratio I/I_0 decreased again with further addition of surfactant, above 60 mM. Since decrease in the ratio means an enhanced collision probability between pyrene and HFDePC, the decrease indicates that the two kinds of micelles come into contact very easily within the life time of exited pyrene or the two kinds of micelles change into one kind of micelle through intermicellar collisions. However, such rapid collisions between the micelles within the lifetime are not reasonable [75,76]. The quenching of pyrene emission by HFDePC is effective only when the probe and quencher are solubilized at the same time in a micelle. In addition, the value of I/I_0 above 60 mM was almost the same as that in the pure SDS system [71]. In contrast to HFDePC, the value of I/I_0 in the presence of CPC increased monotonously with increasing surfactant concentration and approached a constant value without showing a maximum. The decrease in I/I_0 above 60 mM can then be concluded to correspond to demicellization [71]. At surfactant concentrations beyond 60 mM, one of the two kinds of mixed micelles disappears and the other kind of mixed micelles grow up to a larger micelle having a high solubilization power that can solubilize both the hydrocarbon probe and the fluorocarbon quencher.

Although HFDePC had a very similar quenching ability toward pyrene fluorescence in water to CPC, the fluorinated quencher was always less effective in micelle systems than CPC because the fluorinated quencher was partitioned among micelles in a manner deviated from the Poisson distribution, as discussed in (Section V.A). This deviation can be explained as follows. In single surfactant micelles, (1) micelles solubilizing a fluorinated compound are more preferable to a fluorinated quencher than both unoccupied micelles and pyrene containing micelles, or (2) quencher with a fluorocarbon chain is shielded from pyrene in a micelle due to a low mutual miscibility between its fluorinated chain and pyrene. In the present case of mixed micelles, a slight difference in quenching between CPC and HFDePC results from segregation of the fluorocarbon quencher among micelles. CPC is solubilized in hydrocarbon surfactant rich micelles, some of which contain pyrene, while HFDePC is less solubilized in such micelles. In a mixed micelle, CPC and pyrene are solubilized in the hydrocarbon sur-factant rich region and HFDePC is favorably located in the fluorocarbon surfactant rich region of the micelle, respectively (microscopic phase separation). Imperfect depression of quenching around 40–55 mM in Fig. 13, indicates that at least either of the two kinds of mixed micelles solubilizes both pyrene and HFDePC.

The mutual miscibility between a hydrocarbon and a fluorocarbon depends on their alkyl chain length and a similar dependency is found in binary surfactant solutions as discussed before(Section IV.D). Microscopic phase separation in the inside of a micelle occurs at a higher concentration and its separation region is more limited or disappears when a surfactant with a shorter alkyl chain is used. Sufficient addition of salt to the SPFO–sodium decyl sulfate (SDeS) mixed system gave similar CMCs to those in the LiFOS–LiDS mixed system. Then, comparison of quenching behavior between the two mixed systems became possible in concentration ranges similar to each other [71]. CPC was more effective in quenching of pyrene fluorescence than HFDePC but the difference between the two quenchers was smaller in the SPFO–SDeS system containing 0.2 M NaCl than in the LiFOS–LiDS system. In addition, the ratio I/I_0 for HFDePC was 0.5–0.6 in the SPFO–SDeS system, while the ratio was over 0.9 in the LiFOS–SDS system. The small difference in quenching between CPC and HFDePC and the low value of I/I_0 suggest the formation of rather miscible mixed micelles in the SPFO–SDeS system.

Liquid mixtures of hydrocarbon and fluorocarbon become more miscible with rising temperature and give the upper critical solution temperature, which increases with increasing alkyl chain length [4]. A similar miscibility behavior and the upper critical solution temperature were found in binary micelle solutions of hydrocarbon and fluorocarbon surfactants [6]. Two kinds of mixed micelles change to one kind of larger mixed micelle above the upper critical solution temperature. This temperature may thus be better to be termed "the critical demicellization temperature" in mixed micelle systems. This phenomenon can also be examined using a fluorescence probe method [71]. Figure 14 shows the fluorescence intensity of pyrene as a function of temperature in the SPFO–SDS system. In the absence of salt, the difference in quenching between CPC and HFDePC was larger at lower temperature and decreased with increasing temperature. The difference decreased significantly above 40°C. CPC effectively quenched pyrene fluorescence at every temperature and its temperature dependence was slight, while the quenching by HFDePC became more

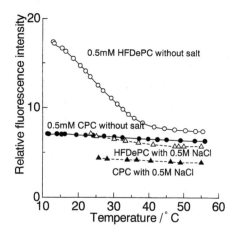

Figure 14 Effect of temperature on fluorescence intensity in SPFO equimolar mixture. (Data of solutions without salt were cited from Ref. 71.)

effective with increasing temperature and reached a level similar to that of CPC. In addition, the difference at high temperatures was at a similar level to that found in single component micelles. The large difference between CPC and HFDePC can be explained by separated solubilization of pyrene (in SDS rich micelles) and HFDePC (in SPFO rich micelles). The small difference in quenching between CPC and HFDePC at high temperatures indicates that both quenchers are in similar environments where the quencher can effectively quench pyrene fluorescence. The quenching behavior above 40°C results from the fusion of two kinds of mixed micelles.

In the presence of 0.5 M NaCl, the quenching behavior of HFDePC was similar to that of CPC and its temperature dependence diminished while it remained in the case where no salt was present as shown in Fig. 14. This is attributable to micellar growth. Strictly speaking, addition of salt induces fusion of two kinds of small micelles into one kind of large micelle in which a SDS rich region and a SPFO rich region coexist (this situation corresponds to "microscopic or nanoscale phase separation in a micelle"). HFDePC solubilized in the SPFO rich region can come into contact with pyrene that is solubilized in the SDS rich region, and quench its emission within its lifetime, but its collision probability is less than CPC, which is solubilized in the same region together with pyrene. A HFDePC molecule should collide with an excited pyrene molecule within its lifetime because the quenching is a diffusion-controlled reaction. Then, long separation between pyrene and HFDePC is disadvantageous for the quenching. However, the larger the micelle size is, the longer becomes the distance between pyrene and HFDePC. Since the increase in micelle size produces elevation in micro-viscosity in a micelle [80], it may make the solubilized molecule difficult to move. However, the perfect inhabitation of quenching has not been found yet. On the other hand, Kamogawa and Tajima [81] proposed a model of a single kind of mixed micelle consisting of two different intra-micellar sites, instead of the model that two kinds of mixed micelles coexist in a solution. However, their model cannot explain the high ratio of I/I_0 (nearly 1) for HFDePC at low temperatures.

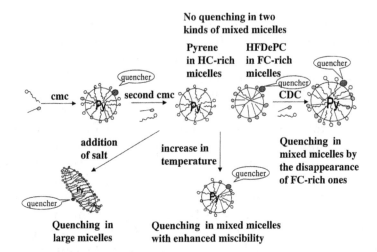

Figure 15 Quenching behavior of a fluorinated quencher and microscopic phase separation in micellar systems. (From Ref. 27.)

In conclusion, the quenching behavior of pyrene fluorescence by HFDePC can be schematically illustrated with a microscopic phase separation model in Fig. 15. This figure clearly demonstrates the high utility of HFDePC in detecting two kinds of micelles and the critical demicellization concentration (CDC) and in examining the miscibility in a mixed micelle and its temperature dependence.

REFERENCES

1. C. Chothia, *Nature* **248**: 338 (1974).

2. T. Asakawa, K. Johten, S. Miyagishi and M. Nishida, *Langmuir* **1**: 347 (1985).

3. J. C. Ravey, A. Gherbi, and M. J. Stebe, *Prog. Colloid Polym. Sci.* **76**: 234 (1988).

4. J. H. Hidebrand, J. M. Prausnitz, and R. L. Scott, *Regular and Related Solutions*, Van Nostrand Reinhold, New York (1970).

5. P. Mukerjee, and A. Y. S. Yang, *J. Phys. Chem.* **80**: 1388 (1976).

6. N. Funasaki, *Mixed Surfactant Systems* (K. Ogino and M. Abe, eds.). Marcel Dekker, New York (1993), pp. 145.

7. R. Zana, *Surfactant Solutions: New Methods of Investigation*, Marcel Dekker, New York (1987).

8. P. Infelta, M. Grätzel, and J. K. Thomas, *J. Am. Chem. Soc.* **78**: 190 (1974).

9. S. Miyagishi, M. Higashide, T. Asakawa, and M. Nishida, *Langmuir* **7**: 51 (1991).

10. T. Asakawa, H. Hisamatsu, and S. Miyagishi, *Langmuir* **11**: 478 (1995).

11. H. C. Evans, *J. Chem. Soc.* 579 (1956).

12. E. Götte, and M. J. Schwuger, *Tenside* **3**: 131 (1969).

13. S. Miyagishi, and M. Nishida, *J. Colloid Interface Sci.* **65**: 380 (1978).

14. M. Takehara, I. Yoshimura, and R. Yoshida, *J. Am. Oil Chem. Soc.* **51**: 419 (1974).

15. K. Shinoda, M. Hato, and T. Hayashi, *J. Phys. Chem.* **76**: 909 (1972).

16. N. J. Turro, M. Grätzel, and A. M. Braun, *Angew. Chem. Int. Ed., Engl.* **19**: 675 (1980).

17. T. Asakawa, M. Mouri, S. Miyagishi, and M. Nishida, *Langmuir* **5**: 343 (1989).

18. D. C. Turner, and L. Brand, *Biochemistry* **7**: 3381 (1968).

19. G. Greiser, and C. J. Drummond, *J. Phys. Chem.* **92**: 5580 (1988).

20. R. C. Mast, and L. V. Haynes, *J. Colloid Interface Sci.* **53**: 35 (1975).

21. T. Asakawa, J. Ikehara, and S. Miyagishi, *JOCS,* **73**: 21 (1996).

22. K. Kalyanasundaram, *Langmuir* **4**: 942 (1988).

23. T. Asakawa, A. Saruta, and S. Miyagishi, *Colloid Polym. Sci.* **275**: 958 (1997).

24. D. F. Evans, and H. Wennerstrom, *The Colloidal Domain*, VCH Publishers Inc., New York (1994), pp. 14.

25. N. Muller, and H. Simsohn, *J. Phys. Chem.* **75**: 942 (1971).

26. J. Ulmius, and B. Lindman, *J. Phys. Chem.* **85**: 4131 (1981).

27. T. Asakawa, and S. Miyagishi, *Hyomen* **39**: 160 (2001).

28. T. Asakawa, T. Kitaguchi, and S. Miyagishi, *J. Surfactants Detergents* **1**: 195 (1998).

29. S. Miyagishi, H. Kurimoto, and T. Asakawa, *Langmuir,* **11**: 2951 (1995).

30. D. Bendedouch, and S. H. Chen, *J. Phys. Chem.* **88**: 658 (1984).

31. R. Nagarajan, and E. Ruckenstein, *Langmuir* **7**: 2934 (1991).

32. T. Asakawa, T. Fukita, and S. Miyagishi, *Langmuir* **7**: 2112 (1991).

33. C. Tanford, *The Hydrophobic Effect.* Wiley, New York (1973), pp. 74.

34. W. Prapaitrakul, and A. D. King, Jr., *J. Colloid Interface Sci.* **118**: 224 (1987).

35. J. N. Israelachvili, D. J. Mitchell, and B. W. Ninham, *J. Chem. Soc. Faraday Trans. I,* **72**: 1525 (1976).

36. H. Hoffmann, J. Kalus, K. Reizlein, W. Ulbricht, and K. Ibel, *Colloid Polym. Sci.* **260**: 435 (1982).

37. S. J. Burkitt, R. H. Ottewill, J. B. Hayter, and B. T. Ingram, *Colloid Polym. Sci.* **265**: 619 (1987).

38. Y. Muto, K. Esumi, K. Meguro, and R. Zana, *J. Colloid Interface Sci.* **120**: 162 (1987).

39. T. Asakawa, and S. Miyagishi (unpublished data).

40. T. Asakawa, H. Sunagawa, and S. Miyagishi, *Langmuir* **14**: 7091 (1998).

41. E. Grieser, and C. J. Drummond, *J. Phys. Chem.* **92**: 5580 (1988) (and references therein).

42. H. Hoffmann, J. Kalus, and H. Thurn, *Colloid Polym. Sci.* **261**: 1043 (1983).

43. K. Fontell, and B. Lindman, *J. Phys. Chem.* **87**: 3289 (1983).

44. K. Wang, G. Karlsson, M. Almgren, and T. Asakawa, *J. Phys. Chem. B* **103**: 9237 (1999).

45. H. Hoffmann, and J. Wurtz, *J. Mol. Liq.* **72**: 191 (1997).

46. M. Monduzzi, *J. Curr. Opin. Colloid Interface Sci.* **3**: 67 (1998).

47. N. Funasaki, and S. Hada, *J. Chem. Soc. Chem. Commun.* 253 (1980).

48. N. Funasaki, and S. Hada, *J. Phys. Chem.* **84**: 342 (1980).

49. T. Asakawa, S. Miyagishi, and M. Nishida, *Langmuir* **3**: 821 (1987).

50. P. Mukerjee, and T. Handa, *J. Phys. Chem.* **85**: 2298 (1981).

51. H. Matsuki, S. Kaneshina, N. Ikeda, M. Aratono, and K. Motomura, *J. Colloid Interface Sci.* **150**: 33 (1992).

52. I. W. Osborne-Lee, R. S. Schechter, and W. H. Wade, *J. Colloid Interface Sci.* **94**: 179 (1983).

53. I. W. Osborne-Lee, R. S. Schechter, W. H. Wade, and Y. Baraket, *J. Colloid Interface Sci.* **108**: 60 (1985).

54. T. Asakawa, K. Johten, S. Miyagishi, and M. Nishida, *Langmuir* **4**: 136 (1988).

55. R. F. Kamrath, and E. I. Franses, *Ind. Eng. Chem. Fundam.* **22**: 230 (1983).

56. C. Treiner, A. A. Khodja, and M. Fromon, *Langmuir* **3**: 729 (1987).

57. J. Weers, *J. Am. Oil Chem. Soc.* **67**: 340 (1990).

58. N. Nishikido, *Langmuir* **7**: 2076 (1991).

59. M. E. Morgan, H. Uchiyama, S. D. Christian, E. E. Tucker, and J. F. Scamehorn, *Langmuir* **10**: 2170 (1994).

60. D. N. Rubingh, in *Solution Chemistry of Surfactant I* (K. L. Mittal. ed.), Plenum Press, New York (1979), pp. 337.

61. K. Shinoda, and T. Nomura, *J. Phys. Chem.* **84**: 365 (1980).

62. D. S. Abram, and J. M. Prausnitz, *AIChE J.* **21**: 116 (1975).

63. A. Fredenslund, R. Johes, and J. M. Prausnitz, *AIChE J.* **21**: 1086 (1975).

64. T. Mafnussen, P. P. Rasumussen, and A. Fredenslund, *Ind. Eng. Chem. Process Des. Dev.* **20**: 331 (1981).

65. J. Gmehling, P. Rasumussen, and A. Fredenslund, *Ind. Eng. Chem. Process Des. Dev.* **21**: 118 (1982).

66. T. Asakawa, K. Ishikawa, and S. Miyagishi, *J. Colloid Interface Sci.* **240**: 365 (2001).

67. K. Motomura, M. Yamanaka, and M. Aratono, *Colloid Polym.* **262**: 948 (1984).

68. R. Nagarajan, *Langmuir* **1**: 331 (1985).

69. N. J. Turro, and A. Yekta, *J. Am. Chem. Soc.* **100**: 5951 (1978).

70. Y. Moroi, *Micelles—Theoretical and Applied Aspect*, Plenum Press, New York (1992), pp. 219.

71. T. Asakawa, and S. Miyagishi, *Langmuir* **15**: 3464 (1999).

72. M. Grätzel, K. Kkyanasundaram, and J. K. Thomas, *J. Am. Chem. Soc.* **96**: 7869 (1974).

73. M. Almgren, K. Wang, and T. Asakawa, *Langmuir* **13**: 4535 (1997).

74. T. Asakawa, K. Amada, and S. Miyagishi, *Langmuir* **13**: 4569 (1997).

75. J. Dederen, M. V. Auweraer, and F. C. D. Schryver, *Chem. Phys. Lett.* **68**: 451 (1979).

76. M. Tachiya, *J. Chem. Phys.* **76**: 340 (1982).

77. T. Asakawa, H. Hisamatsu, and S. Miyagishi, *Langmuir* **121**: 1204 (1996).

78. K. J. Mysels, *J. Colloid Interface Sci.* **66**: 331 (1978).

79. M. Lake, *J. Colloid Interface Sci.* **91**: 496 (1983).

80. S. Miyagishi, H. Kurimoto, and T. Asakawa, *Bull. Chem. Soc. Jpn* **68**: 135 (1995).

81. K. Kamogawa, and K. Tajima, *J. Phys. Chem.* **97**: 9506 (1993).

14

Dynamic Interfacial Properties of Mixed Surfactant Systems Containing Light-Sensitive Surfactants

JASON Y. SHIN and NICHOLAS L. ABBOTT

Department of Chemical and Biological
Engineering, University of Wisconsin,
Madison, Wisconsin, U.S.A.

SYNOPSIS

This chapter addresses the influence of light on dynamic interfacial properties of aqueous solutions containing mixtures of 4,4'-bis(trimethylammoniumhexyloxy) azobenzene bromide (BTHA) and sodium dodecylsulfate (SDS) by analyzing the shapes of pendant bubbles of air. The interfacial states of freshly created surfaces relax to equilibrium through two processes that possess clearly distinguishable characteristic times. The first process, a fast process with a characteristics time <1 s, is independent of the state of illumination of the surfactant system and is associated with the transport of monomeric SDS to the surface. The second relaxation process, which is substantially slower than the first one, exhibits a characteristic time that is a strong function of the illumination of the surfactant system. Without prior or concurrent illumination with UV light, the characteristic time of the second process is ~ 1500 s. With prior or concurrent illumination with UV light, the characteristic time of the second process is reduced to ~ 50 s and ~ 160 s, respectively. The second relaxation process is associated with the transport of BTHA to the interface and regulated through the influence of UV light on the state of aggregation of the surfactants in bulk solution. The characteristic time of the slow interfacial relaxation process is accelerated by five orders of magnitude by the addition of $1:1$ electrolyte to mixtures of BTHA and SDS. Measurement of the intensity of visible light scattered from bulk solutions of BTHA and SDS reveals the presence of large vesicle-sized aggregates in bulk solution, independent of whether the process of assembly is triggered by the mixing of the two surfactants or by illumination with visible light.

I. INTRODUCTION

Surfactants are widely used to manipulate the interfacial and bulk properties of aqueous systems and thereby control phenomena such as wetting, detergency, solubilization, and foaming [1]. The interfacial and bulk properties of a given surfactant system are determined by a number of factors including the molecular structure of the surfactant and solution conditions, the concentration of surfactant, the presence of added electrolytes and temperature [1]. Although manipulation of solution conditions, such as addition of salt, can lead to changes in interfacial properties, these changes are not readily reversed. In contrast, changes in temperature are readily reversed, but the effects of temperature on interfacial properties are generally too small to be broadly useful (although specific exceptions do exist [2,3]). The work described in this chapter addresses the properties of a light-sensitive, mixed surfactant system that permits large and reversible changes in dynamic interfacial properties of aqueous system to be actively controlled by illumination [4]. In this system, illumination of the mixed surfactant system by UV or visible light leads to reversible changes in the molecular structure (isomerization to *cis* or *trans* state, respectively) of the surfactant.

Several past studies [4–12] have reported approaches leading to active control of interfacial properties of surfactant solutions by using redox-active [13–15] and light sensitive surfactants. For example, ferrocene, which can undergo a one electron oxidation to form the ferrocenium cation, has been exploited in a number of past studies of interfacial phenomena [5–8,10,11]. Surfactants with the structure, $Fc(CH_2)_nN$ $(CH_3)_3Br$, where Fc is ferrocene and $n = 8$, 11 and 15, have been used to demonstrate large and reversible *in-situ* changes in equilibrium surface tensions (up to 23 mN/m) upon electrochemical oxidation and reduction of the ferrocenyl group. In addition, electrochemical control of gradients in surfactant-based properties have been used to control the motion of surfaces of liquids [10]. A second class of redox-active surfactant, the bolaform disulfide, $(CH_3)_3N^+C_8H_{16}SSC_8H_{16}N^+$ $(CH_3)_3$, which can be reduced to form two single-headed surfactants, $(CH_3)_3N^+C_8H_{16}SH$, has been used to regulate nonequilibrium interfacial states of surfactant systems. These nonequilibrium interfacial states possess both excess surface concentrations of thiol fragments and surface pressures that are larger than equilibrium values [6]. Finally, we note that light-sensitive, water-soluble surfactants have also been explored as the basis of approaches leading to active control of interfacial properties of aqueous systems. For example, Drummond and coworkers have demonstrated that illumination of spirobenzopyran surfactants, which leads to formation of merocyanin products, can drive changes in surface tension [12]. These surfactants, however, are not stable and possess a short-lived photochromic response. More recently, a light-sensitive, water-soluble surfactant based on the azobenzene group, when mixed with sodium dodecyl sulfate, SDS, has been used to demonstrate control of the dynamic surface tension of freshly created interfaces by using UV light [4]. These mixtures of surfactants are stable over numerous photocycles, and dynamic surface tensions of solutions of these surfactants measured before and after illumination with UV light were shown to differ by as much as 26 mN/m. This chapter describes further the interfacial properties of this novel class of mixed surfactant system.

It is well known that adsorption of surfactant onto a freshly created surface of an aqueous solution results in a time-dependent reduction in surface tension. The rate of arrival of surfactant at the freshly created surface is determined by convective and diffusive transport of surfactant from the bulk of the solution to the sub-surface, and by the sorption kinetics that govern the rate of transfer of surfactant from the sub-surface to the surface of the solution [16]. Whereas relatively little is known about the molecular factors that govern the sorption kinetics of surfactants [16], the effective diffusion coefficient of a surfactant is influenced by its state of aggregation in bulk solution and thus the state of aggregation of surfactant affects the time required to transport surfactant from bulk solution to an interface [17,18]. For example, large aggregates with hydrodynamic diameters of ~150 nm (e.g., vesicles) have diffusion coefficients ($\sim10^{-8}\,cm^2/s$) that are roughly two orders of magnitude smaller than those of the surfactant monomers ($\sim10^{-6}\,cm^2/s$) [16]. Control of the state of aggregation of a surfactant in bulk solution can, therefore, provide a means to control the rate of transport of surfactant to a freshly created interface.

The diffusion-limited model is the simplest description of the processes that underlie the time-dependent lowering of surface tension of a freshly created surface of a surfactant solution [17,18]. In this model, the sub-surface concentration of surfactant is assumed to be in equilibrium with the surfactant adsorbed at the interface. It is accurate in describing the initial decrease in surface tension of a freshly created interface

[17,18]. However, as the excess surface concentration of surfactant increases through the adsorption process, the role of sorption kinetics becomes increasingly important in determining the time-dependence of the state of the interface. Under these conditions, the rate of adsorption of surfactant is described by a combination of diffusive transport and sorption kinetics. In addition, convection and the presence of aggregates can also affect the rate of delivery of surfactant to a surface (as discussed below).

The presence of aggregates of surfactant within the bulk of a solution can influence the rate of adsorption of surfactant to an interface in two ways. First, as described above, the effective diffusion coefficient of the surfactant can be lowered significantly by its state of aggregation. In addition, the adsorption rate can also be limited by the dynamics of the aggregate disassembly processes that form part of the pathway for transfer of surfactant from bulk solution to an interface [19–22]. For example, past studies have described the dynamics of micellar systems in terms of two processes (a fast process and a slow process) [23]. The fast relaxation process corresponds to the redistribution of surfactant between existing micelles by the exchange of surfactant between the monomeric state and micelles. The slow relaxation process corresponds to a change in the number of micelles in the system. For micellar surfactant solutions of SDS, the fast and slow relaxation processes have characteristic times of microseconds and milliseconds, respectively [24]. However, when a surfactant system responds to strong disturbances, the relaxation processes have been reported to be extremely slow with characteristic times of minutes or hours [25]. We also point out that past measurements of the dynamics of solutions containing mixtures of anionic and cationic surfactants have revealed the presence of very slow processes, such as those processes leading to the formation of solutions of vesicles by the mixing of micellar solutions [26,27]. Light scattering measurements have shown that the formation of vesicles from micellar solutions is characterized by a rapid increase in the apparent radii over the first hour followed by a slow increase over three months [26]. In contrast, the reverse process (breakup of vesicles) was found to be rapid. For example, the breakup of vesicles in a solution containing 2.9% by weight of a mixture of cetyltri-methylammonium bromide (CTAB) and sodium octylsulfate (SOS) occurred faster than the dead time of the apparatus used for the measurement (4 ms) [27]. For solutions containing 1% by weight of a mixture of dodecyltrimethylammonium bromide (DTAB) and SDS, vesicle breakup occurs with a characteristic time of 9 s [27].

As mentioned above, we have recently reported on a light-sensitive cationic surfactant and its use in mixtures with SDS to achieve optical control of the dynamic surface tensions of aqueous solutions [4]. The light-sensitive surfactant, 4,4′-bis(trimethylammoniumhexyloxy) azobenzene bromide (BTHA), is a bolaform surfactant that contains the azobenzene functional group. Azobenzene can undergo reversible *trans* to *cis* or *cis* to *trans* isomerization when illuminated by UV or visible light, respectively (Fig. 1). In our past study, we reported

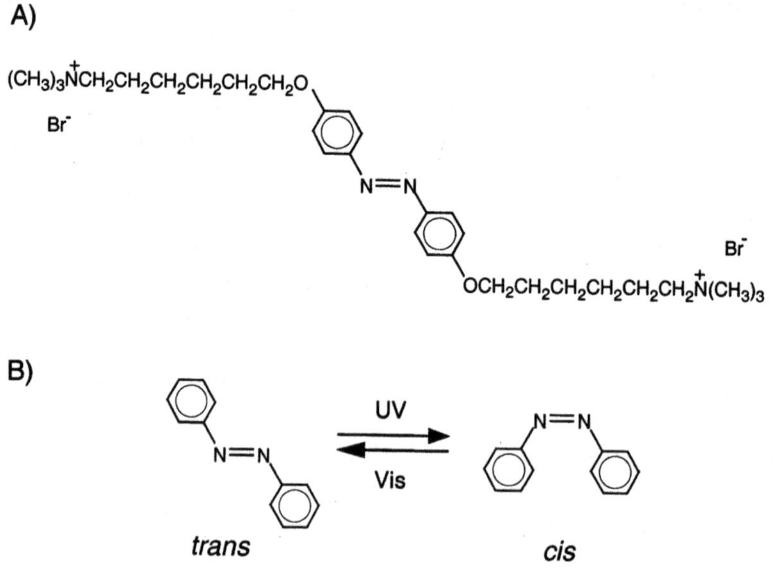

A)

$(CH_3)_3\overset{+}{N}CH_2CH_2CH_2CH_2CH_2CH_2O$

Br⁻

N=N

Br⁻

$OCH_2CH_2CH_2CH_2CH_2CH_2\overset{+}{N}(CH_3)_3$

B)

N=N

UV
⇄
Vis

N=N

trans   *cis*

Figure 1 (A) Structure of BTHA. (B) Schematic illustration of photoisomerization of azobenzene between the *cis* and *trans* states upon illumination with UV and visible light, respectively.

that the mixing of aqueous solutions of *trans*-BTHA and SDS to a concentration of 0.1 mM BTHA and 1.6 mM SDS leads to the formation of large aggregates. By using measurements of dynamic light scattering, the hydrodynamic diameters of these aggregates were estimated to be 150 ± 10 nm. The intensity of scattered light was measured to be 2500 kcnts/s. Following illumination of these solutions of aggregates with UV light and the conversion of ~80% of the *trans*-BTHA into *cis*-BTHA, the concentration of large aggregates was found to decrease [4]. Although the hydrodynamic size of the largest aggregates in solution did not appear to undergo a large change after illumination with UV light (140 ± 10 nm), the static intensity of scattered light decreased by a factor of almost two (to 1480 kcnts/s). Because the intensity of scattered light is dominated by the largest aggregates in solution and is approximately proportional to their number density, the decrease in the intensity of scattered light after UV irradiation indicates that roughly half the aggregates in solution are disrupted into surfactant monomers or smaller aggregates by illumination with UV light.

In our past study, we also investigated the interfacial properties of aqueous solutions containing mixtures of BTHA and SDS prior to and after illumination with UV light [4]. Surface tension measurements were performed using various concentrations of BTHA (1 µM to 1 mM) with the concentration of SDS fixed at 1.6 mM. These measurements of surface tension were performed using a combination of methods (maximum bubble pressure, du Nouy ring and Wilhelmy plate methods) so as to determine the state of the surface at various points in time following the formation of the surface. For example, whereas the maximum bubble pressure method provided measurements of surface tensions approximately ~1 s after formation of the surface, use of the Wilhelmy plate provided measurements of equilibrium surface tensions (minutes to hours after formation of the interface). We mention here three principal observations from these past measurements of surface tension of mixtures of BTHA and SDS that are relevant to the results described in this chapter. First, the dynamic surface tensions of aqueous solutions of BTHA and SDS measured using the maximum bubble pressure method (age of interface ~1 s) were found to be similar for solutions that had or had not been illuminated with UV light prior to measurement of the surface tension. Second, the equilibrium surface tensions (age of interface ~1 hr) of aqueous solutions of BTHA and SDS were found to be similar (<2 mN/m difference) for solutions that had or had not been illuminated by UV light prior to measurement of the surface tension. Third, measurements of surface tension using the du Nouy ring method, which characterizes the state of the surface with an age (~10 s) that is intermediate to the maximum bubble pressure and Wilhelmy methods, revealed the dynamic surface tension of a solution that had not been illuminated by UV light (containing *trans*-BTHA) to be as much as 26 mN/m higher than a solution that had been illuminated by UV light (containing *cis*-BTHA). The influence of illumination by UV light on the dynamic surface tension, as measured by the du Nouy ring, was further evidenced by the qualitative effect of illumination on pendant droplets of the solution. Whereas pendant droplets of aqueous solutions containing 0.04 mM BTHA, 0.16 mM SDS and 0.1 M NaCl could be poised at the tip of a syringe for more than a minute, illumination of the droplets with UV light led to the release of the droplets within 3–5 s of the onset of illumination with UV light.The release of the droplets was triggered by the light-induced lowering of the surface tension below a threshold value, and thus provided qualitative evidence of control of dynamic surface tensions by using light. The experiments described in this chapter further our understanding of the equilibrium and dynamic processes that control the influence of light on the dynamic surface tension of this mixed surfactant system.

Finally, we mention a second recent study of the cationic surfactant, 4-butylazobenzene-4′-(oxyethyl)trimethylammonium bromide (AZTMA), and the anionic surfactant, sodium dodecylbenzenesulfonate (SDBS). This study also reported the influence of UV illumination on the state of aggregation of the mixture of surfactants in bulk aqueous solution [28]. The change in the state of aggregation was followed by using transmission electron microscopy to image carbon-on-platinum replicas of a rapidly quenched solution. The observations reported using AZTMA, a single-headed cationic surfactant, and SDBS differ from the past study using mixtures of BTHA, a bolaform surfactant, and SDS in a number of interesting ways. In particular, an *increase* in the turbidity of mixtures of *trans*-AZTMA and SDBS is reported upon illumination with UV light: TEM images indicate the transformation of vesicle-like structures into "large, elongated molecular aggregates." In contrast, a decrease (by ~50%) in the scattering of light from aqueous solutions of BTHA and SDS is observed to follow illumination with UV light [4]. Second, the mixtures of AZTMA and SDBS form "precipitates" in solutions containing 1:1 molar ratios of AZTMA and SDBS, similar to past studies of mixtures of classical cationic and anionic surfactants [26,29,31]. In contrast, past studies using mixtures of BTHA (a bolaform surfactant) and SDS have not yet yielded evidence of similar precipitates [4]. These results suggests that the phase diagram of mixtures of cationic bolaform surfactants and anionic surfactants (as described in this chapter) may be substantially different from mixtures of classical cationic and anionic surfactants.

The sections presented below in this chapter address several issues that further our understanding of the equilibrium and dynamic properties of aqueous solutions containing mixtures of BTHA and SDS. First, because the dynamic and equilibrium bulk solution properties of mixtures of BTHA and SDS appear to play a central role in determining the influence of light on the dynamic interfacial properties of this system, we address the characteristic time for equilibration of bulk solutions following (I) mixing of the *trans*-BTHA and SDS, (II) illumination of the previously equilibrated mixture of *trans*-BTHA and SDS with UV light, and (III) illumination of the equilibrated solution in (II) with visible light. Second, by analysis of the evolving shape of a floating bubble, we report on the time-dependent evolution of the surface tension of mixtures of BTHA and SDS for times ranging from 1 to 10^3 s. In our past study, conclusions regarding dynamic surface tension were obtained by patching together results obtained by using three separate techniques (maximum bubble pressure, du Nouy ring, and Wilhelmy plate) [4]. By analyzing the shape of a

floating bubble, here we report a complete description of the
evolution of the state of the interface as a function of time. We
can extract from these measurements estimates of the characteristic times of the dynamic processes that lead to the
formation of the equilibrium states of the interfaces. Third, we
report here on the role of fluid convection in determining the
time-dependent states of the surfaces of solutions containing
mixtures of BTHA and SDS. In our previous study, we
hypothesized that the influence of light on the dynamic surface
tension is mediated by the effect of the light on the state of
aggregation of the surfactant in bulk solution and thus its rate of
transport to the surface of the solution. Here we address the role
of transport processes in determining the dynamic surface
tensions of aqueous solutions of mixtures of BTHA and SDS by
comparing the evolution of the state of the solution with and
without forced convection (stirring). Fourth, in this paper, we
report on the dynamic surface tensions *during* the illumination
of the surface of the solution. In our past study, we only
quantified the state of the surface of the solution for solutions
that were illuminated (or not) prior to formation of the interface.
Here we discuss the response of a preformed interface to illumination, and compare the dynamics of the interfacial response to
the dynamics of disassembly of aggregates in bulk solution and
mass transport of surfactant from bulk solution to the interface.
Fifth, and finally, we describe the role of added electrolyte on
the dynamics of surfaces of aqueous solutions containing
BTHA and SDS. We report that the dynamics of the interface
can be tuned over a wide range by the addition of electrolyte.

II. EXPERIMENTAL

The azobenzene-based surfactant used in our study, BTHA,
was synthesized as described in our past study [4]. Here we
simply note that we found it important to obtain high purity in
the intermediate product, 4,4'-dihydroxyazobenzene. In one
synthesis, impurities carried through the alkylation and
quaternization steps were present in the final product and
could not be removed. This material was discarded and the

synthesis repeated. The product was purified by recrystallization from ethanol and characterized by ^{1}H NMR, UV-vis absorption spectroscopy and electrospray ionization mass spectroscopy. Sodium dodecyl sulfate (SDS), was obtained from Aldrich (Milwaukee, WI) and purified by recrystallization in ethanol. Measurements of light scattering were performed using samples that were prepared by dissolving each surfactant in ethanol, passing the solution through a filter with a 0.22 μm pore size (Millipore, Bedford, MA), and then recovering the surfactant by crystallization. Solutions were prepared using water obtained from a water purification system (Millipore, Bedford, MA). The purity of the water was confirmed by measurements of ionic resistivity (18.2 MΩ cm) and surface tension (72 mN/m).

Measurements of the scattering of light were performed using a light scattering apparatus (Brookhaven Instruments, Brookhaven, NY) with an Ar ion laser (488 nm). The sample vial was placed in a bath containing index matching fluid thermostated at 25°C. The static intensity of scattered light was averaged over four minutes for each data point.

Measurements of dynamic surface tension were performed using an FTÅ 200 pendant drop tensiometer (First Ten Angstroms, Portsmouth, VA). The bubble was formed from an inverted, blunt needle (22 gauge) in a 1 cm standard cuvette (Fisher 14-385-985). All solutions were stirred with a 7 mm stir bar (Fisher) unless otherwise indicated. Measurements of surface tension were performed at 25°C. The evolution of the shape of each bubble was recorded using a frame grabber with a maximum grab rate of 60 images per second. Values of the interfacial tension were determined from the instantaneous shape of the bubble using the Bashforth–Adams solution to the Young–Laplace equation [32]. The bubble was illuminated using red light so as to not affect the state of isomerization of BTHA.

Solutions were illuminated using a UV pen lamp (Oriel, model 6035) with a long wave filter (Oriel, model 6042) to convert the BTHA from the *trans* state to a photostationary state composed mostly of the *cis* isomer. In a previous study, we determined that the photostationary states of these solutions contained ~80% of the BTHA in the *cis* state [4]. An Ar ion laser (488 nm) was used to isomerize the BTHA back to the *trans* state. The state of isomerization was followed by using UV-vis spectroscopy. Because factors such as vessel geometry, volume, stirring speed, and concentration of surfactant all affect the rate of isomerization of BTHA within the solution, UV-vis spectroscopy was used to establish the duration of illumination needed to reach the photostationary state for each experimental setup.

III.   RESULTS AND DISCUSSION

A.   Dynamics of Disruption and Formation of Aggregates of Surfactant in Bulk Solution

We first investigated the dynamics of the process of aggregation within the BTHA/SDS surfactant system with a particular focus on identification of the time required to reach an equilibrium population of aggregates. The dynamics of aggregation were followed by measurement of the intensity of light scattered from a bulk solution of the surfactants as a function of time. The BTHA/SDS solution was placed in the light scattering apparatus equipped with an Ar ion laser (488 nm). The intensity of light scattered from the solution, averaged over 4 min intervals, was recorded for several hours. Because the wavelength of the light from the Ar ion laser does not drive the *trans* to *cis* isomerization of BTHA, the dynamics of aggregation of solutions containing *trans*-BTHA and SDS could be measured. Here we report measurements of the dynamics of aggregation (I) after mixing of solutions of *trans*-BTHA and SDS and (II) after a mixed and equilibrated solution of *trans*-BTHA and SDS was illuminated sequentially with UV and then visible light (i.e., sent through a photocycle). We used solutions containing 0.1 mM BTHA and 1.6 mM SDS in these experiments because our past studies [4] showed the effect of light on the dynamic surface tension of these solutions, as measured using a du Nouy ring, to be large.

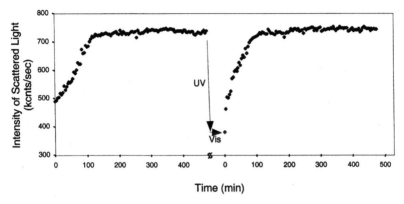

Figure 2 Intensity of light scattered as a function of time from a solution of 0.1 mM BTHA and 1.6 mM SDS. The BTHA and SDS were mixed at $t = 0$. See text for details.

Figure 2 shows a typical set of measurements of scattered light immediately after mixing equal volumes of 0.2 mM BTHA and 3.2 mM SDS to form a 0.1 mM BTHA and 1.6 mM SDS solution. The measurements of scattered light were initiated within 1 min of mixing of the two solutions. Inspection of Fig. 2 shows that the intensity of light scattered from the surfactant solution increases for ~120 min. This increase in intensity is interpreted to be a result of the formation and growth of aggregates containing BTHA and SDS. For times greater than ~120 min and less than ~450 min, the intensity of light scattered from the solution does not change measurably as a function of time. We equilibrated some solutions for as long as one week, and measured no further change in the intensity of light scattered from the solution. From these measurements, we conclude that the equilibrium distribution of aggregates in the solution is likely established (as indicated by the scattering of light) within a few hours of mixing of these solutions of BTHA and SDS.

Following equilibration of the solution of *trans*-BTHA and SDS for ~450 min, we illuminated the solution with UV light for 3 min (while stirring the solution). Measurements reported in the past have revealed that this level of illumination by UV light leads to conversion of ~80% of the *trans*-BTHA to *cis*-BTHA and to a reduction in the scattering of light from the solution by a factor of ~2 [4]. In the experiments reported here, we followed the illumination of the solution with UV light by illumination with visible light. By stirring the solution in the path of the Ar ion laser for 1 min, we determined that the *cis*-BTHA generated upon exposure to the UV light was transformed back to the *trans*-state by illumination with the visible light. Immediately following the isomerization back to the *trans* state, as shown in Fig. 2, the intensity of scattered light is approximately one half the value measured before illumination (equilibrated *trans*-BTHA/SDS solution). This observation is in agreement with previous light scattering measurements of the BTHA/SDS solution before and after illumination with UV light [4]. In our past measurements, the intensity of light scattered from the solution was measured using a laser that did not drive the *cis*-to-*trans* (or reverse) isomerization. The close agreement between the measurements reported previously (of *cis*-rich mixtures of BTHA and SDS) and the measurements reported here *immediately* after one photocycle (*trans*-to-*cis*-to-*trans*) suggests that there is little change in the intensity of light scattered from the surfactant solution after 1 min of illumination by visible light. Inspection of Fig. 2 reveals the change of the intensity of light scattered from the solution following the photocycle to occur over ~100 min. The final value of the intensity of light scattered by the equilibrated solution after one photocycle is indistinguishable from the value of light scattered from the equilibrated solution prior to the photocycle. These observations are consistent with our conclusion reported above that near-equilibrium structures form within the solution in ~2 h (at least, as judged by static light scattering), both after mixing of SDS and BTHA and after one photocycle of the surfactant solution.

We conclude our discussion of Fig. 2 by making three additional observations. First, we note that the dynamics of the assembly process leading to the equilibrium distribution of aggregates in solution is remarkably similar for two processes that have different initial states. The first process of assembly shown in Fig. 2 was initiated by the mixing of aqueous solutions of BTHA and SDS whereas the second process of assembly was initiated by illumination of the *cis*-rich BTHA solution with visible light. This result suggests that the rate-limiting processes governing the formation of the equilibrium aggregates along each pathway may be similar. Second, whereas the overall time taken to establish an equilibrium distribution of aggregates within solution (as measured by the scattering of light) is similar for the two cases, we note, however, that there are differences in the shapes of the two response curves. Inspection of Fig. 2 reveals two principal differences in shape. Whereas the intensity of light scattered from the solution that was prepared by the mixing of BTHA and SDS possesses an inflection point, the process of assembly following the photo-cycle evolves without the presence of an inflection. The second difference that we point out is the abruptness with which the equilibrium intensity of scattered light is reached. Whereas the solution that had passed through the photocycle slowly asymptotes to the equilibrium intensity of scattered light, the solution formed by mixing of BTHA and SDS reaches saturation with a distinct "break" in the plot. Although we do not yet know how to interpret these differences in the features of the response curves, these differences do suggest that the pathways leading to the equilibrium state in these two solutions may, in fact, possess some minor differences. Third, and finally, the light scattering measurements in Fig. 2 also reveal that the light-induced formation (by illumination with visible light) and disruption of the aggregates (by illumination with UV light) occur on different time scales. Whereas aggregates form over ~100 min, the disruption of aggregates occurs during the illumination of the solution with UV light and is complete by the time we have measured the scattering of light from the solution (~4 min, using our procedures). The rates of assembly (initiated by mixing) and disassembly (initiated by dilution) of aggregates within systems containing cationic and anionic surfactants have been reported in the past to differ in a similar manner to that measured by us [27].

B. Measurements of Dynamic Surface Tension

Our past measurements of the dynamic surface tensions of aqueous solutions of mixtures of BTHA and SDS were performed using three different techniques (maximum bubble pressure, du Nouy ring and Wilhelmy plate methods) [27]. These three measurements provided three "snapshots" of the state of the interface at three different ages. These measurements did not, however, provide knowledge of the dynamic surface tension as a continuous function of time [$\gamma(t)$]. Here we report an analysis of floating bubbles so as to measure the evolution of the surface tension as a function of time and thereby identify one or more time-scales (and associated processes) over which the surface tension of the solution is reduced. The solutions were formed from stock solutions of each surfactant and allowed to equilibrate for at least 2 h before measurement of the surface tension. The measurements were performed by creating a bubble and recording the shape of the bubble as a function time. The evolution of the bubble shape was used to calculate the surface tension as a function of time. Dynamic surface tensions of the BTHA/SDS solutions were measured before any illumination with light (i.e., a solution containing *trans*-BTHA) and after illumination with UV light for 1 min (i.e., a solution containing *cis*-rich BTHA). Unless otherwise indicated, all measurements were performed with stirring.

Figure 3 shows the dynamic surface tensions of aqueous solutions of 0.1 mM BTHA and 1.6 mM SDS before and after illumination with UV light. We make three observations from these measurements. First, inspection of Fig. 3 reveals that the initial surface tension of the aqueous surfactant solution (measured within ~1 s of formation of the interface) is ~65 mN/m. We conclude that there exists a process which gives rise to the lowering of the surface tension from ~72 mN/m to ~65 mN/m that is complete within ~1 s of formation of the bubble. Second, for solutions not previously illuminated with UV light, we observe that there exists a second, slow process that begins 600 s after formation of the bubble and concludes when the surface tension has reached the limiting value

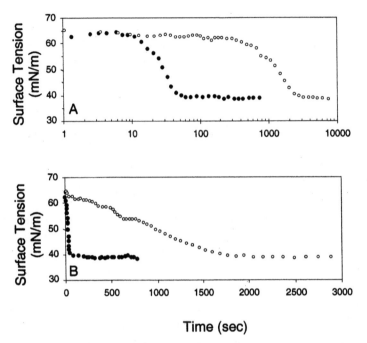

Figure 3 Dynamic surface tension of an aqueous solution of 0.1 mM BTHA and 1.6 mM SDS before (open) and after (filled) illumination with UV light. The two plots show the same data using different axes.

reached after 1500 s. Third, for solutions previously illuminated with UV light, the second slow process begins ~10 s after the formation of the bubble and concludes when the surface tension has reached the limiting value after 50 s. From these observations we conclude that the overall process of lowering of the surface tension of the solution occurs in two distinct steps with distinguishable time-scales (a fast process and a slow process). The slow process depends on whether or not the solution had been previously illuminated with UV light.

We also point out that the limiting values of surface tension measured with the floating bubble apparatus (shown in Fig. 3) before and after illumination agree well with previously obtained values measured using the Wilhelmy plate technique [4]. This agreement supports our conclusion that the values of the limiting surface tensions shown in Fig. 3 are equilibrium values. The values of dynamic surface tensions measured using the floating bubble apparatus at short times ($t < 10$ s) also agree well with values obtained previously using the maximum bubble pressure method.

We propose that the fast process described above (<1 s) is associated with the adsorption of monomeric SDS from bulk solution to the surface. Because the solution contains an excess amount of SDS as compared with BTHA, the concentration of monomeric SDS is expected to be close to the overall concentration of SDS. We estimate the excess concentration of SDS by assuming, for simplicity, that electrically neutral aggregates of BTHA and SDS are formed by these surfactants in solution. Past studies have shown that the aggregates formed in solution by cationic and anionic surfactants are, in fact, not electrically neutral [29]. Here we use this assumption for reasons of simplicity. For a solution containing 0.1 mM BTHA and 1.6 mM SDS, the excess concentration of SDS is 1.4 mM. Two facts support our proposition that the fast process is associated with the adsorption of SDS to the interface. First, the surface tension of the solution at the end of the fast process is the same as the equilibrium surface tension of aqueous solutions of SDS containing 1.4 mM of SDS [30]. Second, the time scale of the fast process is consistent with the time scale of diffusion-limited adsorption of 1.4 mM SDS. The characteristic time for monomeric surfactant to diffuse to the surface is given by [22]

$$t_D = \frac{\Gamma^2}{c^2 D} \tag{1}$$

where Γ is the excess surface concentration of a monolayer of surfactant, c is the bulk concentration of surfactant and D is the diffusion coefficient. Using $c = 1.4$ mM and $D = 10^{-6}$ cm^2/s, we calculate a characteristic time for diffusion-limited adsorption of 0.03 s. We note also that this estimate represents an upper bound since the aqueous solutions of SDS and BTHA were stirred during the measurement of the shape of the floating bubble (see below).

We propose that the slow process in Fig. 3 is associated with the delivery of BTHA to the surface of the solution. The slow rate of delivery of BTHA is likely due to the fact that BTHA is largely hosted within the aggregates of surfactant in the solution and thus it diffuses at the diffusion coefficient of these aggregates.The difference in time scales of the slow process before and after illumination with UV light also suggests that the rate of diffusion is a significant factor. The previously described disassembly of aggregates upon illumination with UV light leads to a larger effective diffusion coefficient and a decrease in the time required to transport the BTHA to the interface. In order to further test the proposed role of diffusion in determining rate of the adsorption of BTHA, the effect of convection on the slow process is tested below.

The proposal that the slow process is governed by diffusion of BTHA is further supported by estimates for the time scale of adsorption of BTHA using Eq. (1). The excess surface concentration of BTHA in the SDS/BTHA monolayer is estimated to be equivalent to an area per molecule of BTHA of $100 \, \text{Å}^2$. This estimate is based on the high areal densities found in mixed cationic/anionic systems [33]. Using a concentration of 0.1 mM and a diffusion coefficient of $10^{-8} \, \text{cm}^2/\text{s}$, the time required for adsorption is calculated to be 300 s. This value corresponds closely with the time at which the slow process reduces surface tension (before illumination with UV light, Fig. 3).

C. Effect of Stirring

The results described above are consistent with the proposition that the characteristic times of the fast and slow processes of adsorption (before and after illumination) are determined by the time required to transport SDS and BTHA to the surface. Here we test further the proposed role of mass transport in determining the characteristic times of the fast and slow processes by exploring the influence of the rate of stirring on the characteristic times. Specifically, we have measured the dynamic surface tensions of aqueous solutions of BTHA and SDS with and without stirring.

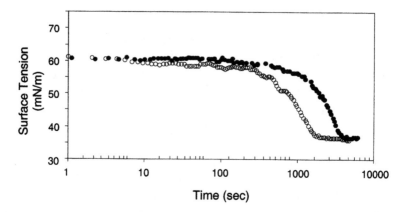

Figure 4 Dynamic surface tension of aqueous solution of 0.1 mM BTHA and 1.6 mM SDS. Measurements were made with stirring (open circles) and without stirring (filled circles).

We used a level of stirring that was sufficiently low such that it did not measurably affect the shape of a bubble used during tensiometry. We observed that high rates of stirring caused the shape of the bubble to oscillate and to introduce uncertainty in estimates of surface tension obtained by analysis of the shape of the bubble. Figure 4 shows the effects of stirring on the dynamic surface tensions of solutions of BTHA and SDS using the same experimental set-up. These measurements were performed by mixing a solution of 0.1 mM BTHA and 1.6 mM SDS and then allowing it to equilibrate for 2 h prior to measurement of the dynamic surface tension. The measured effect of stirring on the dynamic surface tension is consistent with the proposed role of mass-transport in determining the slow process that influences the dynamic surface tension behavior of aqueous solutions of BTHA and SDS. A second conclusion of this experiment is that, even when stirred, the mechanism of reduction of surface tension is dominated by mass-transport across the concentration boundary layer that is influenced by the forced convection. Because convection affects the time-scale of the dynamic surface tension, use of Eq. (1) represents an approximation. These experiments, in combination with dynamic surface tension measurements and previous light scattering measurements lead to the conclusion that the state of aggregation affects the rate of diffusion of surfactant to the interface and the characteristic time of surface tension reduction.

D. Reversibility of Light-Induced Changes in Surface Tension

Measurements of the intensity of light scattered from the bulk of solutions containing mixtures of BTHA and SDS before illumination and after sequential illumination with UV and visible light showed that the intensity of scattered light returned to its initial value after one photocycle (Fig. 2). Here we report the influence of a photocycle on the dynamic surface tensions of aqueous solutions containing mixtures of BTHA and SDS.

We first tested the reversibility of the light-induced changes in dynamic surface tension by isomerization of an UV-illuminated (*cis*-rich) solution of BTHA and SDS back to its *trans*-state using the Ar ion laser. The solution was sequentially illuminated with UV light and then by the Ar laser and allowed to rest 2 h before measurements of the dynamic surface tension were performed (Fig. 5). A comparison of Fig. 5 and Fig. 3 reveal that the dynamic surface tension of the *trans*-rich solution of BTHA and SDS is recovered after one photocycle. These results, when combined with the bulk light scattering experiments shown in Fig. 2, demonstrate that the influence of light on the bulk and interfacial properties of this surfactant system is not a result of an irreversible photochemical reaction and that the initial properties of both the bulk and interface can be recovered after one photocycle.

E. Control of Timing of Reduction in Surface Tension

The experiments described above suggest that the change in dynamic surface tension induced by illumination with UV light is the result of a reversible change in the state of aggregation of the surfactant in the bulk of the solution. Measurements of the intensity of light scattered from the solutions containing

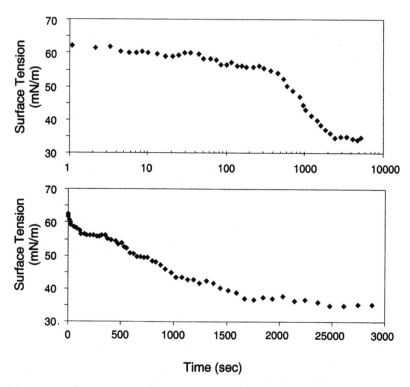

Figure 5 Dynamic surface tension of an aqueous solution of 0.1 mM BTHA and 1.6 mM SDS measured after one photocycle. The two plots show the same data with different axes.

mixtures of BTHA and SDS show that the disruption of aggregates occurs within the time required to illuminate the solution with UV light (3 min) and visible light (1 min) and average the scattered intensity for the first data point (4 min). Figure 6 shows measurements of dynamic surface tensions performed with aqueous solutions containing mixtures of *trans*-BTHA and SDS that were illuminated 0, 60, 300, and 600 s after formation of a bubble within each solution. Inspection of Fig. 6B reveals that the surface tension of a bubble that was immediately illuminated ($t = 0$ s) with UV light reaches its equilibrium value within 160 s of the onset of illumination. In contrast, the surface tension of a solution of BTHA and SDS that was illuminated with UV light prior to formation the

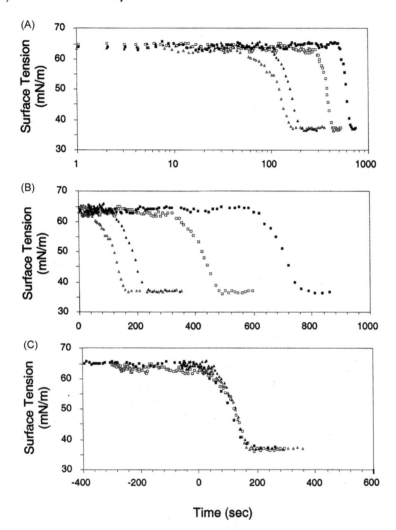

Figure 6 Dynamic surface tension of an aqueous solution of 0.1 mM BTHA and 1.6 mM SDS that was illuminated at $t = 0$ min (open triangles), 1 min (filled triangles), 5 min (open squares), and 10 min. (filled squares) after formation of the bubble. Panels (A) and (B) show the same data with different axes. Panel (C) shows the same data except $t = 0$ is referenced to the time of onset of illumination. See text for details.

bubble is observed to reach its equilibrium value in ~50 s (Fig. 3). The difference between these two time-scales (~110 s) is due to the time required to isomerize the *trans*-rich BTHA solution to a *cis*-rich state as well as the time required for the aggregates in solution to reorganize as a consequence of illumination. Separate experiments showed that these solutions achieved photostationary state after ~4 min of illumination of UV light. These results suggest that it is not necessary to reach the photostationary state in order to affect the slow relaxation process.

Figures 6A and 6B also show the time-dependent surface tensions of bubbles that were illuminated with UV light approximately 60, 300, or 600 s after their formation. Inspection of Fig. 6A and 6B show that the absence of illumination can result in a surface tension of ~65 mN/m for as long as 10 min whereas the onset of illumination with UV light can lower the surface tension to ~35 mN/m within ~160 s of the onset of illumination. We also point out the similarity of the shapes of the four response (to UV illumination) curves by replotting the data in Fig. 6A and 6B with an x-axis calculated as $t-t_{onset}$, where t_{onset} is duration of time between formation of the bubble and the onset of illumination of UV light (see Fig. 6C). The similarity of the four response curves is surprising since the state of the interface immediately after the formation of the bubble and after 10 min of equilibration are expected to be different. Our measurements show, however, that the response of the dynamic surface tension to illumination is largely independent of the age of the interface for times less than 10 min. These results may reflect a strongly non-linear dependence of the surface equation of state on the excess surface concentration of BTHA.

F. Effect of Electrolyte

In our previous study, we demonstrated that it was possible to control the release of droplets of 0.04 mM BTHA, 0.16 mM SDS and 0.1 M NaCl from the tip of a syringe needle upon illumination with UV light [4]. In the absence of illumination (*trans*-rich BTHA), the droplets hung from the needle for ~1–2 min prior to detachment. However, upon illumination with UV light, the drops were released within 3 to 5 s of the *onset* of illumination. The characteristic times of the processes leading to the decrease in surface tension that are implied by the observed release of the droplets are much faster than those reported in Fig. 3 or Fig. 6 of this paper (using 0.1 mM BTHA and 1.6 mM SDS and no added salt). We considered it likely that this difference in behavior resulted from the different salt and surfactant concentrations used in the two sets of experiments. Previous experimental and model-based studies of aqueous solutions of anionic and cationic surfactants have revealed that an increase in the concentration of salt within systems containing unequal bulk concentrations of the anionic and cationic surfactant does lead to a decrease in the characteristic size of the aggregate within the system [24]. In short, the addition of salt to these systems leads to the formation of aggregates with compositions that are closer to the bulk surfactant concentrations of the system, thus resulting in the formation of aggregates with high surface change densities. The high surface charge density promotes the formation of aggregates that possess high curvature and thus are small in size. Here we report an investigation of the influence of added salt on the interfacial dynamics of mixed surfactant systems containing BTHA and SDS. We hypothesized that the addition of salt leads to the formation of small aggregates in these systems and thus rapid interfacial dynamics (because of the role of mass-transport in determining the interfacial properties of these systems).

Figure 7 shows dynamic surface tensions of aqueous solutions containing 0.04 mM BTHA and 0.16 mM SDS (Fig. 7A), 0.04 mM BTHA, 0.16 mM SDS and 0.1 M NaCl (Fig. 7B), and 0.1 mM BTHA, 1.6 mM SDS and 0.1 M NaCl (Fig. 7C) that were measured before and after illumination with UV light. When combined with the measurements of 0.1 mM BTHA and 1.6 mM SDS (Fig. 3), this set of data provides an account of the effects of two surfactant concentrations, both with and without added salt, on the dynamic interfacial properties of this surfactant system.

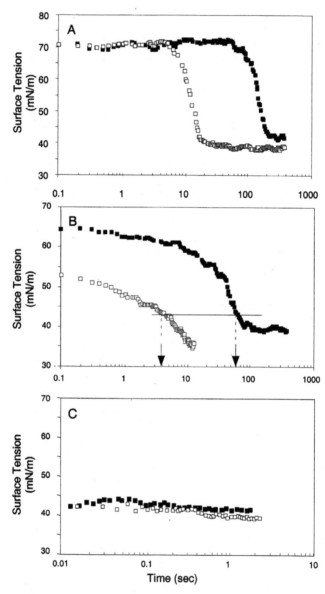

Figure 7 Dynamic surface tensions of aqueous solutions of (A) 0.04 mM BTHA and 0.16 mM SDS, (B) 0.04 mM BTHA, 0.16 mM SDS and 0.1 M NaCl, (C) 0.1 mM BTHA, 1.6 mM SDS and 0.1 M NaCl. Measurements were made before (filled squares) and after (open squares) illumination with UV light.

The initial values of the dynamic surface tension of 0.04 mM BTHA and 0.16 mM SDS (Fig. 7A) are similar to the surface tension of pure water. This result is in agreement with our hypothesis that the fast process discussed in the context of Fig. 3 is, indeed, due to adsorption of the excess SDS within the bulk solution. In the case shown in Fig. 7A, the excess concentration of SDS is too low (0.08 mM) to cause a significant reduction in surface tension below that of water. In contrast to Fig. 7A, the effect of the addition of salt shown in Fig. 7B (0.04 mM BTHA, 0.16 mM SDS and 0.1 M NaCl) leads to a measurable decrease in surface tension at short times ($\sim$64 mN/m). The value of the dynamic surface tension at 0.1 s reflects the adsorption and consequent reduction of surface tension by monomeric SDS. Inspection of Fig. 7B, and comparison to Fig. 7A, also reveals that the characteristic time of the slow process is shortened by the addition of salt. Whereas the slow process did not reduce the surface tension until $\sim$60 s after formation of the surface in the absence of added salt (Fig. 7A), the presence of salt caused the reduction of the surface tension to occur within a few seconds.

The results shown in Fig. 7B are consistent with our previous report of a light-induced release of droplets [4]. In the absence of illumination of the droplets with UV light, we reported the droplets to be released approximately 1–2 min after their formation. By using Fig. 7B, we estimate the threshold value of the surface tension below which the drop was released from the tip of a syringe to be 43 mN/m. After illumination, this value of surface tension is reached within $\sim$4 s. This latter time is consistent with our experimental observations of the light-triggered release of the droplets.

Finally, we mention that the limiting surface tensions of aqueous solutions of 0.1 mM BTHA, 1.6 mM SDS and 0.1 M NaCl are reached within the time required to form the bubble (Fig. 7C). The rapid relaxation of the surface tension to its equilibrium value suggests that the delivery of surfactant to the interface is rapid in this system. Comparison with Fig. 3 reveals that the addition of salt causes the slow process to be accelerated by four orders of magnitude.

IV. CONCLUSIONS

The principal conclusions of the experimental investigation reported in this chapter are four-fold. First, we have identified the interfacial dynamics of aqueous solutions of *trans*-BTHA and SDS (no added salt) to involve two clearly distinguishable relaxation mechanisms. A fast process is measured to be consistent with the transport of the excess monomeric SDS in solution to the surface. A second, slow process is measured to be consistent with diffusion-dominated transport of BTHA to the surface of the solution. Because the *trans*-BTHA is sequestered in large ($\sim$150 nm), slow diffusing aggregates, this second process occurs over $\sim$10 min. Second, we have measured the influence of sequential illumination of aqueous solutions of BTHA and SDS with UV and then visible light to lead to the reversible disruption and formation of vesicle-sized aggregates within the bulk of the aqueous solution. This same photocycle leads to reversible changes in the dynamic interfacial properties of aqueous systems of BTHA and SDS (see next conclusion). Third, we have measured illumination of the surface of an aqueous solution of BHTA and SDS to accelerate the slow relaxation process at the interface by as much as an order of magnitude (from 10 min to $\sim$1 min). Fourth, we have determined that it is possible to tune the time-scale of the slow relaxation process by the addition of salt to aqueous solutions of BTHA and SDS. These conclusions and others provide new principles by which to design mixed surfactant systems with interfacial properties that can be tuned by illumination.

REFERENCES

1. M. J. Rosen, *Surfactants and Interfacial Phenomena*. Wiley, New York (1989).

2. D. Langevin, *Langmuir* **13**: 599–601 (1997).

3. R. Sharma, Z. H. Yang, H. Yu, M. Green, and H. K. Pak, *Abs. Papers ACS*, **217**: 015–PMSE (1999).

4. J. Y. Shin, and N. L. Abbott, *Langmuir* **15**: 4404.2 (1999).

5. B. S. Gallardo, M. J. Hwa, and N. L. Abbott, *Langmuir* **11**: 4209–4212 (1995).

6. D. E. Bennett, B. S. Gallardo, and N. L. Abbott, *J. Am. Chem. Soc.* **118**: 6499 (1996).

7. B. S. Gallardo, K. L. Metcalfe, and N. L. Abbott, *Langmuir* **12**: 4116 (1996).

8. B. S. Gallardo, and N. L. Abbott, *Langmuir* **13**: 203 (1997).

9. L. I. Jong, and N. L. Abbott, *Langmuir* **14**: 2235 (1998).

10. B. S. Gallardo, V. K. Gupta, L. I. Jong, R. Shah, F. D. Eagerton, and N. L. Abbott, *Science* **283**: 57 (1999).

11. N. Aydogan, B. S. Gallardo, and N. L. Abbott, *Langmuir* **15**: 722 (1999).

12. C. J. Drummond, S. Albers, D. N. Furlong, and D. Wells, *Langmuir* **7**: 2409 (1991).

13. T. Saji, K. Hoshino, and S. Aoyagui, *J. Am. Chem. Soc.* **107**: 6865 (1985).

14. T. Saji, K. Hoshino, Y. Ishii, and M. Goto, *J. Am. Chem. Soc.* **113**: 450 (1991).

15. T. Saji, *Chem. Lett.* 693 (1988).

16. C. H. Chang, and E. I. Franses, *Colloids and Surfaces A*, **100**: 1 (1995).

17. J. K. Ferri, and K. J. Stebe, *Adv. Colloid Int. Sci.* **85**: 61.19 (2000).

18. S. S. Datwani, and K. J. Stebe, *J. Coll. Int. Sci.* **219**: 282 (1999).

19. A. Patist, B. K. Jha, S. G. Oh, and D. O. Shah, *J. Surf. Deterg.* **2**: 317 (1999).

20. J. Lucassen, *Faraday Discuss.* **59**: 76 (1975).

21. R. Miller, *Colloid and Polymer Science* **259**: 1124 (1981).

22. V. B. Fainerman, *Colloids and Surfaces* **62**: 333 (1992).

23. E. A. G. Aniansson, *Ber. Bunsenges. Phys. Chem.* **82**: 981 (1978).

24. C. Trachimow, L. DeMaeyer, and U. Kaatze, *J. Phys. Chem. B* **102**: 4483 (1998).

25. L. DeMaeyer, C. Trachimow, and U. Kaatze, *J. Phys. Chem. B.* **102**: 8024 (1998).

26. M. T. Yatcilla, K. L. Herrington, L. L. Brasher, E. W. Kaler, S. Chiruvolu, and J. A. Zasadzinski, *J. Phys. Chem.* **100**: 5874 (1996).

27. A. J. O'Connor, T. A. Hatton, and A. Bose, *Langmuir* **13**: 6931 (1997).

28. H. Sakai, A. Matsumura, S. Yokoyama, T. Saji, and M. Abe, *J. Phys. Chem. B* **103**: 10737 (1999).

29. E. W. Kaler, K. L. Harrington, A. K. Murthy, and J. A. N. Zasadzinski, *J. Phys. Chem.* **96**: 6698 (1992).

30. P. H. Elworthy, and K. J. Mysels, *J. Coll. Int. Sci.* **21**: 331 (1966).

31. K. L. Harrington, E. W. Kaler, D. D. Miller, J. A. Zasadzinski, and S. Chiruvolu, *J. Phys. Chem.* **97**: 13792 (1993).

32. J. F. Padday, *Theory of Surface Tension, In E.Matijevic (ed.), Surface and Colloid Science.* Wiley, New York (1969). Vol. 1, p. 84.

33. P. K. Yuet, and D. Blankchstein, *Langmuir* **12**: 3802 (1996).

15

Control of Molecular Aggregate Formation and Solubilization Using Electro- and Photoresponsive Surfactant Mixtures

HIDEKI SAKAI and MASAHIKO ABE

Faculty of Science and Technology and Institute
of Colloid and Interface Science,
Tokyo University of Science,
Chiba, Japan

SYNOPSIS

Control of reversible formation of surfactant molecular aggregates (micelles, vesicles, etc.) by external stimuli has been a
subject of significant attention because it can be applied to
controlled release of encapsulated drugs and perfumes and

removal of organic impurities dissolved in water. This study dealt with electrochemical and photochemical control of molecular aggregate formation and solubilization of oily substances making use of mixed surfactant systems. Ferrocene-modified and azobenzene-modified surfactants were used as electro-responsive and photo-responsive surfactants, respectively. Electrochemical and photochemical control of vesicle formation was made possible in ternary systems of cationic "switchable" surfactant/anionic surfactant/water. Moreover, uptake (solubilization) in and release from micelles of oily substances were also possible through the use of changes in CMC and solubilizing capacity induced by electrochemical and photochemical "switching."

I. VESICLES FORMED IN AQUEOUS MIXTURES OF CATIONIC AND ANIONIC SURFACTANTS

Liposomes and vesicles are closed bilayer shells that encapsulate an aqueous interior. Liposomes consisting of phospholipids have been extensively studied as biomimetic cell models and carriers in drug delivery systems [1–3]. Kunitake et al. have reported that double-tailed synthetic surfactants such as dialkyldimethylammonium bromides can also form vesicles upon ultrasonic treatment [4]. Since this report appeared, vesicles have gathered much attention with the view to apply them as an ultrasmall reaction field [5,6], sensor technology [7,8], and waste water treatment, etc. [9,10]. However, these vesicles are thermodynamically metastable and require the application of external forces such as ultrasonic treatment and french-press for their preparation. This has so far restricted the practical use of vesicles. For instance, when these metastable vesicles are used as a small reaction field for the preparation of metal oxide, the recycle of the used surfactants is troublesome because they have to be recovered from the product and an external force has to be applied again for re-formation of vesicles. In this respect, self-formation of vesicles is of great interest.

In 1989, Kaler *et al.* [11] first observed the spontaneous formation of stable vesicles in aqueous solution of two ionic surfactants with oppositely charged head groups (cetyltrimethylammonium bromide and sodium dodecylbenzene sulfonate). The key to the formation of this unique phase is nonideal mixing of the two kinds of surfactants in the vesicle [12]. This nonideal mixing is, in turn, dominated by both electrostatic interactions in the head group region and chain-packing considerations for the surfactant hydrophobic tails in the bilayer core. The effective neutralization in the head group plane reduces the repulsion between head groups, and as a result, the effective surfactant packing parameter depends on composition. Subsequent experimental observations show that the range of compositions over which vesicles form can be increased in a variety of ways, including increasing the asymmetry between the two surfactant tails and branching of one of the surfactant chains. Adding a simple electrolyte destabilizes the vesicle phase [13–16]. This work aroused theoretical interest among physicists and especially Safran *et al.* [17] proposed a theoretical interpretation of formation of vesicles in a "catanionic" system. In addition to aqueous mixtures of single-tailed cationic and anionic surfactant [18–21], many other mixed surfactant systems, including mixtures of double-tailed cationic and single-tailed anionic [22], double-tailed and single-tailed cationic [23], and lysolecithin and lecithin [24] have been shown to undergo spontaneous vesicle formation. Furthermore, spontaneous vesicle formation has also been reported in single-component surfactant solutions of didodecyldimethylammonium hydroxide (DDAOH) [25] and a double-tailed fluorinated surfactant [26].

Control of reversible formation of surfactant molecular aggregates (micelles, vesicles, emulsions, etc.) by external stimuli, e.g., thermal, electrical [27–37], and optical ones, has been a subject of significant attention because it can be applied to the controlled release of encapsulated drugs and perfumes and the removal of organic impurities dissolved in water. If vesicles can be disintegrated into monomers by these stimuli (switching on) and the monomers can reform vesicles upon removal of the stimuli (switching off) they can be applied more effectively as a small reaction field and to waste water treatment. Use of spontaneously formed vesicles is most suitable for this purpose. This chapter describes the properties of thermodynamically stable vesicles formed in aqueous mixtures of "electrochemically" and "photochemically" switchable cationic surfactant and simple anionic surfactant. Furthermore, the chapter deals with the control of reversible formation of the vesicles that are "electrochemically" and "photochemicaly" switchable.

II. ELECTROCHEMICAL SWITCHING OF FORMATION AND DISRUPTION OF VESICLES

A. Short Review On Electrochemically "Switchable" Surfactants

Reversible control of the solution properties of surfactant molecular aggregates by *electrochemical* stimulation has been reported extensively [27–40]. Formation-disruption control of surfactant *micelles* utilizing redox active ferrocene-modified surfactants is one of the most promising techniques [27–39]. For instance, Saji et al. [27–30] have reported that micelles formed with cationic 11-ferrocenylundecyltrimethylammonium bromide (FTMA) disrupt into monomers upon oxidation on the electrode surface; they applied this phenomenon in the micelle electrolysis method, which is a promising method for preparing organic thin films.

Abott and coworkers have reported active control of interfacial properties (interfacial tension, surface tension, etc.) using a ferrocene-modified cationic surfactant [31,32]. They also succeeded in controlling the motions and position of aqueous and organic liquids on millimeter and smaller scales. Detailed electrochemical properties of aqueous solutions of ferrocene-modified cationic and nonionic surfactant have also been investigated by Takeoka et al. [34] and Tajima et al. [33].

Formation control of *vesicles*, having a water phase inside their bilayer structure, is also of significant attention, since it has a potential applicability in drug delivery and preparation of inorganic thin films. Medina *et al.* [40] have reported the formation control of vesicles with use of ferrocene-modified surfactants. This system, however, requires ultrasonication treatment for vesicle formation, meaning that reversible control of formation and disruption of vesicles with redox reactions is impossible. If we have a system with "switchable" surfactant, in which vesicles form spontaneously in an equilibrium, then we would be able to reversibly control vesicle formation and disruption reversibly. The following sections describe the spontaneous formation of vesicles in an aqueous mixture of a cationic surfactant modified with "redox switchable" ferrocenyl moiety and a simple anionic surfactant [41]. In addition, these sections deal with the effect of redox reactions of the ferrocene-modified surfactant on the aggregation state of vesicles as studied by means of microscopic observations and electrochemical measurements.

B. Vesicle Formation in FTMA/SDBS/Water Ternary System

In this study, 11-ferrocenylundecyltrimethylammonium bromide (FTMA, chemical structure is shown in Scheme 1) was used as the redox switchable cationic surfactant. Cationic FTMA was mixed with anionic sodium dodecylbenzenesulfate (SDBS, Scheme 1) at various concentrations and compositions in water and a phase diagram of FTMA/SDBS/H_2O ternary system was prepared. Aqueous mixtures of FTMA and SDBS

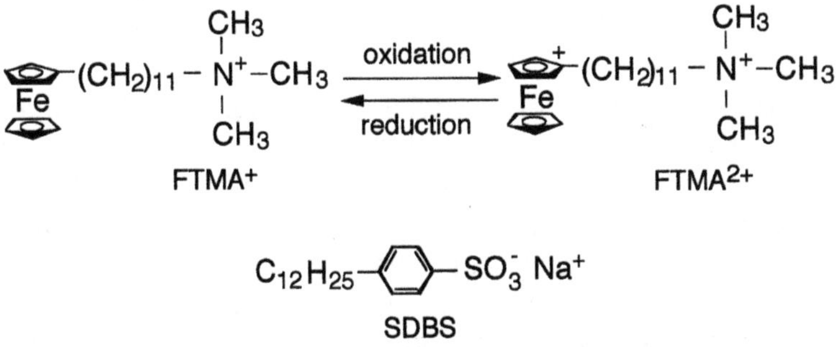

$$\underset{\text{Fe}}{\bigcirc\!\!\!-}(CH_2)_{11} - \underset{\underset{CH_3}{|}}{\overset{\overset{CH_3}{|}}{N^+}} - CH_3 \quad \underset{\text{reduction}}{\overset{\text{oxidation}}{\rightleftharpoons}} \quad \underset{\text{Fe}}{\bigcirc\!\!\!-}(CH_2)_{11} - \underset{\underset{CH_3}{|}}{\overset{\overset{CH_3}{|}}{N^+}} - CH_3$$

$$\text{FTMA}^+ \qquad\qquad\qquad \text{FTMA}^{2+}$$

$$C_{12}H_{25} - \bigcirc - SO_3^- \, Na^+$$

$$\text{SDBS}$$

Scheme 1 Molecular structures of FTMA and SDBS.

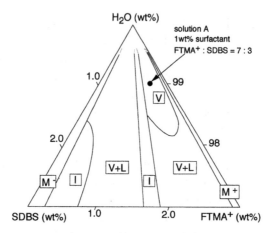

Figure 1 Ternary phase diagram for FTMA$^+$/SDBS/H$_2$O at 25°C. One-phase regions are vesicles (V), FTMA$^+$-rich micelles (M$^+$), and SDBS-rich micelles (M$^-$). Two-phase region is V and lamellar phase (V + L). Three-phase region is V, L, and precipitate (I).

were prepared by first making stock solutions of FTMA and SDBS at the desired concentrations and then mixing the stock solutions gently at the desired ratios.

Figure 1 is the phase behavior of dilute (less than 3%) mixed solutions at 25°C. The symbol M denotes an isotropic solution, which in this case is a micellar solution, since the diameter of the dispersed particles in the solution, as determined with dynamic light scattering, is less than 10 nm. In region Ia, a crystalline equimolar precipitate is formed. Except for these regions, a lamellar phase liquid crystal is observed (L + V, V, and Ib regions). Vesicles are formed spontaneously in region V.

Figure 2(a) is a differential interference optical micrograph of the mixed aqueous solution of FTMA (0.7 wt%) and SDBS (0.3 wt%) at point A in Fig. 1, almost the center of region V. A donut-like image, characteristic of multilamellar vesicles, is observed. This picture suggests that the vesicles are from several hundred nanometer to 10 μm in size, relatively large and polydisperse. Glucose trapping experiments [42] indicate the existence of an inner aqueous phase, confirming vesicle

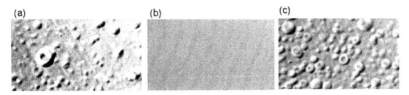

Figure 2 Differential interference optical micrographs of aqueous mixtures of FTMA and SDBS: (a) as prepared solution, (b) after oxidation, and (c) after reduction. Bar length is 17 μm.

formation. These vesicle solutions are stable at 25°C for at least four months. The solution at point A in Fig. 1 is also used in the electrochemical measurements described below.

C. Electrochemical Control of Vesicle Formation

For the oxidation experiments, 1.9 g equivalents (against FTMA) of ceric sulfate are added to solution A. A UV-visible absorption peak around 440 nm, assigned to the ferrocenyl moiety, is reduced remarkably by oxidation, and another peak centered at 628 nm, assigned to the ferricenium ion, appears, indicating oxidation of the ferrocenyl moieties of FTMA surfactants by ceric disulfate. The change in aggregation state accompanying the oxidation is observed with optical microscopy. Figure 2(b) presents an optical micrograph of solution A after oxidation. The donut-like structure characteristic of vesicles completely disappears. This would suggest that the oxidation of FTMA induces disruption of the vesicles into aggregates with a size much smaller than those of the vesicles existing before the oxidation. A differential interference micrograph of solution A after addition of sodium hydrosulfite (reducing reagent) is shown in Fig. 2(c). Donut-like vesicles are seen again. These results imply that vesicle formation, disruption, and reformation are controllable by oxidation and reduction of the amphiphile molecules (FTMA).

Aggregation states of the spontaneously formed vesicles before and after oxidation and the re-reduction of FTMA can be monitored electrochemically with cyclic voltammetry (CV).

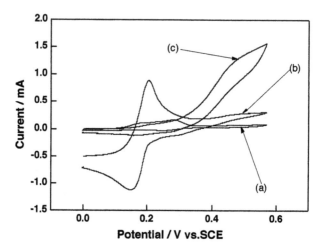

Figure 3 Cyclic voltamograms for aqueous mixtures of FTMA and SDBS (1 wt% surfactant, FTMA:SDBS = 7:3); (a) as prepared, (b) oxidized, and (c) re-reduced solutions at 25°C. Supporting electrolyte 0.02 M NaCl, scan rate; 20 mV s^{-1}, Working electrode area; 1.16 cm^2.

As shown in Fig. 3(a), redox peaks due to the FTMA$^+$/FTMA^{2+} reaction are observed before the oxidation with ceric disulfate. Plural oxidation peaks may suggest the existence of plural aggregation states in the FTMA/SDBS mixture. After oxidation of FTMA with excess cerium sulfate, the oxidation peak current in the cyclic voltammogram shows a drastic increase from 0.2 to 0.9 mA (4.5 times). When the concentration of the redox-active agent is kept constant, the peak current value depends on the diffusion coefficient of the redox species and is known to be proportional to the square root of the diffusion coefficient. Thus, if the efficiency of charge transfer between FTMA and the electrode is assumed to be kept constant before and after oxidation, the diffusion coefficient of FTMA is estimated to be 4.52, a 20-fold increase after oxidation of the ferrocenyl moieties. In other words, the vesicles disrupt into molecular aggregates with a size 20 times smaller than that before the oxidation. When this suspension is re-reduced with excess sodium hydrosulfite, the oxidation current peak in the cyclic voltammogram almost reduces to the initial value [Fig. 3(c)], suggesting re-formation of the mixed vesicles.

The mechanism of vesicle disruption and re-formation with the redox reaction of FTMA molecules is related to changes in the phase behavior induced by changes in the hydrophilic/lipophilic balance of the molecule. When the ferrocenyl moiety of FTMA, which serves as a hydrophobic part of the amphiphile molecule, is oxidized to ferricenium ion, the solution properties of this molecule are known to change drastically; i.e., the hydrophilicity of the molecule drastically increases. For instance, Abbott and coworkers [31,32] reported that the critical micelle concentration (CMC) of FTMA, obtained by surface tension measurement, increases with oxidation through addition of ferric disulfate from 0.07 mM to 0.60 mM. Recent results using light scattering show that even above the cmc, the size of micelles after oxidation is much smaller than before oxidation [39].

Spontaneous formation of vesicles strongly depends on the hydrophilicity and geometry of the surfactant molecules. The ternary phase diagram of FTMA/SDBS/H$_2$O undoubtedly changes significantly with the oxidation of FTMA$^+$ to FTMA^{2+}. Thus, the FTMA/SDBS aqueous solution at position A could no longer form vesicles after the oxidation. As shown in Fig. 4, the oxidized FTMA molecules come from the vesicle surface and reform into other molecular aggregates with much smaller size. The aggregation state after the oxidation is still unclear,

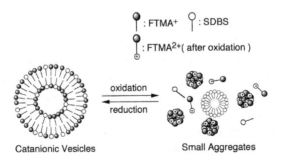

Figure 4 Schematic drawing of vesicle disruption induced by oxidation of FTMA$^+$.

but it may be a mixed micelle, since the trapping efficiency for water-soluble compounds decreases to almost zero after the oxidation treatment.

Control of vesicle formation is made possible by preparing vesicles from a mixture of a "switchable" (redox active) cationic surfactant (FTMA) and an anionic surfactant (SDBS), as described here. This technique would be applicable to controlled release of drugs encapsulated in the aqueous core of the bilayer or to the preparation of a structured thin film on a solid substrate. In addition, redox control of vesicle formation is also possible with the use of direct redox reactions on an electrode surface without the aid of the oxidizing and/or reducing reagents [39].

III. PHOTOCHEMICAL SWITCHING IN FORMATION AND DISRUPTION OF VESICLES

A. Short Review on Photochemically "Switchable" Surfactant

As mentioned in the previous section, the reversible control of formation of surfactant molecular aggregates by external stimuli has been a subject of significant attention with the view to apply it to the controlled release of drugs and perfumes, and to the removal of organic impurities dissolved in water.

Among various ways of control, optical control [43–52] is promising since it requires no addition of a third component to the system. A number of studies including the pioneer works by Kunitake and coworkers [43–46] on vesicles prepared with azobenzene-modified double-tailed amphiphiles have been reported. In those studies, the bilayer properties such as permeability and microviscosity were successfully controlled with *trans/cis* photoisomerization of the azobenzene unit induced by light irradiation, however, no attempt has been made to demonstrate reversible photochemical control of formation and disruption of vesicles. Use of vesicles in a thermodynamic equilibrium formed in an aqueous mixture of cationic and anionic surfactants [11–16] enables such reversible control of vesicle formation and disruption.

This section describes preparation of vesicles that spontaneously form upon mixing a cationic surfactant modified with a "photo switchable" azobenzene moiety and a simple anionic surfactant [52]. The effect of *trans–cis* photoisomerization of the azobenzene-modified surfactant on the aggregation state of vesicles is also studied by means of microscopic observations and spectroscopic measurements.

B. Vesicle Formation in AZTMA/SDBS/H₂O Ternary System

In this study, 4-butylazobenzene-4'(oxyethyl)trimethylammo-trimethylammonium bromide (AZTMA, Scheme 2) is used as the "photoswitchable" cationic surfactant. AZTMA was synthesized and purified as described elsewhere [49,50]. Monomeric and micelle-forming AZTMA molecules have been reported to undergo reversible *trans–cis* photoisomerization [49,50]. Sodium dodecylbenzenesulfonate (SDBS) is used as an anionic amphiphile. Sample preparations are done by first making stock solutions of AZTMA and SDBS at desired concentrations in deionized water. These stock solutions are equilibrated at room temperature and then samples are prepared by vortex-mixing stock solutions of AZTMA and SDBS at desired ratios for 3 s. Except for gentle stirring, the sample solutions were not subjected to any type of mechanical agitation.

A ternary phase diagram of dilute (less than 3%) aqueous mixtures of AZTMA and SDBS at 25°C is shown in Fig. 5. Mixed micellar phases (M) exist on the binary surfactant–water

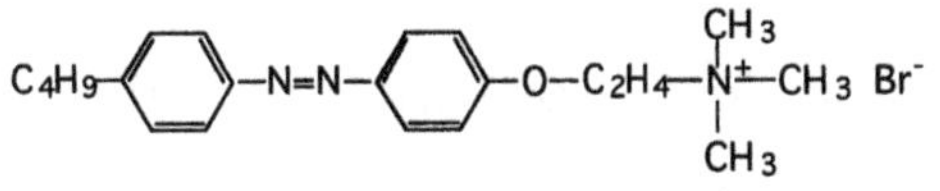

4-butylazobenzene-4' -(oxyethyl)trimethyl
ammonium bromide (AZTMA)

Scheme 2 Molecular structure of AZTMA.

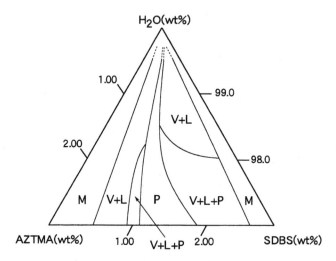

Figure 5 Ternary phase diagram for AZTMA/SDBS/H$_2$O at 25°C: Micelle region (M), precipitate (P), two-phase region (vesicles and lamellar phase (V + L)), three-phase region (vesicles, lameller phase and precipitate, (V + L + P)).

axes in Fig. 5. Precipitates (P) form along the equimolar line (AZTMA : SDBS = 54.7 : 45.3 (wt%). Spontaneously formed vesicles (V) are observed in a relatively wide range of mixing compositions on both cation-rich and anion-rich sides (V + L and V + L + P regions). Aqueous mixtures with the compositions in V + L regions seem to be a single vesicular phase just after sample preparation. However, as the samples age for weeks, small amounts of lamellar clouds form and become visible. These samples are then assigned to a vesicle/lamellar (V + L) phase. Freeze replica TEM observations and glucose dialysis experiments [42] also confirm vesicle formation.

C. Photochemical Control of Vesicle Formation

First, the effect of photo-induced *trans–cis* isomerization of AZTMA on the phase behavior of the catanionic system is described. Figure 6 shows the UV/vis absorption spectra of diluted (total surfactant concentration; 0.05 wt%) aqueous

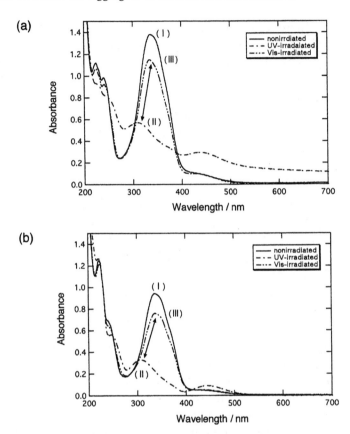

Figure 6 UV/VIS spectra of aqueous AZTMA/SDBS mixtures (total surfactant concentration 0.05 wt%). (a) AZTMA : SDBS = 6 : 4 (wt%), (b) AZTMA : SDBS = 4 : 6.

mixtures of AZTMA and SDBS. The spectroscopic characteristics of AZTMA in aqueous mixtures with SDBS are almost the same as those of AZTMA alone in its aqueous solutions as reported [47,48]. An absorption band characteristic of trans-azobenzene is observed at 344 nm for as-prepared mixed solutions. The absorption of this band decreases after UV-light irradiation (260–390 nm) with a Hg–Xe lamp, and a weak absorption band, due to a n–π^* transition of the *cis*-isomer, appears at about 440 nm. Visible-light irradiation (>410 nm) following UV-light irradiation causes *cis*-AZTMA to revert to the trans form, but not completely [19], indicating that the photo-stationary state of the *cis*-isomer remains to some extent in this system. Repeated UV and visible light irradiation, however, results in a reversible *trans–cis* isomerization that corresponds to the change between the spectra (II) and (III) in Fig. 6. These results show that the AZTMA molecules forming the bilayer structure (vesicles) can undergo reversible *trans–cis* photo isomerization. It should be noted that the turbidity of the cation-rich solution (Fig. 2(a), AZTMA : SDBS = 6 : 4) increases after UV-light irradiation, while such increase in turbidity is not observed for the anion-rich solution (Fig. 2(b), AZTMA : SDBS = 4 : 6). The aggregation state of aqueous *cis*-AZTMA/SDBS mixtures on the cation (AZTMA)-rich side is thus considered to be significantly different from that of *trans*-AZTMA/SDBS mixtures.

The effect of light irradiation on the aggregation state of cation-rich aqueous mixtures of AZTMA and SDBS is then directly observed for the AZTMA/SDBS = 6/4 wt% solution (total concentration; 0.05%) using a transmission electron microscope via freeze replica technique. The samples are frozen in liquid nitrogen and fractured at $-120°C$ using a freeze fracture device. The fractured surfaces are immediately replicated by evaporating platinum at an angle of 45°, followed by carbon at normal incidence, to increase the mechanical stability of the replica. The replicas thus prepared are examined on an electron microscope. In a micrograph of the as-prepared solution (*trans* form, Fig. 7(a)), spherical vesicles with an average size of 50–100 nm are observed. After 2 h UV-light irradiation [Fig. 7(b)], spherical vesicles disappear and large elongated molecular aggregates are observed. Even though the nature of these large molecular aggregates has not yet been identified, they may be lamellar structures dispersed in water, judging from a drastic increase in solution turbidity. Furthermore, subsequent visible light irradiation results in reformation of vesicles with an average size of *ca.* 50 nm [Fig. 7 (c)]. These results clearly demonstrate that vesicle formation and disruption can be reversibly controlled with photo irradiation in the present catanionic AZTMA/SDBS system.

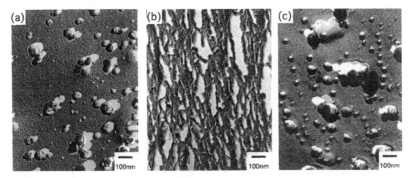

Figure 7 Freeze replica TEM micrographs of aqueous AZTMA/
SDBS mixed solutions. (AZTMA : SDBS = 6 : 4, total surfactant concentration; 0.05 wt%). (a) As prepared, (b) after 2 h UV-irradiation,
(c) after 2 h visible light irradiation.

D. Photochemical Release Control of Drug

Vesicles can encapsulate aqueous compounds (drugs, perfumes,
etc.) in their inner aqueous phase. Hence, if the release of these
encapsulated compounds from the inside of vesicles to the
outside can be controlled by photo irradiation, this should give
a novel functionality to vesicles. The effect of photo irradiation
on the trapping efficiency of AZTMA/SDBS vesicles is investigated with the glucose dialysis technique [42]. Briefly, AZTMA/
SDBS vesicles are prepared in 0.28 mol/L aqueous glucose
solution. The unencapsulated glucose is separated by dialysis
using a cellulose tube. The vesicles inside the tube are then
disrupted by the addition of ethanol. Finally, the amount of
glucose inside the tube is quantified using the mutarotase ·
GOD method [53].

Figure 8 shows the trapping efficiency of AZTMA/SDBS
mixed solutions (total surfactant concentration; 1 wt%) as
a function of their composition. The trapping efficiency of
as-prepared (*trans* form) solutions is high (2–4% against total
glucose amount) at the compositions where vesicles are formed
(AZTMA compositions to the total surfactant concentration are
0.30–0.50 and 0.60–0.75), whereas it is almost zero at the
equimolar composition (the AZTMA ratio is 0.547). After *cis*

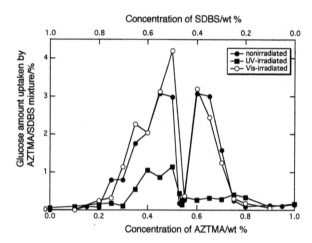

Figure 8   Dependence of trapping efficiency of aqueous AZTMA/
SDBS mixtures on photo irradiation (total surfactant concentration
1.00 wt%).

form formation induced by UV-light irradiation for 12 h, the
trapping efficiency decreases drastically. For instance, it
decreases from 3.1% to 0.3% by UV-light irradiation for an
AZTMA/SDBS = 6/4 wt% solution. Furthermore, subsequent
visible light irradiation for 12 h (in advance of dialysis) causes
a re-increase in the trapping efficiency to 3.2%. These results
confirm the disruption and reformation of vesicles induced by
UV and visible light irradiation, and also suggest that the
release of aqueous compounds encapsulated in the vesicles can
be controlled by the photochemical reaction of AZTMA.

E.   Mechanism of Reversible Formation Control
of Vesicles

The allowed packing of surfactant molecules is governed by
the "surfactant number [54]" $v/a_o l_c$, where v is the volume of
the hydrophobic portion of the surfactant, l_c is the length of
the hydrophobic group, and a_o is the head group area of the
surfactant molecule. In this scheme, when the surfactant
number is less than 1/3, spherical micelles are the preferred
form of aggregation, as the surfactant head group area is large

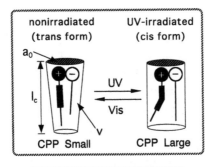

nonirradiated   UV-irradiated
(trans form)   (cis form)

CPP Small   CPP Large

$\dfrac{v}{a_0 \cdot l_c}$	: Critical Packing Parameter (CPP)

$<1/3$	: Spherical micelle
$1/3-1/2$	: Rod-like micelle
$1/2-1$	: Vesicle
-1	: Planar lamellar

Figure 9   Change in packing parameter of AZTMA/SDBS pseudo double-tailed complex induced by photoisomerization of AZTMA.

in comparison to the volume of the surfactant tail. Cylindrical micelles form when the number is between 1/3 and 1/2 and highly curved bilayer vesicles and then flat bilayers are formed when it is greater than 1/2. Regarding the surfactant mixtures of interest here, dynamic ion pairing of ionic single-tailed surfactants apparently brings about a pseudo double-tailed zwitterionic surfactant, which results in a smaller head group and a larger hydrophobic regions than those in the individual surfactants. This ion pairing was confirmed by Kaler *et al.* [15] with surface tension and conductivity measurements. The dynamic pairing of *trans*-AZTMA and SDBS should roughly double the surfactant number, leading to a transition from spherical micelles in the pure component systems to vesicles in the mixed surfactant systems (Fig. 9).

On the other hand, formation of the *cis*-form by UV-light irradiation causes an increase in the critical micelle concentration (cmc). Thus, the CMC of *trans*-AZTMA corresponds to that of alkyltrimethylammonium bromide with a carbon chain length of 16, while the CMC of *cis*-AZTMA corresponds to that of alkyltrimethylammonium bromide with a carbon chain length of 14. Furthermore, the volume of the hydrophobic portion (v) increases and the length of the hydrophobic group (l_c) decreases through *cis*-form formation. This would produce an increase in the surfactant number of AZTMA, thereby causing transformation from vesicles to a planer lamellar structure. Kaler *et al.* [15] have also reported that the stability of catanionic vesicles is higher when the difference in the alkyl tail length between cationic and anionic surfactants is larger. The alkyl chain length of cationic *trans*-AZTMA is significantly longer than that of anionic SDBS, while the alkyl chain length of *cis*-AZTMA is close to that of SDBS. This may cause the dispersion instability of the *cis*-AZTMA–SDBS ion pair.

IV. ELECTROCHEMICAL SWITCHING OF SOLUBILIZATION

A. Solubilization

One of the functions of molecular assemblies such as micelles and vesicles is to solubilize hardly water-soluble and oil-soluble substances in their interior. Solubilization is applied in various fields including those of cosmetics, perfumes, and foods [55].

On the other hand, control of solubilization by means of external stimuli such as pH change, temperature change, redox reaction, and photochemical reaction (Fig. 10) is of significant interest because of the important role of the assemblies in such applications as release rate control of perfumes and drugs held in their interior and targeting of drug delivery systems. Since micelles are in particular a system in thermodynamic equilibrium, unlike liposomes (vesicles), they are a medium most suitable to reversibly control the release of oily substances and the properties of these molecular assemblies. Yet, no report has dealt with control of solubilization of oily substances using "switchable" amphiphilic molecules.

This section describes our recent results on the electrochemical control of release-uptake of oily substance using ferrocene-modified amphiphilic molecules.

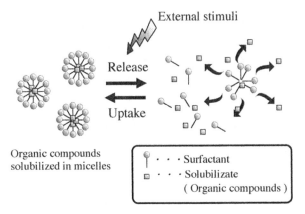

Figure 10 Reversible control of uptake and release of organic compounds by external stimuli.

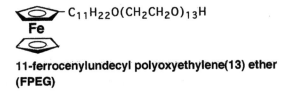

11-ferrocenylundecyl polyoxyethylene(13) ether (FPEG)

Scheme 3 Molecular structure of FPEG.

B. Solution Properties of "Electrochemically Switchable" Surfactants

As the ferrocene-modified nonionic surfactant, 11-ferrocenylundecyl polyoxyethylene ether (FPEG; Scheme 3) is used. Since nonionic surfactants possess ether bond(s) and/or indissociable hydroxyl group(s) as hydrophilic group(s) in their molecule, they can be used over a wide range of pH values and as mixtures with ionic (anionic, cationic) surfactants. In particular, their applications as solubilizing agents of drugs and perfumes are important since their micelles solubilize sparingly water-soluble substances to a higher extent than those of ionic surfactants. The reduced and oxidized forms of the surfactant are abbreviated hereafter to FPEG and FPEG$^+$, respectively. Oxidation of FPEG is carried

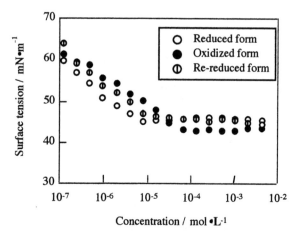

Figure 11 Relationship between the surface tension of FPEG solutions and the surfactant concentration. FPEG was oxidized and re-reduced by electrolysis using a platinum electrode.

out by constant-potential electrolysis of the reduced form in N_2 stream at $+0.6\,V$ vs SCE. Constant-potential electrolysis of a solution of the oxidized form at $0.0\,V$ vs SCE in N_2 stream for 21 hr yields a solution of the re-reduced form of FPEG.

Figure 11 shows the effect of electrochemical reactions on the surface tension-concentration relationship for FPEG solutions. The surface tension decreases with increasing concentration and exhibits a bend at a certain concentration, beyond which it levels off with further increase in the concentration for solutions of the reduced, oxidized and re-reduced forms of FPEG. Oxidation of the ferrocenyl group causes the CMC of FPEG, determined from the bending point on the surface tension-concentration curve, to shift upward from 6×10^{-6} to $5 \times 10^{-5}\,mol/L$, thus slowing down micelle formation. Furthermore, re-reduction of the group produces a downward shift in the CMC from $5 \times 10^{-5}\,mol/L$ to $9 \times 10^{-6}\,mol/L$, thus promoting micelle formation. This indicates that micelle formation of FPEG can be reversibly controlled by electrochemical oxidation-reduction of the ferrocenyl group. The upward shift of the CMC of FPEG observed after oxidation of the ferrocenyl group would be due to an

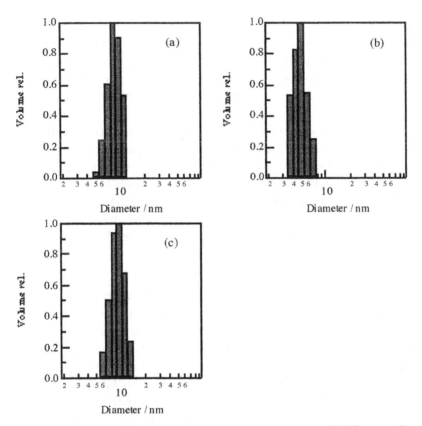

Figure 12 Particle size distribution of 2 mM FPEG micellar solutions measured by dynamic light scattering technique. (a) Reduced form, (b) oxidized form, (c) re-reduced form.

increase in the hydrophilicity of the surfactant molecule caused by the formation of the ferrocenium ion and to the electrostatic repulsion between the ferricinium ions that prevents aggregate formation in the oxidized form.

Dynamic light scattering measurements are performed on solutions of the reduced, oxidized and re-reduced forms of FPEG to examine how the micellar size of the surfactant is affected by its redox reaction on the electrode. The mean micellar sizes of FPEG are found to be 9, 4 and 10 nm for the reduced, oxidized and re-reduced forms, respectively, as shown in Fig. 12. This indicates that oxidation of the ferrocenyl group causes the micellar size to decrease and its re-reduction allows the size to return to the initial value. The decrease in the micellar size brought about by oxidation is due to the increased curvature of micelles because the oxidized form of FPEG has a double hydrophilic group-type structure in which the group is very bulky.

As mentioned so far, the oxidation state of the ferrocenyl group is suggested to strongly affect the properties of the FPEG solution (surface tension, micellar size, area occupied per molecule, etc.). We can then expect to electrochemically control the solubilization of oily substances into FPEG micelles reversibly if we make use of changes in the solution properties produced by the redox reaction of the ferrocenyl group.

C. Electrochemical Control of Solubilization

The effect of the redox reaction of the ferrocenyl group on the solubilization equilibrium constant, K, is examined. The value of K (M^{-1}) can be obtained by the head space method [56,57]. Namely, micellar solutions of FPEG containing a solubilizate in amounts below the solubilization limit are prepared and portions of these solutions are pipetted into GC vials. After solubilization equilibrium is attained, the gas phase in each of the vials is withdrawn with a syringe and the amount of the substance in the gas is determined with a gas chromatograph.

The solubilization equilibrium constant, K (M^{-1}), for the equilibrium of the solubilizate between the micelle and bulk phases is defined by Eq. (1) [58,59]

$$K = \frac{X_{\text{org}}}{C_{\text{org}}} \tag{1}$$

where X_{org} denotes the mole fraction of the solubilizate in micelled and C_{org} the molar concentration of the solubilizate remaining unsolubilized in the bulk phase. Here, the equation predicts that the solubilizate is partitioned (solubilized) more into micelles as the value of K becomes larger.

Benzene as the solubilizate is examined first. Benzene is added to water or aqueous FPEG solutions in various amounts

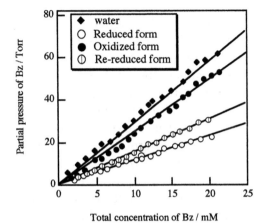

Figure 13 Partial vapor pressures of benzene over aqueous solution and 10 mM FPEG aqueous solution as a function of benzene concentration.

and the partial pressures of benzene in equilibrium with these aqueous phases are measured. After measuring the vapor pressure of pure benzene by gas chromatography, the vapor pressure of benzene in equilibrium with the sample solution in the sample vial is similarly measured. The activity of benzene, a, is calculated using Eq. (2) and the partial pressure of benzene is then estimated through subsitution of a in Eq. (3)

$$a = \frac{P_k}{P_k^0} \tag{2}$$

$$P = P^0 a \tag{3}$$

where P_k and P_k^0 are the peak areas for sample and pure benzene, respectively, and P^0 is the vapor pressure of benzene at 30°C, which is 119 Torr [60].

Figure 13 shows the relationship between the partial pressure of benzene, P, in equilibrium with the aqueous benzene solution or benzene-solubilizing FPEG solution, and the concentration of benzene in the aqueous phase, C_{sol}. The partial pressure of benzene over the solubilized system is found to be lower than that over the aqueous solution at the same benzene concentration. This can be ascribed to the solubilization of benzene dissolved in the bulk phase into micelles. Oxidation of FPEG causes the partial pressure of benzene in equilibrium with the solubilized system to increase over that observed for the reduced form and re-reduction of FPEG produces a decrease in the partial pressure. The observed partial pressure increase brought about by the oxidation is due to the release of a portion of benzene solubilized in micelles into the bulk phase, and the partial pressure decrease produced by the re-reduction can be ascribed to the resolubilization of benzene into micelles from the bulk phase.

Evaluation of the solubilization equilibrium constant is then performed for the solubilized system of benzene on the basis of the results in Fig. 13. The relation between the partial pressure of benzene in equilibrium with aqueous benzene solution, P, and the total concentration of benzene in solution, C_{sol}, is combined with the Hansen–Miller equation [61], yielding Eq. (4)

$$P = 3.3C_{sol} + 2.3 \times 10^{-3}C_{sol^2} \tag{4}$$

which is the relationship between the partial pressure of benzene and the concentration of benzene remaining unsolubilized in the bulk phase. Substitution of the partial pressure of benzene over the solubilized system into Eq. (4) gives the concentration of benzene dissolved in the bulk phase, C_{org}. The difference between C_{sol} and C_{org} is the concentration of benzene solubilized in micelles. Since the CMC values for the reduced, oxidized and re-reduced forms of FPEG are known from the results of surface tension measurements, X_{org} is obtained from Eq. (5)

$$X_{org} = \frac{(C_{sol} - C_{org})}{([FPEG] - CMC) + (C_{sol} - C_{org})} \tag{5}$$

Finally, substitution of X_{org} and C_{org} thus calculated into Eq. (1) gives the solubilization equilibrium constant, K.

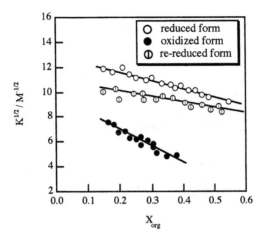

Figure 14 Dependence of solubilization equilibrium constant ($K^{1/2}$) of benzene in FPEG micelles on intramicellar mole fraction (X_{org}).

The following empirical equation is known to hold between the solubilization equilibrium constant, K, and the mole fraction of solubilizate in micelles

$$K^{1/2} = K_0^{1/2} - BK_0^{1/2}X_{org} \qquad (6)$$

where $K^{1/2}$ is the square root of the solubilization equilibrium constant, K_0 is the solubilization equilibrium constant at $X_{org} \to 0$ and B is a constant. Figure 14 shows the relation between the square root of the solubilization equilibrium constant for the solubilized system of benzene in FPEG micelles, K, and the mole fraction of benzene solubilized in FPEG micelles, X_{org}, verifying Eq. (6) to hold in this case as evidenced by the nearly linear decrease of $K^{1/2}$ with increasing X_{org} as shown in the figure.

Reduction in $K^{1/2}$ with increase in X_{org} is observed for all of the reduced, oxidized and re-reduced forms. Furthermore, the value of $K^{1/2}$ is lowered appreciably by oxidation of the ferrocenyl group due to the release of a portion of benzene solubilized in micelles into the bulk. Since the value of $K^{1/2}$ is found to rise again upon reduction of the ferrocenyl group,

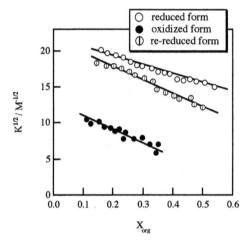

Figure 15 Dependence of solubilization equilibrium constant ($K^{1/2}$) of ethylbenzene in FPEG micelles on intramicellar mole fraction (X_{org}).

re-entry of benzene into FPEG micelles appears to have occurred from the bulk.

The effect of the properties of oily solubilizates on their redox reaction-dependent releasability from micelles is examined. Figure 15 shows how oxidation of the ferrocenyl group affects the solubilization equilibrium constant for the solubilized system of ethylbenzene, which is more hydrophobic than benzene. As in the case of benzene, the solubilization equilibrium constant is reduced by oxidation of the ferrocenyl group and raised by re-reduction of the group. Changes in the solubilization equilibrium constant produced by the redox reaction are larger for ethylbenzene than for benzene, indicating a higher efficiency in electrochemical release control for oily substance of a higher hydrophobicity. This would suggest that ethylbenzene is more accessible than benzene to the reduction in the hydrophobic region of micelles caused by oxidation because it is solubilized in the core (hydrocarbon chains) of micelles.

The results mentioned so far demonstrate that the solubilization of oily substances into FPEG micelles can be reversibly and dynamically controlled through oxidation-reduction of the surfactant.

IV. PHOTOCHEMICAL SWITCHING OF SOLUBILIZATION

A. Solution Properties of "Photochemically Switchable" Surfactants

As mentioned in the previous section, the control of solubilization by means of external stimuli is of significant interest because it can be applied to such applications as control of release rate perfumes and drugs held in their interior and targeting of drug delivery systems. In addition, clean and rapid control of solubilization would become possible without addition of third substances if light can be used as the external stimulus. Amphiphilic molecules with photofunctional groups such as azobenzene [43–50], stilbene [62–65], diphenylazomethine [66–68], and spiropyran [69] have been synthesized as photoresponsive molecules and they have been used in the forms of LB membranes, liquid crystals, and vesicles in attempts to control by light the properties of these molecular assemblies. Yet, no report has dealt with control of solubilization of oily substances using photoresponsive amphiphilic molecules. This section describes photochemical control of the release-uptake of oily substance using azobenzene-modified amphiphilic molecules that undergo significant changes in their geometrical structure and surface chemical properties induced by light irradiation.

In this study, 4-butylazobenzene-4′-(oxyethyl)trimethylammonium bromide (AZTMA), a light-responsive cationic amphiphile, is used. Excitation by ultraviolet irradiation of the azobenzene site produces photoisomerization of AZTMA from the *trans*-form to the *cis*-form and irradiation of visible light of the *cis*-form causes its isomerization to the *trans*-form. As a consequence, the geometrical shape of AZTMA molecules changes reversibly with light irradiation.

Figure 16 shows the relationship between the specific conductivity and the concentration for aqueous AZTMA

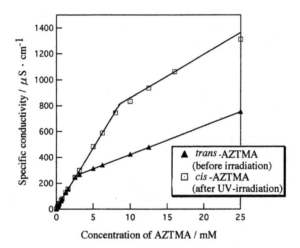

Figure 16 Specific conductivity of aqueous AZTMA solutions as a function of total surfactant concentration.

solutions. The specific conductivity increases linearly with the concentration of AZTMA and the slope of the conductivity curve changes at a certain concentration for both isomers. The CMC values, determined from the bending point on the conductivity curve, are 2.7 mM for the *trans*-isomer and 8.2 mM for the *cis*-isomer, respectively, indicating that photo-isomerization of *trans*-AZTMA produces *cis*-AZTMA, thereby causing a delay in micelle formation. *cis*-AZTMA molecule has a more bulky structure than *trans*-AZTMA and a shorter apparent hydrophobic tail length. The hydrophobicity of *cis*-AZTMA is then lower than that of the *trans* form and the CMC shows a higher value. This result suggests the possibility that formation-disintegration of micelles can be controlled by light using the difference in CMC between the isomers if the concentration of AZTMA is properly chosen. Then, the effect of light irradiation is examined by SHSGC on the solubilization of ethylbenzene in AZTMA micelles.

B. Photochemical Control of Solubilization

Figure 17 shows the vapor pressures of ethylbenzene in equilibrium with micellar AZTMA solutions at various

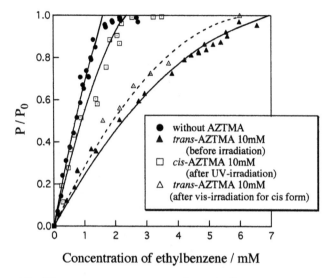

Concentration of ethylbenzene / mM

Figure 17 Vapor pressure curve for ethylbenzene over aqueous AZTMA solution at 30°C.

ethylbenzene concentrations, together with the vapor pressure of ethylbenzene in equilibrium with an aqueous ethylbenzene solution. The concentration of AZTMA is 10 mM, a concentration higher than the CMCs of the *trans*-isomer (2.7 mM) and the *cis*-isomer (8.2 mM). Comparison at the same ethylbenzene concentration shows that the vapor pressures of ethylbenzene in equilibrium with solutions of *trans*-AZTMA (closed triangles) and *cis*-AZTMA (open squares) with the oil solubilized are lower than that over aqueous ethylbenzene solution without surfactant. This verifies solubilization of ethylbenzene in AZTMA micelles.

A rise in the vapor pressure of ethylbenzene is also noticed upon photoisomerization of the *trans*-AZTMA to the *cis*-isomer caused by UV light irradiation (closed triangle → open square). This is brought about by release of ethylbenzene into the bulk due to disintegration of portions of *trans*-AZTMA micelles into monomers caused by the reduction in the cmc (equivalent to 5.5 mM) produced by photoisomerization of AZTMA. In addition, a fall in the vapor pressure to the level before the light

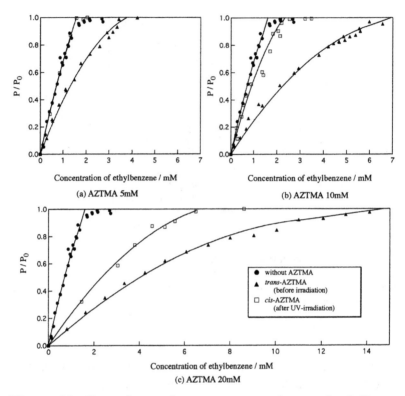

Figure 18 Dependence of vapor pressure change of ethylbenzene induced by photoisomerization of AZTMA on its concentration at 30°C.

irradiation is found when the solution of *cis*-AZTMA is exposed to visible light. These findings demonstrate that uptake by and release from micelles of the substance to be solubilized can be controlled through light irradiation by making use of the difference in CMC between *trans*- and *cis*-isomers.

Figure 18 shows the effect of ethylbenzene concentration on its relative vapor pressure at different AZTMA concentrations. Measurements are carried out using 5 and 20 mM AZTMA solutions, in addition to the 10 mM solution used so far. The vapor pressure of ethylbenzene (open squares) in equilibrium with 5 mM *cis*-AZTMA solutions formed after UV light irradiation is found to be identical to the vapor pressure of ethylbenzene (closed circles) over the solution without surfactant [Fig. 18(a)]. This is because the concentration of 5 mM is higher than the CMC of *trans*-AZTMA (2.7 mM) but lower than the CMC of *cis*-AZTMA (8.2 mM) and therefore no micelle is formed in the solution. Complete release into the bulk solution of ethylbenzene solubilized in *trans*-AZTMA micelles is thus confirmed after UV light irradiation. Although UV light irradiation causes release of ethylbenzene into the bulk solution at 20 mM AZTMA [Fig. 18(c)] as observed with a 10 mM solution, a considerable amount of ethylbenzene is found to remain in *cis*-AZTMA micelles.

If the amount of solubilization per unit concentration of micellized AZTMA (solubilizing capacity) is the same for both the *trans*- and *cis*-isomers, the amount of ethylbenzene released after UV light irradiation corresponds to the difference between their CMCs and an equal amount released per unit volume should be observed with both 10 and 20 mM AZTMA solutions. Contrary to expectation, as shown in Fig. 18, the maximum amount of ethylbenzene released increases to 8.0 mM [the difference between the solubilization limit for the *trans*-isomer (13.0 mM) and that for the *cis*-isomer (5.0 mM)] for a 20 mM AZTMA solution. This would mean that the change in the solubilizing capacity also affects the solubilization in this system, in addition to the change in cmc produced by light irradiation. The solubilizing capacity is then estimated for micelles of the *trans*- and *cis*-isomers.

Figure 19 shows the solubilization limit of ethylbenzene for *trans*-AZTMA and *cis*-AZTMA. The abscissa and ordinate represent AZTMA concentration and solubilization limit, respectively. Solubilization is observed only at surfactant concentrations above the cmc and the solubilization limit increases almost linearly with increasing concentration for both isomers. The solubilization limit is larger for the *trans*-isomer than for the *cis*-isomer. Then, the solubilization limit is plotted against the concentration of AZTMA constituting micelles, to yield a straight line for each isomer (Fig. 19). The slope of the line gives solubilizing capacity and a steeper slope means a higher capacity. Clearly, the slope is steeper for the *trans*-isomer (<slope = 0.74) than for the *cis*-isomer

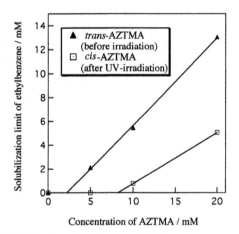

Figure 19 Solubilizing limit of ethylbenzene as a function of AZTMA concentration.

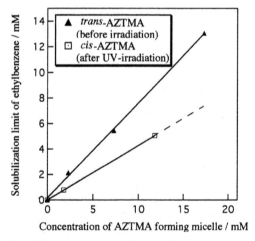

Figure 20 Solubilizing limit of ethylbenzene as a function of concentration of AZTMA forming micelles.

($<$slope $= 0.42$), showing a higher solubilizing capacity for the former. The reason for the higher solubilizing capacity of the *trans*-isomer is ascribed to larger solubilization sites in the hydrophobic moiety of the *trans*-AZTMA micelles since the hydrophobic group of AZTMA is in an extended state in the *trans*-isomer while it is folded in the *cis*-isomer. In addition, a larger hydrophobicity of the interior of *trans*-AZTMA micelles, as suggested by its lower CMC value, enables oily substances to be solubilized in the micelles more easily. These results indicate that the solubilization control using the photoisomerization of AZTMA is possible through light irradiation and this method of solubilization control is based on the differences in both CMC and solubilizing capacity between the *trans*-isomer and the *cis*-isomer.

REFERENCES

1. A. D. Bangham, and R. W. Horne, *J. Mol. Biol.* **8**: 660 (1994).

2. M. Wong, F. H. Anthony, T. W. Tillack, and T. E. Thompson, *Biochemistry* **21**: 4126 (1982).

3. W. K. Surewicz, and R. M. Epand, *Biochim. Biophys. Acta* **856**: 290 (1986).

4. T. Kunitake, and Y. Okahata, *J. Am. Chem. Soc.* **99**: 3860 (1977).

5. M. Shimomura, and T. Kunitake, *J. Am. Chem. Soc.* **104**: 1757 (1982).

6. Y. Murakami, A. Nakano, A. Yoshimatsu, and K. Fukuya, *J. Am. Chem. Soc.* **103**: 728 (1981).

7. N. Nakashima, H. Fukushima, and T. Kunitake, *Chem. Lett.* 1207 (1981).

8. T. Kunitake, S. Nakashima, M. Shimomura, Y. Okahata, K. Kano, and T. Ogawa, *J. Am. Chem. Soc.* **102**: 6642 (1980).

9. S. D. Christian, E. E. Tucker, and J. F. Scamehorn, *American Environmental Laboratory* **2**: 13 (1990).

10. J. F. Scamehorn, and J. H. Harwell (eds.), *Surfactant-Based Separation Processes*, Marcel Decker, New York (1989).

11. E. W. Kaler, A. K. Murthy, B. E. Rodriguez, and J. A. N. Zasadzinski, *Science* **245**: 1371 (1989).

12. L. L. Brasher, and E. W. Kaler, *Langmuir* **12**: 6270 (1996).

13. E. W. Kaler, K. L. Herington, A. K. Murthy, and J. A. N. Zasadzinski, *J. Phys. Chem.* **96**: 6698 (1992).

14. K. L. Herrington, E. W. Kaler, D. D. Miller, J. A. N. Zasadzinski, and S. Chiruvolu, *J. Phys. Chem.* **97**: 13792 (1993).

15. M. T. Yatcilla, K. L. Herrington, L. L. Brasher, E. W. Kaler, S. Chiruvolu, and J. A. Zasadzinski, *J. Phys. Chem.* **100**: 5874 (1996).

16. D. J. Iampietro, L. L. Brasher, E. W. Kaler, A. S. Stradner, and O. Glatter, *J. Phys. Chem. B* **102**: 3105 (1998).

17. S. A. Safran, P. Pincus, and D. Andelman, *Science* **248**: 354 (1990).

18. K. Murthy, N. Easwar, and E. Singer, *Colloid Polym. Sci.* **276**: 940 (1998).

19. J.-B. Huang, B. -Y. Zhu, M. Mao, P. He, J. Wang, and X. He, *Colloid Polym. Sci.* **277**: 354 (1999).

20. B. Jonsson, P. Jokela, A. Khan, B. Lindman, and A. Sadaghiani, *Langmuir* **7:** 889 (1991).

21. A. Khan, and C. Mendonca, *J. Colloid Interface Sci.* **169:** 60 (1995).

22. Y. Kondo, H. Uchiyama, N. Yoshino, K. Nishiyama, and M. Abe, *Langmuir* **11**: 2380 (1995).

23. M. I. Viseu, K. Edwards, C. S. Campos, and S. M. B. Costa, *Langmuir* **16:** 2105 (2000).

24. H. Hauser, *Chem. Phys. Lipids* **43**: 283 (1987).

25. J. E. Brady, D. F. Evans, R. Kacharr, and B. W. Ninham, *J. Am. Chem. Soc.* **106**: 4279 (1984).

26. M. Sagisaka, K. O. Kwon, H. Sakai, and M. Abe, *J. Jpn Soc. Colour Mat.* **74**: 63 (2001).

27. T. Saji, K. Hoshino, and S. Aoyagui, *J. Am. Chem. Soc.* **107**: 6865 (1985).

28. T. Saji, K. Hoshino, and S. Aoyagui, *J. Chem. Soc. Chem. Commun.* 865 (1985).

29. T. Saji, K. Hoshino, Y. Ishii, and M. Goto, *J. Am. Chem. Soc.* **113**: 450 (1991).

30. T. Saji, *J. Jpn. Oil Chem. Soc.* **39**: 717 (1990).

31. B. S. Gallardo, M. J. Hwa, and N. L. Abbott, *Langmuir* **11**: 4209 (1995).

32. D. E. Bennett, B. S. Gallardo, and N. L. Abbott, *J. Am. Chem. Soc.* **118**: 6499 (1996).

33. K. Tajima, T. Huxur, Y. Imai, I. Motoyama, A. Nakamura, and M. Koshinuma, *Colloids Surf. A* **94**: 243 (1995).

34. Y. Takeoaka, T. Aoki, K. Sanui, N. Ogata, and M. Watanabe, *Langmuir* **12**: 487 (1996).

35. Y. Kakizawa, H. Sakai, K. Nishiyama, M. Abe, H. Shoji, Y. Kondo, and N. Yoshino, *Langmuir* **12**: 921 (1996).

36. N. Yoshino, H. Shoji, Y. Kondo, Y. Kakizawa, H. Sakai, and M. Abe, *J. Jpn. Oil Chem. Soc.* **45**: 769 (1996).

37. T. Hayashita, T. Kurosawa, T. Miyata, K. Tanaka, and M. Igawa, *Colloid Polym. Sci.* **272**: 1611 (1994).

38. Y. Kakizawa, H. Sakai, T. Saji, N. Yoshino, Y. Kondo, and M. Abe, *J. Jpn. Soc. Colour Mat.* **72**: 78 (1999).

39. H. Sakai, T. Takei, Y. Kakizawa, Y. Kondo, N. Yoshino, and M. Abe, *Jpn. J. Oil Chem. Soc.* **47**: (1998).

40. J. C. Medina, I. Gay, Z. Chen, L. Echegoyen, and G. W. Gokel, *J. Am. Chem. Soc.* **113**: 365 (1991).

41. H. Sakai, H. Imamura, Y. Kakizawa, K. Yukishige, N. Yoshino, and J. H. Harwell, *Denki Kagaku* **65**: 669 (1997).

42. I. Miwa, J. Okuda, K. Maeda, and G. Okuda, *Clin. Chim. Acta* **37**: 538 (1972).

43. M. Shimomura, and T. Kunitake, *J. Am. Chem. Soc.* **104**: (1982).

44. T. Kunitake, Y. Okahata, M. Shimomura, S. Yasunami, and K. Takarabe, *J. Am. Chem. Soc.* **103**: 5401 (1981).

45. T. Kunitake, J. -M. Kim, and Y. Ishikawa, *J. Chem. Soc. Perkin Trans 2*, 885. (1991).

46. M. Shimomura, R. Ando, and T. Kunitake, *Ber. Bunsenges. Phys. Chem.* **87**: 1134 (1983).

47. T. Hayashita, R. Kurosawa, T. Miyata, K. Tanaka, and M. Igawa, *Colloid Polym. Sci.* **272**: 1611 (1994).

48. L. Yang, N. Takisawa, T. Hayashita, and K. Shirahama, *J. Phys. Chem.* **99**: 8799 (1995).

49. T. Saji, K. Hoshino, Y. Ishii, and M. Goto, *J. Am. Chem. Soc.* **113**: 450 (1991).

50. T. Saji, K. Ebata, K. Sugawara, S. Liu, and K. Kobayashi, *J. Am. Chem. Soc.* **116:** 6053 (1994).

51. Y. Orihara, A. Matsumura, Y. Saito, N. Ogawa, T. Saji, A. Yamaguchi, H. Sakai, and M. Abe, *Langmuir* **17**: 6072 (2001).

52. H. Sakai, A. Matsumura, S. Yokoyama, T. Saji, and M. Abe, *J. Phys. Chem.* **103**: 10737 (1999).

53. J. Okuda, and I. Miwa, *PNE* **17**: 216 (1972).

54. J. N. Israelachvili, D. J. Mitchell, and B. W. Ninham, *J. Chem. Soc. Faraday Trans.* **72**: 1525 (1976).

55. S. D. Christian, and J. F. Scamehorn (eds.), *Solubilization in Surfactant Aggregates*, Marcel Dekker, New York (1995).

56. M. E. Morgan, U. Uchiyama, S. D. Christian, E. E. Tucker, and J. F. Scamehorm, *Langmuir* **10**: 2170 (1994).

57. A. Hewitt, P. Mlyares, D. Leggett, and T. Jenkins, *Environs Sci. Technol* **26**: 1932 (1992).

58. A. J. Bard, and L. R. Faulkner, *Electrochemical Methods*, Wiley, New York (1980), P 136.

59. H. Uchiyama, S. D. Christian, J. F. Scamehorn, M. Abe and K. Ogino, *Langmuir* **7**: 95 (1991).

60. C. L. Yaws, *Handbook of Vapor Pressure*, Vol. 2, Gulf Publishing Co. (1994), P 388.

61. R. S. Hansen, and F. A. Millar, *J. Phys. Chem.* **58**: 193 (1954).

62. D. G. Whitten, I. Furman, C. Geiger, W. Richard, and S. P. Spooner, *Chem. Sci.* **105**: 527 (1993).

63. A. Sokolowski, K. A. Wilk, and A. Masiowska, *World Surfactant Congr. 4th* **2**: 274 (1996).

64. J. C. Russell, S. B. Costa, R. P. Seiders, and D. G. Whitten, *J. Am. Chem. Soc.* **102**: 5678 (1980).

65. P. E. Brown, T. Mizutani, J. C. Russell, B. R. Suddaby, and D. G. Whitten, *ACS Symp. Ser.* **278**: 171 (1985).

66. T. Kunitake, *Angew. Chem. Int. Ed. Eng.* **31**: 709 (1993).

67. Z. Zhang, L. Wu, Y. Liang, and Q. Yin, *J. Colloid Interface Sci.* **188**: 501 (1997).

68. Y. Liang, L. Wu, Y. Tian, Z. Zhang, and H. Chen, *J. Colloid Interface Sci.* **178**: 703 (1996).

69. S. Liu, M. Fujihira, and T. Saji, *J. Chem. Soc. Chem. Commun.* **16**: 1855 (1994).

16

Adsolubilization by Binary Surfactant Mixtures

KUNIO ESUMI

Department of Applied Chemistry and
Institute of Colloid and Interface Science,
Tokyo University of Science,
Tokyo, Japan

SYNOPSIS

Surfactant adsorbed layers formed on particles by adsorption of surfactants exhibit hydrophobic properties where water-insoluble compounds are incorporated in the layers. This phenomenon is called "adsolubilization." Mixed surfactant systems change markedly the adsolubilization behavior. In this chapter, adsolubilization by binary surfactant mixtures and adsorption characteristics by the mixtures are discussed.

I. INTRODUCTION

Surfactants adsorb as layers or micellar-type aggregates onto the surface of solids in water. Some hydrophobic compounds which do not adsorb on these solids can be incorporated in these layers in the presence of small amounts of surfactants. In 1955, Stigter *et al.* [1] observed that orange OT is coadsorbed with a cationic surfactant on glass. They suggested that the dye is located between a double layer of surfactant and coined the term "surface solubilization" to describe the phenomenon. Then, coadsorption of nonpolar oils with surfactants has been reported [2] and it is tempting to postulate that it is solubilized in hemimicelles, by analogy to micellar behavior. Also, it has been demonstrated [3] that pinacyanol is adsorbed into a micelle-like environment on alumina using spectrophotometric analysis. Thus, incorporation of water-insoluble compounds into surfactant adsorbed layers on particles or solid surfaces has been called adsolubilization [4].

Since the mid-1980s all the studies of adsolubilization have actually started. Most of the studies of adsolubilization so far have been limited to the use of single surfactant systems and have been reviewed by O'Haver *et al.* [5].

Adsorption from surfactant mixtures onto solids includes the adsorption of binary surfactant mixtures of anionic surfactants, anionic–nonionic surfactants, cationic–nonionic surfactants, and anionic–cationic surfactants. In particular, it is well known [6] that in the case of ionic–nonionic surfactants, the adsorption of one surfactant is often enhanced by addition of a small amount of the other surfactant. Surfactant mixtures provide several advantages over single surfactants, because the adsorption of surfactants on solids can be controlled using appropriate surfactants and solution properties. From the standpoint of surfactant mixtures, a question of interest is whether the surfactant mixture will solubilize more or less than pure components based on a simple linear weighting fraction. Nishikido [7] presents thermodynamic equations on synergistic solubilization and provides a good point to understanding many solubilization data on the theoretical approaches. On the other hand, adsolubilization studies by binary surfactant mixture are still few, even though they are very important from fundamental and applicable standpoints.

In this chapter, adsolubilization of 2-naphthol, yellow OB, and azobenzene by binary surfactant mixtures is described. Since adsorption of binary surfactant mixtures is directly correlated with the adsolubilization, the adsorption characteristics of binary surfactant mixtures is also discussed.

II. ADSORPTION CHARACTERISTICS OF BINARY SURFACTANT MIXTURES

When adsorption of binary surfactant mixtures on particles is studied, two different procedures for the adsorption are employed. One method is adsorption of binary surfactant mixtures from sequential addition of two surfactants and the other from premixed binary surfactant solutions.

A. Adsorption from Sequential Addition of Two Surfactants

When an ionic surfactant is adsorbed onto oppositely charged particles, the particles flocculate and then redisperse by increasing the concentration of the surfactant. Meguro and Kondo [8] reported this phenomenon in the dispersion of positively charged ferric hydrosols upon addition of anionic surfactants. This flocculation–redispersion process can be correlated with the formation of monolayers and bilayers of the surfactant. To study the interaction of binary surfactant mixtures on particles, adsorption from sequential addition of two surfactants has been investigated.

The alumina particles flocculate upon addition of a certain concentration of anionic surfactant at pH 3.5 where the alumina has an optimum positive ζ potential. Then a nonionic surfactant is added into the dispersion of flocculated alumina. Here, poly(oxyethylene)nonylphenyl ethers (NP) having average oxyethylene chain lengths of 7.5, 10, and 20 are used. After adsorption equilibrium, the adsorbed amount of surfactant has been determined from the difference in the surfactant concentration in the supernatant before and after the adsorption. The optimum flocculation concentration for the alumina (0.3 g/10 ml) is 0.05 mmol dm^{-3} for lithium dodecyl sulfate (LiDS) and lithium perfluorooctane sulfonate (LiFOS) [9]. Figure 1 shows the adsorbed amount of LiDS, LiFOS, and NP onto the alumina as a function of additive concentration of NP. In the LiDS–NP7.5 and LiFOS–NP7.5 systems as shown in Fig. 1(a), the adsorbed amount of LiDS or LiFOS is nearly constant, while that of NP7.5 increases with increasing additive concentration of NP7.5. These results imply that LiDS or LiFOS adsorbing on the alumina, even with the addition of NP7.5, is not desorbed presumably due to strong electrostatic attraction between a positively charged alumina and negatively charged hydrophilic group of LiDS or LiFOS. Upon addition of NP7.5, the adsorption of this surfactant occurs by hydrocarbon to hydrocarbon or hydrocarbon to fluorocarbon chain interaction, resulting in the formation of a mixed bilayer with the hydrophilic groups (oxyethylene chain) of NP7.5 oriented toward the aqueous phase [10,11]. In addition, the adsorbed amount of NP7.5 is greater for the LiDS–NP7.5 system than that for the LiFOS–NP7.5 system. This result is understood by a view that the interaction between hydrocarbon and hydrocarbon is more favorable than that between hydrocarbon and fluorocarbon [12]. Similarly, Figs 1(b) and (c) show that the adsorbed amount of LiDS or LiFOS is nearly constant, whereas that of NP10 or NP20 increases; the adsorbed amount of NP10 or NP20 in the LiDS–NP10 or NP20 system is also greater than that in the LiFOS–NP10 or NP20 system. Further, it is noteworthy that for both systems of LiDS–NP and LiFOS–NP, the adsorbed amounts of NP decreased with increasing oxyethylene chain length of NP. This is interesting since their isotherms essentially follow the adsorption of a nonionic surfactant on a hydrophobic surface [13–15]. A schematic model of a mixed bilayer is shown in Fig. 2. An increase in the length of the oxyethylene chain of NP results in a decrease in the adsorbed amount of NP, presumably because the change in free energy $(-\Delta G)$ of adsorption is decreased in magnitude and the cross-sectional area of the molecule at the LiDS- or LiFOS-covered alumina may increase as the number of oxyethylene units is increased [16]. Thus, the above results indicate that a

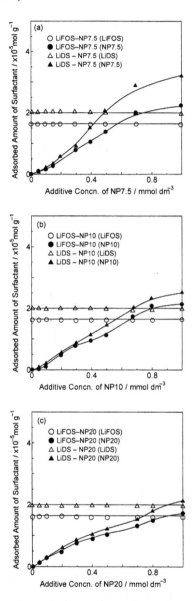

Figure 1 Adsorbed amount of surfactants on alumina: (a) LiDS–NP7.5 and LiFOS–NP7.5 systems; (b) LiDS–NP10 and LiFOS–NP10 systems; (c) LiDS–NP20 and LiFOS–NP20 systems; each mark shows adsorbed amount of surfactant indicated in bracket in each system. (Reproduced from Ref. 9 with permission of Academic Press, Inc.)

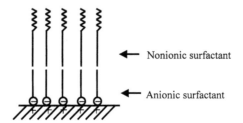

Figure 2 A schematic model of mixed surfactant bilayer on alumina. (Reproduced from Ref. 9 with permission of Academic Press, Inc.)

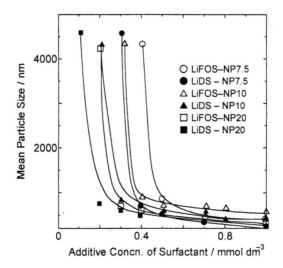

Figure 3 Change in mean-particle size of alumina for systems shown in Fig. 1. (Reproduced from Ref. 9 with permission of Academic Press, Inc.)

decrease in the length of the oxyethylene chain of NP allows the formation of the mixed bilayer to proceed more easily.

The degree of redispersion in the flocculated alumina covered with a LiDS or LiFOS layer upon addition of NP has been evaluated from the measurement of particle size. Figure 3 shows that the mean-particle size of the alumina suspension decreases with increasing concentration of NP and with increasing length of the oxyethylene chain of NP, for both the LiDS and LiFOS systems. It is found that the mean-particle size for the LiDS system decreases with addition of a concentration of NP lower than that for the LiFOS system. Since NP is probably adsorbed on the hydrophobic chain of LiDS or LiFOS while orienting its hydrophilic chain toward the solution as shown in Fig. 2, this adsorption produces a steric repulsive force due to the randomly coiled oxyethylene chains. Similar experiments using three types of iron oxides have been reported by Meguro *et al.* [17] and they have obtained a result similar to that of this study. In addition, the ζ potentials of alumina particles in a redispersion state are almost zero, indicating that the redispersion force between the redispersed alumina particles are not electrostatic force but steric repulsion [18]. The degree of steric repulsion is thought to increase with increasing oxyethylene chain length.

It is very important to characterize the microenvironmental properties of surfactant adsorbed layer onto particles. The fluorescence technique using probes [19] is one of the powerful methods to estimate the polarity of the mixed bilayer on the alumina. Figure 4 shows that the I_1/I_3 value of pyrene in the LiFOS–alumina system in the absence of NP7.5 is about 1.8 which is the same value as that in the supernatant; this ratio decreases steeply upon addition of 0.05 mmol/dm^3 of NP7.5 and then becomes constant (about 1.3) while the ratio in the LiDS–alumina system is nearly constant (1.2), even with different additive concentrations of NP7.5. These results can be interpreted as follows. In the LiFOS–alumina system, pyrene is either adsolubilized into the high polarity part of LiFOS molecules in the absence of NP7.5 or exists in water, and with addition of NP7.5 the pyrene becomes adsolubilized into a hydrocarbon chain of NP7.5 forming the mixed bilayer with LiFOS. On the other hand, in the LiDS–alumina system, in the absence of NP7.5, pyrene is adsolubilized into the hydrocarbon of LiDS; with addition of NP7.5, the pyrene is located in the mixed hydrocarbon chains of LiDS and NP7.5. Further, the difference in the I_1/I_3 ratio of the mixed bilayer between LiDS–NP7.5 and LiFOS–NP7.5 systems probably arises from the polarity difference between the hydrocarbon–hydrocarbon

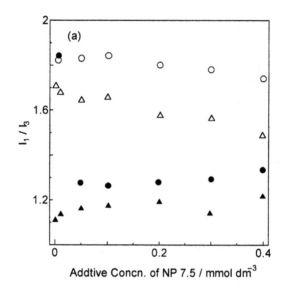

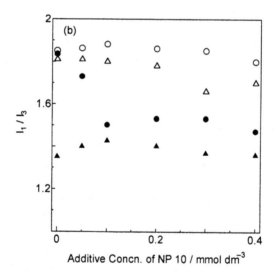

Figure 4   Change in I_1/I_3 ratio as a function of additive concen-
tration of NP: (a) LiDS–NP7.5 ($\triangle$, $\blacktriangle$) and LiFOS–NP7.5 ($\bigcirc$, $\bullet$)
systems; (b) LiDS–NP10 ($\triangle$, $\blacktriangle$) and LiFOS–NP10 ($\bigcirc$, $\bullet$) systems; (c)
LiDS–NP20 ($\triangle$, $\blacktriangle$) and LiFOS–NP20 ($\bigcirc$, $\bullet$) systems. Open mark:
supernatant, closed mark: suspension. (Reproduced from Ref.09 with
permission of Academic Press, Inc.)

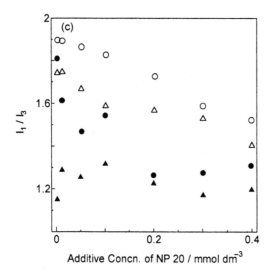

Figure 4 Continued.

chain and fluorocarbon–hydrocarbon chain [20]. The result that the I_1/I_3 ratios for both systems are smaller in the suspension than in the supernatant may indicate that the hydrophobic chains between the surfactants are more compact in the mixed bilayer than in the supernatant. This difference in the compactness would reflect the adsolubilization of water-insoluble compounds in the mixed bilayer [21] and in the mixed micelles. Similar results of the I_1/I_3 ratios are observed in NP10 and NP20 systems. The added concentration of NP to attain a constant I_1/I_3 value decreases with decreasing length of the oxyethylene chain in NP. This corresponds with the change in the adsorbed amount of NP; i.e., a decrease in the oxyethylene chain length of NP increases the adsorbed amount of NP and enhances the formation of a mixed bilayer. Using alumina and monodispersed ferric hydro sols [10,11], sequential adsorption of ionic and ionic surfactants have been reported and found that the hydrophobic interaction between anionic fluorocarbon and anionic hydrocarbon surfactants is weaker than those between anionic hydrocarbon and nonionic fluorocarbon surfactants, and between anionic fluorocarbon and nonionic hydrocarbon surfactants.

Formation of cationic and anionic mixed surfactant bilayers on laponite clay has also been studied [22]. Adsorption of SDS is performed with a suspension containing laponite and cationic surfactant, cetylpyridinium bromide (CP^+) at the cation exchange capacity. Figure 5 shows the result. As illustrated in its adsorption isotherm (curve a in Fig. 5), DS^- is efficiently adsorbed by an aqueous suspension of CP^+-exchanged laponite flocs. For DS^- concentration below 7 mmol/dm^3, the amount of adsorbed DS^- gradually increases as the DS^- monomeric bulk concentration increases and redispersion of the flocs into a colloidal suspension progressively occurs. Curve b in Fig. 5 indicates that in this region CP^+ does not desorb into the aqueous bulk. In the presence of 0.05 mol/dm^3 Na$_2$SO$_4$, adsorption of DS^- by CP^+–laponite flocs and further desorption of CP^+ and DS^- occur in a similar way (curves c and d of Fig. 5). In deionized water and 0.05 mol/dm^3 Na$_2$SO$_4$ aqueous solution, adsorption of DS^- occurs by tails to tails hydrophobic interaction between hydrocarbon of DS^- and the first layer of adsorbed CP^+, resulting in the formation of a cationic/anionic surfactant bilayer.

B. Adsorption from Premixed Binary Surfactants

Adsorption from premixed binary surfactants on particles can be mainly classified as three groups: (1) surfactants with the same head groups; (2) ionic–nonionic surfactants; or (3) oppositely charged surfactants.

Adsorption from binary mixed surfactants with the same head groups such as nonionic–nonionic, anionic–anionic, and cationic–cationic surfactants can be described as ideal mixing of the adsorbed pure components [23–25]. On the other hand, large deviation from ideal mixing has been observed for ionic–nonionic surfactant and opposite charged surfactant adsorption. Here, we describe mainly studies of ionic–nonionic mixed surfactant adsorption.

Figures 6 and 7 show the adsorption of 1 : 1 SDS/C$_n$EO$_8$ ($n = 10, 12, 14, 16$) mixtures on kaolinite [26]. It is interesting to note from Fig. 6 that the isotherms for the adsorption of SDS are identical when the hydrocarbon chain length of C$_n$EO$_8$ is

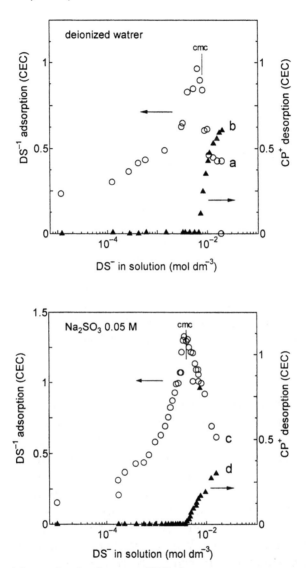

Figure 5 Adsorption isotherms of DS⁻ by an aqueous suspension of $CP^+(1\,cec)$–laponite $(1\,g/dm^3)$: (a) in deionized water and (c) in presence of Na_2SO_4 $0.05\,mol/dm^3$. concentration of desorbed CP^+ as a function of unadsorbed DS⁻ bulk concentration: (b) in deionized water and (d) in presence of Na_2SO_4 $0.05\,mol/dm^3$. The concentration along the y axis are expressed as a percentage of cec (cation exchange capacity). (Reproduced from Ref. 22 with permission of the American Chemical Society.)

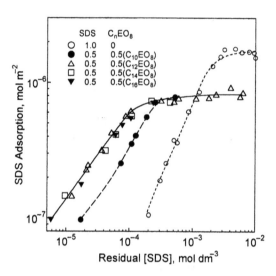

Figure 6 Effect of nonionic surfactant chain length on the adsorption of sodium dodecyl sulfate (SDS) from $1:1$ SDS/C_nEO_8 ($n = 10, 12, 14, 16$) mixtures; 0.03 M NaCl, pH 5, 25°C. (Reproduced from Ref. 26 with permission of Elsevier Science-NL.)

equal to or longer than that of SDS (C12). The presence of SDS is seen in Fig. 7 to enhance the plateau adsorption of the C_nEO_8, and the isotherms are shifted to lower concentration regions. This structural effect on surfactant mixture adsorption is schematically illustrated in Fig. 8. When the hydrocarbon chain length of C_nEO_8 is equal to or longer than that of SDS, the hydrocarbon chains of SDS are equally shielded from hydrophilic environment by the hydrocarbon chains of the co-adsorption C_nEO_8. When the hydrocarbon chain of the C_nEO_8 is shorter than that of SDS, however, part of the SDS hydrocarbon chain is exposed to the hydrophilic environment. The environment for the SDS hydrocarbon chain is, in this case, less hydrophobic and, therefore, the isotherm is shifted less into the low concentration region.

The adsorption of mixtures of a cationic surfactant, tetradecyltrimethylammonium chloride (TTAC) and a nonionic surfactant, pentadecylethoxylated nonylphenol (NP15) at the

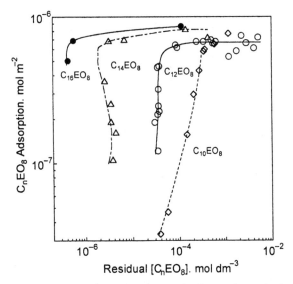

Figure 7 Effect of nonionic surfactant hydrocarbon chain length on the adsorption of C_nEO_8 from $1:1$ SDS/$CnEO_8$ ($n = 10, 12, 14, 16$) mixtures; 0.03 M NaCl, pH 5, 25°C. (Reproduced from Ref. 26 with permission of Elsevier Science-NL.)

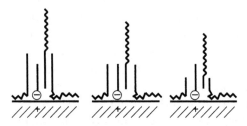

Figure 8 Schematic presentation of the effect of nonionic surfactant hydrocarbon chain length on the adsorption of SDS: (a) nonionic surfactant hydrocarbon chains longer than that of SDS; (b) nonionic surfactant hydrocarbon chain length equal to that of SDS; (c) nonionic surfactant hydrocarbon chain length shorter than that of SDS, partially exposing SDS hydrocarbon chains to the aqueous solution or the hydrophilic ethoxyl chains of the non-ionic surfactant. (Reproduced from Ref. 26 with permission of Elsevier Science-NL.)

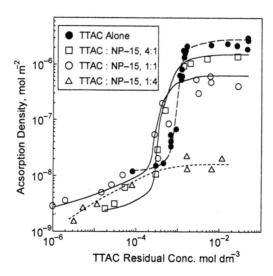

Figure 9 Adsorption of tetradecyltrimethylammonium chloride (TTAC) on alumina in the presence and absence of pentadecylethoxlated-nonylphenol (NP15), pH 10, I.S. $0.03 \, \text{mol/dm}^3$ NaCl. (Reproduced from Ref. 26 with permission of Elsevier Science-NL.)

alumina–liquid is shown in Figs 9 and 10 [26]. At pH 10, TTAC adsorbs on alumina since the alumina surface is negatively charged and hence the electrostatic attraction with TTAC is dominant. On the other hand, NP15 alone hardly adsorbs on alumina. It is seen from Fig. 9 that TTAC hemimicelle formation occurs at lower TTAC concentrations in the presence of NP15 but only at 4 : 1 and 1 : 1 TTAC : NP15 ratios. At the 1 : 4 ratio, however, the sharp increase in adsorption density corresponding to such aggregation is not observed over the entire concentration range studied. In all cases, the plateau adsorption density decreases markedly upon the addition of NP15. Figure 10 shows that with an increase in the TTAC content of the mixtures, the adsorption of NP15 is enhanced significantly, and the adsorption isotherms are shifted to lower concentration ranges. This is attributed to the more effective co-adsorption of NP15 as the number of TTAC hemimicelles at the alumina–water interface increases.

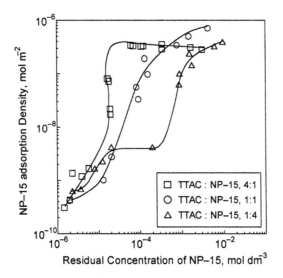

Figure 10 Adsorption of NP15 on alumina in the presence of varying amounts of TTAC, pH 10, I.S. 0.03 mol/dm³ NaCl. (Reproduced from Ref. 26 with permission of Elsevier Science-NL.)

Thus, it is important to study the effect of surfactant structure on the adsorption of surfactant mixtures. In particular, the relationship between the adsorption behaviors of surfactant mixtures and the surfactant adsorbed structures should be established.

III. ADSOLUBILIZATION BY BINARY SURFACTANT MIXTURES

A. Adsolubilization by Sequential Addition of Two Surfactants

Since alumina is positively charged at pH 3.5, anionic surfactants such as LiFOS or LiDS are added to alumina suspension at this pH to obtain optimum flocculation of alumina particles. Then, 1 mmol/dm³ NP aqueous solutions containing various concentrations of yellow OB or azobenzene solubilized are added to the optimum flocculated alumina. In all cases, the alumina becomes redispersed by the addition of NP,

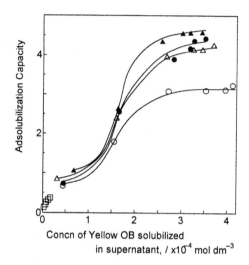

Figure 11 Adsolubilization capacity for yellow OB in surfactant bilayers on alumina as a function of concentration of yellow OB solubilized in supernatant: (△) LiDS–NP7.5; (○) LiFOS–NP7.5; (▲) LiDS–NP20; (●) LiFOS–NP20; (□) LiDS–LiDS. The adsolubilization capacity is calculated by dividing the amount of yellow OB adsolubilized by the amount of surfactants adsorbed. (Reproduced from Ref. 21 with permission of the Chemical Society of Japan.)

indicating that a mixed bilayer is formed on the alumina surface [21]. Figure 11 shows the adsolubilization capacity for yellow OB in bilayers on alumina as a function of concentration of yellow OB solubilized in the supernatant. The adsolubilization capacity can be calculated by dividing the amount of the yellow OB adsolubilized by the total amounts of surfactants adsorbed. It is seen that the adsolubilization capacity for yellow OB in the mixed bilayers increases with an increase in the concentration of yellow OB solubilized in the supernatant. In addition, these results show that the systems containing NP20 have a higher adsolubilization capacity than those containing NP7.5, and that the LiDS–NP systems also have a higher adsolubilization than do the LiFOS–NP system with the same ethylene oxide chain length of NP.

Figure 12 shows the adsolubilization capacity for azobenzene in the LiDS–NP and LiFOS–NP systems. The adsolubilization capacity for azobenzene in the respective

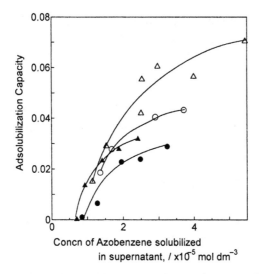

Figure 12 Adsolubilization capacity for azobenzene in surfactant bilayers on alumina as a function of concentration of azobenzene solubilized in supernatant: (△) LiDS–NP7.5; (○) LiFOS–NP7.5; (▲) LiDS–NP20; (●) LiFOS–NP20. The adsolubilization capacity is calculated by dividing the amount of azo benzene adsolubilized by the amount of surfactants adsorbed. (Reproduced from Ref. 21 with permission of the Chemical Society of Japan.)

single bilayers of LiDS and LiFOS is not large enough to measure. It is apparent, however, that the adsolubilization capacity for azobenzene in the LiDS–NP system is greater than that in the LiFOS–NP system, indicating that the adsolubilization of azobenzene is less favorable in the hydrocarbon chain–fluorocarbon chain layer for the LiFOS–NP system than in the hydrocarbon chain–hydrocarbon chain layer for the LiDS–NP system. This result emphasizes the aversion of aromatic compounds for fluorocarbon chains [12,27]. The adsolubilization capacity for azobenzene decreases with the increase in the oxyethylene chain length of NP, contrary to the case of yellow OB. These differences may be attributed to the location of the adsolubilization sites of yellow OB and azobenzene. It is generally said that yellow OB is soluble in the poly(oxyethylene) shell of nonionic surfactant micelles and in the hydrocarbon core of anionic micelles [28], while azobenzene is soluble in the deep region of the poly(oxyethylene) shell for nonionic surfactant micelles and in the hydrocarbon core for the anionic surfactant micelles [29]. The adsolubilization behaviors for yellow OB and azobenzene by these mixed bilayers can be interpreted by a view that the adsolubilized sites are directly correlated with the solubilized sites for the mixed surfactants.

B. Adsolubilization by Addition of Premixed Binary Surfactants

Three binary surfactants systems are selected to study the adsolubilization of 2-naphthol by addition of premixed binary surfactants. They are as follows: SDS–hexaoxyethylenedecylether ($C_{10}E_6$)–alumina [30], hexadecyltrimethylammonium bromide (HTAB)–$C_{10}E_6$–silica, and 1,2-bis(dodecyldimethylammonio)ethane dibromide (2RenQ)–$C_{10}E_6$–silica [31] systems. Since 2-naphthol is soluble in water, adsolubilization experiments are carried out by adding mixed surfactant solutions containing a constant concentration of 2-naphthol into suspensions.

Since aqueous properties of binary surfactant mixtures affect adsorption behavior of mixed surfactants onto particles, mixed critical micelle concentrations for the three systems determined by surface tension measurements are plotted with the molar fraction of ionic surfactant of mixtures (Fig. 13). It is found that the mixed CMCs are very similar to those of the surfactant with lower CMC for the three systems. To estimate the interaction between two surfactants in mixed micelles, the interaction parameter using the regular solution theory [6] is obtained. Their parameters are as follows: −3.4 for the SDS–$C_{10}E_6$ mixed system; −3.4 for the HTAB–$C_{10}E_6$ mixed system; −1.5 for the 2RenQ–$C_{10}E_6$ mixed system. It is suggested that the mutual phobicity between the hydrocarbon chains, as well as the reduction in Coulombic repulsion between the headgroups, dominates the interactions between ionic surfactants and $C_{10}E_6$. In addition, the comparison of the interaction parameter between the HTAB–$C_{10}E_6$ and 2RenQ–$C_{10}E_6$

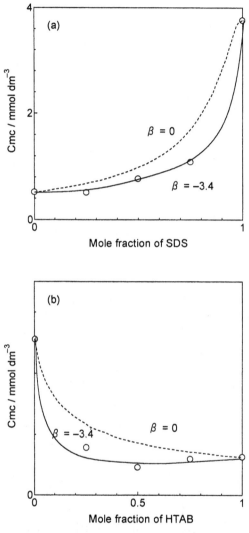

Figure 13 CMCs in for binary surfactant mixtures in 10 mmol/dm^3 NaCl or NaBr at 25°C: (a) SDS/C$_{10}$E$_6$; (b) HTAB/C$_{10}$E$_6$; (c) 2RenQ/ C$_{10}$E$_6$. The plotted points are experimental results; the solid line is the prediction of the regular solution theory; and the dotted line is the prediction for ideal mixing. (Reproduced from Refs. 30 and 31 with permission of the American Chemical Society.)

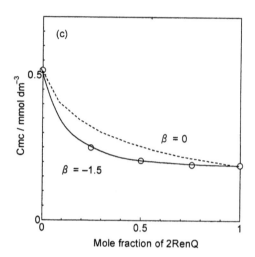

Figure 13 Continued.

systems indicates that the interaction between HTAB and
$C_{10}E_6$ is much stronger than that between 2RenQ and $C_{10}E_6$.

Figure 14 shows the adsolubilized amount of 2-naphthol as
a function of total surfactant equilibrium concentration for the
SDS–$C_{10}E_6$–alumina system and as a function of total initial
surfactant concentration for the HTAB–$C_{10}E_6$–silica and
2RenQ–$C_{10}E_6$–silica systems. Because the adsolubilization of
2-naphthol is not observed both onto alumina and silica
without surfactants, it is apparent that 2-naphthol is incorporated into the surfactant adsorbed layer that exhibits a
hydrophobic property. For the SDS–$C_{10}E_6$–alumina system
the adsolubilized amount of 2-naphthol is very low due to
the adsorption of $C_{10}E_6$ alone, but becomes greater with an
increase in the SDS content of the initial mixtures, from SDS:
$C_{10}E_6 = 1:3$ to $3:1$. The adsolubilized amount of 2-naphthol by
the adsorption of SDS alone ranges between those of
SDS:$C_{10}E_6 = 1:1$ and $3:1$. In the case of the HTAB–
$C_{10}E_6$–silica system, the adsolubilization of 2-naphthol is
appreciable due to the adsorption of $C_{10}E_6$ alone and the
maximum of 2-naphthol adsolubilized is almost the same for all
three different mixed compositions and is slightly greater than
that for HTAB alone. Further, the maximum occurs at the

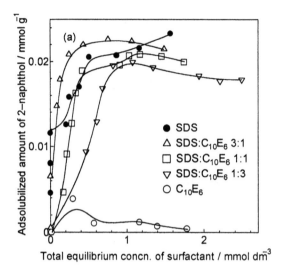

Figure 14 Adsolubilization of 2-naphthol on particles as a function of surfactant concentration: (a) SDS–$C_{10}E_6$–alumina; (b) HTAB–$C_{10}E_6$–silica; (c) 2RenQ–$C_{10}E_6$–silica systems. 0.4 mmol/dm^3 2-naphthol, 10 mmol/dm^3 NaCl or NaBr, 25°C. (Reproduced from Refs 30 and 31 with permission of the American Chemical Society.)

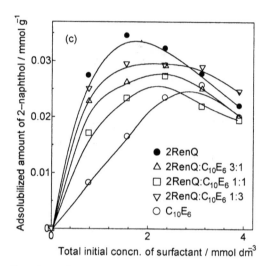

Figure 14 Continued.

lowest initial concentration for $HTAB:C_{10}E_6 = 3:1$ and $1:3$. On the other hand, for the $2RenQ-C_{10}E_6$–silica system, the change in the adsolubilization with the total initial concentration of surfactant from all three different mixed compositions is similar to that of 2RenQ alone, although the magnitude in the adsolubilization is not directly proportional to the mixed compositions of 2RenQ in the mixtures.

To elucidate the effect of mixed surfactant compositions on the adsolubilization of 2-naphthol, the ratios of adsolubilized amount to adsorbed amount of surfactant are plotted with the total adsorbed amount of surfactant in Fig. 15. For the $SDS-C_{10}E_6$–alumina system, the proportions by the adsorption of SDS alone are about 0.1 in a whole SDS adsorption concentration studied, whereas those for $C_{10}E_6$ alone range between 0.02 and 0.06, suggesting that the efficiency of adsolubilization is considerably different between SDS and $C_{10}E_6$ single surfactant adsorption. This difference occurs probably due to the different adsorption states between SDS and $C_{10}E_6$ as well as the different hydrocarbon chain lengths between SDS and $C_{10}E_6$. In the mixed systems, the ratios for the three different mixed compositions are greater than those for SDS alone below 0.2 mmol/g of adsorption. This result can

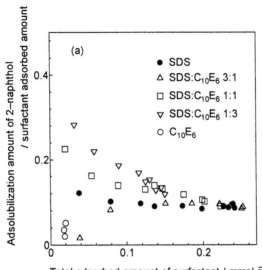

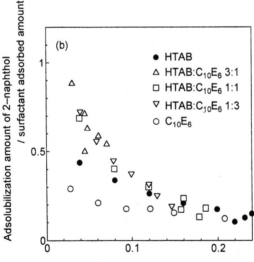

Figure 15 Ratio of adsolubilized amount and surfactant adsorbed amount on particles: (a) SDS–$C_{10}E_6$–alumina; (b) HTAB–$C_{10}E_6$–silica; (c) 2RenQ–$C_{10}E_6$–silica systems. 0.4 mmol/dm^3 2-naphthol, 10 mmol/dm^3 NaCl or NaBr, 25°C. (Reproduced from Refs 30 and 31 with permission of the American Chemical Society.)

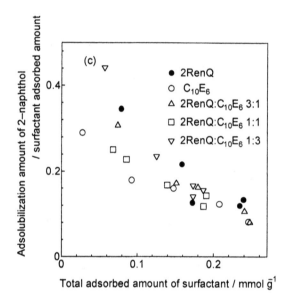

Figure 15 Continued.

be interpreted by a view that the mixed surfactant adsorbed layer is more compact because of a shield of electrostatic repulsion of SDS adsorbed by incorporation of $C_{10}E_6$, in particular for the case of SDS: $C_{10}E_6 = 1:3$. It is interesting to note that the ratio of adsorbed amount of SDS to $C_{10}E_6$ onto alumina is about $1:1$ for SDS: $C_{10}E_6 = 1:3$. In the cases of HTAB–$C_{10}E_6$–silica and 2RenQ–$C_{10}E_6$–silica systems, the ratios of adsolubilized amount of 2-naphthol to adsorbed amount of surfactant for HTAB alone are greater than those for $C_{10}E_6$ alone. Since it is suggested that HTAB adsorbs onto negatively charged silica surface by electrostatic attractive force at low HTAB concentration and as a bilayer due to hydrophobic interaction between HTAB molecules at high HTAB concentration, while $C_{10}E_6$ adsorbs onto silica surface by hydrogen bonding with surface silanol groups, the differences in the ratios may be attributed to the adsorption state and the alkyl chain length between HTAB and $C_{10}E_6$. The ratios for all the three different mixed compositions are greater than that for HTAB alone below 0.15 mmol/g of surfactant adsorption. On the other hand, the ratios for all the three different mixed compositions for 2RenQ–$C_{10}E_6$–silica system are smaller than that for 2RenQ alone. This suggests that the mixed adsorbed layer consisting of 2RenQ and $C_{10}E_6$ is not so compact to enhance the adsolubilization of 2-naphthol. This result coincides with the interaction parameter of the mixed surfactant systems.

Adsolubilization is a very important topic, in particular using mixed surfactants systems. However, to understand these phenomena in detail, many experimental data are still needed in the near future.

REFERENCES

1. D. Stigter, R. J. Williams, and K. J. Mysels, *J. Phys. Chem.* **59**: 330 (1955).

2. A. M. Koganovskii, N. A. Klimenko, and A. A. Tryasorukova, *Colloid J.* **36**: 790 (1974).

3. C. C. Nunn, R. S. Schechter, and W. H. Wade, *J. Phys. Chem.* **66**: 3271 (1982).

4. J. Wu, J. H. Harwell, and E. A. O'Rear, *Langmuir* **3**: 531 (1987).

5. J. H. O'Haver, J. H. Harwell, L. L. Lobban, and E. A. O'Rear, In S. D. Christian and J. F. Scamehorn (eds), *Solubilization in Surfactant Aggregates*, Surfactant Science Vol. 55, Marcel Dekker, New York (1995), Ch. 8.

6. J. H. Harwell, and J. F. Scamehorn, In K. Ogino and M. Abe (eds), *Mixed Surfactant Systems*, Surfactant Science Vol. 46, Marcel Dekker, New York (1993), Ch. 9.

7. N. Nishikido, In S. D. Christian and J. F. Scamehorn, (eds) *Solubilization in Surfactant Aggregates*, Surfactant Science Vol. 55, Marcel Dekker, New York (1995), Ch. 5.

8. K. Meguro, and T. Kondo, *Nippon Kagaku Zasshi* **76**: 642 (1955).

9. K. Esumi, Y. Sakamoto, and K. Meguro, *J. Colloid Interface Sci.* **134**: 283 (1990).

10. K. Esumi, Y. Ono, M. Ishizuka, and K. Meguro, *Colloids Surf.* **32**: 139 (1988).

11. K. Esumi, Y. Sakamoto, K. Yoshikawa, and K. Meguro, *Bull. Chem. Soc. Jpn* **61**: 1475 (1988).

12. P. Mukerjee, *J. Amer. Oil Chem. Soc.* **59**: 573 (1982).

13. J. M. Corkill, J. F. Goodman, and J. R. Tate, *Trans. Faraday Soc.* **62**: 979 (1966).

14. R. Abe, and H. Kuno, *Kolloid-Z.Z. Polym.* **181**: 70 (1962).

15. F. Wolf, and S. Wurster, *Tenside*, **7**: 140 (1970).

16. M. J. Rosen, *Surfactants and Interfacial Phenomena*, Wiley-Interscience, New York (1978), p. 49.

17. K. Meguro, S. Tomioka, N. Kawashima, and K. Esumi, *Prog. Colloid Polym Sci.* **68**: 97 (1983).

18. Th. F. Tadros, *Solid/Liquid Dispersions*, Academic Press, New York/London (1987), p. 266.

19. A. Nakajima, *Bull. Chem. Soc. Jpn* **44**: 3272 (1971).

20. Y. Muto, K. Esumi, K. Meguro, and R. Zana, *J. Colloid Interface Sci.* **120**: 162 (1987).

21. K. Esumi, Y. Sakamoto, T. Nagahama, and K. Meguro, *Bull. Chem. Soc. Jpn* **62**: 2502 (1989).

22. L. Capovila, P. Labbé, and G. Reverdy, *Langmuir* **7**: 2000 (1991).

23. J. F. Scamehorn, R. S. Schechter, and W. H. Wade, *J. Colloid Interface Sci.* **85**: 479 (1982).

24. K. Esumi, Y. Tokui, T. Nagahama, and K. Meguro, *J. Colloid Interface Sci.* **146**: 313 (1991).

25. K. Esumi, H. Otsuka, and K. Meguro, *J. Colloid Interface Sci.* **142**: 582 (1991).

26. P. Somasundaran, and L. Huang, *Adv. Colloid Interface Sci.* **88**: 179 (2000).

27. T. Suzuki, K. Esumi, and K. Meguro, *J. Colloid Interface Sci.* **93**: 205 (1983).

28. F. Tokiwa and K. Tsuji, *Bull. Chem. Soc. Jpn* **46**: 2684 (1973).

29. Y. Nemoto, and H. Funahashi, *Hyomen* **15**: 625 (1977).

30. K. Esumi, N. Maedomari, and K. Torigoe, *Langmuir* **16**: 9217 (2000).

31. K. Esumi, N. Maedomari, and K. Torigoe, *Langmuir* **17**: 7350 (2001).

17

Admicellization and Adsolubilization of Fluorocarbon–Hydrocarbon Mixed Chain Surfactant

MASAHIKO ABE and HIDEKI SAKAI

Faculty of Science and Technology and
Institute of Colloid and Interface Science,
Tokyo University of Science,
Chiba, Japan

SYNOPSIS

Admicellization and adsolubilization of a fluorocarbon/hydro-carbon mixed chain surfactant (hybrid surfactant), 1-oxo-1-[4-(fluoroalkyl)phenyl]-2-alkanesulfonate (FCm–HCn), have been studied. Interfacial properties of FCm–HCn, bearing both a fluorocarbon chain and a hydrocarbon chain in the molecule, have been compared with those of a di-alkyl hydrocarbon surfactant, sodium 1-(4-alkylphenyl)-1-oxo-2-alkanesulfonate (HC*m*–HC*n*). Adsorption isotherms and zeta-potential measurement indicated that FC*m*–HC*n* can adsorb on the aluminum oxide surface at low concentrations compared to the hydrocarbon surfactant. The total amount of FC*m*–HC*n* adsorbed on the aluminum oxide surface was higher than that of HC*m*–HC*n*. In addition, a pyrene fluorescence probe study showed that FC*m*–HC*n* is packed more rigidly in the admicelle than HC*m*–HC*n*.

The FC*m*–HC*n* admicelles were found to adsolubilize fluorocarbons such as fluorobenzene and perfluorobenzene as well as hydrocarbon (2-naphthol). The amount adsolubilized increased in the order, 2-naphthol < fluorobenzene < perfluorobenzene, showing a high ability of FC*m*–HC*n* to solubilize oily compounds containing fluorine. Moreover, FC*m*–HC*n* was found capable of adsolubilizing both hydrocarbon and fluorocarbon simultaneously since both 2-naphthol and perfluorobenzene were solubilized into admicelles of the hybrid surfactant from a mixture of these oily compounds.

I. FLUOROCARBON/HYDROCARBON INTERMOLECULAR MIXED SURFACTANTS

The performance of a mixed surfactant system is often superior to that of a single surfactant system [1,2]. The surfactants used in a multitude of industrial products, processes, and other practical applications almost always consist of a mixture of surfactants. In particular, aqueous mixtures of fluorocarbon and hydrocarbon surfactants have gathered significant attention as a multi-functional system because it can reduce both water/hydrocarbon and water/fluorocarbon interfacial tensions [3–5]. A combination of fluorocarbon surfactant and a suitable hydrocarbon surfactant can produce a degree of wetting which cannot be accomplished by either type of surfactant alone. Normally, in such a combination, it is the fluorocarbon surfactant which reduces the surface tension, while the hydrocarbon surfactant aids in reducing the interfacial tension [3].

On the other hand, unlike mixed hydrocarbon chain surfactants of similar molecular structure, fluorinated surfactant and hydrocarbon chain surfactant do not mix ideally, even when the surfactants have a similar hydrophilic group [6,7]. Mixture of anionic fluorinated surfactant with anionic hydrocarbon surfactant exhibits a positive deviation from the ideal behavior [6]. If positive deviation from ideal mixing in the micelle is significant, the mixed micelles are thermodynamically unstable and separate into two coexisting micelles of different types and compositions.

Recently, investigations have increasingly been conducted on fluorocarbon/hydrocarbon mixed chain surfactants (hybrid surfactants) with both fluorocarbon chain and hydrocarbon chain in the molecule [8–11]. These surfactants are expected to form micelles bearing both lipophilic and fluoro-philic interior in aqueous solutions. In 1992, a new series of hybrid anionic surfactants, such as $C_7F_{15}CH(OSO_3Na)C_7H_{15}$, was synthesized first by Guo *et al.* [8,9]. However, the surfactants were found to slowly hydrolyze in air through the adsorption of moisture and they had to be stored in a desiccator at $-25°C$. They also hydrolyzed slowly in aqueous solution at room temperature, and hence, all physical measurements had to be made within 20 hours after sample preparation. These results, however, gave us good hints for preparation of hydrolysis-resistant hybrid surfactants. Thus, we have synthesized six novel hybrid surfactants (sodium 1-oxo-1-[(4-fluoroalkyl)phenyl]-2-alkane-sulfonate (FCm–HCn; $m = 4, 6$, $n = 2, 4, 6$)) having both a fluorocarbon chain and a hydrocarbon chain with an aromatic ring in a molecule as shown in Fig. 1 [10]. These surfactants are very stable in the presence of moisture due to the resonance between the benzene ring and the carbonyl group and to the bonding structure CSO_3Na at the hydrophilic segment, which is structurally stronger against hydrolysis.

As described in our previous publications [10–17], FCm–HCn shows peculiar solution properties including excellent surface and interfacial tension lowering ability and anomalous concentration- and temperature-dependent viscoelastic behavior observed with their concentrated solutions. These unique surfactants are thus expected to have a unique

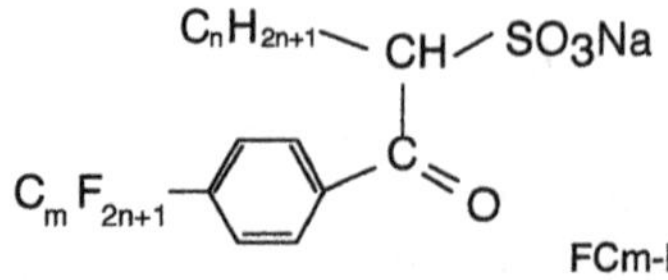

$$C_nH_{2n+1} \diagdown CH \diagup SO_3Na$$

$$C_mF_{2n+1} \diagup \diagdown C \diagdown\!= O$$

FCm-HCn

Sodium 1-oxo-1-[4-(tridecafluorohexyl)phenyl]-2-hexanesulfonate

Figure 1 Chemical structure of 1-oxo-1-[4-(fluoroalkyl)phenyl]-2-alkanesulfonate (FCm–HCn).

adsorption property at solid/liquid interfaces. Furthermore, one can expect that the adsorbed micelles (admicelles) of the novel surfactants will be able to solubilize hydrocarbon oil as well as fluorocarbon oil into their hydrophobic domain because of their two characteristic hydrophobic groups.

This chapter describes the adsorption characteristics of the hybrid surfactant (FCm–HCn) onto a solid/liquid surface and also its solubilization behavior toward organic solutes into the admicelle.

II. ANOMALOUS SOLUTION PROPERTIES OF FLUOROCARBON/HYDROCARBON MIXED CHAIN SURFACTANTS (FCm–HCn)

Many studies on mixed systems of fluorocarbon and hydrocarbon surfactants have shown that extremely heterogeneous micelles are formed in aqueous solution [18–25]. Such heterogeneity may also be reproduced by chemically bound hybrid systems. It is then important to study the aggregation state of the hybrid surfactant micelle as a model of the fluorocarbon-hydrocarbon mixed micelle. The aggregation state of fluorocarbon-hydrocarbon hybrid surfactant micelle has been studied by Guo *et al.* [8] and Inoue *et al.* [20] by means of ^{1}H- and ^{19}F-NMR. Guo *et al.* [8] reported that both fluorocarbon and hydrocarbon chains are incorporated in the interior of the micelle when the hydrocarbon chain bears three or more carbon atoms. Inoue *et al.* [20] studied the surfactants which have a short fluorocarbon chain and a long hydrocarbon

Table 1 Krafft point, CMC, and γ_{CMC} of FCm–HCn.

	Krafft point (°C)	CMC (mol/L)	γ_{CMC} (mN/m)
FC4–HC2	<0	8.2×10^{-3}	24.0
FC4–HC4	<0	3.5×10^{-3}	22.5
FC4–HC6	20	1.2×10^{-3}	18.9
FC6–HC2	<0	8.3×10^{-4}	21.5
FC6–HC4	15	2.3×10^{-4}	20.4
FC6–HC6	48	5.5×10^{-5} (50°C)	16.2 (50°C)

chain, $C_nF_{2n+1}CONHCH_2CH_2N^+(CH_3)_2{}^{C_{16}H_{33}Br}$ ($n = 1, 2, 3$) and showed that the fluoroalkyl groups fully extend their chains to the micellar surface and the terminal CF_3 groups directly face the bulk phase water.

Sodium 1-oxo-1-[(4-fluoroalkyl)phenyl]-2-alkanesulfonate [FCm–HCn; $m = 4, 6$, $n = 2, 4, 6$, (Fig. 1)] with different fluorocarbon and hydrocarbon chains also has very peculiar solution properties. This section briefly introduces these anomalous solution properties of FC6–HC4 before discussion of its adsorption and adsolubilization characteristics. FCm–HCn was synthesized according to our previous reports [10,11].

Table 1 shows the critical micelle concentrations (CMC) of FCm–HCn in aqueous solution obtained by the surface tension method at 25°C. The surface tension of FC6–HC6 solution was measured at 50°C because of the high Krafft point. The FCm–HCn was very effective in lowering surface tension. In addition, the surface tension at CMC (γ_{CMC}) decreased with an increase in fluorocarbon and/or hydrocarbon chain length.

FCm–HCn is also effective in reducing the interfacial tension between water and hydrocarbon oil (octane). As shown in Table 2, the interfacial tension at the CMC ranged from 2.7 mN/m (FC4–HC6) to 9.6 mN/m (FC4–HC2) and decreased with an increase in fluorocarbon and/or hydrocarbon chain length. FCm–HCn also gave lower interfacial tensions between water and a fluorocarbon substance (Table 2). Because of this unique interfacial property, FCm–HCn is able to emulsify three kinds of immiscible liquids (water, hydrocarbon oil, fluorocarbon oil) simultaneously [10]. In addition, aqueous solutions of

Table 2 Interfacial tensions between aqueous FCm–HCn solution and n-octane or between aqueous FCm–HCn solution and perfluoro-n-hexane at 25°C.

	Interfacial tension (mN/m)	
Surfactant aqueous solution	With n-octane (conc of FCm–HCn)	With perfluoro-n-hexane (conc of FCm–HCn)
FC4–HC2	9.6(CMC)	4.0(16 mM)
FC4–HC4	6.2(CMC)	1.7(80 mM)
FC4–HC6	2.7(CMC)	1.2(10 mM)
FC6–HC2	7.7(CMC)	1.5(30 mM)
FC6–HC4	6.4(CMC)	1.9(2 mM)
FC6–HC6	3.4(CMC, 50°C)	

FCm–HCn can float on hydrocarbon substances such as benzene, toluene, xylene, cyclohexane, since the following equation holds due to low the interfacial tension,

$$\gamma_{a/o} > \gamma_{a/w} + \gamma_{o/w} \tag{1}$$

where $\gamma_{a/o}$ and $\gamma_{a/w}$ are the surface tensions of oil and aqueous surfactant solution, respectively, and $\gamma_{o/w}$ is the oil/solution interfacial tension.

Concentrated solutions of FCm–HCn also have attractive solution properties. Above all, aqueous concentrated solution of FC6–HC4 (sodium 1-oxo-1-[4-(tridecafluorohexyl)phenyl]-2-hexanesulfonate) exhibits unusual viscoelastic behavior [14–17]. Figure 2 shows the change observed at the solution surface when the containers with 1, 10, 20 and 30 wt% solutions of FC6–HC4 are tilted from the upright position.

Thus, the surface of all solutions, except for 10 wt% solution, inclined immediately after the containers were tiled. The 10 wt% solution took approximately 5 s before the surface inclines. This solution was also found to have a characteristic of rubber-like elasticity when a glass rod was pulled from the solution. We then analyzed the flow properties and dynamic viscoelasticity of hybrid surfactant solutions using a stress-controlled rheometer. Figure 3 shows the relation between

Figure 2 Change in the surfaces of aqueous 1, 10, 20, and 30 wt% solutions of FC6–HC4 when containers are tilted from a upright position.

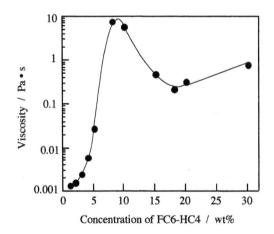

Figure 3 Relationship between viscosity and concentration for FC6–HC4 solution at 25°C.

viscosity and concentration for aqueous FC6–HC4 solutions at 25°C. A maximum appears in the viscosity at a surfactant concentration of about 10 wt% as mentioned previously.

In order to investigate the reason why this hybrid surfactant (FC6–HC4) shows a viscoelasticity when its concentration is around 10 wt%, we have to know changes in

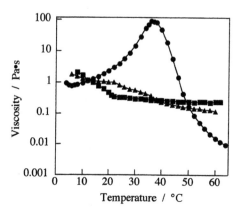

Figure 4 Temperature dependence of the viscosity of an aqueous FC6–HC4 solution: (●) 10 wt%; (■) 20 wt%; (▲) 30 wt%.

aggregating condition of the surfactant. ^{19}F-NMR spectra for FC6–HC4 solutions showed that the four states of association (monomeric, micellar, viscoelastic, and liquid crystalline) are possible for FC6–HC4 molecules in aqueous solutions as their concentration changes and, in particular, in the concentration range where the surfactant exhibits a viscoelastic behavior, it forms molecular assemblies (viscoelastic body) larger in size than micelles in dilute solutions.

The anomalous viscoelasticity behavior of aqueous FC6–HC4 solution observed when the temperature is raised is shown in Fig. 4. The viscosity increased from about 0.9 Pa·s to about 90 Pa·s (10^2 times) as the temperature rose from 6 to 36°C to show a maximum and then decreased to about 0.009 Pa·s at 64°C (10^{-4} times) with further temperature rise when the surfactant concentration was 10 wt%. We called such temperature-dependent viscosity change "thermoresponsive viscoelasticity." The thermoresponsive viscoelasticity was observed with this surfactant at least in the concentration range of 7–15 wt%.

TEM observation of a freeze fracture replica of molecular assemblies in 10 wt% solution of FC6–HC4 showed that small molecular assemblies of about 80 nm in size are clearly observed, in addition to large assemblies of about 500 nm in size. It is a striking fact that molecular assemblies of FC6–HC4 in solution have a nearly spherical shape, instead of thread- or rod-like shape that has been reported to be necessary for molecular assemblies to show a notable viscoelastic behavior [26–28].

FC6–HC4 forms small molecular assemblies (micelles) with its more hydrophobic fluorocarbon chains directing towards the micelle core and its less hydrophobic hydrocarbon chains orienting near the bulk phase in diluted solutions [12]. The 10 wt% concentration of FC6–HC4 is about 1000 times as high as the CMC of the hybrid surfactant obtained in the earlier study [12]. At such high concentrations, strongly hydrophobic fluorocarbon chains are forced to be located at the site far away from the bulk water, i.e., the micelle core, and a large number of small (compact) micelles are formed with their hydrocarbon chains inhabiting outside the benzene rings (in the vicinity of the bulk). As the number of small micelles increases, the frequency of intermicellar collisions also increases to make it easy for small micelles to form large aggregates through the hydrophobic interaction between the intertwined hydrocarbon chains inhabiting near the micelle surfaces.

From what have been described so far, the necessary conditions for double chain hybrid type surfactants to exhibit the thermoresponsive viscoelastic behavior are suggested as follows. Compact micelles of the surfactant are formed with its fluorocarbon chains being incorporated into the micelle core and then the micelles gather to make large molecular assemblies through the hydrophobic interaction between the hydrocarbon chains located near the micelle surface as the surfactant concentration increases. After these conditions are satisfied, the viscoelastic behavior is observed in the process of degradation of the large assemblies when temperature is raised.

III. INTERFACIAL PROPERTIES OF FC*m*–HC*n* AT SOLID/LIQUID INTERFACES

As described in the previous section, fluorocarbon hybrid surfactants (FC*m*–HC*n*) show not only a superior surface

tension reduction capability but also a remarkable interfacial
tension lowering ability. Solubilization into micelles and
adsorption layers (admicelle) of the hybrid surfactants have a
great potential in their new applications in the environment
where organic toxic substances are solubilized into the
hydrophobic domain of the aggregate and removed from an
aqueous stream [29,30]. Having two characteristic hydrophobic
groups, FCm–HCn is expected to be able to solubilize hydrocarbon oil as well as fluorocarbon oil into their hydrophobic
domain. In addition, these unique surfactants are also expected
to have a unique adsorption property at solid/liquid interfaces.

This chapter describes the adsorption characteristics of
the hybrid surfactant (FCm–HCn) onto solid/liquid surface in
detail. We also introduce its solubilization behavior toward
organic solutes into the admicelle.

A. Admicelle Formation of FCm–HCn [31]

1. Adsorption Isotherm

Figure 5 shows the adsorption isotherms of FC6–HC4 and
HC6–HC4 hybrid at 30°C onto the surface of aluminum oxide
powders (Nihon Aerosil, Aluminum Oxide C, specific surface
area $100 \pm 15\,\mathrm{m^2/g}$), giving a comparison between the fluorocarbon hybrid surfactant and the hydrocarbon surfactant of
similar molecular structure (molecular structure of hydrocarbon analogues is shown in Figure 6). The equilibrium concentration used in the figures is equivalent to the concentration
in the supernatant after adsorption. The total amount of
surfactant adsorbed onto the aluminum oxide increased with
increasing equilibrium concentration of the surfactants. In
particular, the adsorption amount of the hybrid surfactant
(FC6–HC4) was much larger than that of hydrocarbon
surfactant (HC6–HC4) at low equilibrium concentrations. It
is well known that anionic surfactant adsorption reaches a
maximum and the surface of alumina is saturated with the
surfactant above its critical micelle concentration [32,33].
However, the adsorption of the hybrid surfactants increased
even in the higher concentration region.

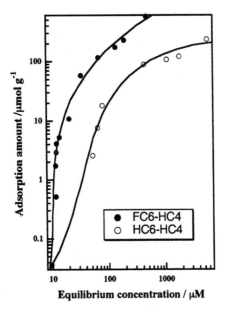

Figure 5 Adsorption isotherms of FC6–HC4 and HC6–HC4 on alumina particles at 30°C.

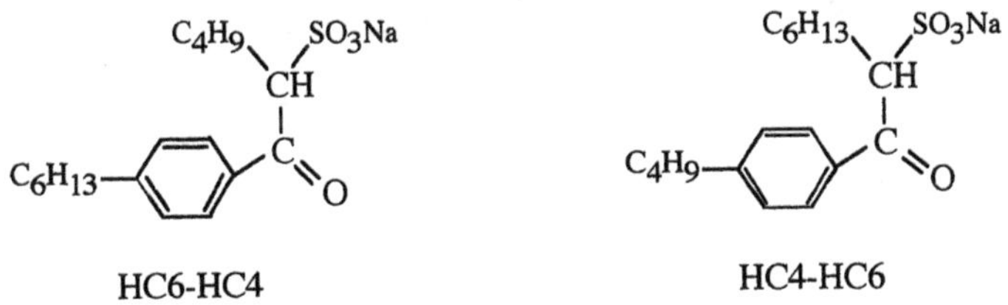

HC6-HC4   HC4-HC6

Figure 6 Chemical structure of FC*m*–HC*n* analogue surfactants (HC*m*–HC*n*).

Figure 7 depicts the zeta-potential of aluminum oxide particles as a function of equilibrium concentration of the surfactants. The zeta-potential in solutions of all surfactants changed its sign from positive to negative one as equilibrium concentration increased. In higher concentration region, the zeta potential reached a constant value and that in solutions of hybrid surfactants was much lower (-34 mV) than that in solutions of hydrocarbon surfactants (-26 mV). The

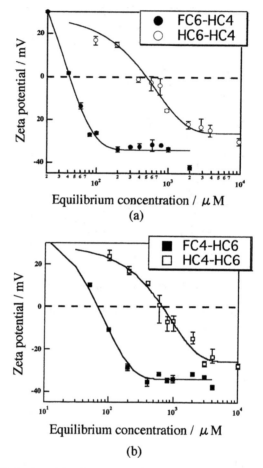

Figure 7 Relationship between the zeta-potential of aluminum oxide particles and equilibrium concentration of FCm–HCn and HCm–HCn at 30°C.

equilibrium concentration for the hybrid surfactants at zero zeta-potential was about one order of magnitude lower than that for the hydrocarbon surfactants. The results are consistent with the fact that both of the hybrid surfactants adsorb on the aluminum oxide surface at lower concentrations than the hydrocarbon surfactants. The fact that the adsorption of the hybrid surfactants continues to increase even above the CMC and shows a constant zeta-potential value indicates that

Table 3 Degree of dissociation of hybrid surfactants (FC-type, HC-type).

FC-type	Degree of dissociation	HC-type	Degree of dissociation
FC6–HC4	0.34	HC6–HC4	0.47
FC4–HC6	0.50	HC4–HC6	0.65

the hybrid surfactant can form multiple layers on the alumina surface and the surfactant molecules orient denser than hydrocarbon surfactants.

The degree of counterion dissociation for FCm–HCn is summarized in Table 3. Both fluorocarbon hybrid surfactants liberated their counterions to a lesser degree than the corresponding hydrocarbon surfactants. Electrostatic repulsion force between adsorbed hybrid surfactant molecules on the aluminum oxide surface will be smaller than that for hydrocarbon surfactants as a result of lower counterion dissociation. Then, the hybrid surfactants can adsorb tightly and the total amount of adsorption is higher than that of the hydrocarbon surfactants. This also explains why the zeta-potential of aluminum oxide surface with hybrid surfactant is more negative than that of hydrocarbon surfactants.

Contact angles of the surfactant solution on an aluminum oxide plate are plotted as a function of surfactant concentration in Figs 8(a) and (b). Contact angles of both fluorocarbon and hydrocarbon surfactants decreased with concentration increase and FCm–HCn surfactants showed much lower contact angles than those of HCm–HCn over the whole concentration range. This may be attributed to the fact that the affinity of fluorocarbon hybrid surfactants to aluminum oxide surface is relatively high and the surfactants can adsorb at lower concentrations.

2. Microenvironment in FCm–HCn admicelles

The fluorescence spectrum of pyrene molecules has five peaks and the ratio of the intensities of the first (375 nm) to the third (386 nm) peaks, I_1/I_3, is well known to be almost proportional to

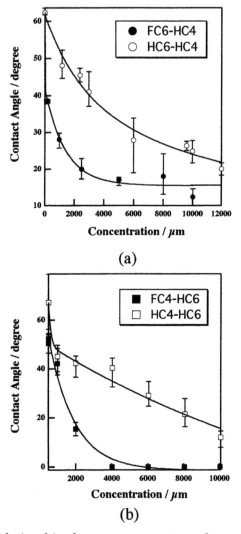

Figure 8 Relationship between contact angle and FC*m*–HC*n* concentration.

the polarity in the region near the pyrene molecules solubilized in the aggregates [34]. The ratio decreases with increase in the hydrophobic environment. Then, the micropolarity in surfactant aggregates such as micelles, vesicles, and liposomes can be monitored by measuring the ratio. In order to confirm that

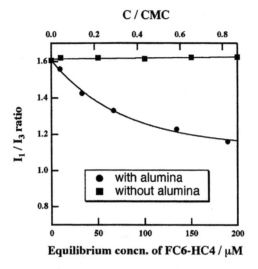

Figure 9 I_1/I_3 ratio of pyrene fluorescence in aqueous FC6–HC4 solution with and without aluminum oxide particles.

pyrene molecules are solubilized into admicelles on the aluminum oxide surface, the experiment was performed below the CMC. Figure 9 is a plot of the I_1/I_3 ratio for pyrene as a function of the equilibrium concentration of FC6–HC4 and the ratio to the CMC, C. Without aluminum oxide particles, the I_1/I_3 ratio was constant, indicating no micelle in the solution. The I_1/I_3 ratio decreased with increase in the surfactant equilibrium concentration in the presence of aluminum oxide particles, showing that pyrene molecules are solubilized in the admicelle and the hydrophobic environment around these pyrene molecules increases with surfactant concentration. The same trend was observed for HC6–HC4 system.

In order to compare the micropolarity in the inside of the membrane, the I_1/I_3 ratios for FC6–HC4 and HC6–HC4 systems were compared with each other as a function of the surfactant amount adsorbed on alumina particles. The I_1/I_3 ratio for the HC6–HC4 system was slightly lower than that for FC6–HC4 system, indicating that the polarity inside the membrane of the fluorocarbon hybrid surfactant is higher than that of the normal hydrocarbon surfactant.

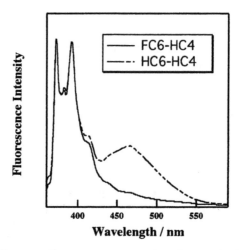

Figure 10 Pyrene fluorescence spectra in surfactant-aluminum oxide solution. (equilibrium concentration is below CMC at 30°C).

Figure 10 shows the pyrene fluorescence spectra in surfactant/aluminum oxide solution for FC6–HC4 and HC6–HC4 systems at 30°C. Typical five peaks in the pyrene fluorescence spectra were seen for the FC6–HC4 system, while the HC6–HC4 system showed an additional peak around 470 nm. This is due to excimer formation of pyrene molecules. No excimer formation of pyrene was observed over a wide range of pyrene concentrations in the FC6–HC4 system. Excimer formation depends on the microfluidity around pyrene molecules. Since pyrene is solubilized into the surfactant membrane on the aluminum oxide surface, excimer formation can be regarded as a result of a higher fluidity inside the membrane of the hydrocarbon surfactant. In contrast, the fluorocarbon hybrid surfactant seems to have a lower microfluidity in the membrane, indicating a packing condition for the fluorocarbon surfactant more rigid than that for the hydrocarbon surfactant. In fact, because the degree of counterion dissociation for the fluorocarbon hybrid surfactant is lower than that for the hydrocarbon surfactant, the packing of fluorocarbon surfactants in the membrane can be tighter on the aluminum oxide surface.

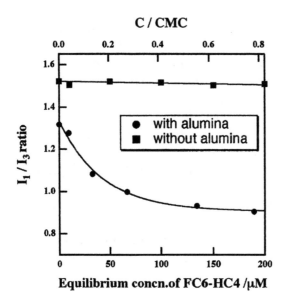

Figure 11 I_1/I_3 ratio of 1-(perfluorooctanoyl)pyrene fluorescence in aqueous FC6–HC4 solutions.

1-Perfluorooctanoylpyrene (abbreviated hereafter to F-Py) is known as a hydrophobic fluorescence probe that allows to evaluate the environment of fluorine atom through changes in the fluorescence intensity ratio, I_1/I_3, in the spectrum since the probe has a pyrenyl group and a fluorocarbon chain in its molecule [35]. Hence, application of this fluorescence probe to admicelles of the hybrid surfactant will provide information about the fluorocarbon region in the interior of the admicelles. Figure 11 shows the results of fluorescence measurements using the fluorescence probe on adsorbed FC6–HC4 admicelles. As a result, no change in the I_1/I_3 ratio for F-Py was observed for solutions containing the surfactant alone, whereas the ratio decreased with increasing equilibrium concentration for surfactant solutions with alumina particles.

Roughly, three differences are found when a comparison is made between the result on the adsorbed system with pyrene in Fig. 9 and that with F-Py in Fig. 11. First, a reduction in the I_1/I_3 ratio was observed in the latter case at zero surfactant concentration in the presence of alumina. This would be brought about by adsorption of F-Py with a carboxyl group on the positively charged alumina surface. Second, the ratio started decreasing at lower surfactant concentrations in the latter case than in the former case. This would suggest a higher solubilization of F-Py into the interior of admicelles than pyrene. Third, the ratio finally reached a lower value in the latter case than in the former case as the equilibrium surfactant concentration was increased, suggesting that F-Py is solubilized in the strongly hydrophobic region (fluorocarbon region) of the admicelles. The experimental results mentioned above imply that admicelles of the hybrid surfactant have both hydrocarbon and fluorocarbon regions in their inside. Hence, the admicelles can adsolubilize both hydrocarbon oils and fluorocarbon oils.

3. Characteristics of FC6–HC4 as a Dispersant

The effect of FC6–HC4 on the dispersibility of alumina particles in an aqueous solution was evaluated. The dispersibility of alumina particles in FC6–HC4 solutions was compared with that in solutions of HC6–HC4 and of a typical hydrocarbon anionic surfactant, SDS. As shown in Fig. 12, alumina particles were dispersed in water without surfactant (a). Increase in the concentration of each surfactant first caused flocculation of the particles (b), then resulted in their redispersion above a certain concentration (c). The concentration range in which the particles are flocculated was narrower and the concentration above which the particles are redispersed was lower for FC6–HC4 than those for HC6–HC4 and SDS (Fig. 13).

Dispersion stability of alumina particles with and without surfactants was evaluated by pursuing the time course of change in turbidity. Alumina particles with FC6–HC4 adsorbed were shown to have an excellent dispersibility in aqueous solution compared to those with adsorbed HC6–HC4 and SDS (Fig. 14). The zeta potentials of alumina particles covered with FC6–HC4 and HC6–HC4 (at saturation, see Fig. 7) were $-34\,\mathrm{mV}$ and $-26\,\mathrm{mV}$, respectively. Thus, more negative surface

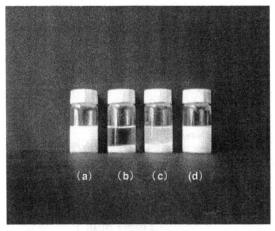

Figure 12 Change in dispersion stability of alumina particles dispersed in FC6–HC4 solutions as a function of concentration. (a) 0 mM, (b) 2.0 mM, (c) 3.0 mM, (d) 4.0 mM.

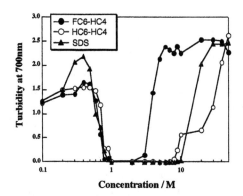

Figure 13 Change in turbidity at 700 nm of alumina suspension in the presense of surfactants as a function of concentration.

charge on the particles with FC6–HC4 caused a better dispersion stability.

B. Adsolubilization of hydrocarbon and fluorocarbon oils

Recently, solubilization by surfactant molecular aggregates, particularly that by admicelles formed at solid–liquid interfaces

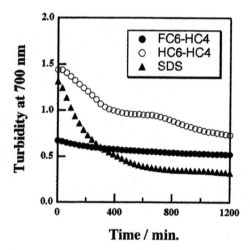

Figure 14 Change in turbidity at 700 nm of alumina particles dispersed in surfactant solutions as a function of time.

using the hydrophobic region in the interior of admicelles has attracted significant attention because it can be used in the recovery of organic wastes from aqueous solutions and as an ultrasmall reaction field for thin film preparation. Fluorinated surfactants are generally good in lowering the surface tension of water while few of them reduce the interfacial tension between oil and water to a great extent [44]. By contract, fluorocarbon–hydrocarbon hybrid surfactants can remarkably lower not only the surface tension of water but the interfacial tension between oil and water and have an excellent property of coemulsifying hydrocarbon oil and fluorocarbon oil. We can then expect cosolubilization of hydrocarbon oil and fluorocarbon oil to occur in admicelles of the hybrid surfactant. In this section, we describe the adsolubilization characteristics of hydrocarbon oil and fluorocarbon oil from their mixture into admicelles of FC6–HC4.

First, solubilization of 2-naphthol into FC6–HC4 admicelles was investigated. Figure 15 shows the relation between the adsorbed amount of FC6–HC4 and the adsolubilized amount of 2-naphthol. With increasing adsorbed amount of the hybrid surfactant (FC6–HC4), the adsolubilized amount of

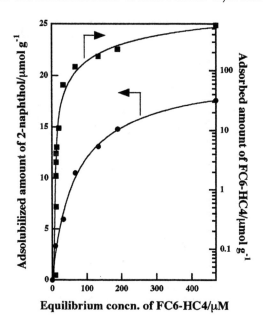

Figure 15 Changes in adsolubilized amount of 2-naphthol and adsorbed amount of FC6–HC4 as a function of equilibrium concentration.

2-naphthol increased, indicating that the solubilized amount of 2-naphthol increases as admicelles are formed. On the other hand, 2-naphthol was also found to be solubilized in admicelles of HC6–HC4, which is similar in chemical structure to FC6–HC4. In view of this, a comparison was made on the adsolubilized amount of 2-naphthol between FC6–HC4 and HC6–HC4. The comparison revealed that FC6–HC4 solubilizes 2-naphthol less than HC6–HC4 at the same adsorbed amount. This is reasonable because fluorocarbon chains in molecules of the hybrid surfactant are lipophobic.

Solubilization of fluorocarbon oils was examined into admicelles of the hybrid surfactant using fluorobenzene with a fluorine atom in the molecule and perfluorobenzene having six fluorine atoms in the molecule as the oily compounds. As a result, both of the compounds were found to be solubilized into admicelles of the hybrid surfactant, showing that the

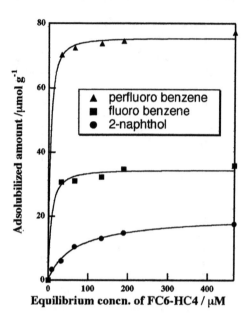

Figure 16 Adsolubilized amounts of hydrocarbon and fluorocarbon compounds in FC6–HC4 admicelles.

surfactant can efficiently adsolubilize hydrocarbon oils as well as fluorocarbon oils (Fig. 16). Comparison of saturated amounts of adsolubilization showed that fluorobenzene is solubilized about two times more than 2-naphthol and perfluorobenzene is solubilized over four times more than 2-naphthol. Thus, the adsolubilized amount for these oily compounds increased in the order, 2-naphthol < fluorobenzene < perfluorobenzene, being indicative of a high ability of the hybrid surfactant to solubilize fluorine-containing oily compounds. In addition, the adsolubilization amount of fluorocarbon oils with the FC6–HC4 reached a plateau value at the concentration below its CMC, while that of hydrocarbon oils did not. This result suggests that fluorocarbon chains of FC6–HC4 molecules form fluorocarbon–philic environment even at lower concentrations than its CMC.

Solubilization of fluorocarbon oils into HC6–HC4 admicelles was then examined to see whether the high ability of the

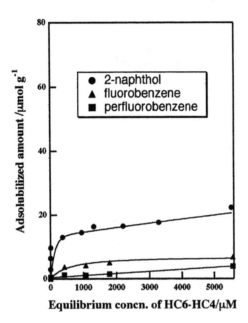

Figure 17 Adsolubilized amounts of hydrocarbon and fluorocarbon compounds in HC6–HC4 admicelles.

hybrid surfactant to solubilize fluorine-containing oils is due to the fluorine environment of the surfactant or due to very low solubilities of fluorocarbon oils in water. As a result, either flurobenzene or perfluorobenzene was found to be hardly adsolubilized as shown in Fig. 17. In fact, the solubilized amounts of the fluorocarbons in HC6–HC4 admicelles were about one tenth of those in adsorbed FC6–HC4 admicelles (Fig. 16). This would suggest that the high ability of the hybrid surfactant to solubilize fluorocarbon oils arises from high affinity of the fluorocarbon oils to the fluoroalkyl chains of the hybrid surfactant rather than the low solubility in water of these oils.

An attempt was made to see if admicelles of the hybrid surfactant can cosolubilize 2-naphthol, a hydrocarbon, and perfluorobenzene, a fluorocarbon, since the surfactant was found to solubilize both hydrocarbon oils and fluorocarbon oils as shown in Fig. 16. Then, an interesting result was obtained

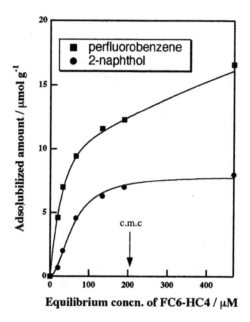

Figure 18 Cosolubilized amounts of 2-naphthol and perfluorobenzene in FC6–HC4 admicelles.

showing that the two oily compounds are actually cosolubilized
by admicelles of the hybrid surfactant (Fig. 18). The simultaneously adsolubilized amounts of 2-naphthol and perfluorobenzene were respectively about half and fifth the separately
adsolubilized amounts of the oily compounds (see Fig. 18). This
would suggest that the solubilization sites for 2-naphthol and
those for perfluorobenzene overlap each other in the inside of
admicelles of the hybrid surfactant. This would mean that the
hydrocarbon and fluorocarbon regions are somewhat close to
each other in admicelles of the hybrid surfactant.

     Comparison of the adsolubilization curves in Fig. 18 for
the single solubilizates and those in Fig. 18 for the mixed
solubilizates reveals that even though a large increase in the
adsolubilized amount of perfluorobenzene started from a low
equilibrium surfactant concentration in the single solubilizate
system, only moderate increases were observed in the mixed
solubilizate system, while no appreciable difference in the

adsolubilized amount of 2-naphthol was seen between the two systems. This would imply that 2-naphthol (lipophilic) is preferably solubilized and perfluorobenzene (lipophobic) is much less solubilizable in admicelles of the hybrid surfactant. The experimental findings mentioned so far would indicate that admicelles of the hybrid surfactant have both hydrocarbon and fluorocarbon regions, thereby enabling them to cosolubilize hydrocarbon oil and fluorocarbon oil.

REFERENCES

1. J. F. Scamehorn (ed.), *Phenomena in Mixed Surfactant Systems*, ACS Symposium Series 311, American Chemical Society, Washington, D.C. (1986).

2. K. Ogino, and M. Abe (eds), *Mixed Surfactant Systems*, Marcel Dekker, New York (1993).

3. N. Funasaki, In K. Ogino and M. Abe, (eds), *Mixed Surfactant System*, Marcel Dekker, New York (1993), p. 145.

4. P. Mukerjiee, and A. Y. S. Yang, *J. Phys. Chem.* **80**: 1388 (1976).

5. G. Perron, R. De Lisi, I. Davidson, S. Gerreux, and B. T. Ingram, *J. Colloid Interface Sci.* **79**: 432 (1981).

6. E. Kissa, *Fluorinated Surfactants, Synthesis, Properties, Applications*. Marcel Dekker, New York (1993).

7. P. Mukerjee and K. J. Mysels, In K. L. Mittal (ed.), *Colloidal Dispersions and Micellar Behavior*, p. 239. ACS Symposium Series, No. 9, American Chemical Society, Washington, D.C. (1975).

8. W. Guo, Z. Li, B. M. Fung, and E. A. O'Rear, and J. H. Harwell, *J. Phys. Chem.* **96**: 6738 (1992).

9. W. Guo, B. M. Fung, and E. A. O'Rear, *J. Phys. Chem.* **96**: 10068 (1992).

10. N. Yoshino, K. Hamano, T. Omiya, Y. Kondo, A. Ito, and M. Abe, *Langmuir* **11**: 466 (1995).

11. M. Abe, and N. Yoshino, In D. R. Karsa (ed.), *Fluorocarbon-Hydrocarbon Hybrid Surfactants*, New Products and Applications in Surfactant Technology. Academic Press, London (1998).

12. A. Ito, K. H. Sakai, Y. Kondo, N. Yoshino, and M. Abe, *Langmuir* **12**: 5768 (1996).

13. A. Ito, K. Kamogawa, H. Sakai, K. Hamano, Y. Kondo, N. Yoshino, and M. Abe, *Langmuir* **13**: 2935 (1997).

14. M. Abe, K. Tobita, H. Sakai, Y. Kondo, N. Yoshino, Y. Kasahara, H. Matsuzawa, M. Iwahashi, N. Momozawa, and K. Nishiyama, *Langmuir* **13**: 2932 (1997).

15. K. Tobita, H. Sakai, Y. Kondo, N. Yoshino, M. Iwahashi, N. Momozawa, and M. Abe, *Langmuir* **13**: 5054 (1997).

16. K. Tobita, H. Sakai, Y. Kondo, N. Yoshino, K. Kamogawa, N. Momozawa, and M. Abe, *Langmuir* **14**: 4753 (1998).

17. K. Tobita, H. Sakai, Y. Kondo, N. Yoshino, K. Kamogawa, N. Momozawa, and M. Abe, *Material Technology* **16**: 202 (1998).

18. P. Mukerjee, and A. Y. S. Yang, *J. Phys. Chem.* **80**: 1388 (1976).

19. K. Shinoda, and T. Nomura, *J. Phys. Chem.* **84**: 365 (1980).

20. N. Funasaki, and S. Hada, *J. Phys. Chem.* **84**: 736 (1980).

21. J. Carlfors, and P. Stilbs, *J. Phys. Chem.* **88**: 4410 (1984).

22. T. Asakawa, S. Miyagishi, and M. Nishida, *J. Colloid Interface Sci.* **104**: 279 (1985).

23. Y. Muto, K. Esumi, K. Meguro, and R. Zana, *J. Colloid Interface Sci.* **120**: 162 (1987).

24. K. Kalyanasundaram, *Langmuir* **4**: 942 (1988).

25. S. J. Burkitt, B. T. Ingram, and R. H. Ottewill, *Prog. Colloid Polym. Sci.* **76**: 247 (1988).

26. T. Imae, R. Kamita, and S. Ikeda, *J. Colloid Interface Sci.* **99**: 300 (1984).

27. T. Imae, R. Kamita, and S. Ikeda, *J. Colloid Interface Sci.* **108**: 215 (1985).

28. K. Hashimoto, and T. Imae, *Langmuir* **7**: 1734 (1991).

29. S. D. Christian, and J. F. Scamehorn (eds), *Solubilization in Surfactant Aggregates*. Marcel Dekker, New York (1995).

30. E. A. Lissi, E. B. Abuin, and A. M. Roacha, *J. Phys. Chem.* **84**: 2406 (1980).

31. M. Abe, A. Saeki, K. Kamogawa, H. Sakai, Y. Kondo, H. Uchiyama, and J. H. Harwell, *Ind. Eng. Chem. Res.* **39**: 2697 (2000).

32. M. Fujie, *J. Jpn. Oil Chem. Soc.* **45**: 181 (1996).

33. D. Bitting, and J. H. Harwell, *Langmuir* **3**: 500 (1987).

34. H. Terada, and T. Yoshimura, *Liposome*. Springer Verlag, Tokyo (1992), p. 136.

35. M. Li, M. Jiang. Y. Zhang, and Q. Fang, *Macromolecules* **30**: 470 (1997).

36. A. Saeki, K. Kamogawa, Y. Kondo, N. Yoshino, H. Sakai, and M. Abe, *Material Technology* **18**: 158 (2000).

37. A. Saeki, H. Sakai, K. Kamogawa, Y. Kondo, N. Yoshino, H. Uchiyama, J. H. Harwell, and M. Abe, *Langmuir* **16**: 9991 (2000).

38. V. Monticone, and C. Treiner. *J. Colloid Interface Sci.* **166**: 394 (1994).

39. K. Esumi, M. Goino, and Y. Koide, *J. Colloid Interface Sci.* **183**: 539 (1996).

40. K. Esumi, S. Uda, M. Goino, K. Ishiduki, T. Suhara, H. Fukui, and Y. Koide, *Langmuir* **13**: 2803 (1997).

41. K. Esumi, M. Goino, S. Uda, H. Fukui, and Y. Koide, *J. Jpn Soc. Colour Mater.* **71**: 494 (1998).

42. K. Esumi, S. Uda, T. Suhara, H. Fukui, and Y. Koide, *J. Colloid Interface Sci.* **166**: 394 (1994).

43. K. Esumi, W. Uda, T. Suhara, H. Fukui, and Y. Koide, *J. Colloid Interface Sci.* **193**: 315 (1997).

44. M. Abe, and N. Yoshino, *J. Jpn. Oil Chem. Soc.* **45**: 991 (1996).

18

Precipitation of Surfactant Mixtures

JOHN F. SCAMEHORN and
JEFFREY H. HARWELL
Institute for Applied Surfactant Research,
University of Oklahoma, Norman,
Oklahoma, U.S.A.

SYNOPSIS

Precipitation of surfactants in aqueous solution is a phenomenon of great practical importance in such applications as detergency and petroleum production using surfactants. This chapter discusses both ionic surfactant precipitation by electrolytes and precipitation of anionic surfactants by cationic surfactants. Precipitation of anionic surfactants by added monovalent cations and divalent cations or by decreasing temperature is discussed. In all of these precipitation reactions, the presence of micelles decreases the tendency for precipitation to occur since precipitation can be viewed as a competition for monomeric surfactant between micelles and precipitate. Thus, the tendency of anionic surfactant to precipitate is reduced when nonionic surfactant is added to the system. The concentration of cationic surfactant required to precipitate an anionic surfactant (or vice versa) is shown to increase above the critical micelle concentration of the system. Addition of nonionic surfactant to these anionic–cationic surfactant mixtures reduces the tendency for precipitation to occur. Precipitation rates are also discussed. The rate of precipitation is shown to be on the order of several minutes to reach equilibrium for pure anionic surfactant precipitated by calcium and on the order of 30 min for anionic–cationic surfactant mixtures. In both cases, there is a period of inhibition during which little precipitation occurs. The more supersaturated a system is, the faster the precipitation reaches its equilibrium level. Mixtures of anionic surfactants can take substantially longer to precipitate than single surfactants.

I. INTRODUCTION

A. Overview

Surfactant precipitation from aqueous solutions can be deleterious in applications such as detergency. Examples are shampoos [1], or powdered laundry detergents which can be composed of builder to sequester multivalent cations, preventing these ions from precipitating anionic surfactant present [2]. On the other hand, precipitation of anionic and cationic surfactants in oil reservoirs is the basis of an improved waterflooding technology to enhance crude oil recovery [3]. Also, recovery of surfactants from surfactant-based separation processes can be achieved by precipitating the surfactant [4,5]. In concentrated surfactant systems the transition from liquid crystal to gel is related to the precipitation boundary in such a way that decreasing the tendency to precipitate decreases the tendency to form gels. In any of these and other technologies, understanding and manipulating surfactant precipitation is of great industrial importance. The use of surfactant mixtures permits considerable manipulation of precipitation behavior.

More generally, surfactant mixtures can have a number of synergistic advantages over the use of a single surfactant type [6].

This chapter is a qualitative discussion of the basic principles of precipitation in mixed surfactant systems. Mathematical models to describe these systems are given in original literature articles and are not repeated here. The purpose of this chapter is to give the reader an understanding of the basic principles of surfactant precipitation, not to present an exhaustive review of the literature on the topic.

B. Phase Boundaries and Krafft Points

There are two general approaches to presenting surfactant precipitation data in the literature: phase boundaries or Krafft points (sometimes called Krafft temperatures). The term "surfactant solubility" is used here to refer to the tendency of a charged surfactant to precipitate. Laughlin [1] uses the term more broadly to also include the tendency of surfactants to form liquid crystalline phases. Sometimes, dissolution of solid nonionic surfactant is included under the term surfactant solubility. While this is not unreasonable, we also do not include these phenomena in this discussion.

A phase boundary represents the minimum or maximum concentration of an additive required to form an infinitesimal amount of precipitate in the aqueous surfactant solution at constant temperature at various surfactant concentrations. The boundary separates concentration regimes in which precipitation occurs at equilibrium from regimes where no precipitate is present.

The Krafft point is the temperature at which the solubility of hydrated surfactant crystals increases sharply with increasing temperature. This increase is so sharp that the solid hydrate dissolution temperature is essentially independent of concentration above the critical micelle concentration (CMC) and is, therefore, often called the Krafft point without specifying the surfactant concentration. In theory, the Krafft point is at the CMC exactly, but one frequently used arbitrary concentration which is recommended for measuring Krafft points is 1 wt% surfactant.

The phase boundary approach involves forcing a surfactant to precipitate by adding another compound to solution, whereas the Krafft point involves forcing the precipitation by cooling the solution. Historically, the Krafft temperature was the most popular means of quantifying the tendency of surfactants to precipitate. Phase boundaries are a more popular method of data presentation in recent literature because isothermal results are easier to model and are more useful, particularly in mixed surfactant systems. Temperature effects can then be incorporated in model parameters derived from isothermal phase boundaries. However, Krafft points are still widely reported.

An important point about equilibrium phase boundaries in precipitating surfactant systems is illustrated by Fig. 1 which shows the precipitation phase boundary of an anionic

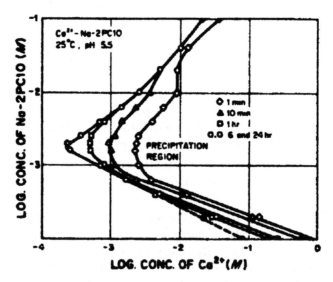

Figure 1 The Ca^{2+}/sodium p-(1-methylnonyl)benzene sulfonate precipitation domain at pH 5.5 and 25°C, 1 min ($\diamond$), 10 min ($\triangle$), 1 hr ($\square$), and 6 and 24 hr ($\bigcirc$) after mixing the reacting components ($\bullet$). Data corrected for activities of Ca^{2+} ion. (From Ref. 7.)

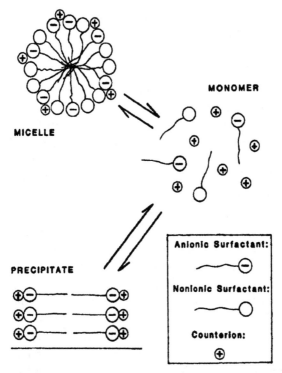

MONOMER

MICELLE

PRECIPITATE

Anionic Surfactant:

Nonionic Surfactant:

Counterion:

Figure 2 Schematic of the equilibrium in a micellar system containing an anionic and nonionic surfactant, where the anionic surfactant is being precipitated by monovalent counterion. (From Ref. 12.)

surfactant [sodium p-(methylnonyl)benzene sulfonate] with calcium as a function of time [7]. Clearly, the phase boundary changes substantially over an extended period of time, indicating how slowly these systems can take to reach equilibrium. There is evidence that surfactant solutions can be clear for long periods of time at constant temperature before precipitate begins to form [8–11]. In our laboratory, to avoid this, we cool a surfactant solution to force it to precipitate, then place it in a heated water bath and observe if the crystals dissolve after a few days to map a phase boundary [12–18]. Section IV expands this discussion of kinetics of surfactant precipitation.

II. ANIONIC SURFACTANT PRECIPITATION AT EQUILIBRIUM

A. Introduction

Hardness tolerance of an anionic surfactant is defined as the minimum concentration of multivalent cation necessary to cause precipitation of the surfactant. Salinity tolerance is defined as the minimum concentration of monovalent cation necessary to cause precipitation of the surfactant. Hardness tolerance of a detergent mixture can limit effective cleaning using anionic surfactants in hard water (water containing a high concentration of calcium and/or magnesium), often requiring the addition of a builder to a detergent formulation [2]. Builders are compounds that reduce the activity of hardness ions by complexing them. For example, sodium tripolyphosphate is a cheap, effective builder. Salinity tolerance is generally of less importance than hardness tolerance, but surfactant flooding of extremely saline oil reservoirs can be limited by precipitation of the sodium salts of the surfactant. If an additive increases the salinity or hardness tolerance of an anionic surfactant solution (a higher concentration of added electrolyte required to cause precipitation), this also means that the Krafft point decreases, which is to say that it requires cooling to a lower temperature at constant electrolyte concentration to initiate precipitation.

Consider a completely disassociated monovalent anionic surfactant such as dodecyl sulfate. If the sodium concentration in solution is increased enough at constant temperature, hydrated crystals of sodium dodecyl sulfate will precipitate from solution. This minimum sodium concentration is the salinity tolerance under those conditions. If a divalent cation (e.g., calcium) were added to the solution, the minimum calcium concentration required to cause the surfactant to precipitate as calcium dodecyl sulfate is the hardness tolerance. This would be true if the anionic surfactant were not completely disassociated (e.g., a fatty acid) as long as the counterion precipitating with the surfactant anion were monovalent for salinity tolerance and divalent for hardness tolerance.

If precipitation were forced by reducing the temperature, the salinity tolerance at the Krafft point would be the sodium concentration in solution if sodium dodecyl sulfate were the precipitate and the hardness tolerance would be the calcium concentration if calcium dodecyl sulfate were the precipitate.

B. Salinity Tolerance in Anionic–Nonionic Surfactant Systems

Figure 2 is a schematic of an anionic–nonionic surfactant system where a monovalent electrolyte is forming precipitate with the anionic surfactant [12]. There is an equilibrium between the monomeric or unassociated anionic surfactant, the anionic surfactant in micelles, and the precipitated anionic surfactant. Likewise, there is an equilibrium between the unbound or unassociated counterion, the bound counterion, and the counterion in the precipitate. The nonionic surfactant is present as a monomer or in micelles.

The surfactant anion activity is often approximately proportional to the surfactant monomer concentration. Likewise, the activity of the counterion is approximately proportional to the unbound counterion concentration in solution. At equilibrium, surfactant precipitation will occur if the product of the surfactant anion activity and the counterion activity equals the solubility product of the surfactant salt (it cannot exceed the solubility product since additional surfactant will precipitate to prevent this). Whether precipitation occurs or not depends on monomer-micelle equilibrium since, to a good approximation, the surfactant present as micelles does not contribute to surfactant anion activity and the bound counterion does not contribute to counterion activity. Qualitatively, the micelles can be considered to be in competition with precipitate for monomer and counterions. At constant total surfactant concentration, the more easily micelles form and the higher counterion binding is on the micelles, the lower the surfactant and counterion activity or concentration is, and the more difficult (higher added salt or lower temperatures) it is for precipitation to occur. This simultaneous equilibrium is shown in Fig. 2. It is a serious mistake to treat a surfactant solution above the CMC as if the surfactant and counterion were present in a single homogenous state.

The key to increasing salinity tolerance is to enhance micelle formation, which decreases monomer concentration. The addition of nonionic surfactant to the system can achieve this synergism because anionic–nonionic mixed micelles form more easily than pure anionic surfactant micelles of the same anionic surfactant. This is because the nonionic surfactant hydrophilic groups insert themselves between the anionic surfactant hydrophilic groups without the electrostatic repulsion that exists between these charged head groups. This reduces the charge density and electrical potential at the micelle surface thus reducing the work required to bring the surfactant ion to the surface of the micelle. The fractional counterion binding on anionic–nonionic mixed micelles (bound counterions/anionic surfactants in micellar form) decreases slowly with increasing nonionic content [19,20], so the addition of low mole fractions of nonionic surfactant does not substantially increase the concentration of unbound counterions; in fact, these may decrease due to increased concentration of micellized anionic surfactant. Cationic surfactants can achieve these effects in mixed micelles with anionic surfactants to an even greater degree, but the anionic–cationic surfactant pair can precipitate easily (as will be discussed in Section III), so the addition of cationic surfactants to increase salinity tolerance of anionic surfactants is risky and only a small concentration of cationic surfactant can be used.

The CMC is the minimum surfactant concentration at which micelles form. The addition of nonionic surfactants (which can have very low CMC values) to anionic surfactants can reduce the CMC of the surfactant mixture to levels much lower than that of the pure anionic surfactant. This is illustrated in Fig. 3, where the CMC of a mixture of sodium dodecyl sulfate (SDS) and a nonylphenol polyethoxylate or $NP(EO)_{10}$ (with a degree of polymerization of 10) is shown in a 0.03 M NaCl solution [12]. If the monomer mole fraction of the SDS in equilibrium with micelles is reduced from 1.0 to 0.9, the CMC is reduced by about an order of magnitude. This would reduce the SDS monomer concentration, increasing the concentration

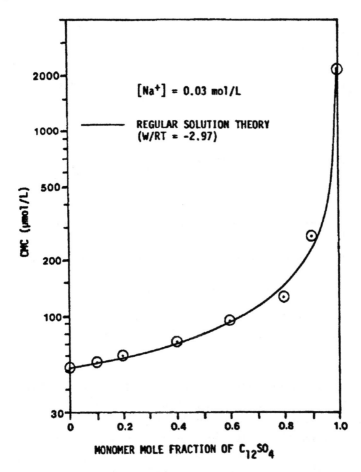

Figure 3 The CMC of an anionic–nonionic surfactant mixture (SDS/NP(EO)$_{10}$). (From Ref. 12.)

of added sodium required to cause precipitation. In a qualitative view, in the diagram shown in Fig. 2, the enhancement of the mixed micelle formation by insertion of the nonionic surfactant shifts the equilibrium from monomer to micelle and from precipitate to monomer. The micelle is better able to compete for monomer with precipitate.

The salinity tolerance of the SDS/NP(EO)$_{10}$ system is shown in Fig. 4 [12] for NP(EO)$_{10}$ mole fractions (not

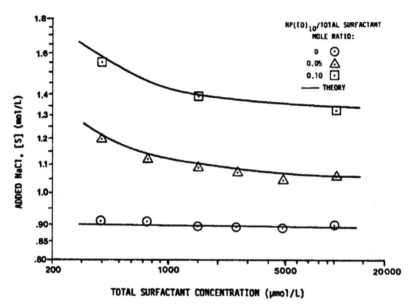

Figure 4 Salinity tolerance phase diagram of $(SDS/NP(EO)_{10})$ at different surfactant compositions. (From Ref. 12.)

necessarily in monomer or in micelles) of 0, 0.05, and 0.1 in the system. Note that the salinity tolerance of the pure anionic–surfactant system is almost independent of surfactant concentration above the CMC, just as is the Krafft point. This is because the surfactant monomer concentration is almost constant above the CMC for single-surfactant systems for nonionic surfactants or ionic surfactants in high electrolyte concentrations, as described by the pseudophase separation model of micelle formation [21].

The addition of nonionic surfactant clearly increases salinity tolerance, by as much as a factor of 2. The theoretical model developed to predict these phase boundaries is also shown in Fig. 4. The theory used a simple solubility product to describe the equilibrium between monomeric surfactant and unbound counterion, and regular solution theory to describe the mixed micelle formation between the anionic and nonionic surfactant. The reader is referred to the original article [12] for details, but it should be pointed out that the model can describe

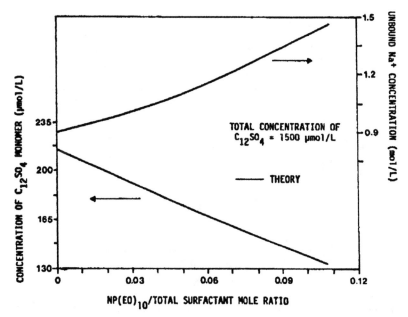

Figure 5 Monomer anionic surfactant composition and unbound sodium concentration on the precipitation phase boundary for the (SDS/NP(EO)$_{10}$) system. (From Ref. 12.)

the data very well, as seen in Fig. 4. This indicates that a simple equilibrium as shown in Fig. 2 provides a valid framework for understanding precipitation of surfactants. Figure 5 shows the calculated SDS monomer concentration in the system shown in Fig. 4 when precipitation occurs (on the phase boundary). As expected from the above discussion, the SDS monomer concentration decreases as the proportion of nonionic surfactant in the system increases because of mixed micelle formation—about a factor of 2 decrease when the nonionic surfactant content of the system is increased to 10% of the total surfactant present.

The synergism of nonionic surfactants in reducing the tendency of anionic surfactants to be precipitated by a monovalent counterion can also be reflected by Krafft point reductions as shown in Fig. 6 for the SDS/NP(EO)$_{10}$ system at several salinities. In the absence of added salt,

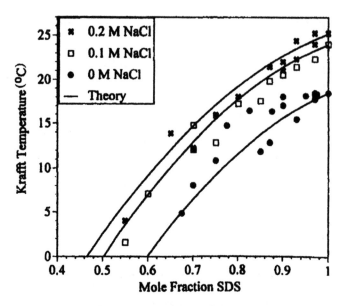

Figure 6 Krafft temperatures for 0.04 M sodium dodecyl sulfate (SDS)/nonylphenol polyethoxylate with an average degree of polymerization of 10 (NP(EO)$_{10}$) at 0, 0.1, and 0.2 M NaCl. (From Ref. 17.)

adding 0.3 mole fraction nonionic surfactant reduces the Krafft point by over 10°C.

It is interesting to note that the reason that the nonionic surfactant can increase salinity tolerance or decrease the Krafft point is that it synergizes the formation of mixed micelles, but not the formation of the precipitate which is determined solely by the solubility product between counterion and monomer ion. This is because only one surfactant is present in a precipitate from a solution of dissimilar surfactants [12,22,23] (excluding, of course, anionic–cationic surfactant systems), particularly if crystals from mixed ionic surfactant systems are permitted to ripen [17]. It is this peculiar nature of the crystalline precipitate which allows tremendous flexibility in manipulating the tendency for precipitation to occur. This manipulation is much more difficult in some other cases; for example, in surfactant adsorption on clay, the adsorbed aggregate formation is synergized by the addition of a mixed anionic–nonionic surfactant compared to either pure surfactant, as in micelle formation. Whether the relative enhancement of one aggregate formation is greater than the other determines whether adsorption at concentrations above the CMC will increase or decrease and cannot be easily predicted. In fact, either increases or decreases in adsorption upon introducing a second surfactant to the system have been observed [24].

While most of the surfactant precipitation literature deals with the effects of addition of nonionic surfactants, it should be noted that addition of amphoteric surfactants will also improve salinity and hardness tolerance of ionic surfactants. In fact, since the deviation from ideal mixing behavior with amphoteric surfactants is greater than with nonionics, they should be likewise more effective at increasing hardness and salinity tolerance just as they are more effective at lowering the CMC of the mixture.

C. Mixtures of Anionic Surfactants

If a mixture of anionic surfactants is present in solution, one might intuitively think that the salinity tolerance, hardness tolerance, or Krafft point might be between those of the individual surfactants present since the enhancement of micelle formation observed with nonionic surfactant addition to an anionic surfactant solution would not exist in this case. In fact, this argument is valid for formation of aggregates (such as adsorbed surfactant on minerals [25]) where the distributions of surfactant components between the mixed aggregate in question and in the mixed micelles are similar. However, for system containing surfactant precipitate, the fact that mixed micelles form which are composed of all the surfactant components present, but only one surfactant component is present in the precipitate when the first precipitate forms (on a phase boundary or at the Krafft temperature) make this intuitive argument invalid. Surfactants precipitate separately but form micelles cooperatively.

In mixed-anionic surfactant systems, precipitation of the surfactant component which first meets the aforementioned solubility product criterion dictates the phase boundary or Krafft temperature. The other surfactant components may be close to or far away from precipitation at this point since each surfactant-counterion pair has its own solubility product. If the anionic surfactants in a mixture have similar structures, the thermodynamics of the mixed micelle formed at concentrations above the CMC can be described by ideal solution theory [6].

Even though the mixing in the micelle is ideal, the monomer concentration of each surfactant component is less than the pure component CMC value under those conditions (added electrolyte, temperature, etc.). Hence, even a mixture of similar surfactants would exhibit a reduced tendency for any individual surfactant to precipitate simply by being present in a mixture. Note that the micelle composition can be substantially different than the monomer composition. Of course, systems of practical importance (e.g., a commercial linear alkylbenzene sulfonate) may contain hundreds of components, none of which may be present at more than a few mol% of the total surfactant present. Therefore, the salinity or hardness tolerance is increased and Krafft point decreased in such mixtures. Rosen [26] observes that mixtures of isomeric materials generally have Krafft points that are considerably lower than those of individual compounds.

Figure 7 shows the Krafft points of a binary mixture of sodium dodecyl sulfate (SDS)/sodium octylbenzene sulfonate (SOBS) as a function of total solution composition (not necessarily equal to the monomer or to micelle composition) [17]. In the absence of added NaCl, an equimolar mixture of the two surfactants had a Krafft point at least 10°C below that of each pure component. At mole fractions of SDS of less than approximately 0.5, the SOBS is the precipitating surfactant, whereas at higher mole fractions, the SDS is precipitating. This eutectic type of behavior is commonly observed for binary mixtures of precipitating surfactants. The minimum Krafft point doesn't necessarily correspond to an equimolar mixture as shown by the systems with added salt in Fig. 7. Added non-ionic surfactant depresses the Krafft temperature of mixtures

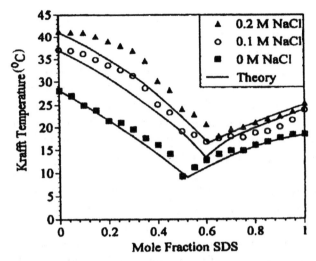

Figure 7 Krafft temperatures for sodium dodecyl sulfate (SDS)/ sodium 4-*n*-octylbenzene-1-sulfonate (SOBS) system at 0.04 M total surfactant concentration and 0, 0.1, and 0.2 M NaCl concentrations compared with theory. Activity-based solubility product (K_{SP}) adjusted for pure SDS and pure SOBS at each NaCl concentration. (From Ref. 17.)

of anionic surfactants [17], just as with single anionic surfactant systems. The Krafft points of these anionic/anionic (Fig. 7), anionic/nonionic (Fig. 6), or anionic/anionic/nonionic [17] surfactant systems can be predicted successfully using a relatively simple thermodynamic model which combines monomer-micelle equilibrium compositions (from ideal or regular solution theory to describe mixed micelle formation) with solubility products of the anionic surfactant components [17].

The greater the number of surfactant components present, the more difficult it is for any one component to exceed its solubility product (depressed Krafft point or increased hardness tolerance). So, a practical consequence is that increased heterogeneity in an anionic surfactant mixture is beneficial in avoiding surfactant precipitation. Commercial LAS, with many components of different hydrophobe length (homologues) and point of attachment of benzene ring on the linear alkyl chain (isomers), has a much lower tendency to precipitate than a pure isomer with the average properties of the LAS. An important development in the last decade is a class of new catalysts which can adjust the isomer distribution of LAS, with as much as 95% of the isomers in the 2 or 3 position [27,28], sometimes called superhigh-2-phenyl LAS or modified LAS. The closer to the terminal position the benzene is attached, the higher the surface activity and detergency as long as surfactant precipitation is avoided, but the tendency to precipitate increases. By introducing a random methyl branch along the otherwise linear alkyl chain, this increased tendency to precipitate can be offset by the decreased tendency to precipitate associated with branching. The resulting modified LAS can have superior detergency properties with no substantial increase in Krafft temperature [27,28]. This illustrates how surfactant precipitation characteristics continues to drive performance of surfactant systems.

D. Hardness Tolerance

1. Introduction and General Comparison to Salinity Tolerance

An atomic force micrograph of precipitated crystals of calcium dodecyl sulfate is shown in Fig. 8 [29]. The head groups of the surfactant exhibit hexagonal packing with a center-to-center distance of 5.9 Å, the expected spacing based on Van der Waal radii and are about 20% more compact than the head group packing in a Gibbs close packed monolayer at the air–water interface due to the more organized structure and more effective charge neutralization [29]. An optical image of the calcium dodecyl sulfate crystal is shown in Fig. 9 [30] 4 min after mixing the counterion and surfactant. Well structured trapezoidal, rhombic, and hexagonal shaped crystals can be seen in Fig. 9. The size and shape of crystal varies with aging, type of pure anionic surfactant, and degree of supersaturation [30]. In general, precipitated surfactant crystals tend to be smaller when formed in the presence of other surfactants [31,32]. When more than one surfactant is initially supersaturated, an optically amorphous precipitate will tend to

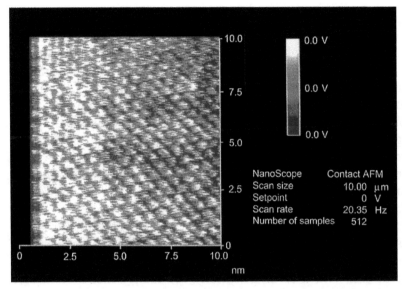

Figure 8 Three-dimensional (3-D) atomic force microscope (AFM) graph of calcium dodecyl sulfate (Ca(DS)₂) head groups in crystal lattice. (From Ref. 29.)

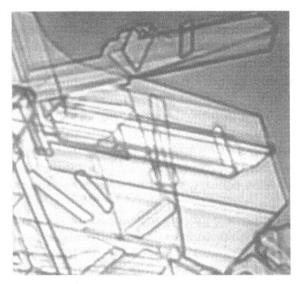

Figure 9 Image analysis picture of Ca(DS)₂ crystals at 40× precipitated from a 0.010 M sodium dodecyl sulfate (SDS)/0.008 M CaCl₂ solution; taken 4 min after mixing at 30°C. (From Ref. 30.)

separate into the different pure surfactant crystals upon ripening [31].

The addition of nonionic surfactants to anionic surfactants to enhance hardness tolerance in practical applications is receiving increasing attention. Nonbuilt heavy-duty liquid laundry detergents may utilize mixtures of anionic and nonionic surfactants [2,33]. Mixed surfactant systems employing nonionic surfactant with anionic surfactant have been proposed for use in enhanced oil recovery [34]. There have been a number of studies investigating the precipitation of anionic surfactants by calcium [7,13,14,18,23,27–31,35–49], indicative of the importance of this phenomena.

Figure 10 shows a schematic of the equilibrium in a system containing anionic–nonionic surfactant at concentrations above the CMC with a precipitate of the anionic surfactant with a divalent cation [14]. Added monovalent salt is omitted from the diagram for clarity. The anionic surfactant is present as a monomer, in micelles, and as precipitate. The nonionic surfactant is present as a monomer or in micelles. The divalent or hardness cation is present in the unbound state, as bound counterion on the micelles, or in the precipitate. Any monovalent cation present is either unbound or bound onto micelles. At equilibrium, surfactant precipitation will occur if the product of the surfactant anion activity squared and the divalent counterion activity equals the solubility product of the surfactant salt.

Whether precipitation occurs or not depends on monomer-micelle equilibrium since the surfactant present as micelles does not contribute to surfactant anion activity and the bound counterion does not contribute to counterion activity. The earlier discussion concerning salinity tolerance and the effect of added nonionic surfactant also holds qualitatively for hardness tolerance. Enhancement of micelle formation reduces the tendency of the surfactant to precipitate since the micelle and precipitate can be viewed as being in competition for monomer. Also, binding of the divalent cation to the micelle (either anionic or mixed anionic–nonionic micelle) can reduce the concentration of unbound hardness ion, reducing the tendency for the divalent cation/surfactant anion to precipitate.

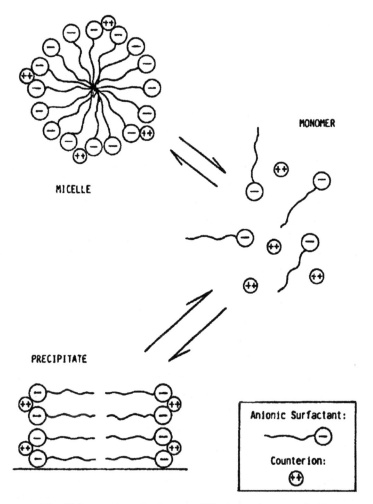

MONOMER

MICELLE

PRECIPITATE

Anionic Surfactant:

Counterion:

Figure 10 Schematic of the equilibrium in a micellar system containing anionic–nonionic surfactant where the anionic surfactant is being precipitated by divalent counterion. (From Ref. 14.)

There are, however, several important differences between salinity tolerance and hardness tolerance adjustment in mixed surfactant systems. First, it requires much lower concentrations of divalent cations to precipitate a given anionic surfactant than when monovalent cations are responsible for precipitation. For example, for SDS at 30°C, it requires

$0.9\,M\ Na^+$ to initiate precipitation (see Fig. 4), whereas less than $10^{-4}\,M$ of Ca^{2+} can cause precipitation of this surfactant under the same conditions [13]. This is why hardness tolerance is generally of much greater concern than salinity tolerance in many applications of anionic surfactants.

Second, in systems in which hardness cations are causing surfactant precipitation, there are normally monovalent cations also present, both counterions from dissolution of the surfactant salt and from other sources. For example, if the anionic surfactant is sodium dodecyl sulfate or sodium dodecylbenzene sulfonate, sodium is added to solution upon surfactant disassociation. Even naturally occurring water, which is considered "hard," may often have an order-of-magnitude higher monovalent cation concentration than the divalent cation concentration. Even when the divalent cation is the cause of precipitation, the monovalent cation can affect the surfactant CMC values, compete for binding sites on the charged micelles, and have other effects which must be considered.

A third difference between salinity and hardness tolerance considerations is that a substantial fraction of divalent ions present in systems can be bound to micelles under reasonable conditions, whereas salinity tolerances are generally so high that a much smaller fraction of the monovalent cation is bound to the micelles. For example, in the system shown in Figs 3–5, the highest fraction of Na^+ bound on the micelles is about 8% (and generally much less) on the phase boundaries; we will present data for Ca^{2+} where binding is 20% of the total calcium in solutions over a wide range of conditions. Hence, counterion binding is generally much more important when considering hardness tolerance than when considering salinity tolerance.

2.   Hardness Tolerance in Monoisomeric Anionic
      Surfactant Systems

Before discussing mixed surfactant systems, an understanding of precipitation of anionic surfactant by divalent cations with no other surfactant present is necessary. Figure 11 shows the phase boundary for SDS being precipitated by Ca^{2+} at 30°C [13]. At low surfactant concentrations below the CMC, the

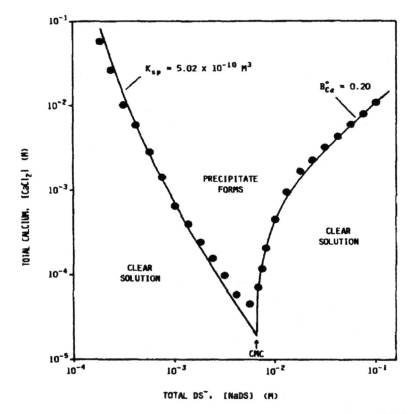

Figure 11 Hardness tolerance phase diagram of SDS without added NaCl. (From Ref. 13.)

minimum Ca^{2+} concentration required to cause precipitation decreases as the surfactant concentration increases since all the surfactant and calcium are unassociated and unbound. However, there is a minimum in the phase boundary at the CMC; the hardness tolerance (calcium concentration to be on the phase boundary) increases with increasing SDS concentration above the CMC. This is due to two reasons. First, the micelles formed above the CMC bind calcium, lowering the calcium activity, and making it unavailable for precipitation. The micelles are acting as sequestering agents or builders. The fractional counterion binding is 0.2 for the calcium in the system shown in Fig. 11 [13]. This means that for every additional mole of micellized surfactant added to the system, an additional 0.2 mole or 0.4 equivalents of calcium are bound to micelles.

A second reason is that as more SDS is added to the system, a higher concentration of unbound sodium is present in solution, lowering the CMC [13], increasing the fraction of surfactant which is present in micellar instead of monomeric form. The increase in surfactant concentration required to stay on the precipitation phase boundary as calcium concentration increases at surfactant concentrations above the CMC could also be interpreted as an increasing Krafft point with increasing calcium concentration at constant surfactant concentration, as confirmed by the data shown in Fig. 12 [36].

Following this logic, it would seem that the addition of a monovalent counterion from other sources would likewise lower the CMC and improve hardness tolerance. This is illustrated in Fig. 13 where 0.02 and 0.1 M NaCl is added to the SDS system. The minimum in the precipitation phase boundary is shifted to lower surfactant concentrations as the addition of the monovalent counterions lowers the CMC; thus,

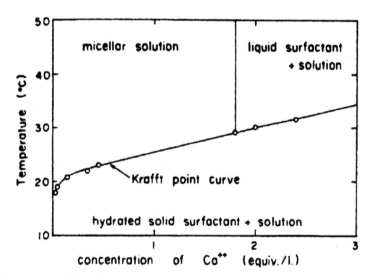

Figure 12   The effect of added $CaCl_2$, on the Krafft point of 1.0 wt% $C_{12}H_{25}OCH_2CH_2SO_4$ 0.5 Ca in water. (From Ref. 36.)

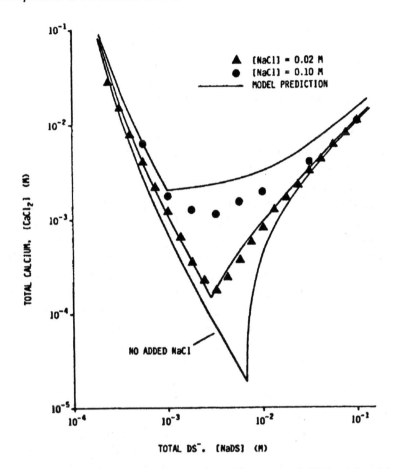

Figure 13 Hardness tolerance phase diagram of SDS with added NaCl. (From Ref. 13.)

the hardness tolerance at concentrations around the CMC of the system with no additives is substantially increased, by almost two orders of magnitude in the extreme case shown. At very high surfactant concentrations, almost all of the surfactant is present in micellar form already, so the added salt has little effect. This is a phenomenon which is known and used in detergency, where approximately 0.02 M sodium as sodium sulfate is a common additive in non-compact, heavy-duty powdered detergents [40] where it acts to improve hardness tolerance by lowering the CMC, as well as aiding processing and acting as a filler. As powdered laundry detergents have largely transitioned to compact form in the past decade, there has been a tendency to remove or reduce the filler [50], increasing the challenge of avoiding anionic surfactant precipitation in these systems.

It might seem that the addition of a cation to solution would make the anionic surfactant more likely to precipitate. However, salinity tolerance is generally much higher than the hardness tolerance, so that very high concentrations of added monovalent cation are required to form the monovalent cation/surfactant anion salt (salinity tolerance). For example, for the surfactant system shown in Figs 4 and 13, the salinity tolerance is $0.9 \, M \, Na^+$, whereas only $0.1 \, M$ NaCl achieved the nearly two orders of magnitude in hardness tolerance enhancement. If the added monovalent cation concentration is high enough, salinity tolerance dictates the tendency to precipitate rather than hardness tolerance as illustrated in Fig. 14, where the Krafft point is plotted as a function of surfactant composition for a mixture of a sodium salt of a given anion and the calcium salt of this same anion for three systems [23]. A eutectic-type relationship is observed, just as when two different anionic surfactants with the same counterion are mixed together (see Fig. 7). In Fig. 14, the compositions to the left or sodium side of the minimum in the Krafft point curve result in precipitation of the sodium salt of the anionic surfactant, whereas the compositions on the other side of the minimum result in precipitation of the calcium salt of the anionic surfactant. In such a system, both salinity and hardness tolerance are of importance, although the calcium salt forms over a much wider range of compositions that the sodium salt in Fig. 14.

3. Hardness Tolerance in Anionic–Nonionic Surfactant Systems

As we have already discussed in Section II.B, the addition of nonionic surfactant to anionic surfactant solutions reduces the CMC, reduces the anionic surfactant monomer concentration,

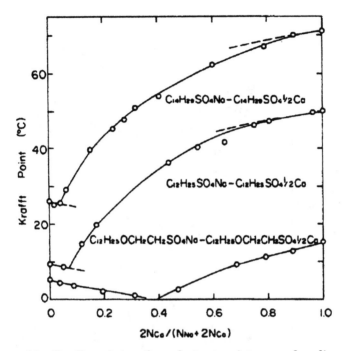

Figure 14 Krafft points of surfactant mixtures of sodium and calcium salts as a function of the fraction of calcium ions in the system. (From Ref. 23.)

and increases salinity tolerance. Exactly the same logic applies to hardness tolerance as illustrated in Fig. 15 [14]. Figure 15 shows that the addition of nonionic surfactant $NP(EO)_{10}$ at a level of 10% of the total surfactant present increases hardness tolerance by as much as an order of magnitude. There are a number of articles in the literature which demonstrate that nonionic surfactants can increase hardness tolerance or reduce the Krafft point of anionic surfactant solutions [13,14,17,34,37–39,51].

Since the addition of either monovalent electrolyte or nonionic surfactant can decrease the CMC and synergize micelle formation, would even greater hardness tolerance occur if both additives were used? Figure 15 shows that, indeed, better hardness tolerance is obtained either by the addition of salt at constant added nonionic surfactant concentration

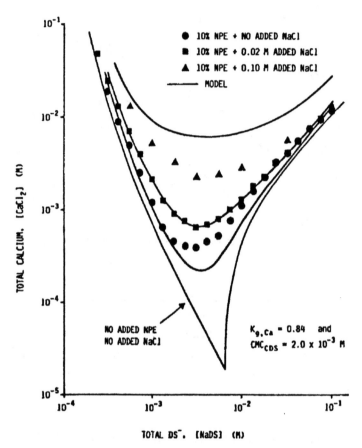

Figure 15 Hardness tolerance phase diagram of SDS with added $NP(EO)_{10}$ and added NaCl. (From Ref. 14.)

or by the addition of nonionic surfactant at constant added monovalent cation concentration, demonstrating that this qualitative additivity of effects from added monovalent salt and nonionic surfactant does indeed occur. Over an order of magnitude increase in hardness tolerance is observed when 0.1 M NaCl and nonionic surfactant are added to the SDS system. Again, at high anionic surfactant concentrations, the added salt or nonionic surfactant have little effect on hardness tolerance since almost all the anionic surfactant is in micellar form anyway.

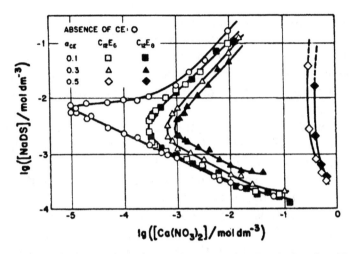

Figure 16 Precipitation boundaries of calcium dodecyl sulfate in the absence and in the presence of nonionic surfactants at 20°C. Nonionic surfactant/total surfactant = α: O, $\alpha = 0$; □, $\alpha = 0.1$; △, $\alpha = 0.3$; ◇, $\alpha = 0.5$. Open symbols are for $C_{12}E_6$, and solid symbols for $C_{12}E_8$. (From Ref. 35.)

Figure 16 shows the effect of nonionic surfactant structure in the precipitation of SDS by calcium [35]. A polyethoxylate with eight ethylene oxides resulted in slightly better hardness tolerance enhancement than one with six ethylene oxides. This is probably due to the greater degree of nonideality of the former system in forming mixed micelles than the latter system. The longer polyethoxylate group can more effectively shield anionic hydrophilic groups from each other, reducing charge density at the micelle surface more, resulting in a greater enhancement of micelle formation and lower CMC.

In Figs 11, 13, and 15, the solid lines shown are from the predictive model developed to describe these systems. This model included solubility product relationships, regular solution theory to describe mixed micelle formation, and new correlations to describe counterion binding on mixed micelles and CMC values of anionic surfactants with both monovalent and divalent cations present. The reader is referred to the original articles for details [13,14]. Except at very high electrolyte concentrations where the Debye–Huckel theory used to describe electrolyte activity coefficients in solutions is not satisfactory, the theory predicted the data extremely well. A predictive model can minimize the need for experimental data when mapping a phase boundary for a new anionic–nonionic surfactant system.

When surfactant precipitation occurs from a mixture of surfactants of which at least one is amphoteric (charge depends on pH), the precipitation phase boundary (or Krafft point) depends on pH, even if the precipitating surfactant is not amphoteric, as long as the system is above the CMC due to mixed micelle formation. Soaps are the most widely used amphoteric surfactant since the carboxylate group can be protonated. For example, the pK_a of sodium octanoate is 5.0 below the CMC [18]. The pK_a is higher above the CMC due to a decreased tendency to protonate when the surfactant is in micellar form [18]. At the pK_a, the soap is present as equal concentrations of the protonated (neutral) form and unprotonated (anionic) form. At a low enough pH, the soap is all protonated and at a high pH, it is all anionic. At intermediate pH, the effect of the soap on the tendency of another anionic surfactant to precipitate is equivalent to adding a mixture of a dissimilar anionic surfactant and a nonionic surfactant to the precipitating surfactant. This is illustrated in Fig. 17 for precipitating anionic surfactant SDS with sodium octanoate (SO) in both protonated and unprotonated form [18]. The hardness tolerance of the SDS is shown in Fig. 18 for different ratios of SDS/SO at pH of 5 [18]. Above the CMC, the hardness tolerance of the SDS (at a given SDS concentration) increases when SO is added due to synergism in mixed micelle formation. Since it is a mixed micelle effect, there is no effect of added SO on the hardness tolerance below the CMC (although the CMC is affected). Of course, if the SO/SDS ratio is high enough, the SO will preferentially precipitate (calcium precipitate of soap is called soap scum). Furthermore, at low pH, the protonated soap comes out of solution as a precipitate (higher molecular weight soaps) or as an oily liquid (low molecular weight soaps like SO).

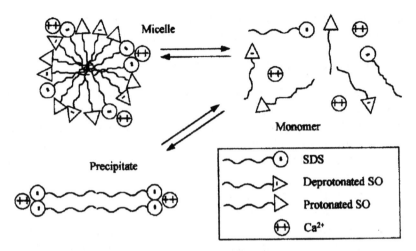

Figure 17 Micelle-monomer-precipitate equilibrium diagram for a sodium dodecyl sulfate/sodium octanoate mixture (SDS/SO) in the presence of calcium ions with $Ca(DS)_2$ precipitating. (From Ref. 18.)

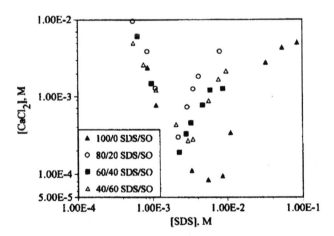

Figure 18 Comparison of precipitation phase boundaries (hardness tolerance) for $100:0$, $80:20$, $60:40$, and $40:60$ sodium dodecyl sulfate/sodium octanoate mixtures (SDS/SO) at a pH level of 5.0 and 30°C. (From Ref. 18.)

III. PRECIPITATION OF ANIONIC–CATIONIC SURFACTANT MIXTURES AT EQUILIBRIUM

A. Anionic–Cationic Surfactant Mixtures

Aqueous solutions of anionic and cationic surfactants are useful in detergency [52–54], pharmaceutical applications [55], analytical chemistry [9,56,57], and enhanced oil recovery [3], among other applications. The reader is referred to an excellent recent review of anionic–cationic surfactant mixture behavior and applications [58].

Figure 19 shows the equilibrium present in a solution containing anionic and cationic surfactants under conditions where the anionic and cationic surfactant form a precipitate with each other and micelles are present in solution [15]. Each charged surfactant may be present as a monomer, in micelles, or in precipitate. As in previous discussions of salinity and

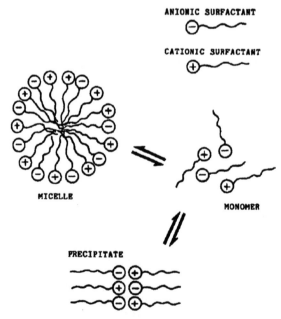

Figure 19 Schematic of the equilibrium in a micellar system containing anionic surfactant and cationic surfactant which is forming precipitate. (From Ref. 15.)

hardness tolerance, the micelles can be thought of as competing with the precipitate for surfactant monomer. Changes in the system that enhance micelle formation will therefore inhibit precipitation. It has been shown that the precipitate contains a 1:1 ratio of anionic and cationic surfactant in a sodium dodecyl sulfate/dodecylpyridinium chloride (DPC) system [15].

A phase boundary for the SDS/DPC system at 30°C is shown in Fig. 20 [16]. At equilibrium, surfactant precipitation will occur if the product of anionic-surfactant activity (roughly proportional to its monomer concentration) and the cationic-surfactant activity (roughly proportional to its monomer concentration) equals the solubility product of the anionic–cationic

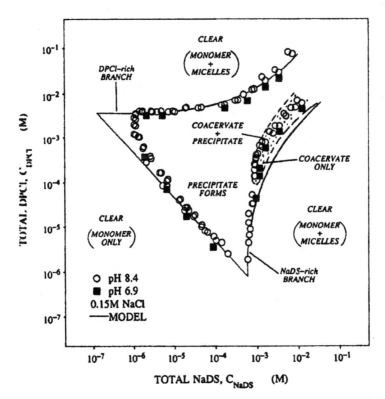

Figure 20 Effect of pH on precipitation phase boundary for sodium dodecyl sulfate(NaDS)/dodecylpyridinium chloride (DPCl) mixture. (From Ref. 16.)

surfactant salt. At low concentrations of either surfactant, where no micelles are present, the cationic surfactant concentration decreases as the anionic surfactant concentration in solution increases, in order to satisfy the solubility product and remain on the phase boundary. This is the region in Fig. 20 where the straight line is drawn through the data. Then, as the anionic surfactant concentration continues to increase, a concentration is reached on the precipitation phase boundary where anionic-rich mixed micelles are formed. At higher anionic surfactant concentrations, an increasing concentration of cationic surfactant is necessary to cause precipitation to occur due to competition for monomer with mixed micelles. As the cationic surfactant concentration increases along the phase boundary where no micelles are present, a concentration is reached where cationic-rich micelles are formed. In similarity to the anionic-rich micelle region of the phase boundary, it requires a higher concentration of the anionic surfactant to cause precipitation as cationic surfactant concentration increases along this cationic-rich micelle region due to mixed micelle formation. As with salinity and hardness tolerance, it is the formation of micelles which decreases the tendency for precipitation to occur.

There is a region noted in Fig. 20 on the anionic-rich portion of the phase boundary where coacervate forms. Coacervate is a dispersion of a second liquid phase in another liquid phase, the second liquid phase generally being very concentrated in surfactant compared to the external liquid phase. This is the same phenomenon as the cloud point in nonionic surfactant solutions, a temperature above which two liquid phases are in equilibrium and below which the solution is isotropic [59,60].

Figure 21 shows the "cloud point" for an alkylpoly (oxyethylene) sulfate/tetradecyltrimethylammonium bromide mixture as a function of surfactant composition at constant total surfactant concentration [61]. Mehreteab and Loprest [61] have referred to this anionic–cationic surfactant behavior as "pseudononionic" and attribute it to uncharged ion pairs formed between the oppositely charged surfactants. In Fig. 20, it was not possible to separate clearly a coacervate from a

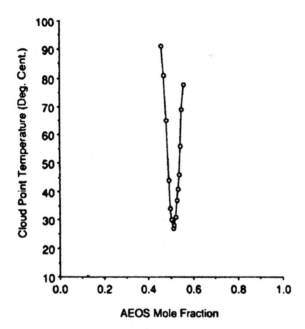

Figure 21 Cloud point temperature of the anionic–cationic surfactant system alkylpoly-(oxyethylene) sulfate (AEOS)/tetradecyltrimethylammonium bromide. (From Ref. 61.)

precipitation region since both appeared to be present at the same time [15,16]. Hoyer *et al.* [62] have observed such coacervate in a SDS/dodecyltrimethylammonium chloride solution. The point of this discussion is to warn the reader that phase behavior in anionic–cationic surfactant systems can be very complex and a clear delineation on the phase boundary between isotropic solutions and precipitation may not always be possible.

Other surfactant aggregates can form in anionic/cationic surfactant mixtures, such as vesicles or liquid crystals [63–70]. For example, Fig. 22 shows ternary phase diagrams for three different anionic/cationic surfactant pairs in heavy water [63]. As illustrated by the sodium dodecanoate/dodecytrimethylammoniumchloride system (SD/DoTAC), the precipitation region tends to dominate when the anionic/cationic surfactant molar ratio is 1/1 over a wide range of concentrations with charged

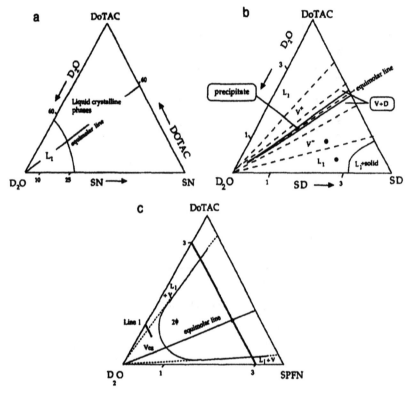

Figure 22 Isothermal ternary phase diagram of mixed surfactant–water systems at 298 K: (a) dodecytrimethylammoniumchloride–sodium nonanoate–water (DoTAC–SN–D$_2$O); (b) dodecytrimethylammoniumchloride–sodium dodecanoate–water (DoTAC–SD–D$_2$O); (c) dodecytrimethylammoniumchloride–sodium perfluorononanoate–water (DoTAC–SPFN–D$_2$O). Notation: L$_1$, Isotropic phase; V$^+$ and V$^-$, positively and negatively charged vesicles; Ves, vesicles; 2Φ, two-phase region. (From Ref. 63.)

vesicles bordering this region of precipitation. It is beyond the scope of this chapter, but there has been much progress over the past decade in understanding the relationship between surfactant structure and other variables on the type of aggregate the anionic/cationic surfactant mixture will form—e.g., whether a lamellar liquid crystal or vesicle will form [66]. Scientifically interesting as these non-solid phases are,

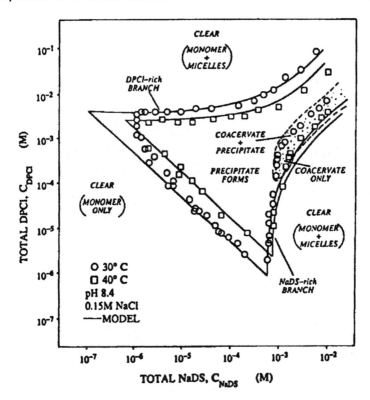

Figure 23 Precipitation phase diagram for the anionic–cationic surfactant system SDS/DPC at two different temperatures. (From Ref. 16.)

the predominant nonisotropic phase formed when traditional, commercial anionic and cationic surfactants are mixed together in aqueous solution is solid precipitate.

The effect of temperature on the phase boundary of the SDS/DPC system is shown in Fig. 23 [16]. At higher temperatures, below the CMC and on the cationic-rich micelle region of the phase boundary, precipitation requires higher concentrations of oppositely charged surfactant (precipitation is inhibited). In the anionic micelle-rich region, precipitation is slightly enhanced with increasing temperature.

The effect of anionic surfactant alkyl chain length for *n*-sodium alkyl sulfate/DPC on the precipitation phase

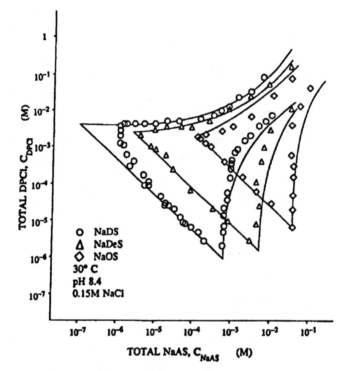

Figure 24 Precipitation phase diagram for the anionic–cationic surfactant system sodium alkyl sulfate/DPC at three different anionic surfactant alkyl chain lengths. (From Ref. 16.)

boundary shown in Fig. 24 [16]. Decreasing the alkyl chain length reduces the tendency to precipitate in all regions of the phase boundary. This is because the solubility product of the cationic–anionic surfactant salt increases substantially with decreasing alkyl chain length. Adjusting the alkyl chain length of either the anionic or cationic surfactant is a useful way of manipulating the phase behavior of these systems.

The effect of branching the hydrophobic group of the anionic surfactant for the anionic/DPC system is shown in Fig. 25 where sodium dihexyl sulfosuccinate (SDHS) and sodium hexadecyl diphenyloxide disulfonate (SHDPDS) are used as the anionic surfactant [71]. The SHDPDS is also a gemini surfactant. The SHDPDS has a much larger hydrophobe than

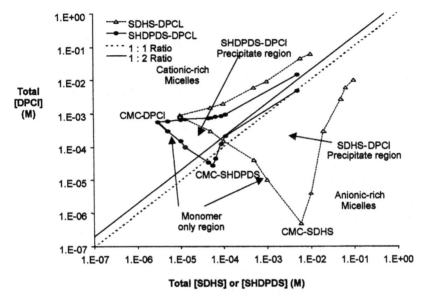

Figure 25 Oil-free precipitation phase boundary of sodium dihexyl sulfosuccinate(SDHS)/dodecylpyridinium chloride (DPCl), and sodium hexadecyl diphenyloxide disulfonate (SHDPDS)/dodecylpyridinium chloride (DPCl) systems at 25°C. (From Ref. 71.)

SDS, so has low CMC. Yet, it does not fit well into a crystal lattice structure, so the solubility product with the DPC is about the same as for SDS. The result is that above the CMC, the precipitation region is much smaller than for SDS. Note that the SHDPDS is divalent. Other authors [58,63,66,69,70] have observed that anionic/cationic surfactant precipitate formation is inhibited by using surfactants with increased branching or bulky groups in the hydrophobe or by chain-length asymmetry between the precipitating surfactants. The SDHS is branched, but has about the same total number of carbons in the hydrophobe as the SDS. So, the CMC is higher than for SDS, so the surfactant concentration at which the micelles start to form is increased. The increased CMC results in an expanded precipitation region for SDHS vs SDS. The increased symmetry between hydrophobes also synergizes precipitation. The reason that the octyl sulfate has a much smaller precipitation region in Fig. 24 compared to the dodecyl sulfate (with DPC) despite the much higher CMC of the former may be due to the greater asymmetry of the anionic/cationic surfactant hydrophobes. So, while branching can reduce the tendency to precipitate, it must be accompanied by appropriately balanced alkyl chain lengths to achieve desired synergisms. Precipitation is generally most likely when the cationic/anionic surfactant molar ratio is stoichiometric (1/1 for monovalent surfactants).

A theory has been developed to predict anionic–cationic surfactant phase boundaries [15,16]. It includes the solubility product of the anionic and cationic surfactant monomers and regular solution theory to describe mixed micelle formation. The results of the model are shown by the solid lines in Figs 20, 23, and 24 and it can be seen to describe the data quite well.

Sometimes, one knows that an anionic–cationic surfactant system will precipitate: The question is how much will precipitate? Combination of material balance equations with the aforementioned phase boundary model allows calculation of the amount of each surfactant added to solution which will precipitate and how much will remain dissolved in solution [15]. The resulting predictions compare very well to experimental results [15]. This type of calculations of amount of precipitate formed have also been used in studies of precipitation of anionic surfactants by inorganic cations [40,46]. These calculations also predict the surfactant composition in both monomer and micelles (if present) when equilibrium is attained.

B. Effect of Added Nonionic Surfactant

Nonionic surfactant can enhance micelle formation with either anionic surfactants or with cationic surfactants [6]. It seems reasonable that nonionic surfactants could also enhance the formation of micelles in mixed anionic–cationic systems. When the aforementioned mathematical model to describe precipitation in anionic–cationic surfactant systems is modified to include the effect of nonionic surfactants on mixed micelle formation, the phase diagram shown in Fig. 26 results [32].

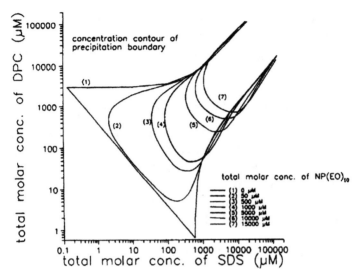

Figure 26 Effect of added nonionic surfactant $(NP(EO)_{10})$ concentration on precipitation phase diagram for the anionic–cationic surfactant system SDS/DPC. (From Ref. 32.)

As the concentration of added nonionic surfactant increases, the tendency to precipitate is reduced, except at high concentrations of anionic or cationic surfactant. Short chain alcohols reduce the tendency of anionic–cationic surfactant systems to precipitate [64], probably because of synergism in mixed micelle formation; these alcohols are likely to be less effective in preventing precipitation than nonionic surfactants since they have much less tendency to micellize.

IV. KINETICS OF SURFACTANT PRECIPITATION

In surfactant systems of practical importance, not only is equilibrium information important, but rate or kinetic considerations may sometimes dominate. For example, a fabric wash cycle in the United States may last about 10 min. The surfactant technologist is much more interested in whether surfactant precipitation will occur in this 10 min time interval than if it will occur at equilibrium, a state which may take days or weeks to be attained. Figure 1 shows how precipitation phase boundaries can shift substantially over hours or days.

A. Anionic Surfactants Precipitated by Calcium

Consider the SDS/Ca^{2+} system shown in Fig. 10. The supersaturation (S_0) indicates the thermodynamic driving force for precipitation and reflects the difference between the appropriate products of initial surfactant monomer concentration and unbound calcium concentration and those at equilibrium as defined in Eqs (1) and (2) [30]

$$K_{SP} = (([S^-]_{mon})^2 [Ca^{2+}]_{un} f_S^2 f_{Ca})_{eq} \tag{1}$$

$$S_0 = (([S^-]_{mon})^2 [Ca^{2+}]_{un} f_S^2 f_{Ca} / K_{SP})^{1/3})_{init} \tag{2}$$

where $[S^-]_{mon}$ is the anionic surfactant monomer concentration, $[Ca^{2+}]_{un}$ is the calcium concentration which is unbound to micelles, f_S is the activity coefficient of the anionic surfactant monomer, f_{Ca} is the activity coefficient of the unbound calcium, and K_{SP} is the solubility product of the calcium dodecyl sulfate. The subscripts refer to the equilibrium condition (eq) and to the initial conditions (init). The K_{SP} can be obtained from precipitation phase boundaries below the CMC and activity coefficient correlations [13]. If the system were at equilibrium, the solubility product (Eq. (1)) would be satisfied. So, the higher the value of S_0, the greater the thermodynamic driving force for precipitation. It is very important to note that above the CMC, the monomer–micelle equilibrium needs to be measured or modeled to obtain the monomeric surfactant concentration and the calcium counterion binding needs to be measured or modeled to deduce the unbound calcium concentration. For mixed surfactant systems, for example, regular solution theory is often used to deduce monomer–micelle equilibrium from CMC measurements [14,17]. Ion specific electrodes [19,20] or the dependence of the CMC on added salt concentration [13] are commonly used to deduce counterion binding, although the

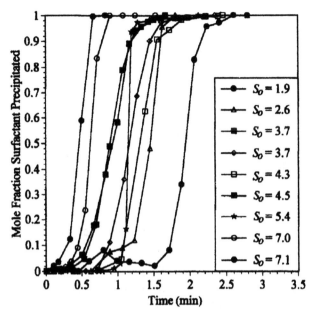

Figure 27 Precipitation rate curves of sodium dodecyl sulfate (SDS) with $CaCl_2$ at 30°C. S_0, supersaturation ratio. (From Ref. 30.)

pure surfactant precipitation phase boundary with the counterion of interest can be used to deduce counterion binding [13]. At equilibrium, $S_0 = 1$, so the degree to which S_0 exceeds unity reflects how supersaturated the solution is.

A number of concentrations inside the precipitation phase boundary both above and below the CMC were chosen as initial conditions for the SDS/Ca^{2+} system. Figure 27 shows the rate of precipitation of these systems and the value of S_0 for each initial condition. In Fig. 28, precipitation rates for this same system are shown at four different levels of supersaturation using a different experimental technique by different investigators [72]. It is remarkable how similar Figs 27 and 28 appear. Three general conclusions can be reached from these results. First, the general time frame required for equilibrium to be attained in the precipitation of anionic surfactants by multivalent cations is on the order of a few minutes. Second, the rate of precipitation is not uniform throughout the time

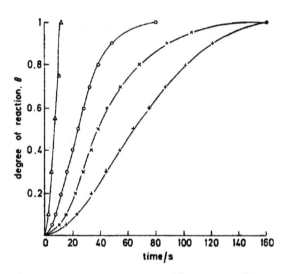

Figure 28   Kinetics of precipitation of SDS by Ca^{2+}. The degree of supersaturation of the initial composition increases in the order: $(+) > (\times) > (\bigcirc) > (\triangle)$. (From Ref. 72.)

period; there is an initial inhibition period during which the precipitation occurs slowly, followed by a period during which the system rapidly approaches equilibrium. Third, the greater the initial degree of supersaturation, the more rapidly the system approaches equilibrium.

Clarke, Lee, and Robb [72–74] have found that, in the precipitation of calcium alkyl sulfates, diffusion in bulk solution is the rate-limiting process in the early stages of precipitation, whereas first-order polynuclear growth is likely the rate-limiting step in later stages. In anionic surfactant-only systems, the presence of micelles reduces the rate of precipitation, presumably by reducing S_0 due to micellar binding of calcium.

As a way of normalizing the data, the extent of precipitation can be plotted against reduced time as shown in Fig. 29 for the data in Fig. 27. Reduced time is time/half life. The half life is the time when 1/2 of the surfactant is precipitated. It is remarkable that all the initial conditions (below and above the CMC) coincide when plotted as a universal curve in Fig. 29.

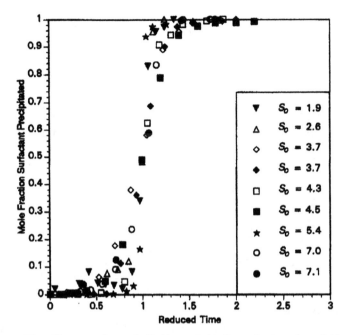

Figure 29 Extent of precipitation plotted against reduced time for sodium dodecyl sulfate (SDS) with CaCl$_2$ at 30°C. S_0, supersaturation ratio. (From Ref. 30.)

So, if the half life is known for a system, the shape of the precipitation rate curve appears to be universal.

B. Anionic–Cationic Surfactant Mixtures

Figure 30 shows kinetic precipitation data for the cationic-anionic surfactant system whose phase boundary is shown in Fig. 20 [30]. Similar to the case of anionic surfactant/calcium, supersaturation for the cationic/anionic system is defined by Eqs (3) and (4) [30]

$$K_{SP} = ([S^-]_{mon}[DP^+]_{mon})(f_\pm)^2 \qquad (3)$$

$$S_0 = ((([S^-]_{mon}[DP^+]_{mon})(f_\pm)^2/K_{SP})^{1/2})_{init} \qquad (4)$$

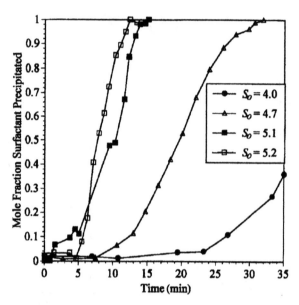

Figure 30 Precipitation rate curved of sodium dodecyl sulfate (SDS) with dodecyl pyridinium chloride (DPC) above the CMC at 30°C. S_0, supersaturation ratio. (From Ref. 30.)

where $[DP^+]_{mon}$ is the cationic surfactant monomer concentration and $f_{\pm}$ is the mean activity coefficient of the ions in solution. All the initial conditions in Fig. 30 are above the CMC and regular solution theory is used to model monomer–micelle equilibrium to obtain S_0. The time frame for equilibrium to be attained is as much as 30 min for the conditions chosen, much longer than for the SDS/Ca^{2+} system. The other conclusions reached for the SDS/Ca^{2+} system are also observed here: the presence of an inhibition period and the higher precipitation rate at greater degrees of supersaturation. It may be that crystal structure for the anionic–cationic surfactant system is less compact than the SDS/Ca^{2+} system, requiring more time for the molecules to arrange themselves geometrically for precipitation to occur. Since it is expected that the anionic/cationic surfactant pair will synergistically adsorb on the precipitate solid, it may be that an adsorbed surfactant layer further inhibits growth of the crystal.

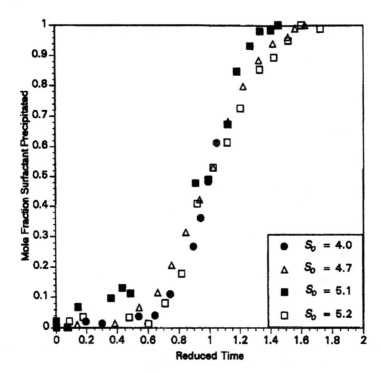

Figure 31 Extent of precipitation plotted against reduced time for SDS with DPC above CMC at 30°C. (From Ref. 30.)

In Fig. 31, the extent of precipitation is shown as a function of reduced time. As with the SDS/Ca^{2+} system, a universal shape is formed by the systems with different values of S_0. One can construct a plot of the half life as a function of the initial value of S_0 as seen in Fig. 32 for SDS/Ca^{2+}, SDS/DPC, and sodium octylbenzene sulfonate (SOBS)/Ca^{2+}. The two anionic surfactant/Ca^{2+} systems coincide in Fig. 32. It may be that a wide range of anionic surfactants obey this correlation, but since only two such systems have been studied, this is speculation. For all systems in Fig. 32, the half life of precipitation decreases with increasing supersaturation. This points out the importance of being able to calculate the surfactant monomer concentrations above the CMC to be able to calculate S_0.

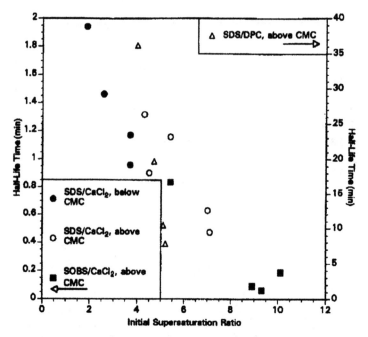

Figure 32 Half-life as a function of initial supersaturation ratio for all systems studied at 30°C. (From Ref. 30.)

C. Anionic Surfactant Mixtures Precipitated by Calcium

Figure 33 shows the rate of precipitation curves for pure anionic surfactants SDS and SOBS with calcium and for various SDS/SOBS concentration ratios [31]. As already shown, the single component systems reach equilibrium within a few minutes. However, at certain mixture compositions, substantially longer times can be required to reach equilibrium; for example, for a SDS/SOBS ratio of 60/40, it takes nearly an order of magnitude increase in time to attain a certain extent of precipitation. As also illustrated in Fig. 33, when more than one anionic surfactant is supersaturated, stepwise precipitation can occur where crystals of different composition can form with different induction times; these crystals do not necessarily contain a single surfactant component [31].

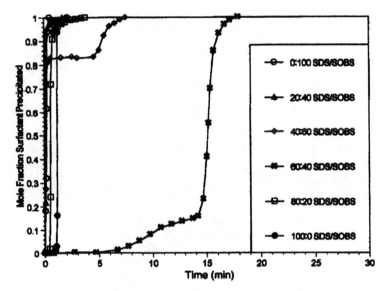

Figure 33 Precipitation rate curves for 0.0192 M total surfactant concentration and varying sodium dodecyl sulfate (SDS)/sodium octyl benzene sulfonate (SOBS) mole fractions with 0.01 M CaCl$_2$ at 30°C. (From Ref. 31.)

In order to better understand the crystallization process, atomic force micrographs (AFM) of precipitated crystals are shown in Figs 34 and 35 [29]. In Fig. 34, the calcium precipitate of SDS, SOBS, and a mixture thereof is shown (where both surfactants are supersaturated). The SDS exhibits well ordered layering or spiral growth patterns, consistent with a screw dislocation crystallization mechanism. The Ca-SOBS shows very jagged, uneven growth patterns. When both surfactants are supersaturated, the precipitate crystals show the layered patterns characteristic of Ca-SDS in the background in Fig. 34c, while in the foreground, the jagged crystals characteristic of Ca-SOBS form. One potential mechanism explaining the reduction in precipitation rate when mixture of SDS and SOBS are present is that Ca-SOBS crystals use the Ca-SDS crystals as nucleation sites, interrupting the growth of the Ca-SDS crystals, presumably followed by Ca-SDS interrupting growth of the Ca-SOBS crystals, ad infinitum. This explanation

Figure 34 (a) Three-dimensional (3-D) atomic force microscope (AFM) graph of $Ca(DS)_2$ crystals precipitated from a 0.020 M (SDS)/ 0.010 M $CaCl_2$ solution at 30°C; 50 × 50 μm. DS, dodecyl sulfate. (b) 3-D AFM graph of $Ca(OBS)_2$ crystals from a 0.020 M SOBS/0.010 M $CaCl_2$ solution at 30°C; 50 × 50 μm. OBS, octyl benzene sulfonate. (c) 3-D AFM graph of crystals precipitated from 0.020 M total surfactant solution with 60 : 40 SDS/SOBS mole ratio and 0.010 M $CaCl_2$ at 30°C; 50 × 50 μm. (From Ref. 29.)

is consistent with Fig. 34c. Another mechanism is the formation of holes in the crystals formed in mixed systems as illustrated in Fig. 35. Such holes result from strains in the crystal structure which are not observed when only one surfactant component is precipitating. Presumably, the holes can fill with mother liquor and get covered up with further

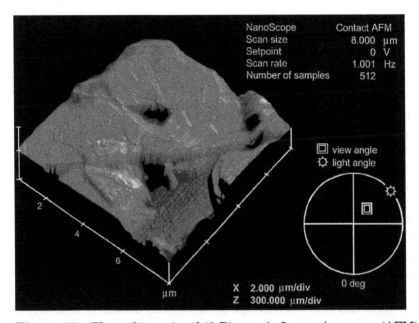

NanoScope   Contact AFM
Scan size   8.000 μm
Setpoint   0 V
Scan rate   1.001 Hz
Number of samples   512

□ view angle
☼ light angle

0 deg

μm   X 2.000 μm/div
      Z 300.000 μm/div

Figure 35 Three-dimensional (3-D) atomic force microscope (AFM) graph of crystals precipitated from a 0.020 M total surfactant solution with 40:60 sodium dodecyl sulfate (SDS)/sodium 1-octyl benzene sulfonate (SOBS) mole ratio and 0.010 M CaCl$_2$ at 30°C; 8 × 8 μm. (From Ref. 29.)

crystallization and the trapped liquid takes an extended period of time to diffuse through the crystal and equilibrate with the bulk solution.

An anionic surfactant can adsorb on the crystalline precipitate of a dissimilar anionic surfactant [31]. This can dramatically slow the rate of precipitation as adsorption blocks sites for continuing crystal growth (and generally result in smaller crystals). This is not surprising, since surfactants are commonly used to adjust crystal habits and crystallization rates for simple inorganic salts. It is interesting that addition of nonionic surfactant to calcium/dodecyl sulfate systems had no effect when micelles were not present [72–74], implying that the nonionic surfactant does not adsorb significantly on the anionic surfactant precipitate. However, the nonionic surfactant reduces the rate of calcium dodecyl sulfate precipitation when mixed anionic–nonionic micelles are present, perhaps due to reduced anionic surfactant monomer concentration and resulting reduction in the value of S_0.

There is anecdotal evidence in the surfactant industry that industrial surfactants can remain supersaturated for extended periods of time (weeks or months). These industrial surfactants are almost always complex mixtures of surfactant components. The results shown in this section for well defined binary surfactant mixtures are consistent with these supersaturation effects being caused by the heterogeneous nature of the industrial surfactants. Much of this chapter has discussed the fairly well researched advantages of surfactant mixtures in reducing the tendency of surfactants to precipitate at equilibrium. The limited results here indicate that there are also substantial synergisms in using mixtures to reduce the rate of precipitation. The kinetics of precipitation is clearly where the need for more research is most acute within the topic of surfactant precipitation. Scientists are able to quantitatively model or even predict phase boundaries, Krafft temperatures, hardness tolerances and other thermodynamic properties related to surfactant precipitation as seen in a number of figures in this chapter. However, our understanding of, let alone ability to model, the rate of precipitation is still in its infancy— in comparing this chapter to its earlier version from a decade ago [75], this section on kinetics has been altered far more than the other sections on more established topics.

ACKNOWLEDGMENTS

Financial support for this work was provided by the industrial sponsors of the Institute for Applied Surfactant Research at the University of Oklahoma, including Akzo Nobel Chemicals Inc, Albemarle Corporation, Amway Corporation, Clorox Company, Colgate-Palmolive, Dial Corporation, Dow Chemical Company, DowElanco, E. I. DuPont de Nemours & Co., Halliburton Services Corp., Henkel Corporation, Huntsman Corporation, ICI Americas Inc., Kerr-McGee Corporation, Lever Brothers,

Lubrizol Corporation, Nikko Chemicals, Phillips Petroleum Company, Pilot Chemical Company, Procter & Gamble Company, Reckitt Benckiser North America, Schlumberger Technology Corp., Shell Chemical Company, Sun Chemical Corporation, Unilever Inc. and Witco Corporation. John Scamehorn holds the Asahi Glass Chair and Jeff Harwell holds the Conoco/DuPont Professorship in Chemical Engineering at the University of Oklahoma.

REFERENCES

1. R. G. Laughlin, In G. Broze (ed.), *Handbook of Detergents, Part A: Properties*. Marcel Dekker (1999), p. 99.

2. D. Coons, M. Dankowski, M. Diehl, G. Jakobi, P. Kuzel, E. Sung, and U. Trabitzsch, In J. Falbe (ed.), *Surfactants in Consumer Products, Theory, Technology, and Applications*. Springer-Verlag, Berlin (1987), Ch. 5.

3. S. A. Arshad, and J. H. Harwell, SPE paper No. 14291, presented at the *60th Annual Technical Conference of the Society of Petroleum Engineers*, Las Vegas, September (1985).

4. L. W. Brant, K. L. Stellner, and J. F. Scamehorn, In J. F. Scamehorn, and J. H. Harwell (eds), *Surfactant-Based Separation Processes*. Marcel Dekker, New York (1989), Ch. 12.

5. B. Wu, S. D. Christian, and J. F. Scamehorn, *Progr. Colloid Polym. Sci.* **109**: 60 (1998).

6. J. F. Scamehorn (ed.), *Phenomena in Mixed Surfactant Systems*. ACS Symposium Series 311, American Chemical Society, Washington, D.C. (1986), p. 1.

7. J. M. Peacock, and E. Matijevic, *J. Colloid Interface Sci.* **77**: 548 (1980).

8. B. W. Barry, and G. F. J. Russel, *J. Pharm Sci.* **61**: 502 (1972).

9. R. V. Scowen, and J. Leja, *Can. J. Chem.* **45**: 2821 (1967).

10. R. L. Reeves, and S. A. Harkaway, In K. L. Mittal (ed.), *Micellization, Solubilization and Microemulsions*. Plenum Press, New York (1977), Vol. 2, p. 819.

11. H. W. Hoyer, and I. L. Doerr, *J. Phys. Chem.* **68**: 3494 (1964).

12. K. L. Stellner, and J. F. Scamehorn, *J. Am. Oil. Chem. Soc.* **63**: 566 (1986).

13. K. L. Stellner, and J. F. Scamehorn, *Langmuir* **5**: 70 (1989).

14. K. L. Stellner, and J. F. Scamehorn, *Langmuir* **5**: 77 (1989).

15. K. L. Stellner, J. C. Amante, J. F. Scamehorn, and J. H. Harwell, *J. Colloid Interface Sci.* **123**: 186 (1988).

16. J. C. Amante, J. F. Scamehorn, and J. H. Harwell, *J. Colloid Interface Sci.* **144**: 243 (1991).

17. C. H. Rodriguez, and J. F. Scamehorn, *J. Surf. Deterg.* **2**: 17 (1999).

18. C. H. Rodriguez, C. Chintanasathien, J. F. Scamehorn, C. Saiwan, and S. Chavadej, *J. Surf. Deterg.* **1**: 321 (1998).

19. J. F. Rathman, and J. F. Scamehorn, *J. Phys. Chem.* **88**: 5807 (1984).

20. J. F. Rathman, and J. F. Scamehorn, *Langmuir* **3**: 372 (1987).

21. K. Shinoda, In K. Shinoda, B. Tamamushi, I. Nakagawa, and T. Isemura (eds), *Colloidal Surfactants*. Academic Press, New York (1963), Ch. 1.

22. K. Tsujii, N. Saito, and T. Takeuchi, *J. Phys. Chem.* **84**: 2287 (1980).

23. M. Hato, and K. Shinoda, *J. Phys. Chem.* **77**: 378 (1973).

24. J. F. Scamehorn, R. S. Schechter, and W. H. Wade, *J. Colloid Interface Sci.* **85**: 494 (1982).

25. J. J. Lopata, J. H. Harwell, and J. F. Scamehorn, In D. H. Smith, (ed.), *Surfactant Based Mobility Control*. ACS Symposium Series 373, American Chemical Society, Washington, D.C. (1988), p. 205.

26. M. J. Rosen, *Surfactants and Interfacial Phenomena*, 2nd edn. J. Wiley and Sons, New York (1989), p. 215.

27. J. B. St. Laurent, D. Connor, T. Cripe, K. Kott, J. Scheibel, R. Stidham, and R. Reilman, *5th World Surfactants Congress, CESIO 2000: Proceedings*, Vol. 1, Federchimica Assobase—P.I.T.I.O., Milan, Italy (2000), p. 716.

28. K. L. Kott, T. W. Gederle, G. Baillely, J. J. Scheibel, and R. J. Lawson, *Proceedings of New Horizons: Detergents for the New Millenium*, AOCS Press, Champaign, IL (2001), Ch. 11.

29. C. H. Rodriguez, W. L. Yuan, J. F. Scamehorn, and E. A. O'Rear, *J. Surf. Deterg.* **5**: 269 (2002).

30. C. H. Rodriguez, L. H. Lowery, J. F. Scamehorn, and J. H. Harwell, *J. Surf. Deterg.* **4**: 1 (2001).

31. C. H. Rodriguez, and J. F. Scamehorn, *J. Surf. Deterg.* **4**: 15 (2001).

32. B. J. Shiau, J. H. Harwell, and J. F. Scamehorn, *J. Colloid. Interface Sci.* **167**: 332 (1994).

33. M. F. Cox, N. E. Borys, and I. P. Matson, *J. Am. Oil Chem. Soc.* **62**: 1139 (1985).

34. J. Novosad, B. Maini, and J. Batycky, *J. Am. Oil Chem. Soc.* **59**: 833 (1982).

35. X. J. Fan, P. Stenius, N. Kallay, and E. Matijevic, *J. Colloid Interface Sci.* **121**: 571 (1988).

36. K. Shinoda, and H. Tsuyoshi, *J. Phys. Chem.* **81**: 1842 (1977).

37. M. E. Cox, and K. L. Matheson, *J. Am. Oil Chem. Soc.* **62**: 1396 (1985).

38. Y. Chiu, In K. L. Mittal (ed.), *Solution Behavior of Surfactants*. Plenum Press, New York (1980), Vol. 2, p. 1415.

39. W. E. E. Gerbacia, *J. Colloid. Interface Sci.* **93**: 556 (1983).

40. K. L. Matheson, M. F. Cox, and D. L. Smith, *J. Am. Oil Chem. Soc.* **62**: 1391 (1985).

41. P. Somasundaran, K. P. Ananthapadmanabhan, M. S. Celik, and E. D. Manev, *Soc. Pet. Eng. J.* **24**: 667 (1984).

42. M. Baviere, B. Bazin, and R. Aude, *J. Colloid Interface Sci.* **92**: 580 (1983).

43. C. Noik, M. Baviere, and D. Defives, *J. Colloid Interface Sci.* **36**: 115 (1987).

44. S. I. Chou, and J. H. Bai, *J. Colloid Interface Sci.* **96**: 192 (1983).

45. J. Bozic, I. Krznaric, and N. Kallay, *Colloid Polym. Sci.* **257**: 201 (1979).

46. K. L. Matheson, *J. Am. Oil Chem. Soc.* **62**: 1269 (1985).

47. N. Kallay, M. Pastuovic, and E. Matijevic, *J. Colloid Interface Sci.* **106**: 452 (1985).

48. M. S. Celik, E. D. Manev, and P. Somasundaran, *AlChE Symp. Ser.* **78**: 86 (1982).

49. S. Miyamoto, *Bull. Chem. Soc. Jpn* **33**: 371 (1960).

50. L. H. T. Tai, *Formulating Detergents and Personal Care Products.* AOCS Press, Champaign, IL (2000), pp. 145–147.

51. N. Nishikido, H. Akisada, and R. Matuura, *Mem. Fac. Sci. Kyushu Univ., Ser. C* **10**: 91 (1977).

52. D. L. Smith, M. F. Cox, G. L. Russell, G. W. Earl, P. L. Lacke, R. L. Mays, and J. M. Fink, *J. Am. Oil Chem. Soc.* **66**: 718 (1989).

53. M. J. Schwuger, *Kolloid Z* **243**: 129 (1971).

54. D. N. Rubingh, In D. N. Rubingh, and P. M. Holland (eds), *Cationic Surfactants: Physical Chemistry.* Marcel Dekker, New York (1990), Ch. 7.

55. B. K. Sadhukkan, and D. K. Chattoraj, In K. L. Mittal (ed.), *Surfactants in Solution.* Plenum Press, New York (1984), Vol. 2, p. 1249.

56. B. J. Birch, and R. N. Cockroft, *Ion-Sel Electrode* Rev. **3**: 1 (1981).

57. W. Lin, M. Tang, J. J. Stanahan, and S. N. Deming, *Anal. Chem.* **55**: 1872 (1983).

58. A. Mehreteab, In G. Broze (ed.), *Handbook of Detergents, Part A: Properties.* Marcel Dekker (1999), p. 133.

59. D. J. Mitchell, G. J. T. Tiddy, L. Waring, T. Bostock, and M. P. McDonald, *J. Chem. Soc. Faraday Trans.* **179**: 975 (1983).

60. O. E. Yoesting, and J. F. Scamehorn, *Colloid Polym. Sci.* **264**: 148 (1986).

61. A. Mehreteab, and E. J. Loprest, *J. Colloid Interface Sci.* **125**: 602 (1988).

62. H. W. Hoyer, A. Marmo, and M. Zoellner, *J. Phys. Chem.* **65**: 1804 (1961).

63. O. Regev, and A. Khan, *J. Colloid Interface Sci.* **182**: 95 (1996).

64. C. Wang, S. Tang, J. Huang, X. Zhang, and H. Fu, *Colloid Polym. Sci.* **280**: 770 (2002).

65. X. R. Zhang, J. B. Huang, M. Mao, S. H. Tang, and B. Y. Zhu, *Colloid Polym. Sci.* **279**: 1245 (2001).

66. E. F. Marques, O. Regev, A. Khan, and B. Lindman, *Adv. Colloid Interface Sci.* (in press).

67. E. W. Kaler, K. L. Herrington, A. K. Murthy, and J. A. Zasadzinski, *J. Phys. Chem.* **96**: 6698 (1992).

68. E. W. Kaler, A. K. Murthy, B. E. Rodriguez, and J. A. Zasadzinski, *Science* **245**: 1371 (1989).

69. K. L. Herrington, E. W. Kaler, D. D. Miller, J. A. Zasadzinski, and S. Chinuvolu, *J. Phys. Chem.* **97**: 13792 (1993).

70. M. T. Yatcilla, K. L. Herrington, L. L. Brasher, E. W. Kaler, and S. Chinuvolu, *J. Phys. Chem.* **100**: 5874 (1995).

71. T. Doan, E. Acosta, D. A. Sabatini, and J. F. Scamehorn, *J. Surf. Deterg.* **6**: 215 (2003).

72. D. E. Clarke, R. S. Lee, and I. D. Robb, *Faraday Discus. Chem. Soc.* **61**: 165 (1976).

73. R. S. Lee, and I. D. Robb, *Faraday Discus. Chem. Soc.* **75**: 2116 (1979).

74. R. S. Lee, and I. D. Robb, *Faraday Discus. Chem. Soc.* **75**: 2126 (1979).

75. J. F. Scamehorn, and J. H. Harwell, In K. Ogino and M. Abe (eds), *Mixed Surfactant Systems.* Marcel Dekker (1993), Ch. 10.

19

Surfactant–Polymer Interactions

**MARINA TSIANOU and
PASCHALIS ALEXANDRIDIS**
Department of Chemical and
Biological Engineering, University at Buffalo,
The State University of New York, Buffalo,
New York, U.S.A.

SYNOPSIS

We review the self-assembly of surfactants and polymers in aqueous solutions, with a focus on polymers that are hydrophobically-modified. We start from dilute systems, where we examine association between polymer chains and formation of mixed micelles between the surfactant and the polymer hydrophobes, we explain the viscosifying effects that surfactants can have on polymer solutions, and then move on to consider surfactant–polymer systems that may undergo macroscopic phase separation, and concentrated systems where the polymer is embedded into ordered structures formed by the surfactant. Finally, we present examples on how surfactants affect the behavior of water-borne systems containing various types of polymers.

I. INTRODUCTION

We present here the association behavior and corresponding properties of mixtures of surfactants and polymers in aqueous solutions. Such mixtures are ubiquitous in nature (where surface active species such as lipids, and polymers such as DNA, proteins, and extracellular matrix components are basic ingredients of all live organisms) and are commonly encountered in several consumer products (e.g., foods, pharmaceuticals, cosmetics, inks for ink-jet printers) and industrial applications (coatings, paints, enhanced oil recovery). More recently, surfactants and polymers are actively being explored as building blocks for the nanotechnology toolbox. Add to all the above the inherent friendliness of water-borne systems toward the environment, compound the fascinating (and often unanticipated) manifestations of self-assembly, and you have systems that deserve all the attention that they have attracted in the scientific literature, and more.

Our review is organized as follows. We first present the salient features of the interactions occurring between surfactants and polymers, where we focus on the formation of mixed micelles between surfactants and amphiphilic copolymers, on the rheological behavior of aqueous surfactant–polymer systems, on surfactant–polymer phase behavior and phase separation, on surfactant–polymer complexes that are formed at low water contents, and on the various theoretical approaches used to understand surfactant–polymer association. We then examine interactions of surfactants with specific classes of well-studied polymers that are of relevance to various applications: polyether block copolymers, hydrophobically end-capped ethoxylated urethane polymers, cellulosic polymers, hydrophobically graft-modified polyacrylate polymers, and hydrophobically-modified alkali-soluble emulsion polymers.

II. SELF-ASSEMBLY OF SURFACTANTS IN AQUEOUS SOLUTIONS

Surfactant molecules consist of both hydrophilic and hydrophobic parts (covalently joined together) and thus have the tendency, when introduced in water, to self-assemble and form domains that shield the hydrophobic parts from the aqueous milieu [1–3]. The hydrophilic parts of surfactants are usually charged/ionized groups and/or polar non-ionic groups such as poly(ethylene oxide) (PEO) or glucose, while the hydrophobic parts are usually aliphatic hydrocarbons. The self-assembled domains have characteristic dimensions of the same magnitude as the extended length of the surfactant (nanometer-scale), and can take different shapes/forms (e.g., spheres, rods, lamellae). The ability of surfactants to self-assemble finds numerous applications and has fascinated many researchers. The understanding of surfactant self-assembly has advanced significantly following decades of investigations [2,3].

The surfactant self-assembly is dictated by thermodynamic considerations. As such, it is very sensitive to a number of variables, some inherent to the surfactant (e.g., type and size of hydrophilic and hydrophobic parts, architecture) and some external (e.g., ionic strength and pH of the aqueous solution, temperature). These variables affect to various extents the subtle free energy contributions that dictate the mode of self-assembly. The sensitivity of the surfactant self-assembly to multiple variables creates many opportunities for the use of surfactants at different functions and in different applications. At the same time, the ability to tune the self-assembly in a judicious manner remains a challenge, from both a fundamental and an application point of view.

A number of additives have been tested as means of modulating the self-assembly of surfactants in water, the most common of which are simple salts [3–6]. For example, the addition of alkali-halide salts such as LiCl, KCl, NaCl, and NaBr increases the tendency of surfactants to self-assemble [5], whereas addition of sodium thiocyanate (NaSCN) [5] and urea [6] lead to opposite effects. The surfactant self-assembly properties can also be altered by the addition to water of

polar organic solvents such as alcohols [7], hydrazine, formamide, glycerol, propylene glycol, and ethylene glycol [8,9].

The presence of water-soluble polymers can have a much
more dramatic effect on the surfactant self-assembly than the
low-molecular weight additives outlined above [10,11]. Water-
soluble polymers, when added to surfactant solutions, may
modify the free energy contributions during the surfactant
micellization process in a number of ways. Ionic polymers with
charge similar to that of the surfactant have a moderate
electrolyte effect, whereas, polymers of opposite charge act as
multivalent electrolytes and cause more dramatic effects. A
nonionic polymer with more or less amphiphilic character will
tend to localize at the micellar surface, thus influencing the
surfactant association process. On the basis of the above
effects, water-soluble polymers are often used in conjunction
with surfactants. The structure-property relationships that
originate from interactions between surfactants and polymers
play an important role in these cases, and have motivated
unwavering and concerted academic and industrial research
and development efforts.

III. INTERACTIONS IN SURFACTANT–HOMOPOLYMER AQUEOUS SOLUTIONS

It is well established by now that, in aqueous solutions,
surfactant molecules interact with homopolymers above a
critical aggregation concentration (CAC) forming micellar-like
aggregates with them [10,11]. The CAC is typically considerably lower than the critical micellization concentration (CMC)
exhibited by the corresponding surfactant in the absence of
polymer. The interactions between surfactants and polymers
are usually expressed in terms of surfactant binding isotherms.
These are obtained by a variety of methods, such as surfactant
ion-selective electrodes, surfactant self-diffusion, equilibrium
dialysis, spectroscopy, etc. [10–13]. A typical binding isotherm
of a surfactant to a polymer (having no distinct hydrophobic
moieties), expressed as the concentration of bound surfactant
vs that of the free surfactant, is illustrated in Fig. 1. At very

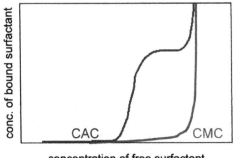

Figure 1 Schematic representation of the binding isotherm of a surfactant to a homopolymer. The critical aggregation concentration (CAC) and the critical micelle concentration (CMC) (in the absence of the polymer) are indicated.

low surfactant concentrations (and at any polymer concentration), no significant interaction can be observed. A strong, co-operative binding starts at CAC, followed by a plateau as the surfactant concentration increases. At even higher concentrations, the binding becomes similar to that reflected by the surfactant CMC curve in the absence of the polymer (see Fig. 1). At these high surfactant concentrations, free surfactant micelles coexist with surfactant aggregates bound to the polymer chains. The size and association number of the polymer-bound micelles are typically similar to those of micelles forming in the absence of the polymer [11,13–15]. The most widely accepted model for the morphology of the complexes formed between the polymer and the surfactant molecules in such cases was proposed by Cabane [16,17] on the basis of his pioneering studies on poly(ethylene oxide) (PEO)–sodium dodecyl sulfate (SDS) interactions. The picture that emerges is that of surfactant micelles adsorbed on PEO segments, thus converting the polymer chain to a "necklace" decorated with "pearls," the surfactant micelles (see schematic of Fig. 2) [16,17].

The surfactant–polymer interaction has been shown to depend on the relative charge and the hydrophobicity of the surfactant–polymer pair. Anionic surfactants exhibit a very

Figure 2 The pearl-necklace model for the polymer–surfactant complexation.

distinct interaction with most homopolymers, while cationic surfactants usually show a weak interaction. When the homopolymers are "slightly hydrophobic" (i.e., they contain hydrophobic groups, but are not hydrophobic enough to associate by themselves), such as poly(ethylene oxide) (PEO), poly(vinyl pyrrolidone) (PVP), poly(N-isopropylacrylamide) (PNIPAm), and ethyl hydroxyethyl cellulose (EHEC), then a hydrophobic attraction sufficient for binding will exist between the polymer and surfactant molecules. Nonionic surfactants (with much lower CMCs than ionic surfactants of comparable hydrophobic chain length) and zwitterionic surfactants rarely show an interaction with homopolymers. In the case of mixtures comprising ionic polymers and charged surfactants, the interactions are predominately electrostatic in nature and can be strong compared to those occurring when uncharged polymers are used. In systems containing polyelectrolytes and oppositely-charged surfactants, where the polymer acts as a multivalent electrolyte, the complexes formed are stabilized by both electrostatic attractions and cooperative hydrophobic effects, thus leading to a CAC which is lower than the CMC by several orders of magnitude. On the contrary, associations between polyelectrolytes and surfactants of the same charge are expected to be weak or even absent due to unfavorable electrostatic repulsions, and can only occur when the polymer has a very pronounced hydrophobic nature [10,11,18–22].

IV. INTERACTIONS IN SURFACTANT–AMPHIPHILIC POLYMER SYSTEMS

A. Association Properties of Amphiphilic Polymers in Aqueous Solutions

A significant number of studies over the past decade have addressed interactions occurring in aqueous solutions containing surfactants and amphiphilic polymers. In the case of amphiphilic polymers that bear both hydrophilic and hydrophobic parts, a hydrophobic attraction will exist between the polymer and the surfactant molecules. Such interactions are expected to be strong for block copolymers having hydrophilic and hydrophobic blocks, and for water-soluble polymers endcapped or graft-modified with hydrophobic groups, where there is sufficient segregation between the different parts of the polymer molecule [10,22–27].

The difference in polarity between the hydrophilic and hydrophobic groups in amphiphilic polymers drives their self-association in aqueous solutions even in the absence of surfactant. For example, block copolymers of $(EO)_x(PO)_z(EO)_y$ structure (where EO stands for ethylene oxide and PO for propylene oxide) are characterized by a considerable tendency to self-assemble in water. This leads to the formation of micelles, as well as several lyotropic liquid crystalline phases (cubic, hexagonal, or lamellar), in analogy with surfactant systems that exhibit the same types of phases and phase sequence [28]. The composition and temperature stability ranges of the various phases formed can be controlled by varying the polymer molecular weight and ratio between the more polar EO and the less polar PO groups [28].

The modification of homopolymers by hydrophobic groups ("hydrophobes") also leads to amphiphilic polymers that have a tendency to self-associate by hydrophobic interactions [13, 23–27]. These polymers, called hydrophobically modified polymers (HM-polymers or HMPs), are essentially hydrophilic water-soluble polymers (ionic or nonionic) to which a small number of charged or non-charged groups (typically alkyl chains) have been chemically attached. The level of incorporation of these groups is limited by the desired solubility of the resulting polymer in water, and it can be only up to a few mol%, except for hydrophobically modified polyelectrolytes (HMPEs) where the charges render the polymers highly water-soluble. HMPs can be end-modified (when the hydrophobes are attached to the two ends of the polymer), or graft-modified (when the hydrophobes are grafted along the polymer backbone). In aqueous media, the hydrophobic moieties of HMPs tend to self-associate and form intra- and inter-molecular aggregates, hence HMPs are also referred to as "associative polymers." In dilute solutions, and when the fraction of the hydrophobes is high, these associations are mainly intra-molecular thus causing the polymer chain to adopt a more compact conformation as compared to the homopolymer (random) coil. On the other hand, inter-chain associations are favored in semidilute solutions leading to a reversible cross-linking of the system. An illustration of the structural changes in solutions of end- and graft-modified polymers is presented in Figs 3 and 4, respectively. On the basis of this transient network formation, HMPs show interesting rheological features (i.e., unique response to shear rate and solvent quality), rather different from those of the corresponding unmodified polymers. Graft-modified HMPs exhibit viscosities much higher than those of flexible, linear, unmodified or even end-modified analogues under the same conditions. In the case of HMPEs, these associative interactions between hydrophobic moieties differentiate them from regular polyelectrolytes as they demonstrate certain anti-polyelectrolyte behavior (e.g., salt induced increase of HMPE solution viscosity) [11,14,23]. The intriguing rheological properties of HMPs have led to their increasing commercial applications as rheology modifiers and thickeners in several industrial processes and products (such as paints, food, pharmaceutics, cosmetics and skin care products, fluids for enhanced oil recovery), and to their utilization as dispersing/stabilizing agents for colloidal dispersions and emulsions [24–26].

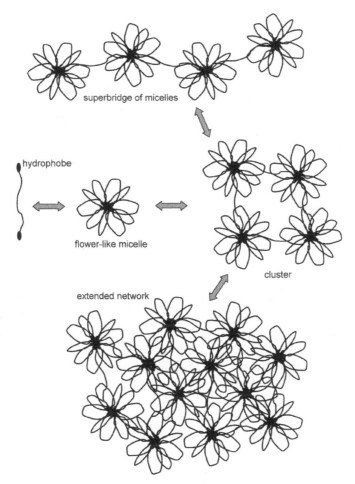

Figure 3 Depiction of the association between end-modified hydrophobically modified polymers (HMPs) in aqueous solutions. Isolated polymer chains exist at low concentrations, while flower-like micelles are formed in dilute solutions. Extended networks can be created via bridging of the flower-like micelles in the semidilute regime.

B. Surfactants and Hydrophobically Modified Polymers: Mixed Micelle Formation

As HMPs comprise both hydrophilic and hydrophobic parts and combine both polymer and surfactant properties (thus are often

(a)   (b)

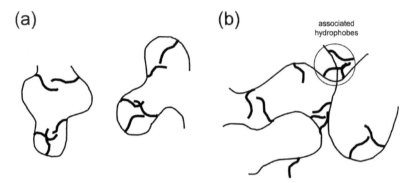

associated
hydrophobes

Figure 4 Association structures in aqueous solutions of graft-modified HMPs (hydrophobes are indicated in bold). (a) *Intra*-molecular associations in dilute solutions. (b) *Inter*-molecular associations and network formation in semidilute solutions.

regarded as "polymeric surfactants" or "polymer-modified surfactants"), they can show an increased surface activity, and a strong tendency to interact with surfactants, surfaces, and colloidal (inorganic or organic) particles [10,11,20,22,25]. The general picture for surfactant–HMP interactions that is emerging from recent investigations is somehow different from that of the non-modified polymers; in particular, there is no CAC [12,14,15,23]. Instead, as it is now realized, surfactant self-assembly is not a prerequisite for surfactant–HMP complexation. The HM-polymers are sufficiently hydrophobic and tend to self-associate and form micellar-type aggregates in aqueous media, even in the absence of surfactants. In doing so, they create hydrophobic domains in which individual surfactant molecules can be bound or solubilized and, as a consequence, "mixed micelles" are formed, involving both surfactant hydrophobic tails and polymer hydrophobes. In fact, because of this mixed micelle formation, hydrophobically modified polyelectrolytes can associate even with similarly charged surfactants [22,23]. Moreover, surfactants can render HMPs with high degrees of hydrophobic modification (that are otherwise insoluble in pure water) water-soluble [14].

In this context, it has proven fruitful to view the surfactant–HMP complexation in a way similar to the mixed micellization between two surfactants [11,12,14,15]. Such a

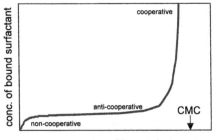

Figure 5 Schematic representation of the binding isotherm of an ionic surfactant to a nonionic hydrophobically modified polymer. The critical micelle concentration (CMC) (in the absence of the polymer) is indicated.

viewpoint rationalizes the complex binding isotherms with regions of non-cooperativity and anti-cooperativity observed in surfactant–HMP and in ionic–nonionic surfactant micellization. A typical binding isotherm of an ionic surfactant to a nonionic HMP is shown in Fig. 5. Three concentration regions can be distinguished: (i) in the first region, non-cooperative binding of individual ionic surfactant molecules to the hydrophobic domains created by the HMP hydrophobes takes place; (ii) as the concentration of the surfactant increases, the binding becomes anti-cooperative because of the unfavorable incorporation of a new surfactant molecule into a similarly charged micelle; (iii) in the last region, as the free surfactant concentration approaches the CAC (which signals the onset of the polymer–surfactant complexation in the case of the parent homopolymer), the binding becomes highly cooperative due to the self-assembly of the surfactant into micelles (at the polymer chains initially, and then as free micelles in solution) [11,12,14,15].

Recent studies on the interactions between surfactants and HMPs have resulted in a deeper understanding of the origin of their rheological and phase behavior on the basis of the molecular picture of the aggregates formed [10,13,23]. As expected from the discussion above, the nature of the mixed aggregates will vary with the polymer/surfactant mixing ratio. This transition from micelles consisting of polymer hydrophobes alone (at the limits of large polymer dominance), to mixed micelles involving both surfactant and polymer hydrophobic groups, to coexistence between mixed micelles and micelles formed by surfactant alone (occurring at the limits of very large surfactant dominance), is presented schematically in Fig. 6. The composition of the mixed micelles and the ratio of the polymer to the surfactant hydrophobes in these will change progressively with increasing surfactant concentration, but it will not necessarily be the same as the global polymer/surfactant ratio. This happens because nonassociated polymer hydrophobes or free surfactant molecules that do not participate in micelles may also exist. The exact mixed micellar compositions are difficult to determine experimentally, and data on the binding isotherms of surfactants to HMPs are rather limited [12,13,23].

V. RHEOLOGICAL BEHAVIOR OF SURFACTANT–POLYMER SYSTEMS

Mixed polymer–surfactant systems are most commonly employed in industrial practice to achieve desired rheological properties that depend on specific applications. These can range from simple thickening effects to producing viscoelastic liquids, or "gels", that exhibit specific response to shear or display solid-like characteristics [10,20,25].

A surfactant–promoted inter-polymer interaction usually gives rise to a viscosity enhancement that can become more pronounced with increasing polymer concentration or temperature [10,11]. In mixtures of surfactants and clouding polymers (which contain groups with a polarity that can decrease with increasing temperature), thermoreversible gelation can often occur. On increasing temperature, stronger gels are formed, and upon cooling these gels melt. A similar effect may also be achieved in mixtures of nonionic surfactants and hydrophobically modified water-soluble polymers by a temperature-induced micellar growth or a transition from micelles to other self-assembled structures (e.g., vesicles) [10,11].

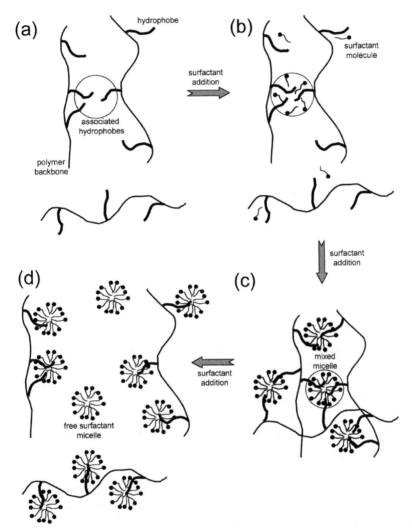

Figure 6 Schematic of interactions between surfactants and hydrophobically modified polymers (HMP). (a) Association between polymer hydrophobes takes place in the absence of surfactant. (b) Added surfactants bind to hydrophobic domains created by the polymer hydrophobes. (c) Upon further addition of surfactant, mixed micelles between surfactant and polymer hydrophobes form, which constitute the network connecting points. (d) Eventually, the number of micelles becomes greater than the number of hydrophobes and the cross-linking of the polymer chains is lost.

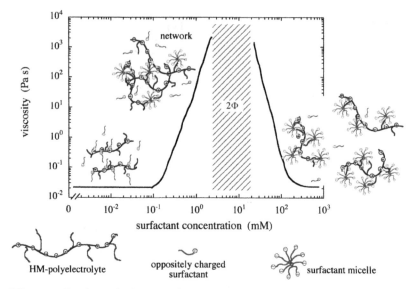

Figure 7 Association and network formation in mixtures of oppositely-charged surfactant and hydrophobically-modified polymer.

A remarkably different rheological behavior is observed for aqueous solutions of hydrophobically-modified polymers when small amounts of surfactant are added to them. This viscosity enhancement has been ascribed to the ability of the surfactants to assemble at the hydrophobic moieties of the polymers and to the rearrangement of the solution structure due to the formation of the mixed micelles [10,12,13,22,23]. We exemplify this behavior with a schematic representation of the viscosity changes in a system containing a hydrophobically-modified polyelectrolyte and an oppositely charged surfactant (Fig. 7). Upon addition of surfactant, the viscosity of the system can increase by several orders of magnitude (with the viscosity peak generally occurring in the vicinity of CAC), followed by a decrease at higher surfactant concentrations. This non-monotonic variation of the viscosity is related to the composition of the mixed micelles formed during the polymer–surfactant association. In order to have cross-linking in the system and, thus, a viscosifying effect, a sufficiently high number (equal to or higher than 2) of polymer hydrophobes must be present in each micelle. This happens at relatively low surfactant concentrations, and the mixed micelles act as the connecting points between different polymer chains in the transient network formed. However, at the final stage of surfactant addition, the stoichiometry of the micelles becomes such that they are dominated by the surfactant. As the number of polymer hydrophobes in each micelle decreases to one, the connectivity between polymer chains is lost, and so is the enhanced system viscosity [10,13,23,25].

In the specific case presented here, involving a polyelectrolyte and a surfactant of opposite charge, the interactions usually result in an associative phase separation. The behavior is generally characterized by three distinct regions with increasing surfactant concentration: (i) a single-phase region where the polymer is in excess; (ii) a precipitation region, where the surfactant-to-polymer ratio (in terms of charges) is close or equal to charge neutralization (i.e., equal amounts of polymer and surfactant charges); and (iii) a single-phase region again. In this last region, redissolution of the complex can take place, a phenomenon that has been ascribed to the charge reversal of the complex by an excess of surfactant. This region though, can be unattainable if the charge density of the polymer is too high. The phase separation presented here occurs close to the viscosity maximum. This is the case when the polymer is not sufficiently charged and, as a consequence, charge neutralization takes place at relatively low surfactant concentrations. For highly charged HMPEs, the phase separation can occur at much higher surfactant concentrations, often after the disruption of the network and the viscosity decrease (depending on the micellar association number) [13,19,29].

The rheological behavior of the mixed surfactant–polymer systems will depend on the polymer concentration and its modification (type and degree), and on the type, concentration and micellar association number of the surfactant, and will be greatly influenced when electrostatic effects come into play. Generally, for a nonionic polymer nonionic surfactants have a weak influence on the system viscosity, while anionic surfactants have a stronger effect than cationics. For charged HMPEs, the strongest viscosity effects are caused by the addition of oppositely charged surfactants [10,11,25].

VI. PHASE BEHAVIOR OF SURFACTANT–POLYMER SYSTEMS

In industrial applications of surfactant–polymer systems, the polymer is usually added with an aim to impart desired rheological properties, while the surfactant offers a hydrophobic microenvironment suitable for solubilization. When surfactant and polymer are mixed, macroscopic phase separation may occur, which is often an unwanted outcome. There are cases, however, where surfactants are deliberately added to polymers in order to cause phase separation. This is relevant to applications like deposition and two-phase partitioning.

The phase behavior of a ternary surfactant–polymer–solvent system will depend on the balance between the interactions of the species involved. Two main classes of biphasic demixing may be distinguished with regard to phase separation phenomena: "associative" phase separation (involving one dilute phase in equilibrium with a phase concentrated in both the polymer and the surfactant), and "segregative" phase separation (each of the two phases is enriched in one of the solutes). These two types of separation are shown schematically in Fig. 8. This terminology was first introduced by Piculell and Lindman [30] who emphasized the idea that a micelle is also characterized by high molecular weight, and made an analogy of the surfactant–polymer behavior to that of

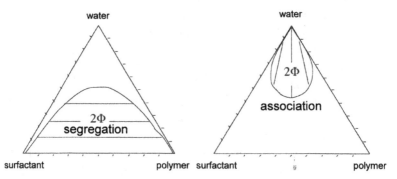

Figure 8 Phase separation in ternary water/surfactant/polymer systems.

Table 1 Phase behavior of aqueous polymer–surfactant mixtures

Polymer/ surfactant system[a]	Type of separation[b]
P^0/S^0	Segregation
	Possible association for less polar polymers (at high temperatures)
P^0/S^+ or P^0/S^-	Miscibility
	Segregation or association with added electrolyte
P^+/S^0 or P^-/S^0	Miscibility
	Segregation or association with added electrolyte
P^+/S^+ or P^-/S^-	Segregation
P^+/S^- or P^-/S^+	Association
	Miscibility with added electrolyte
	Segregation at high electrolyte concentrations

[a]Superscript (+, −, or 0) indicates the charge of the polymer (P) or the surfactant (S).
[b]The type of separation will depend on the specific system studied and may be different than these general observations given here.

polymer–polymer mixtures. Here, we present an overview of the phase separation in different systems (for further information and a description of experimental ternary polymer–surfactant–water phase diagrams the readers are referred to previous excellent reviews on the subject [14,18,19,21,29–31]).

The general phase behavior of aqueous mixtures of a surfactant (S) and a polymer (P) is summarized in Table 1. The kind and extent of the phase separation is determined by external conditions (e.g., temperature, addition of electrolyte), as well as by the "internal" properties of the polymer and the surfactant involved in the system. These include the charge, the degree of polymerization, and the nature of the polymer on one hand, and the association number of the surfactant micelles, and the chemical nature and charge of the surfactant head-group on the other. Segregative phase separation is commonly observed for mixtures of a nonionic polymer and a nonionic surfactant, and also for mixtures of similarly charged polymer and surfactant. Associative phase separation generally takes place in oppositely charged systems where the polymer and the surfactant often form insoluble complexes. Transition, in the same system, from one type of phase separation to the other (through conditions of miscibility) by addition of electrolyte, is also possible. The phase behavior of surfactant–polymer mixtures can also be affected by specific interactions such as the hydrophobic interactions in the case of HMPs. This is demonstrated by an enhanced separation for nonionic systems, and an increased miscibility for oppositely charged mixtures [14,29,30].

VII. SURFACTANT–POLYMER COMPLEXES IN CONCENTRATED SYSTEMS

As alluded to in the previous section, ionic surfactants and oppositely charged polyelectrolytes can spontaneously form complexes that precipitate from aqueous solutions, typically with 1 : 1 stoichiometry. These complexes combine in a unique way the amphiphilic properties of the surfactants and the mechanical strength and stability of polymeric materials. In the solid state, the polyelectrolyte–surfactant complexes can form highly ordered phases [32–43], thus offering the opportunity of producing a variety of new "solid" materials with interesting applications that may range from switchable membranes with controlled permeability to materials with non-wetting properties. Water insoluble complexes containing conventional synthetic polyelectrolytes [34–40], or charged synthetic polypeptides [41–43] have been investigated. Noteworthy are the contributions originating from Antonietti and co-workers [34–40] who have characterized a number of lamellar or cylindrical phases in such complexes; the type of structure formed depended on the type of the surfactant (length and shape) and the polymer (chemical nature and charge density) used (see Fig. 9). Stoichiometric complexes can be easily processed to films by casting from organic solvents. However, melt processability is limited due to the high glass transition temperature (T_g) of the complexes (caused by the high density of ionic groups), which usually exceeds the decomposition temperature. The incorporation in the ionic layers of non-charged, polar polymeric units (use of a HMPE) has been proposed as an approach to lowering the T_g of the complexes [36].

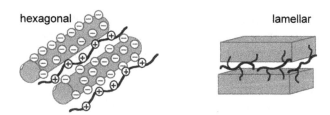

hexagonal

lamellar

Figure 9 Incorporation of polymers in organized (lyotropic liquid crystalline) surfactant assemblies: (left) cationic polymer in hexagonal phase formed by anionic surfactant; (right) hydrophobically-modified polymer in lamellar phase formed by nonionic surfactant.

The alkyl groups of the HMPE act as internal plasticizers and, as a result, highly ordered, solid mesomorphous materials with properties and thermomechanical behavior different from those of the reference polyelectrolyte–surfactant systems are produced. Incorporation of functional groups into a polyelectrolyte–surfactant complex has also lead to materials responsive to external stimuli such as UV-irradiation [38]. Furthermore, complexes of polyallyldimethylammonium chloride (pDADMAC) and SDS, systems that combine mesoscopic order, appropriate charge density and good binding properties with good swellability and solubility, have been recently employed as hosts for the structure-directed polymerization of certain monomers used as guests [39].

The variable water content of polyelectrolyte–surfactant complexes poses a challenge in a number of practical applications. The role of the water in the complexes has been recently examined in detailed studies of mixtures of a polyelectrolyte, sodium polyacrylate (NaPA), a cationic surfactant, cetyltrimethylammonium bromide (CTAB), and water in the high concentration regime, just before the solid state material is reached [44–46]. Concentrated phases with different water content and a range of lyotropic structures (micellar, hexagonal, cubic) were characterized depending on the polyelectrolyte/surfactant ratio and the overall water content of the initial mixture. The concentrated phases may be readily produced by dilution of stoichiometric polyelectrolyte–surfactant mixtures with water. In this context, an interesting alternative approach for the representation of phase diagrams has been proposed [29,44,45,47]. It is suggested that a three-dimensional phase diagram of a pyramid shape be used instead of the classical triangular one, since the polyelectrolyte/surfactant/water system is not truly ternary (more than three species are involved). The phase behavior is better depicted in a pyramidal representation with the water on top and the four possible combinations of the four ions into neutral salts (NaPA, CTAB, NaBr-simple salt and CTAPA-complex salt) at the corners of a square basis.

VIII. THEORETICAL APPROACHES ON SURFACTANT–POLYMER COMPLEXATION

The practical importance of surfactant–polymer systems has stimulated significant efforts to study their behavior theoretically [11,15,48–79]. Theoretical models range from simple equilibrium models to more advanced thermodynamic and detailed mean-field lattice models, as well as recent Monte Carlo simulations. The surfactant–polymer interaction is a complex process, which involves many degrees of freedom and is controlled by hydrophobic and electrostatic interactions, while, at the same time, both a surfactant micelle and a polymer molecule have large possibilities of internal responses. For these reasons the modeling of such systems is not trivial, and even more, computationally-demanding simulations are required for the advanced molecular models. Therefore, all approaches have to rely on certain assumptions based on the nature of the polymer and the surfactant, and often, models that succeed in reproducing certain systems sufficiently well and are in agreement with available experimental data, fail to describe in a satisfactory way the complexation process in other systems. There are two alternative pictures for the description of the polymer–surfactant complexation, a "polymer-centered" and a "surfactant-centered." In the "polymer-centered" viewpoint, the interactions are described in terms of a strongly co-operative association or binding of the surfactant molecules to certain sites of the polymer. According to the "surfactant-centered" picture, the surfactant–polymer complexation is essentially the micellization of surfactant onto or in the vicinity of the polymer chain. The critical aggregation concentration (CAC) is then considered as the surfactant CMC in the surfactant–polymer solution, albeit lowered compared to the CMC by the presence of the polymer. While these two descriptions are largely overlapping, their implications for the complexation and the aggregate structures may be different for different situations. For hydrophilic homopolymers, the "surfactant-centered" picture has distinct advantages, whereas the binding approach has proven very fruitful for hydrophobically modified polymers where no threshold concentration for surfactant binding is observed (as discussed in Section IV.B).

The thermodynamic models are based on the calculation of the free energy contributions for several simple steps during the surfactant self-association process in the presence of the polymer. In one theoretical treatment, it is assumed that polymer segments penetrate into the region occupied by the surfactant heads and shield partially the contact area between the micellar hydrocarbon core and water, thus decreasing the free energy of formation of the micelle–water interface on one hand, while increasing the free energy change arising from the steric repulsions at the micellar surface on the other. Whether the complexation process actually occurs is determined by the relative importance of the two competing effects. In another approach, the polymer adsorption at the micelle surface induces a change in the microenvironment surrounding the micelle and affects the association process. The interfacial tension between the hydrocarbon core of the micelle and the solvent becomes smaller, but that between the surfactant headgroups and the solvent increases (thus opposing surfactant–polymer complexation).

These two approaches have been applied to the self-assembly of surfactant molecules into micelles and to the complexation between surfactant micelles and nonionic polymers (e.g., PEO, PPO) to predict the CMC and micelle association numbers [48–52]. Nikas and Blankschtein [55] have used a theoretical formulation which combines a thermodynamic description of polymer–surfactant solutions with a molecular model of polymer–surfactant complexation, and have incorporated explicitly the effects of the solvent quality, the polymer hydrophobicity and flexibility, as well as specific interactions between the polymer segments and the surfactant hydrophilic moieties on the complexation behavior. The theory can be utilized to predict, among others, the CMC (for polymer-free solutions), the CAC, the number of bound micelles per polymer chain and their association numbers, in reasonable agreement with available experimental data for aqueous solutions of PEO or PVP mixed with various sodium alkyl sulfates. Additionally, the theory predicts that poor solvent quality and small sizes of the surfactant hydrophilic moieties may also promote complexation. Borisov and Halperin [56] have considered the self-assembly of polysoap-type polymers and the modifications to both the configuration and the elasticity of the polymer chain by the presence of surfactants. Moreover, the interactions between end-capped associating polymers and surfactants have been investigated by Balazs and coworkers [57–60] by solving simple and lattice based models by Monte Carlo simulations. They predict that the surfactant tail length and the ratio of the polymer to surfactant concentration affect the aggregates formed, and they are able to correlate the microscopic geometry of the aggregates to the macroscopic rheological behavior observed in such systems.

   In recent years, the theoretical investigations have extended to the complexation between polyelectrolytes and charged surfactants [61–79]. Wallin and Linse [61] used the mean-field lattice theory to study qualitatively the association between polyelectrolytes and oppositely charged surfactants and predict the distribution of the polyelectrolyte chain around the micelle. In a comparison between different polyelectrolytes, they find that the hydrophilic polyelectrolyte strongly adsorbs onto the micelle surface replacing the surfactant counterions, while more hydrophobic polyelectrolytes penetrate into the headgroup region and eventually into the micellar core. Monte Carlo simulations have also been employed to obtain structural and thermodynamic aspects of the polyelectrolyte–macroion complexation. Schiessel *et al.* [66], Chodanowski and Stoll [67–69], Linse and coworkers in a series of papers [62–64,70–72,78,79] and Nguyen and Shklovskii [73–77], have investigated the effect on the complexation of parameters such as the polyelectrolyte chain flexibility and charge density, the surfactant tail length, and the macroion charge. Interestingly, it is found that, for flexible chains, a compact complex involving a collapsed polyelectrolyte wrapping around the macroion is formed, whereas, as the stiffness of the chain is increased, a range of different structures such as "tennis ball seam"-like, solenoid, "rosette" (multiloop), single loop, and structures involving only one contact region between the polyelectrolyte and the macroion are obtained [66,70–72,74]. Moreover, the effects of salt addition to the system [67,68,76,78], the overcharging of a macroion by an oppositely charged polyelectrolyte, the complexation and charge inversion [68,73–76,78], as well as the phase separation and redissolution of the polyelectrolyte–macroion complex [78,79] have been successfully simulated.

IX. CASE STUDIES ON SURFACTANT–POLYMER INTERACTIONS

We review here recent results on the association in aqueous solutions of surfactants with amphiphilic polymers of industrial relevance belonging to the following five classes: amphiphilic polyether block copolymers (Poloxamers), hydrophobically end-capped ethoxylated urethane (HEUR) polymers, cellulosic polymers, hydrophobically-modified polyacrylate polymers, and hydrophobically-modified alkali soluble emulsion (HASE) polymers.

A. Interactions between Surfactants and Amphiphilic Polyether Block Copolymers

Considerable research efforts have been directed to the study of the self-assembly properties of amphiphilic polymers and, in particular, block copolymers with the structure poly(ethylene oxide)-block-poly(propylene oxide)-block-poly(ethylene oxide) (PEO-PPO-PEO) that are commercially available as Poloxamers or Pluronics [28,80]. The solvent quality and quantity are controlling factors in the self-assembly and structure of such amphiphiles and provide extra degrees of freedom in tailoring their properties for specific applications [81]. In the case of water, selective solvent for PEO, variation of the solvent quality can be achieved by changes in the temperature [82,83], by the addition of cosolvents miscible (or partially miscible) with water, such as formamide, ethanol, propanol, or glycerol [84–88], and/or by the dissolution in water of solutes such as urea [6] and salts [5]. More dramatic effects on the self-assembly properties of amphiphilic polymers in aqueous solutions can be achieved by the addition of typical (low-molecular weight) surfactants, as these interact specifically with different blocks of the amphiphilic polymer and modulate their segregation [89–103].

Several studies have examined the effect of surfactants on PEO-PPO-PEO block copolymer micelle [89–99] and lyotropic liquid crystalline phase [100–103] formation and structure in aqueous solutions. Here we focus on the salient aspects of surfactant effects on PEO-PPO-PEO block copolymer micelles. Anionic surfactants are those most commonly considered [89–96,100,101], while some studies used cationic [97] and nonionic [98,99,102,103] surfactants. Various experimental techniques have been utilized: NMR chemical shift measurements and fluorescence quenching [89], differential scanning calorimetry [90,95,98], surface tension [90,91], use of selective electrodes via electromotive force, and isothermal titration calorimetry [92,94,98], rheology [99,103], small-angle neutron scattering [94,100], and small-angle X-ray scattering [102,103]. Usually, a specific sample (system) needs to be interrogated with more than one experimental techniques in order to draw conclusions [82]. For the case of PEO-PPO-PEO block copolymers in aqueous solutions, for which the micellization is an endothermic process and can be induced both by an increase in copolymer concentration (at fixed temperature) and by an increase in temperature (for a given copolymer concentration) [104], isothermal titration calorimetry (ITC) and differential scanning calorimetry (DSC) have been shown very useful and practical [5,82,95,96,98,103]. In what follows we highlight findings on the self-assembly in aqueous solution of mixtures containing (i) the ionic surfactant SDS and Pluronic F127 block copolymer of relatively high PEO-content and molecular weight

($EO_{97}PO_{69}EO_{97}$) [92–95], (ii) SDS and Pluronics P123 or L121 (both of relatively low PEO-content and molecular weight) [95], (iii) the cationic surfactant tetradecyltrimethylammonium bromide (TTAB) and Pluronic F127 [97], and (iv) the nonionic surfactant hexaethylene glycol mono-n-dodecyl ether ($C_{12}EO_6$) and Pluronic F127 [98].

The binding of SDS to Pluronic F127 was studied using a SDS-selective electrode via electromotive force, ITC, and light scattering [92]. At 35°C, aqueous 0.5% w/v Pluronic F127 forms micelles in equilibrium with unassociated polymer molecules [104]. When SDS was gradually added to this solution, binding of SDS to the Pluronic F127 micelles took place at SDS concentrations as low as 1×10^{-5} mol/L [92]. Mixed F127/SDS micelles were formed, with a size that remained constant until 5×10^{-5} mol/L SDS when each micelle contained (on the average) 6 SDS and 69 Pluronic F127 molecules. Further addition of SDS resulted in a breakdown of the F127-rich mixed micelles into smaller SDS-rich micelles. This process continued until 3.5×10^{-4} mol/L SDS, when SDS micelles started binding unassociated F127 molecules. When the total SDS concentration reached 3×10^{-3} mol/L, all the F127/SDS mixed aggregates were broken down. Addition of SDS in excess of 3×10^{-3} mol/L resulted in the formation of more and larger SDS micelles bound to individual F127 molecules until (at about 0.1 mol/L added SDS) all the unassociated F127 molecules available for binding became fully saturated with bound SDS micelles [92]. In addition to the above effects, SDS was found to induce Pluronic F127 micelle formation at 3°C below the temperature at which micelles would form in the absence of SDS [93]. In the concentration range where mixed SDS–Pluronic F127 micelles were formed, the interaction between the two molecules was synergistic [94].

The interactions between SDS and $EO_nPO_mEO_n$ block copolymers having the same PPO block length (68 PO segments) but different PEO block lengths (5, 20, and 97 EO segments, corresponding to Pluronics L121, P123, and F127, respectively) have been studied in dilute aqueous solution [95]. Addition of SDS to Pluronics P123 and L121 increased the temperature at which micelles started forming (as detected by

DSC). The effect on F127 was different; after an initial decrease in the micellization temperature, the DSC peak flattened out and disappeared at low SDS concentration. Successive addition of SDS led to the disruption of the copolymer micelles/ aggregates and accompanying hydration of the PPO block (as detected by ITC). At 40°C, about 20 SDS molecules per copolymer molecule were required to dissolve the Pluronic P123 micelles, while about 10 sufficed to disrupt the Pluronic F127 micelles [95].

The interaction between the cationic TTAB and Pluronic F127 was reported in [97]. TTAB formed a polymer/micellar TTAB complex with monomeric F127. Similar to SDS, TTAB bound to F127 micelles leading to the transformation of the micellized F127 into mixed F127/TTAB micelles that became smaller as more TTAB was added, until all the micellized F127 was dissociated. DSC measurements have shown that small amounts of TTAB (typically 10^{-4} mol/L) can decrease the critical micellization temperature of F127 [97]. Binary mixtures of the nonionic surfactant hexaethylene glycol mono-n-dodecyl ether ($C_{12}EO_6$) and Pluronic F127 were examined by calorimetry [98]. In all cases considered mixed micelles were formed, and the interaction between the two amphiphiles showed synergistic behavior [98].

B. Interactions Between Surfactants and Hydrophobically–modified Ethoxylated Urethane Polymers

Hydrophobically-modified ethoxylated urethane (HEUR) polymers are prepared by connecting blocks of poly(ethylene oxide) with isophorone diisocyanate, followed by capping with either hydroxyl, dodecyl, or hexadecyl linear alkyl end-groups; this results in telechelic molecules where the hydrophobic groups are located at the two ends. In another synthetic route, excess isophorone diisocyanate is reacted with PEO to prepare an isocyanato functional precursor, followed by reaction of the terminal isocyanate group with nonylphenol. Viscous aqueous solutions are obtained with only partial terminal modification of PEO, while swollen gels result from full terminal modification [25,27,105–108]. Model HEUR polymers with alkyl end-groups interact at very dilute concentrations in aqueous solutions to generate a pronounced increase in reduced viscosity [108]. The rheological behavior exhibited by aqueous solutions of model HEUR polymers has been studied as a function of concentration, molecular weight, temperature, and hydrophobic end-cap length [109–112]. The results are consistent with a transient network [109]. The more hydrophobic the end-groups are, the stronger the cross-link points in the network, and the more elastic is the solution [112]. HEUR polymers form spherical flower-like micelles consisting of looped chains; larger structures are formed through secondary aggregation of the polymer micelles at higher polymer concentrations (see Fig. 3) [113–115].

The addition of surfactant has pronounced effects on the viscosity of aqueous HEUR solutions [112,116–118]. The plateau modulus of HEUR polymer solutions in the presence of added surfactant (SDS or nonylphenol ethoxylates) was found to exhibit the following features: (i) at low polymer concentration, the modulus was significantly augmented at low to moderate surfactant concentration; (ii) at higher surfactant concentration, the modulus decreased monotonically; (iii) at high polymer concentrations, no significantly detectable peak in modulus was observed; (iv) the position of the modulus peak shifted to lower surfactant concentration as the polymer concentration was lowered [116]. Dynamic light scattering data also reflect this non-monotonic behavior [119]. These effects have been attributed to two competing principal factors, the number and the composition of micelles (network points of cross-link) [116]. Higher surfactant concentrations increase the number density of the micelles in the system, which, in turn, causes an entropically driven transition from a network composed predominantly of loops (individual polymer molecules) to one in which links prevail. A higher portion of chains in link configurations and a reduction in the number of undesirable uni- and bifunctional micelles both serve to increase the elastic modulus of the network [116]. Against this however, there is a second factor associated with the surfactant addition: the formation of mixed micelles between surfactant and HEUR end-groups. From the point of the HEUR chains, these mixed micelles will tend to have a lower average functionality. As the surfactant concentration is increased, the probability of finding many end-groups simultaneously within the same micelle decreases rapidly. Thus, this second factor promotes a higher number of uni- and bifunctional micelles, resulting in a drop in the modulus. The net effect of the above two opposing factors is a maximum in the modulus as a function of surfactant concentration. Clearly, at high polymer concentrations, where loops are rare, the addition of surfactant has a rather limited influence in turning loops to links. It is then likely that the second factor described above dominates quickly, resulting in an almost immediate drop in the modulus. Any maximum in the modulus, if it exists at all, would become too small to detect at these higher polymer concentrations [116].

The magnitude of the variation in viscosity is dependent on the chemical structures of both the polymer and the surfactant. The observed differences can be explained qualitatively by stoichiometric considerations of the mixed micelles that are formed between the hydrophobic chains of the polymer and the surfactant [112]. A nonionic surfactant was found to bind by a non-cooperative process only to the hydrophobic aggregates of the HEUR [112], whereas SDS was bound both to the hydrophobic domains and to the PEO blocks of the HEUR polymers [112,117]. SDS binds to HEUR with greater affinity than dodecyltrimethylammonium bromide (DTAB) [117].

C. Interactions between Surfactants and Cellulosic Polymers

Water-soluble polymers derived from cellulose, one of nature's most abundant polysaccharides, find significant commercial utilization due to their biological compatibility and because they exhibit great ecological advantages. Various cellulose derivatives can be synthesized by post-functionalization techniques in which hydrophobic and/or ionogenic groups are introduced into the structure of an already existing nonionic polymer. Cellulosic polymers are used as emulsifiers, stabilizers, and thickening agents in food, pharmaceutical and cosmetic products, in paint formulations, for coating, cementing, oil-well fracturing and drilling applications. A substantial body of research by several groups [13,18–20,22,29,120–163] has been devoted to the study of cellulose ethers such as hydroxyethyl cellulose (HEC), ethyl hydroxyethyl cellulose (EHEC), and hydrophobically modified derivatives of them, hydrophobically modified hydroxyethyl cellulose (HM-HEC), hydrophobically modified ethyl hydroxyethyl cellulose (HM-EHEC), and their interactions with surfactants.

An inherent feature of EHEC (which can be considered as a slightly hydrophobic polymer) and HM-EHEC polymers is an inverse temperature dependence of solubility in water. Upon heating, an increased turbidity is first observed (denoted as cloud point temperature), followed by a macroscopic phase separation into polymer-rich and polymer-lean phases [11,29]. The addition of surfactants can have a significant (depending on the type of surfactant) influence on the cloud point of these polymers, with practical implications [120–137]. The interaction between the EHEC polymers and surfactants gives rise to the formation of polymer–surfactant clusters at low levels of surfactant addition. Such interaction is enhanced at higher temperatures when the EHEC polymers approach their cloud point. The cluster size increases at low surfactant concentrations and passes through a maximum slightly above the critical aggregation concentration. The position of the maximum is shifted toward lower surfactant concentrations with increasing temperature. At moderate levels of surfactant in the system, addition of salt is found to strengthen the intermolecular associations. The level of chain association or entanglement is higher for the EHEC/SDS system than for the EHEC/CTAB system (as determined from light scattering). The degree of surfactant binding (determined by NMR) to EHEC, at a given total surfactant concentration, is higher in the presence of CTAB than with SDS. In both cases, the amount of surfactant bound to the polymer is roughly independent of temperature [120–131]. While only small temperature effects can be observed in the dilute regime, thermo-reversible gelling can occur in semidilute EHEC solutions with appropriate amounts of surfactant [121,122,133,134]. This thermal gelation has been correlated (through small angle neutron scattering experiments) with a microphase separation in the system. As proposed by Cabane *et al.* [133,134], addition of surfactant causes a reduction in the size of "lumps" of polymers (formed as the polymer molecules become strongly associating at higher temperatures) and a swelling of the system; these lumps provide the mechanical rigidity of the network and lead to gel formation.

A large number of studies have focused on the interactions of HM-EHEC with surfactants [123–127,130,135–137]. Compared to the behavior of the non-modified analogue EHEC, the presence of hydrophobes (consisting of a low number of nonylphenol groups grafted to the polymer backbone) on HM-EHEC led to a lowering of the cloud point at low SDS concentrations, and a broadening and deepening of the observed cloud point minimum. The qualitative phase behavior, however, was the same for both polymers. The observed behavior upon addition of surfactant is non-monotonic: the tendency to phase separate first increases (the mixed micelles formed lead to an increased inter-polymer attraction), and then decreases at higher surfactant concentrations (the connectivity is lost as the surfactant dominates in the mixed micelles) [123–127]. Both dynamic light scattering and rheological measurements revealed enhanced surfactant–polymer interactions at SDS concentrations of 8–10 mM and 4–5 mM for the systems EHEC/SDS and HM-EHEC/SDS, respectively. An increase in the dynamic viscosity of approximately fivefold for EHEC and fifteen times for the HM-EHEC was also reported [123,125–127]. Moreover, addition of small amounts of SDS was found to induce segregative phase separation to monophasic samples of EHEC/HM-EHEC mixtures as the interchain attractions become significantly enhanced with the binding of SDS to the hydrophobic moieties of HM-EHEC [138]. The effect of an added "spacer" (a hydrophilic segment of four EO groups) between the hydrophobes and the polymer backbone of the HM-EHEC polymer has also been considered [139]. It was evident that the polymer with the spacer interacted stronger with SDS; the spacer renders the polymer hydrophobe more flexible and effectively longer and thus facilitates interaction with the surfactant molecules.

Interactions between surfactants and hydroxyethyl cellulose (HEC) or several HEC derivatives modified with hydrophobic groups have been investigated by means of phase behavior and rheology [140–150]. When anionic surfactants were added to HEC or HM-HEC, polymer–surfactant interactions were observed in the vast majority of systems at surfactant concentrations lower than the CMC of free surfactant in water. Addition of a nonionic surfactant (Triton X-100) to HEC resulted in segregative phase separation above a certain surfactant concentration. However, phase separation was suppressed for HM-HEC compared to the non-modified analogue, and only occurred at much higher surfactant concentrations. Certain amount of surfactant is consumed in the formation of mixed micelles with the polymer hydrophobes before pure surfactant micelles can form and, as an outcome, segregation is delayed [143]. For any individual surfactant, the observed rheological effects on HM-HEC were found to be dependent on the level of surfactant addition; an increase of the system viscosity happened at first, followed by a decrease at higher surfactant concentrations and finally a breakdown of the network structure. The magnitude of the observed rheological effects was also found to be dependent on the alkyl chain length and the nature of the hydrophilic head group of the added surfactant [145–148].

The loss of viscosity from the break-up of associations at high surfactant concentrations, due to masking of the polymer hydrophobes by excess surfactant, generates problems in practical applications. A desire for better viscosity control has directed attention to other surfactant morphologies that can enhance the bridging of hydrophobic domains and expand the range of added surfactant, but without the adverse effects that small spherical micelles may cause [149–151]. In one study involving interactions between HM-HEC and various surfactants, anionic SDS, cationic CTAB, and nonionic cetyldimethylamine oxide (CDMAO), the HM-HEC solution viscosity was maintained high by transforming surfactant micelles from spheres to rods (by addition of salt or other solvent) [149].

In another approach, the presence of mixed micelles (consisting of SDS and dodecyltrimethylammonium chloride, DTAC) had an interesting effect on the viscosity of HM-HEC polymer. It was shown that, in solutions of low viscosity where the ratio between polymer hydrophobes and mixed micelles is low, a high viscosity can be recovered by increasing the surfactant association number in the mixed micelles (either by addition of a screening electrolyte, or by changing the ratio between the two surfactants used) [150,151].

The macroscopic properties of aqueous solutions of cationic hydroxyethyl cellulose (catHEC) or cationic hydrophobically modified hydroxyethyl cellulose (catHM-HEC) with anionic or cationic surfactants were initially studied by rheological and phase separation means by Goddard and coworkers [18–20, 152,153] and later on by other groups [13,22,154–161]. The general features observed in these studies were quite similar. On addition of an oppositely charged surfactant (SDS in most cases) catHEC and catHM-HEC polymer solutions show similar phase behavior: associative phase separation at certain amounts of surfactant, followed by redissolution at excess surfactant concentrations. The cross-links between the polymer chains (originating from surfactant micelles electrostatically bound to the oppositely charged polyelectrolyte) lead in a three-dimensional network creation verified by high solution viscosities, or even formation of "gels." A viscosity enhancement by almost three orders of magnitude in the case of catHEC/SDS system and by over four orders of magnitude for catHM-HEC/SDS can be observed by addition of SDS to these polymers [19,20,158, 160,161]. Evidence for the intermolecular interactions between the polymer chains via the bound surfactant has been provided from solubility data [19], electrophoresis and surface tension [19], dye solubilization [152], small-angle neutron scattering [152], and, more recently, from cryo-TEM (transmission electron microscopy) experiments [159] where bilayer fragments, small vesicles, and disc-like aggregates were identified. Winnik and coworkers [22,154–157] carried out extensive studies in these systems using fluorescence techniques. Information on the molecular level was obtained by random labeling of the catHEC and catHM-HEC polymers with pyrene.

An important work, which provided key information on the interactions of hydrophobically modified polyelectrolytes and oppositely charged surfactants, was published by Guillemet and Piculell [158]. They rationalized the macroscopic behavior (phase separation and rheology) of the catHM-HEC/SDS system by taking into account the general features of the surfactant binding isotherm. They were able to explain the associative phase behavior, the redissolution at high surfactant concentrations and the viscosity maximum, and correlate them with the nature of the aggregates formed, by making an analogy with the formation of mixed micelles in simple surfactant systems. Kästner *et al.* [145–148] studied the interactions between surfactants and two types of catHM-HEC derivatives, one with the cationic and the hydrophobic groups separately attached onto the polymer backbone, and one with both cationic and hydrophobic entities at the same substituent and observed differences in the rheological properties, phase behavior, and surface tension of these derivatives in the presence of surfactants. More recently, Chronakis and Alexandridis [160] studied the effect of the polymer charge density and molecular weight on the rheological properties of catHECs upon addition of anionic surfactants with different architecture, SDS, sodium dodecylbenzenesulfonate (SDBS), and sodium bis(2-ethylhexyl) sulfosuccinate (AOT). The use of cyclodextrins has been proposed by Tsianou and Alexandridis [161–163] as a novel approach for the modulation of the rheological properties of surfactant–polymer systems. Cyclodextrins (cyclic oligomers of α-D-glucose) can form inclusion complexes with the hydrophobic moieties of surfactants and polymers. Thus, they are able to disrupt the physically cross-linked network formed between polymers and oppositely charged surfactants as in the case of catHEC/SDS and catHM-HEC/SDS systems (see Fig. 10).

Other water-soluble derivatives of naturally occurring polymers such as sodium carboxymethyl cellulose (Na-CMC) [164,165] and chitosan (non-modified and hydrophobically modified) [166] and their interactions with surfactants have also been explored.

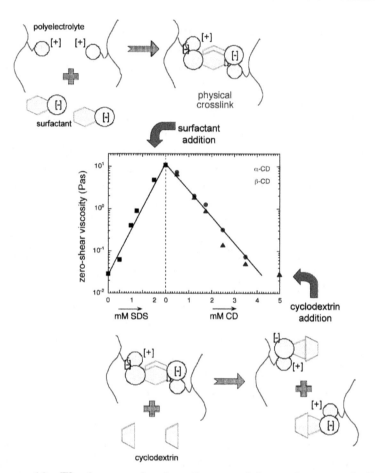

Figure 10 The increase in viscosity caused by surfactant micelles acting as cross-links between polymer chains in mixtures of an anionic surfactant and a cationic polymer, can be completely reversed by the addition of cyclodextrins that bind to the hydrophobic part of the surfactant and "dissolve" the micelles.

D. Interactions between Surfactants and Hydrophobically-modified Polyacrylate Polymers

The interactions between anionic, cationic, and non-ionic surfactants with a series of hydrophobically-modified poly(sodium acrylate) (HM-NaPA) polymers, obtained by modification of a precursor poly(acrylic acid), have been comprehensively investigated by Iliopoulos and coworkers [23,167–172]. The NaPA derivatives contained 1 or 3 mol% of N-dodecyl (or octadecyl) acrylamide side-groups randomly distributed along the polymer backbone. These polymers, above a certain polymer concentration, exhibit much higher viscosities (the level of which depends on the kind and degree of modification) than the precursor non-modified poly(sodium acrylate) [167,168]. On addition of similarly charged SDS surfactant (up to concentrations close to CMC), a viscosity increase of several orders of magnitude was observed, followed by a decrease at higher surfactant concentrations; this has been attributed to variations in the composition of the mixed micelles. The strength of interaction increased with the polymer hydrophobicity [167]. In another study [168], the effects of cationic surfactants DTAC and DTAB were investigated by rheological, and steady-state and time-resolved fluorescence measurements (for the determination of composition and the association number of the micelles). This work has proven influential in elucidating the nature of the interactions between surfactants and HMPs and the importance of inter- and intrachain contributions to the formation of the mixed micelles. It was found that the total number of surfactant molecules and alkyl groups in the mixed micelles is close to the association number of the free surfactant micelles, and it was possible to correlate the observed rheological behavior to the number of polymer hydrophobes in the mixed micelles [10,168]. In these systems, involving a HM-polyelectrolyte and oppositely charged surfactant, associative phase separation takes place. Given that the polyelectrolyte (i.e., polyacrylate) is highly charged, the observed phase separation occurred at very high surfactant concentrations, well after the network disruption and corresponding viscosity decrease took place [14,168]. Hydrophobically-modified poly(sodium acrylate) derivatives bearing perfluorinated side alkyl chains instead of the hydrogenated ones and their interactions with hydrogenated or perfluorinated anionic surfactants have also been studied [169].

HM-NaPA polymers were also found to associate with nonionic surfactants of oligoethylene glycol monododecyl ether type ($C_{12}EO_m$) when the former were introduced in small amounts to the micellar and lamellar phases of binary nonionic surfactant/water systems [171–173]. The association has a large impact on the rheological properties of the mixtures, and moreover was found to stabilize the micellar phase. The principle of viscosity control by HMPs was demonstrated in systems containing $C_{12}EO_4$ surfactant and HM-NaPA polymer where an interesting thermo-reversible gelation was found. By raising the temperature, the system changed from a viscous solution to a stiff gel within a narrow temperature range. The effect was attributed to the transformation, with increasing temperature, of the surfactant aggregates from small micelles (in which polymer hydrophobes participate and thus crosslink the system) to giant vesicles (where the polymer adsorbs and forms bridges between neighboring vesicles) [171,172].

The interactions between surfactants and non-hydrophobically modified sodium polyacrylate (NaPA) or polyacrylic acid (PA) have also been investigated [174–181].

E.   Interactions between Surfactants and Hydrophobically-modified Alkali-soluble Emulsion Polymers

Hydrophobically-modified alkali-soluble emulsion (HASE) polymers result from the copolymerization of methacrylic acid, ethyl acrylate, and macromonomers which contain hydrophobes with an alkyl chain ranging in length from 12 to 20 carbon atoms [182–184]. These polymers possess a comb-like architecture with a polyelectrolyte backbone (ethyl acrylate-co-methacrylic acid) and hydrophobes tethered to the backbone via polyether side chains. As the pH is increased (by, e.g., addition of ammonia) to greater than 5, the carboxyl groups ionize to carboxylate ions and the polymers become water-soluble. The HASE polymers thicken mainly via hydrophobic association. The viscosity can increase by two to three orders of magnitude as the pH is raised to 9. At this pH, the polymers form a network of temporarily associating hydrophobic junctions, causing an increase of the viscosity; this increase is attenuated by an increase of the hydrophobe chain length [182–184]. The structure attained by HASE polymer in water consists of various hydrophobic domains of different composition, varying from nonionic micelle-like structures containing upward of 60–80 hydrophobic groups, to mixed structures containing both hydrophobic and ethyl acrylate groups from the polymer backbone [185,186].

Systematic studies have probed the effects of polymer chemical composition (e.g., length of hydrophobic groups) [182, 184] or additives, such as salts [187–190] and surfactants (ionic [191–194] and nonionic [195–198]), on the aqueous solution properties of HASE polymers. In the absence of any added surfactant, the intrinsic viscosity was found to decrease as the length of the hydrophobes increased. The interpretation is that stronger intra-molecular association of the hydrophobic macro-monomer leads to smaller polymer molecular coils [182]. The degree of ethoxylation in the macromonomer controls the nature of the hydrophobic association junctions by altering the flexibility and hydrophobicity of the macromonomer. Optimum thickening efficiency was observed in a system with a spacer having approximately 10 EO segments between the polymer backbone and the macromonomer [183,184]. An increase of salt content in the solution caused the reduced viscosities of the HASE polymers to decrease. This is attributed to a reduction of the mutual repulsion of the charges along the polymer backbone. Model polymers with stronger hydro-phobic associations were more susceptible to the presence of added salt, however such effects were weakened at high shear stresses [190].

The rheological properties of 0.5 wt% aqueous HASE solutions at around pH 9 in the presence of anionic, cationic, and nonionic surfactants were examined [199]. The presence of surfactant below a certain critical concentration (comparable to the CMC of the surfactant) strengthened the associating network through the formation of mixed polymer hydrophobic group/surfactant aggregates [199]. The increased strength was due to an increase in the number of mechanically active, inter-molecular hydrophobic junctions and lifetime of the average junctions. Above this concentration, the results showed that the presence of anionic or cationic surfactant disrupted and weakened the associating network caused by a decrease in number of these junctions and strength of the overall network structure [194,199]. At the same time, free cations of surfactant molecules screened the carboxyl-anion charges on the polymer backbone, which decreased the electrostatic repulsion between polymer chains [199]. The binding behavior of SDS onto several model HASE polymers with C-1, C-8, C-12, C-16, and C-20 hydrophobes was investigated with isothermal titration calorimetry [192]. The binding of SDS to the hydrophobic junction was found entropy-driven; the critical aggregation concentration and the negative enthalpy of aggregation decreased with increasing polymer hydrophobicity [192].

A further reinforcement of the associating network was observed in the presence of nonionic surfactant [195,196,199]. In particular, a 10 000-fold synergistic increase in the HASE viscosity was reported when the $C_{12}EO_4$ surfactant concentration was increased to 0.1 M, due to the formation of mixed micelles of surfactant hydrophobic tails and polymer hydrophobic groups [14]. In the intermediate surfactant concentration range (0.001–0.01 M), the microstructure of the network was affected by stress even in the low-shear plateau region of the flow curves. At high surfactant concentrations (>0.01 M), polymer hydrophobes interact with the surfactant to form a stiff gel-like structure. Apparently, the $C_{12}EO_4$ molecules formed a mixed bilayer structure with the polymer hydrophobes. These stronger network junctions caused a larger increase in both the viscosity and the longest relaxation time of the system [195].

In the presence of the nonionic surfactant $C_{12}EO_{23}$ that has a high PEO content, a different behavior was observed [197]. The low shear viscosity and dynamic moduli increased at surfactant concentrations above CMC until they reached a maximum at a concentration of approximately 1 mM (which is about 17 times the CMC of free surfactant) and then decreased [197]. The dominant mechanism at in the concentration range between the CMC and 1 mM was an increase in the number of intermolecular hydrophobic junctions and a strengthening of the overall associative network structure. Above 1 mM $C_{12}EO_{23}$, the disruption of the associative network caused a reduction in the number of junctions and the strength of the overall network structure. The binding of $C_{12}EO_{23}$ to HASE before the CMC could not be detected by rheology, but was revealed by isothermal titration calorimetry. The $C_{12}EO_{23}$ aggregation in water and 0.1 wt% HASE polymer solutions was found to be entropically driven. A reduction in the CAC confirmed the existence of polymer–surfactant interactions [197].

SYMBOLS

C_nEO_m	$CH_3(CH_2)_{n-1}(OCH_2CH_2)_mOH$
C_nTA	$CH_3(CH_2)_{n-1}N^+(CH_3)_3,$
CAC	critical aggregation concentration
CMC	critical micelle concentration
catHEC	cationic hydroxyethyl cellulose
catHM-HEC	cationic hydrophobically modified hydroxyethyl cellulose
EHEC	ethyl hydroxyethyl cellulose
EO	ethylene oxide
HASE	hydrophobically-modified alkali soluble emulsion
HEC	hydroxyethyl cellulose
HEUR	hydrophobically-modified ethoxylated urethane
HM	hydrophobically modified
HMP	hydrophobically modified polymer
HMPE	hydrophobically modified polyelectrolyte
HM-EHEC	hydrophobically modified ethyl hydroxyethyl cellulose
HM-HEC	hydrophobically modified hydroxyethyl cellulose
HM-NaPA	hydrophobically modified sodium polyacrylate
NaPA	sodium polyacrylate
PEO	poly(ethylene oxide)
PNIPAm	poly(N-isopropylacrylamide)
PO	propylene oxide
PPO	poly(propylene oxide)
PVP	poly(vinyl pyrrolidone)
SDS	sodium dodecyl sulfate

ACKNOWLEDGEMENTS

We are grateful to the National Science Foundation (grant CTS-0124848/TSE) for supporting our research on surfactant–polymer interactions. Additional funding from Xerox Foundation and Bausch & Lomb is greatly appreciated.

REFERENCES

1. M. J. Rosen, *Surfactants and Interfacial Phenomena*, 2nd edn., Wiley, New York (1989).

2. R. G. Laughlin, *The Aqueous Phase Behavior of Surfactants*, Academic Press, London (1994).

3. D. F. Evans, and H. Wennerström, *The Colloidal Domain*, 2nd edn., Wiley-VCH (1999).

4. T. R. Carale, Q. T. Pham, and D. Blankschtein, *Langmuir* **10**: 109–121(1944).

5. P. Alexandridis, and J. F. Holzwarth, *Langmuir* **13**: 6074–6082 (1997).

6. P. Alexandridis, V. Athanassiou, and T. A. Hatton, *Langmuir* **11**: 2442–2450 (1995).

7. J. Armstrong, B. Chowdhry, J. Mitchel, A. Beezer, and S. Leharne, *J. Phys. Chem.* **100**: 1738–1745 (1996).

8. D. F. Evans, *Langmuir* **4**: 3–12 (1988).

9. M. Sjöberg, and T. Wärnheim, *Surfactant Sci. Ser.* **67**: 179–205 (1997).

10. J. C. T. Kwak, (ed.), *Polymer–Surfactant Systems*; Marcel Dekker, Inc., New York (1998).

11. B. Jönsson, B. Lindman, K. Holmberg, and B. Kronberg, *Surfactants and Polymers in Aqueous Solutions*, Wiley (1998).

12. P. Hansson, and B. Lindman, *Curr. Opin. Colloid Interface Sci.* **1**: 604–613 (1996).

13. L. Piculell, K. Thuresson, and B. Lindman, *Polym. Adv. Technol.* **12**: 44–69 (2001).

14. L. Piculell, B. Lindman, and G. Karlström, In Polymer–Surfactant Systems, J. C. T. Kwak (ed.), Marcel Dekker Inc., New York (1998), pp. 65–141.

15. P. Linse, L. Piculell, and P. Hansson, In Polymer–Surfactant Systems, J. C. T. Kwak (ed.), Marcel Dekker, Inc., New York (1998), pp. 193–238.

16. B. Cabane, and R. Duplessix, *J. Physique* **43**: 1529–1542 (1982).

17. B. Cabane, and R. Duplessix, *J. Physique* **48**: 651–662 (1987).

18. E. D. Goddard, *Colloids Surfaces* **19**: 255–300 (1986).

19. E. D. Goddard, *Colloids Surfaces* **19**: 301–329 (1986)

20. E. D. Goddard, *J. Am. Oil Chem. Soc.* **71**: 1–16 (1994).

21. P. L. Dubin, Y. Li, and P. Bahadur, In Polymeric Materials Encyclopedia. J. C. Salamone (ed.), CRC Press (1996), Vol. 6, pp. 4294–4304.

22. F. M. Winnik, and S. T. A. Regismond, *Coll. Surf. A* **118**: 1–39 (1996).

23. I. Iliopoulos, *Curr. Opin. Colloid Interface Sci.* **3**: 493–498 (1998).

24. D. N. Schultz, and J. E. Glass, (eds), *Polymers as Rheology Modifiers*. Advances in Chemistry Series No. 462. American Chemical Society, Washington, DC (1991).

25. J. E. Glass, (ed.), *Associative Polymers in Aqueous Media* ACS Symposium Series No. 765. American Chemical Society, Washington, DC (2000).

26. C. A. Finch (ed.), *Industrial Water Soluble Polymers* Special Publication No. 186. The Royal Society of Chemistry (1996).

27. M. A. Winnik, and A. Yekta, *Curr. Opin. Colloid Interface Sci.* **2**: 424–436 (1997).

28. P. Alexandridis, and B. Lindman, (eds), *Amphiphilic Block Copolymer: Self-Assembly and Applications*. Elsevier Science, Amsterdam (2000).

29. B. Lindman, and K. Thalberg, In Interactions of Surfactants with Polymers and Proteins. E. D. Goddard, K. P.

Ananthapadmanabhan, (eds), CRC Press, Boca Raton (1993), pp. 203–276.

30. L. Piculell, and B. Lindman, *Adv. Colloid Interface Sci.* **41**:149–178 (1992).

31. P. Dubin, J. Bock, R. M. Davies, and D. N. Schulz, C. Thies, *Macromolecular Complexes in Chemistry and Biology*. Springer-Verlag: Berlin (1994).

32. C. K. Ober, and G. Wegner, *Adv. Mater.* **9**: 17–31(1997).

33. W. J. MacKnight, E. A. Ponomarenko, and D. A. Tirrell, *Acc. Chem. Res.* **31**: 781–788 (1998).

34. M. Antonietti, J. Conrad, and A. Thünemann, *Macromolecules* **27**: 6007–6011 (1994).

35. M. Antonietti, D. Radloff, U. Wiesner, and H. W. Spiess, *Macromol. Chem. Phys.* **197**: 2713–2727 (1996).

36. M. Antonietti, and M. Maskos, *Macromolecules* **29**: 4199–4205 (1996).

37. M. Antonietti, M. Neese, G. Blum, and F. Kremer, *Langmuir* **12**: 4436–4441 (1996).

38. M. Antonietti, R. Kublickas, O. Nuyken, and B. Voit, *Macromol. Rapid Commun.* **18**: 287–294 (1997).

39. C. Faul, M. Antonietti, R. Sanderson, and H.-P. Hentze, *Langmuir* **17**: 2031–2035 (2001).

40. M. Antonietti, H.-P. Hentze, B. Smarsly, M. Löffler, and R. Morschhäuser, *Macromol. Mater. Eng.* **287**: 195–202 (2002).

41. E. A. Ponomarenko, D. A. Tirrell, W. J. MacKnight, *Macromolecules* **29**: 8751–8758 (1996).

42. E. A. Ponomarenko, A. J. Waddon, K. N. Bakeev, D. A. Tirrell, and W. J. MacKnight, *Macromolecules* **29**: 4340–4345 (1996).

43. E. A. Ponomarenko, D. A. Tirrell, and W. J. MacKnight, *Macromolecules* **31**: 1584–1589 (1998).

44. P. Ilekti, L. Piculell, F. Tournilhac, B. Cabane, *J. Phys. Chem. B* **102**: 344–351 (1998).

45. P. Ilekti, T. Martin, B. Cabane, and L. Piculell, *J. Phys. Chem. B* **103**: 9831–9840 (1999).

46. A. Svensson, L. Piculell, B. Cabane, and P. Ilekti, *J. Phys. Chem. B* **106**: 1013–1018 (2002).

47. K. Thalberg, B. Lindman, and G. Karlström, *J. Phys. Chem.* **95**: 6004–6011 (1991).

48. R. Nagarajan, *Langmuir* **1**: 331–341(1985).

49. R. Nagarajan, *Colloids Surf.* **13**: 1–17(1985).

50. R. Nagarajan, *Adv. Colloid Interface Sci.* **26**: 205–264 (1986).

51. E. Ruckenstein, G. Huber, and H. Hoffmann, *Langmuir* **3**: 382–387 (1987).

52. E. Ruckenstein, *Langmuir* **15**: 8086–8089 (1999).

53. T. Odijk, *Langmuir* **7**: 1991–1992 (1991).

54. T. Gilányi, *J. Phys. Chem. B* **103**: 2085–2090 (1999).

55. Y. J. Nikas, and D. Blankschtein, *Langmuir* **10**: 3512–3528 (1994).

56. O. V. Borisov, and A. Halperin, *Curr. Opin. Colloid Interface Sci.* **3**: 415–421 (1998).

57. A. C. Balazs, and J. Y. Hu, *Langmuir* **5**: 1230–1234 (1989).

58. A. C. Balazs, and J. Y. Hu, *Langmuir* **5**: 1253–1255 (1989).

59. A. C. Balazs, M. C. Gempe, and A. P. Lentvorski, *J. Chem. Phys.* **95**: 8467–8473 (1991).

60. A. C. Balazs, K. Huang, and T. Pan, *Colloids Surfaces A* **75**: 1–20 (1993).

61. T. Wallin, and P. Linse, *Langmuir* **14**: 2940–2949 (1998).

62. T. Wallin, and P. Linse, *Langmuir* **12** 305–314 (1996).

63. T. Wallin, and P. Linse, *J. Phys. Chem.* **100**: 17873–17880 (1996).

64. T. Wallin, and P. Linse, *J. Phys. Chem. B* **101**: 5506–5513 (1997).

65. C. Y. Yong, and M. Muthukumar, *J. Chem. Phys.* **109**: 1522–1527 (1998).

66. H. Schiessel, R. F. Bruinsma, and W. M. Gelbart, *J. Chem. Phys.* **115** 7245–7252 (2001).

67. P. Chodanowski, and S. Stoll, *J. Chem. Phys.* **115:** 4951–4960 (2001).

68. P. Chodanowski, and S. Stoll, *Macromolecules* **34:** 2320–2328 (2001).

69. S. Stoll, and P. Chodanowski, *Macromolecules* **35:** 9556–9562 (2002).

70. M. Jonsson, and P. Linse, *J. Chem. Phys.* **115:** 3406–3418 (2001).

71. M. Jonsson, and P. Linse, *J. Chem. Phys.* **115:** 10975–10985 (2001).

72. A. Akinchina, and P. Linse, *Macromolecules* **35:** 5183–5193 (2002).

73. T. T. Nguyen, and B. I. Shklovskii, *Phys. Rev. E* **64:** no. 041407 (2001).

74. T. T. Nguyen, and Shklovskii, B. I. *Physica A* **293:** 324–338 (2001).

75. T. T. Nguyen, and B. I. Shklovskii, *J. Chem. Phys.* **114:** 5905–5916 (2001).

76. T. T. Nguyen, and B. I. Shklovskii, *Phys. Rev. E* **66:** no. 021801 (2002).

77. T. T. Nguyen, and B. I. Shklovskii, *Phys. Rev. E* **65:** no. 031409 (2002).

78. M. Skepö, and P. Linse, *Phys. Rev. E* **66:** no. 051807 (2002).

79. M. Skepö, and P. Linse, *Macromolecules* **36:** 508–519 (2003).

80. P. Alexandridis, *Curr. Opin. Colloid Interface Sci.* **1:** 490–501 (1996).

81. P. Alexandridis, and R. J. Spontak, *Curr. Opin. Colloid Interface Sci.* **4:** 130–139 (1999).

82. P. Alexandridis, T. Nivaggioli, and T. A. Hatton, *Langmuir* **11:** 1468–1476 (1995).

83. L. Yang, P. Alexandridis, D. C. Steytler, M. J. Kositza, and J. F. Holzwarth, *Langmuir* **16:** 8555–8561 (2000).

84. L. Yang, and P. Alexandridis, *Langmuir* **16:** 4819–4829 (2000).

85. P. Alexandridis, and L. Yang, *Macromolecules* **33**: 5574–5587 (2000).

86. Y. Lin, and P. Alexandridis, *Langmuir* **18**: 4220–4231 (2002).

87. Y. Lin, and P. Alexandridis, *J. Phys. Chem. B* **106**: 12124–12132 (2002).

88. R. Ivanova, B. Lindman, P. Alexandridis, *Adv. Colloid Interface Sci.* **89**: 351–382 (2001).

89. M. Almgren, J. van Stam, C. Lindblad, P. Y. Li, P. Stilbs, and P. Bahadur, *J. Phys. Chem.* **95**: 5677–5684 (1991).

90. E. Hecht, and H. Hoffmann, *Langmuir* **10**: 86–91 (1994).

91. K. Contractor, and P. Bahadur, *Eur. Polym. J.* **34:** 225–228 (1998).

92. Y. Li, R. Xu, D. M. Bloor, J. F. Holzwarth, and E. Wyn-Jones, *Langmuir* **16**: 10515–10520 (2000).

93. Y. Li, R. Xu, S. Couderc, D. M. Bloor, E. Wyn-Jones, and J. F. Holzwarth, *Langmuir* **17**: 183–188 (2001).

94. T. Thurn, S. Couderc, J. Sidhu, D. M. Bloor, J. Penfold, J. F. Holzwarth, and E. Wyn-Jones, *Langmuir* **18**: 9267–9275, (2002).

95. R. Cardoso da Silva, G. Olofsson, K. Schillen, and W. Loh, *J. Phys. Chem. B* **106**: 1239–1246 (2002).

96. R. De Lisi, S. Milioto, and N. Muratore, *Macromolecules* **35**: 7067–7073 (2002).

97. Y. Li, R. Xu, S. Couderc, D. M. Bloor, J. F. Holzwarth, and E. Wyn-Jones, *Langmuir* **17**: 5742–5747 (2001).

98. S. Couderc, Y. Li, D. M. Bloor, J. F. Holzwarth, and E. Wyn-Jones, *Langmuir* **17**: 4818–4824 (2001).

99. L. Guo, R. H. Colby, M. Y. Lin, and G. P. Dado, *J. Rheol.* **45:** 1223–1243 (2001).

100. E. Hecht, K. Mortensen, M. Gradzielski, and H. Hoffmann, *J. Phys. Chem.* **99**: 4866–4874 (1995).

101. K. W. Zhang, B. Lindman, and L. Coppola, *Langmuir* **11**: 538–542 (1995).

102. R. Ivanova, P. Alexandridis, and B. Lindman, *Coll. Surf. A* **183–185**: 41–53 (2001).

103. P. Alexandridis, D. Zhou, and A. Khan, *Langmuir* **12**: 2690–2700 (1996).

104. P. Alexandridis, J. F. Holzwarth, and T. A. Hatton, *Macromolecules* **27**: 2414–2425 (1994).

105. J. P. Kaczmarski, J. E. Glass, *Macromolecules* **26**: 5149–5156 (1993).

106. J. P. Kaczmarski, and J. E. Glass, *Langmuir* **10**: 3035–3042 (1994).

107. D. J. Lundberg, R. G. Brown, J. E. Glass, and R. R. Eley, *Langmuir* **10**: 3027–3034 (1994).

108. R. D. Jenkins, D. R. Bassett, C. A. Silebi, and M.S. El-Aasser, *J. Appl. Polym. Sci.* **58**: 209–230 (1995).

109. T. Annable, R. Buscall, R. Ettelaie, and D. Whittlestone, *J. Rheol.* **37**: 695–726 (1993).

110. K. C. Tam, R. D. Jenkins, M. A. Winnik, and D. R. Bassett, *Macromolecules* **31**: 4149–4159 (1998).

111. W. K. Ng, K. C. Tam, and R. D. Jenkins, *J. Rheol.* **44**: 137–147 (2000).

112. M. Hulden, *Coll. Surf. A* **82**: 263–277 (1994).

113. A. Yekta, B. Xu, J. Duhamel, H. Adiwidjaja, and M. A. Winnik, *Macromolecules* **28**: 956–966 (1995).

114. B. Xu, A. Yekta, L. Li, Z. Masoumi, and M. A. Winnik, *Coll. Surf. A* **112**: 239–250 (1996).

115. E. Alami, S. Abrahmsen-Alami, M. Vasilescu, and M. Almgren, *J. Colloid Interface Sci.* **193**: 152–162 (1997).

116. T. Annable, R. Buscall, R. Ettelaie, P. Shepherd, and D. Whittlestone, *Langmuir* **10**: 1060–1070 (1994).

117. K. W. Zhang, B. Xu, M. A. Winnik, and P. M. Macdonald, *J. Phys. Chem.* **100**: 9834–9841 (1996).

118. J. P. Kaczmarski, M.-R. Tarng, Z. Ma, and J. E. Glass, *Coll. Surf. A* **147**: 39–53 (1999).

119. S. Dai, K. C. Tam, and R. D. Jenkins, *Macromolecules* **34**: 4673–4675 (2001).

120. N. Kamenka, I. Burgaud, R. Zana, and B. Lindman, *J. Phys. Chem.* **98**: 6785–6789 (1994).

121. B. Nyström, H. Walderhaug, F. K. Hansen, and B. Lindman, *Langmuir* **11**: 750–757 (1995).

122. H. Walderhaug, B. Nyström, F. K. Hansen, and B. Lindman, *J. Phys. Chem.* **99**: 4672–4678 (1995).

123. B. Nyström, K. Thuresson, and B. Lindman, *Langmuir* **11**: 1994–2002 (1995).

124. K. Thuresson, B. Nyström, G. Wang, and B. Lindman, *Langmuir* **11**: 3730–3736: (1995).

125. B. Nyström, A.-L. Kjøniksen, and B. Lindman, *Langmuir* **12**: 3233–3240 (1996).

126. K. Thuresson, B. Lindman, and B. Nyström, *J. Phys. Chem. B* **101**: 6450–6459 (1997).

127. K. Thuresson, and B. Lindman, *J. Phys. Chem. B* **101**: 6460–6468 (1997).

128. C. Holmberg, S. Nilsson, and L.-O. Sundelöf, *Langmuir* **13**: 1392–1399 (1997).

129. H. Evertsson, and S. Nilsson, *Macromolecules* **30**: 2377–2385 (1997).

130. A.-L. Kjøniksen, B. Nyström. and B. Lindman, *Langmuir* **14** 5039–5045 (1998).

131. H. Evertsson, S. Nilsson, C. J. Welch, and L.-O. Sundelöf, *Langmuir* **14**: 6403–6408 (1998).

132. S. K. Singh, and S. Nilsson, *J. Colloid Interface Sci.* **213**: 133–151 (1999).

133. B. Cabane, K. Lindell, S. Engström, and B. Lindman, *Macromolecules* **29**: 3188–3197 (1996).

134. K. Lindell, and B. Cabane, *Langmuir* **14**: 6361–6370 (1998).

135. S. Nilsson, K. Thuresson, B. Lindman, and B. Nyström, *Macromolecules* **33**: 9641–9649 (2000).

136. R. Lund, R. A. Lauten, B. Nyström, and B. Lindman, *Langmuir* **17**: 8001–8009 (2001).

137. E. Hoff, B. Nyström, and B. Lindman, *Langmuir* **17**: 28–34 (2001).

138. M. Tsianou, K. Thuresson, and L. Piculell, *Colloid Polym. Sci.* **279**: 340–347 (2001).

139. A.-L. Kjøniksen, S. Nilsson, K. Thuresson, B. Lindman, and B. Nyström, *Macromolecules* **33**: 877–886 (2000).

140. A. J. Dualeh, and C. A. Steiner, *Macromolecules* **23**: 251–255 (1990).

141. A. J. Dualeh, and C. A. Steiner, *Macromolecules* **24**: 112–116 (1991).

142. R. Tanaka, J. Meadows, P. A. Williams, and G. O. Phillips, *Macromolecules* **25**: 1304–1310 (1992).

143. I. D. Robb, P. A. Williams, P. Warren, and R. Tanaka, *J. Chem. Soc., Faraday Trans.* **91**: 3901–3906 (1995).

144. L. Piculell, S. Nilsson, J. Sjöström, and K. Thuresson, *ACS Symp. Ser.* **765**: 317–335 (2000).

145. U. Kästner, H. Hoffmann, R. Dönges, and R. Ehrler, *Coll. Surf. A* **82**: 279–297 (1994).

146. U. Kästner, H. Hoffmann, R. Dönges, and R. Ehrler, *Progr. Colloid Polym. Sci.* **98**: 57–62 (1995).

147. U. Kästner, H. Hoffmann, R. Dönges, and R. Ehrler, *Coll. Surf. A* **112**: 209–225 (1996).

148. H. Hoffmann, U. Kästner, R. Dönges, and R. Ehrler, *Polym. Gels Networks* **4**: 509–526 (1996).

149. S. Panmai, R. K. Prud'homme, and D. G. Peiffer, *Coll. Surf. A* **147**: 3–15 (1999).

150. S. Nilsson, K. Thuresson, P. Hansson, and B. Lindman, *J. Phys. Chem. B* **102**: 7099–7105 (1998).

151. S. Nilsson, M. Goldraich, B. Lindman, and Y. Talmon, *Langmuir* **16**: 6825–6832 (2000).

152. P. S. Leung, E. D. Goddard, C. Han, and C. J. Glinka, *Colloids Surfaces* **13**: 47–62 (1985).

153. E. D. Goddard, and P. S. Leung, *Langmuir* **8**: 1499–1500 (1992).

154. F. M. Winnik, S. T. A. Regismond, and E. D. Goddard, *Coll. Surf. A* **106**: 243–247 (1996).

155. F. M. Winnik, S. T. A. Regismond, and E. D. Goddard, *Langmuir* **13**: 111–114 (1997).

156. S. T. A. Regismond, K. D. Gracie, F. M. Winnik, and E. D. Goddard, *Langmuir* **13**: 5558–5562 (1997).

157. S. T. A. Regismond, F. M. Winnik, and E. D. Goddard, *Coll. Surf. A* **141**: 165–171 (1998).

158. F. Guillemet, and L. Piculell, *J. Phys. Chem.* **99**: 9201–9209 (1995).

159. M. Goldraich, J. R. Schwartz, J. L. Burns, and Y. Talmon, *Coll. Surf. A* **125**: 231–244 (1997).

160. I. S. Chronakis, and P. Alexandridis, *Macromolecules* **34**: 5005–5018 (2001).

161. M. Tsianou, and P. Alexandridis, *Langmuir* **15**: 8105–8112 (1999).

162. P. Alexandridis, S. Ahn, and M. Tsianou, *ACS Symp. Ser.* **737**: 187–198 (2000).

163. M. Tsianou, and P. Alexandridis, *Polym. Mat. Sci. Eng.* **87**: 23–24 (2002).

164. P. Hansson, and M. Almgren, *J. Phys. Chem.* **100**: 9038–9046 (1996).

165. U. Kästner, H. Hoffmann, R. Dönges, and J. Hilbig, *Coll. Surf. A* **123–124**: 307–328 (1997).

166. A.-L. Kjøniksen, C. Iversen, B. Nyström, T. Nakken, and O. Palmgren, *Macromolecules* **31**: 8142–8148 (1998).

167. I. Iliopoulos, T. K. Wang, and R. Audebert, *Langmuir* **7**: 617–619 (1991).

168. B. Magny, I. Iliopoulos, R. Zana, and R. Audebert, *Langmuir* **10**: 3180–3187 (1994).

169. F. Petit, I. Iliopoulos, R. Audebert, and S. Szonyi, *Langmuir* **13**: 4229–4233 (1997).

170. G. Bokias, D. Hourdet, I. Iliopoulos, G. Staikos, and R. Audebert, *Macromolecules* **30**: 8293–8297 (1997).

171. A. Sarrazin-Cartalas, I. Iliopoulos, R. Audebert, and U. Olsson, *Langmuir* **10**: 1421–1426 (1994).

172. K. Loyen, I. Iliopoulos, U. Olsson, and R. Audebert, *Progr. Colloid Polym. Sci.* **98**: 42–46 (1995).

173. B. S. Yang, J. Lal, P. Richetti, C. M. Marques, W. B. Russel, and R. K. Prud'homme, *Langmuir* **17**: 5834–5841 (2001).

174. K. Thalberg, B. Lindman, and K. Bergfeldt, *Langmuir* **7**: 2893–2898 (1991).

175. J. J. Kiefer, P. Somasundaran, and K. P. Ananthapadmanabhan, *Langmuir* **9**: 1187–1192 (1993).

176. K. Bergfeldt, and L. Piculell, *J. Phys. Chem.* **100**: 5935–5940 (1996).

177. L. Piculell, K. Bergfeldt, and S. Gerdes, *J. Phys. Chem.* **100**: 3675–3679 (1996).

178. P. Hansson, and M. Almgren, *J. Phys. Chem.* **99**: 16684–16693 (1995).

179. S. Ranganathan, and J. C. T. Kwak, *Langmuir* **12**: 1381–1390 (1996).

180. J. Fundin, P. Hansson, W. Brown, and I. Lidegran, *Macromolecules* **30**: 1118–1126 (1997).

181. P. M. Macdonald, and A. Tang, Jr. *Langmuir* **13**: 2259–2265 (1997).

182. V. Tirtaatmadja, K. C. Tam, and R. D. Jenkins, *Macromolecules* **30**: 3271–3282 (1997).

183. K. C. Tam, M. L. Farmer, R. D. Jenkins, and D. R. Bassett, *J. Polym. Sci. B: Polym. Phys.* **36**: 2275–2290 (1998).

184. S. Dai, K. C. Tam, R. D. Jenkins, and D. R. Bassett, *Macromolecules* **33**: 7021–7028 (2000).

185. K. Horiuchi, Y. Rharbi, J. G. Spiro, A. Yekta, M. A. Winnik, R. D. Jenkins, and D. R. Bassett, *Langmuir* **15**: 1644–1650 (1999).

186. E. Araujo, Y. Rharbi, X. Y. Huang, M. A. Winnik, D. R. Bassett, and R. D. Jenkins, *Langmuir* **16**: 8664–8671 (2000).

187. L. Guo, K. C. Tam, and R. D. Jenkins, *Macromol. Chem. Phys.* **199**: 1175–1184 (1998).

188. K. C. Tam, L. Guo, R. D. Jenkins, and D. R. Bassett, *Polymer* **40**: 6369–6379 (1999).

189. S. Dai, K. C. Tam, and R. D. Jenkins, *Macromolecules* **33**: 404–411 (2000).

190. H. Tan, K. C. Tam, and R. D. Jenkins, *J. Appl. Polym. Sci.* **79**: 1486–1496 (2001).

191. W. P. Seng, K. C. Tam, R. D. Jenkins, and D. R. Bassett, *Macromolecules* **33**: 1727–1733 (2000).

192. W. P. Seng, K. C. Tam, R. D. Jenkins, and D. R. Bassett, *Langmuir* **16**: 2151–2156 (2000).

193. H. Tan, K. C. Tam, and R. D. Jenkins, *Langmuir* **16**: 5600–5606 (2000).

194. H. Tan, K. C. Tam, and R. D. Jenkins, *J. Colloid Interface Sci.* **231**: 52–58 (2000).

195. V. Tirtaatmadja, K. C. Tam, and R. D. Jenkins, *AICHE J.* **44**: 2756–2765 (1998).

196. V. Tirtaatmadja, K. C. Tam, and R. D. Jenkins, *Langmuir* **15**: 7537–7545 (1999).

197. K. C. Tam, W. P. Seng, R. D. Jenkins, and D. R. Bassett, *J. Polym. Sci. B: Polym. Phys.* **38**: 2019–2032 (2000).

198. R. J. English, J. H. Laurer, R. J. Spontak, and S. A. Khan, *Ind. Eng. Chem. Res.* **41**: 6425–6435 (2002).

199. W. P. Seng, K. C. Tam, and R. D. Jenkins, *Coll. Surf. A* **154**: 365–382 (1999).

20

Ultrasonic Relaxation Studies of Mixed Surfactant Systems—Analysis of Data to Yield Kinetic Parameters

T. THURN, D. M. BLOOR, and E. WYN-JONES

School of Sciences, Chemistry, University of
Salford, Salford, United Kingdom

SYNOPSIS

The aim of this chapter is to describe how ultrasonic relaxation measurements can be used to determine the kinetic parameters (rate constants) associated with the fast exchange process between surfactant monomers and micelles in micellar solutions of single component surfactants and also mixed surfactant systems containing additives. In relaxation studies all systems are at equilibrium and the above aims can only be met where there are reliable equilibrium data available on the analytical concentration of the species in equilibrium.

The first part of this chapter gives an outline of the seminal theoretical treatment by Aniansson and Wall on the kinetics of micelle formation in nonionic surfactants. In this work we will focus on the fast relaxation time as measured by ultrasonics with particular reference to the various relaxation equations that have been derived to relate the fast relaxation time to the kinetics and equilibrium parameters of the micellar system. With the exception of some recent work, the majority of ultrasonic studies have been carried out on ionic surfactants. The fast relaxation equations that are considered involve the original Aniansson and Wall equation and modifications to account for ionic systems as well as other equations which have been independently derived. In addition we will also describe how an alternative phenomenological treatment can be successfully used with particular reference to the mixed micellar systems. For single component ionic micellar systems, it is shown that when due allowance is made for the decrease of monomer surfactant concentration in the micellar range as measured by surfactant selective electrodes, all the independent relaxation equations yield the same constants and adequately describe the fast relaxation process.

Finally, when mixed micelle systems are considered, the phenomenological treatment has a distinct advantage over other approaches in the sense that reaction rates can be extracted directly from the experimental data and allow the operator to apply normal kinetic concentrations. Data will be presented for mixed surfactant systems containing surface active drugs, alcohols, polymers and also cyclodextrins. In all cases the success of the treatments have relied on the determination of reliable equilibrium concentrations.

I. INTRODUCTION

This chapter describes how ultrasonic relaxation data on mixed surfactant systems have been used to evaluate kinetic data in the form of rate constants for the exchange of surfactant monomers in bulk solutions and those complexed to various additives ranging from other surfactants to polymers. In practise, the ultrasonic method is one of several different chemical relaxation techniques (e.g., temperature-jump, pressure-jump, electric field jump, concentration jump, etc.) which can be used to study the kinetics of very fast processes in solution [1–8]. In the context of these studies, the phrase Chemical Relaxation Spectrometry is used [1,2] to cover a variety of relaxation methods that when used together can cover the time range 10–10^{-10} s. Each method is limited in its time resolution and also the type of equilibrium systems that can be studied. The ultrasonic data described in this work is limited to systems that have relaxation times in the range 0.5×10^{-6} to 1.6×10^{-9} s—fortunately this covers a diverse range of exchange processes mostly involving the almost diffusion controlled rates of surfactant monomers in solution. In practice, there are descriptions of several different ultrasonic methods available which can considerably extend this time range [9,10]. Unfortunately many of these techniques are not commercially available, can be user unfriendly and demand technical expertise and maintenance which are way outside the accepted skills of a normal chemistry postgraduate student. However, two of these techniques, the Eggers resonance and pulse method, are considered as fairly versatile and in the hands of a competent operator can almost be used in routine instruments. Recently, Eggers, Kaatze [11] and co-workers have developed elegant modifications of these systems which produce excellent relaxation data in the time range 10^{-5}–10^{-10} s.

It can be strongly argued that, following the joint award of the Nobel Prize to Eigen in 1967 for his contributions on Chemical Relaxation Studies of fast reactions in Chemistry and Biochemistry, the most significant and noteworthy breakthrough has been the success achieved in the application of these techniques to study micellar kinetics [4,5,12–15]. Another beneficial outcome from this success story has been the way theoretical treatments of the kinetics of multi-step equilibria have been developed so that useable relaxation equations for complex systems have emerged which allow the kinetics of these processes to be accessible to the experimentalist. This is particularly true in the context of the present chapter especially since many rate constants, or more correctly coefficients, depend on solution composition [14]. The other significant underlying feature is the realization that relaxation experiments on their own must be underpinned with reliable equilibrium data in the form of the analytical concentrations of the species in equilibrium before any reliable conclusions can be drawn. Before proceeding with the main theme of this chapter, we will first consider some important aspects in the developments of the kinetics of micelle formation—this we consider to be a prerequisite before we can consider any mixed surfactant system. In this chapter we will attempt to avoid detailed mathematical and experimental details—rather we will concentrate on general aspects of mixed micelles and refer the reader to the detailed treatments which are published.

II. INTRODUCTION TO CHEMICAL RELAXATION

The kinetics of extremely fast reactions in solution can be studied by the use of chemical relaxation methods, as well as continuous and stopped-flow techniques [1–8]. As a result it is possible to time resolve various processes which occur in the time domain $10-10^{-10}$ s. In principle this means that the rates of processes as fast as the diffusion controlled recombination of hydrogen ions with hydroxyl ions with rate constants of the order $10^{10}-10^{11}$ dm^3 mol^{-1} s^{-1} to the rearrangement and breakdown of large molecular aggregates which take place in around 1–10 s can be investigated.

Temperature-jump, pressure-jump, and ultrasonics form a group of techniques known as relaxation methods which were developed primarily by Eigen and De Meyer for kinetic studies of fast reactions in solution [1–8]. The general principle of the jump methods is that we start with a reaction at equilibrium. This system is then rapidly perturbed by means of a temperature or pressure change and the equilibrium constant is changed to its new position in accordance with the van't Hoff

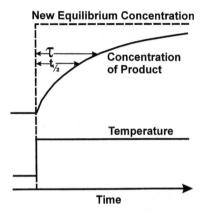

Figure 1   The principle of the temperature-jump method.

equations. The relaxation which subsequently takes place is the adjustment of reaction and product concentrations to the new equilibrium values. The process of readjustment of concentrations is monitored by a fast response method such as optical spectroscopy, light scattering, conductivity, etc. The magnitudes of the temperature or pressure jump are of the order of a few degrees and up to 1000 atmospheres respectively. The principle of this method demands that the rate of change of the temperature or pressure should be rapid compared to the reaction rate. In practice, temperature jumps occurring within 1 ns can be achieved with an iodine laser [16] whereas in pressure jump experiments the limit is in region of $5\,\mu s$. Figure 1 shows the ideal time dependence of the temperature or pressure and the corresponding quantity representing the extent of the reaction such as the concentration of the product.

     In practice, the displacement is small, so that a first-order exponential is obtained. This is important in connection with the evaluation of rate constant from mechanisms [6–8].

     Simple mathematics shows that when a small perturbation occurs in the equilibrium $A + B \underset{k^-}{\overset{k^+}{\rightleftharpoons}} C$, the system moves to a new equilibrium position according to the first-order rate law with a rate constant k, given by $k = k^+(a+b) + k^-$ where k^+ and k^- are the equilibrium forward and backward rate constants, and a and b are the concentrations of A and B at the new equilibrium. The first order rate constant k is often described as τ^{-1}, where τ is the relaxation time and is related to the half life of the reaction by

$$t/\tau = kt = \ln 2 = 0.7$$

In the use of the ultrasonic method the perturbation of the equilibrium is achieved by means of the periodic variation of pressure and temperature accompanying the passage of the sound wave through a system. When conditions are such that the frequency of the sound wave is of order τ^{-1}, effects due to relaxation of the equilibrium give rise to characteristic changes in the quantity α/f^2, where α is the sound absorption coefficient measured at frequency f. The behaviour of α/f^2, the sound velocity, and the absorption per wave length, $\alpha\lambda$ are shown in Fig. 2.

The relaxation time is obtained from the inflection point or the maximum in the absorption curve at frequency f_c using the relationship $1/2\pi\tau$ with $\tau = 1/2\pi f_c$. The time scales for which these techniques operate over are indicated in Fig. 3.

The mechanism of some processes involve several elementary equilibria, some of which are coupled [6–8]. In this case, the relaxation is characterized by a spectrum involving n

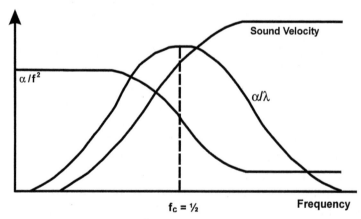

Figure 2 Variation of α/f^2, $\alpha\lambda$, and sound velocity with frequency for a single ultrasonic relaxation.

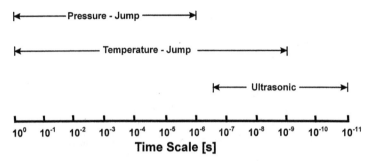

Figure 3 Time scale of pressure jump, temperature jump, and ultrasonic relaxation techniques.

relaxation times where n is the number of independent steps in the mechanism. Each relaxation time is some function of all the rate constants. In the relaxation experiment, the relaxations of individual steps can only be resolved if their times are separated by a factor of four or more, and their amplitudes are reasonable. Despite this limitation, experiments have shown that several complex systems involving multi-step equilibria are nevertheless characterized, within experimental error, by a single well defined relaxation time or two relaxations which can be resolved.

III. SURFACTANTS

The fundamental characteristics of surface active molecules or ions is amphiphility, that is the presence in the molecule or ion of both polar and nonpolar moieties [17]. In aqueous solution the polar head groups are hydrophilic and the nonpolar moieties are usually long hydrocarbon chains, making them hydrophobic. There are several types of surface active agents, including ionic, cationic, amphoteric, and nonionic varieties.

In aqueous solution, these molecules tend to collect at interfaces and their surface activity is illustrated in Fig. 4.

Usually the hydrophilic head groups remain in the aqueous phase whereas the hydrophobic moieties can be partially or completely removed from the aqueous phase to be

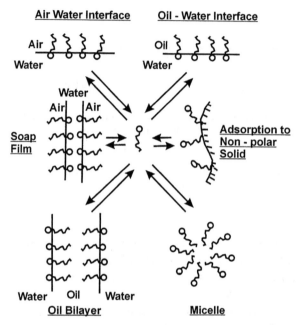

Figure 4 An illustration of the surface active properties of surfactants.

in contact with a more hydrophobic environment. The dual properties of these surfactants also allow them to form organized structures such as soap films, bilayers, and micelles.

In general, when water is added to solid surfactants, three types of behavior can occur:

1. Some of the surfactant dissolves to form an aqueous micellar solution.
2. The surfactant is practically insoluble, and remains as a solid crystal plus an aqueous solution of surfactant monomer.
3. A lyotropic liquid crystal is formed, which may dissolve when diluted, to form an aqueous micellar solution.

When measurable amounts of aggregated material are present, micellar solutions, as distinct from other types of

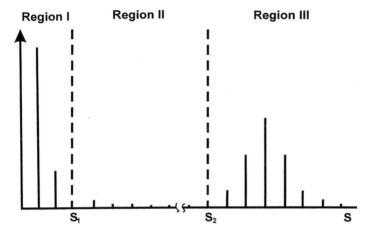

Figure 5 The equilibrium distribution of aggregates in a typical micellar system.

aggregating system, are characterized by an aggregate population of the type shown in Fig. 5 where the concentrations of the least probable species are very small indeed. Even dimers and trimers may be quite rare. Typical average aggregation numbers are of order 50–100. It is easily shown that for systems of this kind, the onset of aggregation occurs suddenly at a concentration known as the critical micelle concentration [18] (CMC) and that once this concentration is reached, the surfactant activity increases very little with further increase in concentration. The onset of micellization is reflected strongly in the variation of solubility with temperature. When micelles are absent, this behavior is like that of a normal dilute solution. However once micelles start to form in significant quantities, the solubility increases dramatically over a range of a few degrees [19]. The temperature at which this effect first becomes noticeable is known as the Krafft point and may be regarded as the temperature at which the solubility is equal to the CMC.

The accepted structure of an ionic micelle in aqueous solution when the surfactant concentration slightly exceeds the CMC is shown in Fig. 6. A variety of evidence in support of the ideal spherical shape has been accumulated for anionic, cationic, and nonionic micelles. The hydrophobic part of the

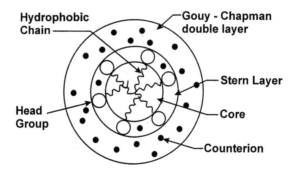

Figure 6 A two-dimensional schematic representation of a typical micelle.

aggregate forms the core of the micelle which resembles the liquid hydrocarbon. The polar head groups are located at the micelle/water interface in contact with hydrated water. This diagram also shows the relative position of the Stern and Gouy/Chapman layers.

IV. KINETIC STUDIES IN MICELLAR SYSTEMS

The formation of micelles from surfactant monomers takes place by a series of stepwise bimolecular equilibria according to the aggregation scheme

$$A_1 + A_{n-1} \xrightleftharpoons[k^-]{k^+} A_n \quad n = 2, 3, 4 \tag{1}$$

where A_1 denotes monomer and A_n an aggregate made up of n monomers and the ks are rate constants. In this scheme, only equilibria which involve an interaction between aggregates and monomers are considered to make a contribution. This means that the bimolecular equilibria involving only aggregates are neglected. In micellar solutions it is generally regarded that monomers and micelles are present in substantial quantities which can be determined analytically. In comparison, the concentrations of the intermediate species are negligibly small and cannot be detected. On the other hand their presence is required in order for the above mechanism to operate. It is interesting to note that in discussing the applications of the law of mass action to micelle formation, Goodeve [20] in 1935 pointed out that the simultaneous coming together of many molecules to produce a micelle in an improbable process and that the stepwise addition of monomer molecules to existing micelles in more likely. A comprehensive thermodynamic treatment of micelle formation has been given by Hall [21]. The micelles whose aggregation numbers are large (>20) are usually polydisperse. In practice, mechanisms can be tested experimentally through kinetic measurements. The above reactions involving the formation of micelles are very fast $(10^{-2}\text{--}10^{-10}\,\text{s})$ and their kinetics can be studied using chemical relaxation techniques. In order to cover the above time spectrum, more than one relaxation method is required and in general both the method of perturbation and detection can vary. Perturbation of fast equilibria takes place in the form of a rapid temperature, pressure, electric field, pH, or concentration jump and the method of detection can be specific to one of the stable species in equilibrium or can be an experimental parameter which is a measure of a particular property of the solution. This means that one of the basic prerequisites before successfully using any relaxation method to study chemical equilibria is to be able to confidently identify the molecular process which is being perturbed in the relaxation experiment. In some cases, this exercise requires imaginative manipulation of the system followed by the relaxation experiment, as well as a process of elimination of all the possible mechanisms. In practise, when a complex multistep system is studied by several independent investigators, one of the inherent consequences can be disagreement between the various operators concerning the interpretation and/or analysis of the relaxation data. This was certainly the case following the pioneering chemical relaxation studies of micellar kinetics. This situation was however resolved due to two reasons. First, a breakthrough was achieved following the seminal theoretical treatment of Aniansson and Wall which presented a novel way of looking at the multi-step problem described above and predicted the two relaxation times which were measured experimentally [4,5,22]. Second, a timely platform which allowed a forum for the early investigators to meet was provided by a NATO Advanced Summer School held at the University of Salford in 1974 [4].

A summary of the main considerations in the Aniansson and Wall treatment now follows. When micellar solutions are subjected to a rapid perturbation as in a relaxation experiment, the concentrations of monomer and of the various aggregate species shift from their equilibrium values after the perturbation. Clearly, in general, this leads to changes both in the monomer concentration and the concentration of micelles. Although the overall process is complex and involves many steps, only two relaxation times are observed experimentally for solutions of a single surfactant—these have been described as the fast time τ_1 and the slow time τ_2. Typically these times are separated by a factor of order 10^3. This behavior can be interpreted in terms of a stepwise association model equation (1). According to this model, reactions between aggregates are sufficiently rare as to be negligible and the only reactions which matter are of the kind

$$A_r + A_1 \rightleftharpoons A_{r+1}$$

where r may be any positive integer greater than one.

For this model as applied to micellar solutions, Aniansson and Wall [4,5,22–24] showed that the fast relaxation time τ_1 can be attributed to a change in surfactant monomer concentration at constant concentration of micelles (see Fig. 7)

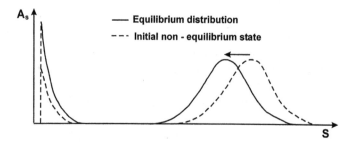

Figure 7 Schematic representation of the fast relaxation process τ_1 in a micellar system.

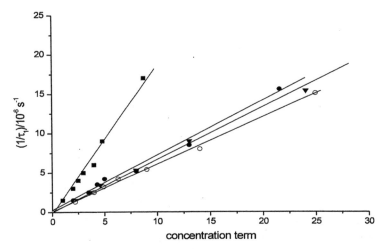

Figure 8 Plot of $1/\tau_1$ vs the concentration term for SDeS in $10^{-4}\,\mathrm{mol\,dm^{-3}}$ NaBr; (■) analysis I; (▼) analysis II; (○) analysis III; (●) analysis IV.

and that the slow relaxation time τ_2 can be attributed to changes in the concentration of the micelles which remain effectively in equilibrium with monomer throughout the slow process.

Qualitatively, the large difference between τ_1 and τ_2 arises from the fact that the monomer concentration can vary through reactions with all micellar species present whereas the concentration of micelles can only change via reactions involving the rare species at and near the minimum in Fig. 5. Quantitatively, the Aniansson–Wall theory gives

$$\frac{1}{\tau_1} = \sum_r k_r^- c_r \left[\frac{1}{\sigma^2 c_m} + \frac{1}{m_1} \right] \tag{2}$$

and

$$\frac{1}{\tau_2} = \frac{1}{Rc_m} \left| \frac{m_1 + N^2 c_m}{m_1 + \sigma^2 c_m} \right| \tag{3}$$

where k_r^- is the backward rate constant of the reaction referred to above, c_r is the equilibrium concentration of species r, $c_m = \Sigma_r c_r$, m_1 is the monomer concentration

$$\bar{N}^2 = \sum_r r^2 c_r/c_m, \quad \sigma^2 = \frac{(\bar{N}^2 - \bar{N}^2)}{N^2},$$

$$\bar{N} = \sum_r r c_r/c_m \quad \text{and} \quad R = \sum [k_s^- c_2]^{-1}$$

where the summation over s covers the rare intermediate micelles depicted by the region II of Fig. 5. When the only mechanism whereby c_m can change is through the stepwise formation or breakdown of aggregates, the slow relaxation time provides information about the rare intermediate species in region II of Fig. 5. This information is contained in the term R in Eq. (3) [23]. If however, c_m can change through the association and disproportionation of aggregates, this is no longer so.

The above "slow" relaxation time is normally slow when compared to the time resolution of ultrasonics and as such, will not be considered further in the chapter. We are however interested in the fast time τ_1 because it relates to monomer/micelle exchange which is active in the ultrasonic relaxation range, especially with reference to the kinetics of exchange processes in mixed surfactant systems. If we now define k^- by

$$k^- = (C - m_1) = \sum_r k_r^b c_r \tag{4}$$

where C denotes total concentration so that $(C - m_1) = \bar{N} c_m$, Eq. (2) may be written as

$$\frac{1}{\tau_1} = k^- \left[\frac{\bar{N}}{\sigma^2} + \frac{C - m_1}{m_1} \right] \tag{5}$$

Equation (5) is then rearranged to its useable form

$$1/\tau_1 = k^-/\sigma^2 + (k^-/n)(C - m_1)/m_1 \tag{6}$$

where n is the micellar aggregation number. This equation has been used extensively by several workers to study the kinetics of the monomer–micelle exchange process in surfactants. During the initial period, when this equation was used extensively, it became apparent, mainly due to experimental/ instrumental reasons that the fast kinetics of *only* ionic surfactants were accessible to study by relaxation techniques. This was unfortunate in the sense that the Aniansson and Wall treatment was based on nonionic micellar systems. A further drawback of the Aniansson–Wall theory is the assumption that the forward and backward rate coefficients of the various reaction steps to not depend on solution composition. For this assumption to be valid, the surfactant monomer concentration must increase monotonically with increasing total concentration above the CMC. However at the time it was recognized from theory and experiment that for ionic surfactants, the concentration of surfactant monomer ions decreases above the CMC [23,25,30]. It follows that the Aniansson–Wall theory cannot apply strictly to ionic surfactants. Three attempts have been made to overcome this difficulty [26–29]. The independent approaches by Lessner *et al.* [28] and Elvingson and Wall [29] were based on Aniansson type kinetics and led to the relaxation equation.

$$1/\tau_1 = k^-/\sigma^2 + (k^-/n)[1 + (1 - \alpha)^2/(1 + \alpha a)]a \qquad (7)$$

where $a = (C - m_1)/m_1$ and α is the degree of micellar dissociation of the ionic micelles.

The third and more rigorous treatment of the fast relaxation in ionic micelles was given by Hall [26,27] and is based on the notion that the rate coefficients depend on solution composition when the forward and backward rates are written in terms of concentrations. This treatment led to the relaxation equation

$$1/\tau_1 = k^-/\sigma^2 + (k^-/n)[1 + (1 - \alpha)^2/(1 + \alpha a) + 2\gamma^1 m_1]a$$
$$(8)$$

where $\gamma^1 = -[0.587/b(1+b)^2]$, $b = (m_1 + m_2)^{1/2}$, and m_3 is the concentration of added salt. It should be noted here that Eq. (8) is not in the original paper [26]. Here it is stated in a form suitable for the graphical analysis of the relaxation data and is derived by algebraic manipulation [30] of Eq. (23) in the original paper [26].

We now consider how the above relaxation equations can be used to determine rate constants and other related micellar parameters from the relaxation experiments. The three relaxation equations (6), (7), and (8) have been arranged in a form that they can be used graphically to analyze relaxation data on the fast relaxation process which usually involves the measurement of ultrasonic relaxation frequency (f_c) and amplitude (μ_m) at different surfactant concentrations. In order to utilize all the experimental data, a linearized graphical procedure is desirable in the analysis using equations (6), (7), and (8) and the above equations have been arranged in such a way that a plot of $1/\tau$ against the coefficient of k^-/n should give a straight line of slope k^-/n and intercept k^-/σ^2. An inspection of equations (6), (7), and (8) shows that in order to evaluate the coefficient of k^-/n in each equation, vital equilibrium data in the form of the monomer concentration of the surfactant in the micellar range and also the degree of micellar dissociation for Eqs (7) and (8) is required.

Following the acceptance of the Aniansson and Wall kinetic treatment around 1974, a major effort was directed at using Eq. (6) to study the kinetics of monomer/micelle exchange in surfactants. As we have mentioned previously, the treatment was based on nonionic micelles whereas all the relaxation experiments were conducted on ionic micelles—this was largely a consequence on the scope and limitation of the various relaxation techniques used. Unfortunately during this active and productive period, no worthwhile quantitative information was available on the variation of the monomer concentration of ionic surfactants in the micellar range. Although it was acknowledged that the monomer concentration decreased with increasing surfactant concentration, there was no reliable method to measure this quantity. Since the kinetic treatment was based on nonionic micelles, it was generally assumed that ionic surfactants behaved much in the same way as nonionic surfactants above the CMC with their monomer concentrations being assumed constant and equal to the CMC. This approximation and other treatments [23] involving extrapolation of various parameters to the CMC were used in those early studies in the analysis of Eq. (6), which basically led to solution of the relaxation equation (6) which at best could be considered semi-quantitative. In the 1980s, considerable progress was made in the development of surfactant membrane selective electrodes. In particular the work of Hall and Mears [31] led to the development of a modified poly vinyl chloride (PVC) membrane which could be chemically treated so as to be selective to specific ionic surfactants [30,32–36]. Subsequent development at Salford have led to versatile routine devices, the latest model being the coated wire (CW) version which can be used successfully by a competent operator [37]. As a result, it became possible to measure simultaneously the monomer concentration of surfactants in the micellar range as well as the degree of micellar dissociation [38].

We then used these electrodes to evaluate monomer concentrations and degree of micellar dissociation for a number of ionic surfactants for which ultrasonic relaxation data were available over a substantial concentration range in the micellar phase. Typical graphical plots resulting from the analyses of Eqs (6), (7), and (8) are in Figs 8 and 9.

Also included in each diagram is the analysis of Eq. (6) with the assumption that the monomer concentration is constant and equal to the CMC. These data show that when monomer surfactant and α are taken into account, Eqs (6), (7), and (8) produce a family of almost identical straight lines having similar slopes but slightly different intercepts. Using a constant monomer concentration ($= \text{CMC}$) on the other hand still produces a straight line with a slope that differs by a factor of 2 from the family of almost identical straight lines. The additional parameters required to carry out these analyses are m_1 and α. m_1 is not often accessible to most experimentalists but α which can be measured using a diverse range of experiments is often known for ionic surfactants. In these

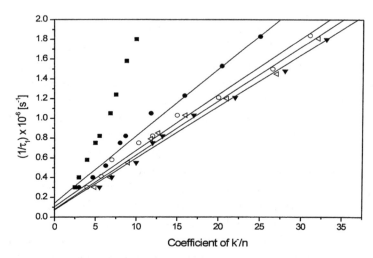

Figure 9   Plot of $1/\tau_1$ vs the coefficient of k^-/n for SDS at 25°C. (○)
Eq. (4); (▼) Eq. (5); (∇) Eq. (6); (●) Eq. (4), m_1 from Eq. (9); (■)
Eq. (4), $m_1 = \text{CMC}$.

circumstances m_1 can be estimated from the mass balance
equation

$$m_1 = \text{CMC}\left[1 + \frac{\alpha(C - m_1)}{m_1}\right]^{-(1-\alpha)/(2+1-\alpha)} \qquad (9)$$

where z is the ratio between the valencies of the counterions
and the amphiphilic ions. The plots using m_1 from Eq. (9) is
also in Fig. 9 and is nearer to reality than when using the
assumption $m_1 = \text{CMC} = \text{constant}$. These results show

1.   In order to fully utilize the relaxation data for ionic
     surfactants in the absence of added salt, the variation
     in monomer concentration in the micellar range
     must be taken into account if reliable kinetic data is
     required.
2.   The closeness in the slopes of the family of curves
     suggest that when the variation of m_1 is taken into
     account, all the relaxation equations gave equally
     reliable backward rate constants which are very close.

Furthermore, the linearity of the plots in the graphical analysis suggests that k^- is constant for pure surfactant micelles.

3. Since $k^+/k^- = m_1$, one of the consequences implicit in 2, above is that k^+ is dependent on concentration. Strictly speaking, the variation of rate constants (a better term is coefficient) with concentration is an obvious consequence of the fact that the monomer concentration of surfactants varies in the micellar range in the absence of salt. As far as we are aware, this has been ignored in the majority of publications dealing with the fast relaxation process in ionic micellar systems [15].

4. The difference between the close straight lines using Eqs (6), (7), and (8) when m_1 is taken into account, shows that, as far as the backward rate is concerned, the treatments almost give identical k^- values. The fine tuning resulting in the expressions arising from consideration of ionic micelles and also ionic strength is, however, reflected in the intercepts and thus k^-/σ^2 values.

In reality, the expressions (6), (7), and (8) for the concentration dependence of the relaxation time are essentially equivalent and do not differ greatly from each other. This similarity is highlighted even more in the graphical analysis of experimental data. This closeness also explains, in part, why the Aniansson and Wall treatment for nonionic micelles works so well for ionic systems when the variation of m_1 in taken into account. The similarity in these expressions and also their closeness in providing consistent backward rate constant data during graphical analysis now lead us to reconsider the merits of a simple two state model [39] for the fast process, which was introduced in 1972 and was the first quantitative relaxation expression to successfully account for the concentration dependence of the fast relaxation time. The concept of the two state model was based on experimental observations that many complex multi-step mechanisms were characterized by a very well defined single relaxation in the ultrasonic experiment.

Furthermore, if this process could be described by a two state, one step, general equilibrium, then an expression for the relaxation time could be evaluated by conventional means. The fast relaxation process associated with monomer/micelle exchange was described as [39]

$$\text{monomer} + (\text{micelle})_1 \rightleftharpoons (\text{micelle})_2$$

which satisfies the above criteria and the following relaxation time was derived

$$\frac{1}{\tau} = k_f C - k_b$$

where k_f was the forward rate constant and k_b the backward rate constant. This form of the equation was used in the original applications and unfortunately led to some misunderstandings [4,15] which resulted in the model being unfairly criticized. However it is worth noting that the above equation can be rearranged as [12]

$$\frac{1}{\tau} = \frac{k^-}{n}\left(\frac{C - m_1}{m_1}\right) \tag{10}$$

where $k^-/n = k_b$ and $k^+/n = k_f$, and is now in a form which is equivalent to Eqs (6), (7), and (8).

This equation is similar to Eq. (6) but without the term k^-/σ^2. At this stage it is perhaps worth remarking that this expression may be regarded as the kinetic equivalent of the equilibrium phase model [40] for micelles whereas Eq. (6) is the corresponding equivalent of the multiple equilibrium model [41]. The deviations from phase behavior are more apparent in the kinetic expressions than in the equilibrium expressions because n/σ^2 is usually considerably greater than $1/n$. Furthermore the term k^-/σ^2 is very small. In the graphical analysis of the above two-state equation (10) and the Aniansson and Wall equation (6) using relaxation and m_1 values from electrode data, exactly the same straight lines are produced— the only difference being that Eq. (10) predicts that the line goes through the origin whereas Eq. (6) demands an intercept of k^-/σ^2. In reality, k^-/σ^2 is very small and often negligible in the context of the analysis.

In summary, it can be stated that when appropriate experimental data is available, all the four independent relaxation equations (6), (7), (8) and (10) give exactly the same backward rate constant for the dissociation of a monomer from the micelle. The forward rate constant varies with concentration and can be evaluated from $k^+ = k^- m_1$. In these circumstances, it should be regarded as a rate coefficient. The most reliable estimate for k^-/σ^2 comes from Eq. (8) where the fine tuning to account for ionic effects has been most rigorous.

In micellar solutions of surfactants which have short hydrocarbon chains ($\sim C_6$), it is well known that the CMC is not as well defined as for the corresponding longer chains [18]. The aggregation numbers for the micelles in these solutions are also very small, $n < 20$. This means that in these systems, the number of aggregates intermediate between the monomer and micelles is much less than for the longer chain length surfactants. Relaxation experiments carried out in micellar solutions of surfactants containing short hydrocarbon chains have shown that only the fast relaxation time, τ_1 is observed— we are not aware of any experiments in which the slow relaxation τ_2 has been measured for these systems. As a result, the Aniansson and Wall treatment may be expected to break down in these cases, because they assign the observed kinetic behavior to two types of processes which are then treated separately. Furthermore, the concentration dependence of τ^{-1} shows a pronounced curvature around the CMC, where its value increases rapidly (by a factor of ~ 20) as the surfactant concentration decreases as shown in Fig. 10.

On the other hand, at concentrations well within the micellar range, τ^{-1} increases linearity as predicted by Eqs (6)–(10). The fast process is considered to be an exchange of monomer with aggregate as depicted by Eq. (1), with n restricted to values within the micellar range. This is the process which has been observed experimentally. The slow process, τ_2 involves the complete formation or dissolution of micelles via a sequence of subsequent reaction steps. When mean aggregation numbers are small, a substantial part of this

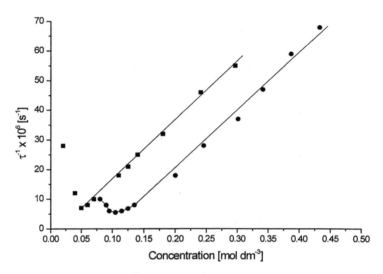

Figure 10 A plot of τ_1^{-1} against concentration for micellar systems having small aggregation numbers. (■) chlorcycline hydrochloride. The solid line here represents the calculated parameters from the numerical analysis; (●) sodium octyl sulphate.

slow process is expected to consist of reaction steps which may also be assigned to the fast process. Indeed, in a recent modification of the Aniansson and Wall theory, Telgmann and Kaatze [42] used a treatment by Teubner and Kahlweit [43] to take into account the possibility of oligomers (i.e., intermediate aggregates between monomer and micelles) participating in the reaction scheme as described above. In these circumstances, it was shown that τ_1 and τ_2 can merge. In addition these authors also showed that the increase of $1/\tau_1$ with decreasing concentration is a result of monomers and aggregates participating in a random scheme of aggregation

$$A_m + A_n \rightleftharpoons A_{m+n} \quad m, n = 1, 2, 3, \text{ etc.}$$

Another interesting feature for micellar systems with small aggregation numbers is that the total number of steps can be limited to the range 5–20. This means that the number of differential equations following the perturbation of the system are small enough to be amenable for solutions using numerical analysis. Indeed, for a series of antihistamine drugs which showed a CMC in light scattering experiments despite having micelles with mean aggregation numbers in the range 5–20, a numerical analysis was carried out in an attempt to test the validity of applying the Aniansson and Wall fast relaxation time equation (6) for these systems. In this analysis [44], the values of k^+ and σ^2 were adjusted to improve the agreement between the calculated relaxation times and those observed experimentally. The predominant process was assumed to be that associated with the monomer exchange in all the steps. One surprising feature of this treatment was that in each of the drugs studied, a value of σ^2 could be chosen which yielded one relaxation time having an amplitude wall in excess of the others. In each case, the value of σ^2 was well within the tolerance of the value found by the application of the Aniansson and Wall equation provided only relaxation data in which $1/\tau_1$ increases linearly when concentration was considered, i.e., well within the micellar range. The agreement between analyzing the kinetic data for these systems using the Aniansson and Wall approach and the numerical method was excellent and consistent with the mechanism of aggregation being a stepwise association phenomena equation (1).

Recent measurements have been carried out in this laboratory on the kinetic and equilibrium measurements associated with surfactant aggregation in nonaqueous solvents [45] and also reverse micelle aggregation of nonionic surfactants in heptane [46]. In these systems, small aggregates are present and in solutions where the amount of aggregated material is substantial, the concentration dependence of the relaxation times and amplitudes were found to be consistent with Eqs (6) and (10) despite the fact that no equivalent CMC was observed in the reverse micellar systems.

In the application of relaxation techniques to study the kinetics of chemical equilibria, it is traditional to assume a detailed model for the system being studied in terms of a simple equilibrium or multiple equilibria and then attempt to obtain an analytical expression for the relaxation time or times as simple functions of rate constants and equilibrium concentrations by solving the appropriate differential equations. In this approach, exact solutions to the differential equations, as elegantly demonstrated by the Aniansson and Wall treatment, are sometimes possible and in other cases various degrees of approximations are necessary. In this conventional approach, the concentration dependence of rate constants as is clearly evident in micellar systems, is a very difficult concept to handle. An alternative approach, which we consider to be a more versatile method, is provided by the phenomenological treatment. The basis of this approach is the linear formalism of irreversible thermodynamics [47,48]. At this point, it is worth remarking that in favorable cases the relaxation methods discussed above yield not only a relaxation time, but also the amplitude of the relaxation process. A good example is the ultrasonic relaxation technique. In general, the prerequisites for the application of the general phenomenological treatment are as follows: (i) the process being perturbed by the sound wave can be described by a general one-step equilibrium; (ii) ultrasonic relaxation arising from the perturbation of such an equilibrium is a well-defined single process; and (iii) equilibrium data are available concerning the concentrations of monomer and aggregated species.

In the case of the monomer/micelle equilibrium, all the above conditions are met, and the equilibrium can be defined as

$$\text{monomer} + (\text{micelle}) \; \rightleftharpoons \; (\text{micelle})_2 \tag{11}$$

In these circumstances, the phenomenological equations of interest to the analysis of ultrasonic relaxation data are

$$1/\tau_1 = (R_1/RT)(\partial A/\partial \xi)_a \tag{12}$$

$$\mu_m = \pi(\Delta V)\rho\sigma^2/2(\partial A/\partial \xi)_a \tag{13}$$

where A is the reaction affinity, ξ is the extent of the reaction, a denotes that the derivatives are evaluated in a closed system. R^+ is the forward (= backward) rate of the process. Combining

Eq. (12) with Eq. (13), we obtain

$$\mu_m/\tau_1 = [(\pi(\Delta V)^2/2RT)\rho v^2]R^+ \tag{14}$$

In the last equation, μ_m and τ_1 on the left side are known from the ultrasonic relaxation experiment, and the thermodynamic term in brackets on the right side follows from the amplitude analysis of Eq. (13) described above. In these circumstances, we can evaluate the equilibrium forward (=backward) rate R^+ associated for the aggregation process described generally in (11). In most cases the ΔV term can be found from the concentration dependence of the ultrasonic amplitude μ_m. In micellar systems, this takes the form of

$$\mu_m = (\pi\rho v^2/2RT)\{(\Delta V)^2 m_1(\sigma^2/n)a/[1+(\sigma^2/n)a]\} \tag{15}$$

where ρ is the density of the solution and ΔV the volume change for the reaction. It is a relatively easy procedure to carry out a curve fitting exercise to evaluate ΔV.

Once the equilibrium rate R^+ has been evaluated in this manner, we can then revert to the principles conventional kinetics in the sense that, if we consider the backward rate process R^- involving the dissociation of monomers from the micelles, then we expect this to be a first-order rate process depending on the concentration of the micelles ($=(C-m_1)/n$). Thus

$$R^+ = R^- = (k^-/n)(C-m_1) \tag{16}$$

the plots of R^- evaluated from the procedure described above against $(C-m_1)$ for various surfactants are shown in Fig. 11.

In all cases, they are straight lines passing through the origin clearly showing that Eq. (16) is appropriate, and the values of k^- are in Table I. It is very encouraging to note that these k^- values agree well from those derived by using the relaxation equations (6)–(8) and (10). The phenomenological treatment gives the same backward rate constant as the other models for the monomer/micelle exchange process [30,34–36,49]. This treatment has the added advantage that

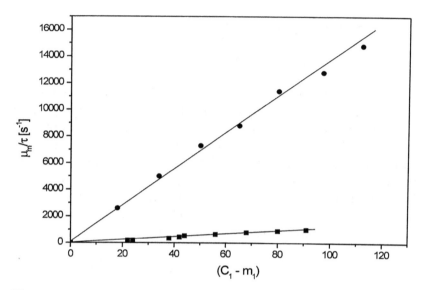

Figure 11   Plot of μ_m/τ_1 vs $(C - m_1)$ according to Eq. (8). (■) SDS; (●) SDHS.

no "a priori" model describing the elementary step of the system is equilibrium is required. In practice a general description of the two state one-step process is only required. From the ratio of the measured relaxation parameters in the form (μ_m/τ) and also ΔV, the equilibrium forward ($=$ backward) rate is evaluated at each concentration. This allows one to treat the forward and/or backward step in the context of conventional kinetics—the only other key requirement being equilibrium concentrations which are always required in order to test for unimolecular or bimolecular processes. If these data are available, equilibria involving ionic or nonionic micellar systems, in which rate coefficient depends on composition, can be handled. Another beneficial outcome in the use of a phenomenological model concerns circumstances where the ultrasonic relaxation lies just outside the operating frequency range. Here it is difficult to estimate the relaxation times τ, and amplitudes μ_m, accurately. We have found that using the phenomenological treatment can, in favorable circumstances, overcome this problem. In this treatment, the quantity μ_m/τ is used because it is a direct measure of the forward (= backward) rate of the process being perturbed. When $f > f_c$ and the tail end of the relaxation is measured experimentally, it is almost always the case that a series of amplitude and relaxation times can give acceptable fits to the measured data. For each curve fitting result, there is always a compensating effect occurring, in the sense that an increase (or decrease) in μ has to be compensated by an increase (or decrease) in τ.

Despite the uncertainties in the individual parameters, the quantity μ/τ is often constant as a consequence of this compensating effect. Therefore, whatever the molecular origin of the relaxation in ultrasonics, we can get a direct handle on the forward (= backward) rates of the equilibrium being disturbed directly from the ultrasonic experiment. Once this is achieved, then it is a matter of applying the same principles as in conventional kinetics to examine the nature of the process.

The first example of the use of this method is in alcohol/ molecular systems which are the basis of oil in water microemulsions.

V. MIXED MICELLES

Strictly speaking, a mixed micelle should be considered as an aggregate of two or more surfactant molecules and/or ions. Indeed in, the context of detergency, the most successful practical applications of surfactants, invariably utilize binary surfactant mixtures. In practice, these surfactant blends exhibit much improved performance in comparison to single components and this phenomenon is often referred to as synergy. This has led to the concept of quantifying the mixing characteristics of binary mixed micelles using regular solution theory, where a mixing parameter is introduced as a measure of synergistic or antagonistic mixing [50,51]. Unfortunately, attempts to study the kinetics of such mixed micelles have not been successful, mainly due to problems involved in resolving the two relxation [52] times associated with the monomer exchange process of each surfactant, and also measuring the monomer concentration of each surfactant. It is possible to overcome these problems by choosing a surfactant blend in which the monomer exchange process of one component is frozen, and on the ultrasonic time scale and/or using an ionic/nonionic mixture. At present, we are not aware that such studies have been carried out.

As a result we have confined this chapter to mixed surfactant systems in which an additive (usually surface active) interacts with a single component surfactant. We recognize of course, that this type of definitive overlaps phenomenon, such as mixed micelles, solubilization, and surfactant macromolecule complexes.

A. Triblock Copolymers

Recently, the kinetics of micellization in nonionic triblock copolymers of the type $EO_nPO_mEO_n$ where EO and PO represent polyethylene oxide and polypropylene oxide blocks have been studied. The studies have been extended to mixed micelles of some triblock copolymers with sodium dodecyl sulfate, and the relaxation time amounts with the exchange process of the triblock copolymer monomer with some mixed micelles have been measured, using temperature jump relaxation [53]. Unfortunately, the results cannot be analyzed because of the lack of equilibrium data on the monomer concentrations.

B. Alcohol/Surfactant Systems

Microemulsions are four component systems containing surfactant, co-surfactant (usually an alcohol), oil and water. There are generally two types:

1. Oil in water—in which the surfactant aggregates are micellar with the oil dissolved in the hydrocarbon-like interior of the micelle and co-surfactant distributing itself between the aqueous phase and the micellar aggregate.
2. Water in oil—the aggregates are essentially reversed micelles with a water pool in the interior which contains the hydrophilic head groups of the

surfactant: the hydrophobic chains being dissolved in the oil-like solvent. Again, when the co-surfactant is present, it is usually distributed between the oil phase and the aggregate.

Aqueous solutions of surfactant micelles containing alcohols such as n-butanol and n-pentanol are model systems for oil in water microemulsions. These solutions contain mixed aggregates composed of surfactant and alcohol which are in dynamic equilibrium with alcohol and surfactant monomers in the aqueous phase (Scheme 1). These processes should be characterized by two fast relaxation times and these have been observed experimentally. Using the ultrasonic technique, it is possible to study the alcohol exchange processes exclusively by choosing a surfactant monomer whose chain length is large enough so as to freeze out its τ_1 in relation to the time scale of the ultrasonic measurements. Examples of such systems are mixed micelles of cetyltrimethylammonium bromide and hexadecylpyridinium chloride containing n-pentanol, where each exchange process between pentanol and the micellar aggregate is a well-defined single relaxation process. The exact mechanism for the formation of such mixed micelles involves a series of stepwise bimolecular reactions, with a micellar aggregate having a distribution of both alcohol and surfactant monomers. Such a mechanism contains many unknown kinetic and equilibrium parameters [54,55]. We describe here how we have used the phenomenological treatment to analyze the ultrasonic relaxation data [56]. As stated previously, one of the key requirements is equilibrium data. In these systems,

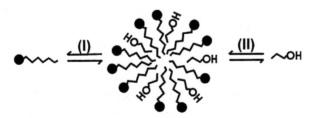

Scheme 1 Micelle.

however, the equilibrium behavior can be determined readily using head space gas chromatography analysis.

C. Head Space Analysis

Quantitative information concerning the partitioning of the alcohol between the bulk phase and the micelles has been obtained using the head space analysis method involving gas chromatography. In this method, solutions of alcohol and alcohol/micellar solutions are placed in a conical flask with fitted septum and placed in a thermostatted bath. After appropriate equilibration times (approximately 24 hr), vapor samples are withdrawn from the head space using a gas-tight syringe and are analyzed by gas chromatography for the partial pressure of alcohol in the vapor over each solution. The underlying principle of this method is that the concentration of alcohol per unit volume of head space over a solution in equilibrium with its vapor is the same for any two solutions in which the alcohol chemical potential is the same. From the head space data, we can evaluate the monomer alcohol concentration, m_2. In all those systems studied, the range of total alcohol and surfactant concentrations were the same as those used in the ultrasonic experiments. The results in Fig. 12 are displayed as plots of the concentration of solubilized alcohol (C_2^m) as a function of the monomer alcohol concentration (m_2).

The total alcohol concentration $C_2 = C_2^m + m_2$. In all cases, C_2^m is a linear function of m_2 and in these circumstances we define the partition coefficient (K) as follows in Eq. (17).

$$K = \frac{C_2^m}{m_2 M} \qquad (17)$$

Here M is the concentration of micellar surfactant expressed as moles monomer dm^{-3}. For the present work, $M = (C - \text{CMC})$ where C is the total surfactant concentration and the critical micellar concentration (CMC) is taken to be the monomer surfactant concentration. As we have mentioned previously, this is not strictly correct but the total surfactant concentration used in many of the surfactants is high enough for this approximation to be valid. Central to the approach in using the

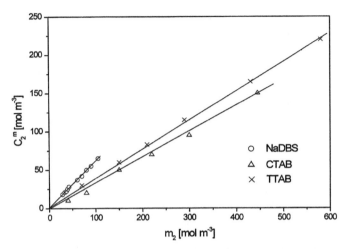

Figure 12 Concentration of solubilized alcohol (C_2^m) as a function of aqueous alcohol concentration (m_2).

phenomenological treatment to evaluate the kinetics of complex systems for which both equilibrium and relaxation data are available is to investigate the behavior of the ultrasonic relaxation parameters μ_m and μ_m/τ. We have shown previously that for an exchange process such as equilibrium (11) μ_m is given by Eq. (18).

$$\mu_m = \frac{\Delta V^2}{2RT\kappa_s} \frac{m_1 C_2^m}{C_2} \tag{18}$$

where ΔV is the reaction volume change and κ_s the adiabatic compressibility. From Eq. (18) it follows that a plot of μ_m against $m_2 C_2^m / C_2$ should be linear, passing through the origin as shown in Fig. 13.

$|\Delta V|$ can be evaluated from the slope of this line. By working along the lines outlined in Eqs (13)–(16), the backward rate R^- was calculated for a mixed alcohol/surfactant micelle. This dissociation step should be a first-order rate process whose rate depends on the alcohol concentration in the mixed micelle phase. Thus

$$R^- = k^- C_2^m \tag{19}$$

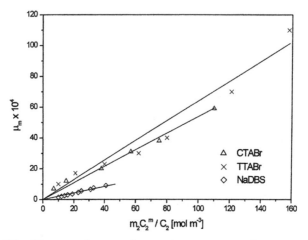

Figure 13 The maximum absorption per wavelength (μ_m) as a function of $m_2 C_2^m / C_2$.

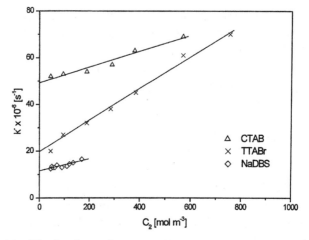

Figure 14 The backward rate coefficient (k^-) as a function of total alcohol concentration (C_2).

In practice, it was found that [56–58] k^- increases with C_2 (Fig. 14) and C_2^m according to the relationship $k^- = k_0(1 + aC_2^m)$.

The increase of k^- with C_2^m implies that the intrinsic rate with which alcohol molecules are released from the mixed micelle increases as the proportion of alcohol increases. This is not surprising since the environment of the micellar alcohol molecules is expected to change substantially with C_2^m.

We now consider the forward rate $R^+ = R^-$. The forward step is biomolecular and involves the association of an alcohol monomer from the bulk phase to the mixed micelle. Thus the rate associated with this step (R^+) is proportional to m_2 and M accordingly

$$R^+ = k^+ m_2 M \qquad (20)$$

where m_2 is the monomer alcohol concentration and M the mixed micellar concentration. At any alcohol concentration, the system is always at equilibrium, thus

$$k^+ m_2 M = k^- C_2^m \qquad (21)$$

or

$$k^+ = k^- \left[\frac{C_2^m}{m_2 M} \right] \qquad (22)$$

If we take this equation to the limit as $C_2^m \to 0$ then

$$k^+ = k_0 Kn \qquad (23)$$

where n is the micellar aggregation number. In some cases, K also varies linearly with C_2^m. In practice, both k_0 and K are known and also in many cases, n. Therefore, k^+ can be evaluated and some data are listed below.

	k^+ $\mathrm{dm^3\,mol^{-1}\,s^{-1}}$	k^- $\mathrm{s^{-1}}$
0.10 SDeS—n-pentanol	3.7×10^9	1.2×10^7
0.15 SDeS—n-pentanol	4.6×10^9	1.3×10^7
0.05 DTAB—n-pentanol	8.0×10^9	1.9×10^7
0.10 DTAB—n-pentanol	6.0×10^9	1.4×10^7
0.10 CTAB—n-pentanol	7.3×10^9	8.4×10^7

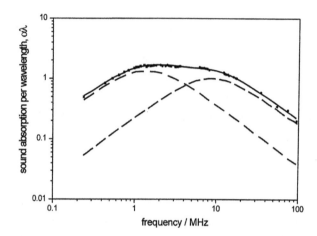

Figure 15 The sound absorption per wavelength, $\alpha\lambda$ as a function of frequency for $0.1\,mol\,dm^{-3}$ SDeS–$0.177\,mol\,dm^{-3}$ n–pentanol.

In the case of ultrasonic studies on the binary systems sodium decyl sulfate (SDeS) and n-pentanol and also dodecyltrimethylammonium bromide (DTAB) and n-pentanol, two reasonably well-separated relaxation processes were observed experimentally as shown in Fig. 15.

These relaxations were assigned respectively to the exchange processes involving the surfactant monomer (1) and n-pentanol (2) from the bulk phase to the mixed micellar phase according to Scheme 1.

For these systems, the monomer alcohol concentration was determined using head space analysis as described above and the monomer surfactant concentration was measured using specially constructed DeS^- and DTA^+ membrane selective electrodes.

The rate constants for the alcohol exchange process were determined using the proceduress described in Eqs (12)–(19) above. In the consideration of the surfactant exchange process, the equilibrium rate at all concentrations was determined using Eqs (12) and (14) separately. In the use of Eq. (12), the quantity $(dA/d\xi)$ was determined from the equilibrium data obtained using the surfactant electrodes to measure the composition of the mixed micellar phase. The rate data using both approaches were consistent. We would also like to point out that the kinetics of both alcohol and surfactant exchange processes (Scheme 1) have also been studied using relaxation equations derived by Aniansson and based on Aniansson and Wall kinetics [54,55,59]. These relaxation equations were derived by solving the differential equations associated with the kinetics of mixed micelle formation. The theory [54] predicts the existence of two fast relaxation processes as shown below for a mixed micelle (S_nS_a), containing neutral surfactant (S) and alcohol (A)

$$A + S_nA_{a-1} \underset{k_2^+}{\overset{k_2^-}{\rightleftharpoons}} S_nA_a \underset{k_1^+}{\overset{k_1^-}{\rightleftharpoons}} S_{n-1}A_a + S \tag{24}$$

where k_1^+ and k_1^- are the forward and backward rate constants, respectively, for the exchange of surfactant with the mixed micelle, and k_2^+ and k_2^- are the forward and backward rate constants, respectively, for the exchange process of the alcohol with the mixed micelle. Equations (7) and (8) are the relations derived by Aniansson [54] which link the relaxation times to rate constants and micellar distribution parameters

$$1/\tau_{1,S} = k_1^- \sigma_S^2[1 + \sigma_S^2 C_m/\text{CMC}$$
$$- \sigma_A^2 \sigma_S^2/\sigma_{AS}^4(1 + \sigma_A^2 C_m/m_2)^{-1}] \tag{25}$$

$$1/\tau_{1,A} = k_2^- \sigma_A^2[1 + \sigma_A^2 C_m/m_2] \tag{26}$$

where σ_A^2 and σ_S^2 are the variances in the mean alcohol aggregation number ($\tilde{a}$) and the mean surfactant aggregation number ($\tilde{n}$), respectively, σ_{AS} is a measure of the correlation (or covariance) between σ_A^2 and σ_S^2, C_A^f is the concentration of alcohol not micellized, CMC is the critical micelle concentration in the presence of the alcohol, and C_m is the concentration of micelles, given by

$$C_m = (C_S - CMC)/\tilde{n} \tag{27}$$

where C_S is the total surfactant concentration.

In the first application of these equations, Yiv *et al.* [59] used the pressure (shock) relaxation technique to study the surfactant exchange process (1) and ultrasonic relaxation to measure the relaxation parameters for step (2). Mixed micelle system of tetradecyl (TTAB) and hexadecyl trimethyl ammonium bromide (CTAB) with the alcohol n-butanol to n-hexanol were studied with the most extensive data reported on TTAB-n-pentanol. The independent relaxation experiments were conducted at two different temperatures and the equilibrium data were limited mainly to aggregation numbers with no measurements made on the monomer surfactant concentration. Although the subsequent analysis required approximations and assumptions, some very useful qualitative characteristics of the exchange processes emerged [54].

Verral *et al.* [55] have also studied the double relaxation observed in ultrasonics associated with exchange processes (1) and (2) in the system DTAB with the alcohols n-propanal$\rightarrow$ n-pentanol and also n-butoxyethanol. This analysis was based on Aniansson's equations (25) and (26) with subsequent approximations. The data was again analyzed to give rate constants for both exchange processes. In general, the difference between the approaches described above is quite reasonable in the sense that the overall qualitative picture remains the same but quantitative details concerning the kinetic data are inevitable. In all cases, k^+ for the association of alcohol to the mixed micelle is an almost diffusion controlled process with values in the range 10^9–$10^{10}\,\mathrm{mol}^{-1}\,\mathrm{dm}^3\,\mathrm{s}^{-1}$. On the other hand, the k^- values are $\sim 10^{-3}\,\mathrm{s}^{-1}$ from the phenomenological treatment and $\sim 10^{-10}$ from the data possessed using the Aniansson and Wall kinetics. Here it must be stated that for the alcohol exchange process the k^- values from the Aniansson approach k_A^- and the phenomenological k_p^- treatment are related as:

$$k_A^- = \frac{k_p^-}{n}$$

The kinetic data for the exchange of surfactant to and from the mixed micelle show more quantitative variations of k^+ and k^- values from the different approaches.

It must be stressed here that the approach of Yiv *et al.* [59] and Verral [55] using Aniansson and Wall type kinetics are based on the assumption that rate constants do not depend on the micellar composition—an assumption which is clearly at odds with the results from the phenomenological approach. In addition the amount of equilibrium data available to Yiv *et al.* [59] and Verral [55] was very limited especially with regard to the partition of the alcohol and the lack of direct information on monomer surfactant concentration as a function of mixed micellar composition.

D. Cationic drugs/Surfactant Mixed Micelles

The transport of drugs through membranes and their subsequent release can be mediated by surfactant vesicles and also di and triblock copolymers which are basically nonionic surfactants. A simple model for such a drug carrier is a mixed micelle containing the drug and a surfactant. Many heterocyclic based drugs are essentially cationic and several behave as surfactants in the sense that they aggregate in solution, sometimes to form drug micelles. In the presence of the cationic surfactant CTAB drugs such as propanolol hydrochloride (PM), penthianate methobromide (PMB), and chlorocyclizine hydrochloride (CCH) whose structures are shown in Scheme 2, can form mixed cationic micelles. In solution, the exchange process of surfactant and drug monomers with the mixed micelle can be represented by Scheme 3.

This model system is ideal for ultrasonic relaxation studies in the sense that the CTAB monomer exchange with the mixed micelle is "frozen" out of the experimental ultrasonic frequency range and the system is characterized by a well-defined single ultrasonic relaxation associated with the exchange of drug monomer from bulk solution to the mixed micelle. The kinetic data were analyzed using the phenomenological equations and the pertinent kinetic parameter associated with the drug release from the mixed micelle is the backward rate, R^-. This rate can be evaluated from the ultrasonic data via Eqs (12)–(23). As in the case of alcohol/surfactant micellar systems,

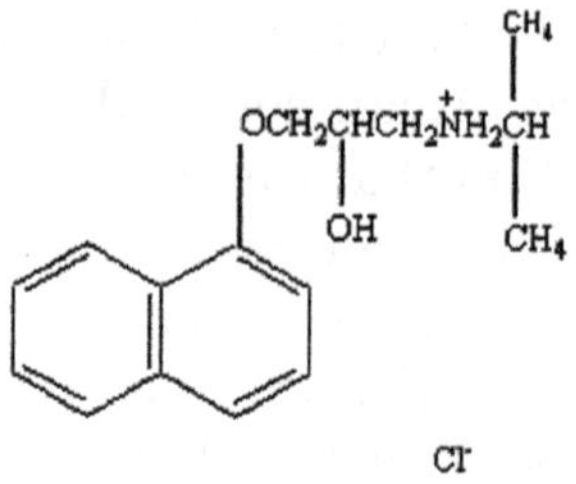

OCH₂CHCH₂NH₂CH

$OCH_2CHCH_2\overset{+}{N}H_2CH$

with CH₄ groups and OH

(PH)

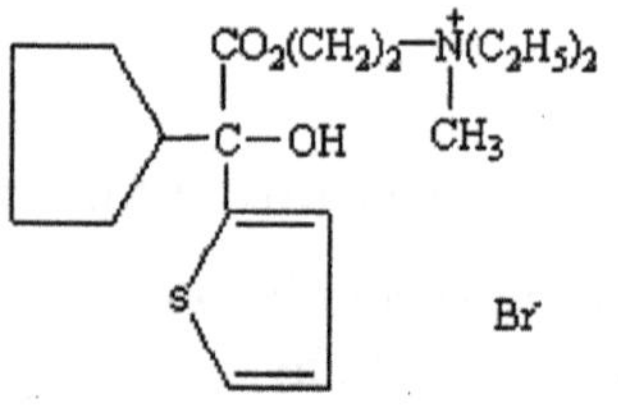

$CO_2(CH_2)_2\overset{+}{N}(C_2H_5)_2$

C—OH   CH₃

Br⁻

(PMB)

CH_3 — N   N — CH – HCl

with Cl-substituted phenyl rings

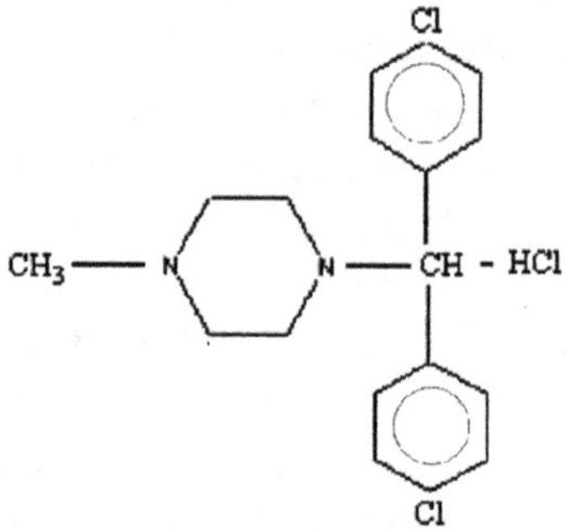

(CCH)

Scheme 2 Structure.

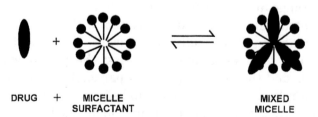

DRUG   +   MICELLE
SURFACTANT

MIXED
MICELLE

Scheme 3 Drug and surfactant micelle.

this backward rate for a drug/surfactant mixed micelle is proportional to the concentration of drug to the mixed micelle, C_2^m according to first-order rate principles, Eq. (19).

$$R^- = k^- C_2^m$$

The rate constant k^- can be evaluated if C_2^m is known. The composition of drug incorporated in the mixed micelle C_2^m, was evaluated from ultrasonic velocities using a novel procedure introduced by Sharma and Gormally [60].

E. Equilibrium Composition of Solutions

The concentration of CTAB used in all experiments which contained the drug was $0.1\,\text{mol}\,\text{dm}^{-3}$. As the critical micelle concentration of CTAB is below $10^{-3}\,\text{mol}\,\text{dm}^{-3}$, it can be safely assumed that this surfactant is almost totally in the micellar form. In order to understand the behavior of a cationic drug in these solutions, it is necessary to determine how the drug is partitioned between the aqueous and micellar phases. In Fig. 16, plots of sound velocity vs concentration for solutions containing CTAB alone, a drug alone, and the drug in the presence of $0.1\,\text{mol}\,\text{dm}^{-3}$ CTAB are shown.

It can be seen that the sound velocity varies little with concentration for CTAB solutions. As these changes in sound velocity are largely due to changes in the solution compressibility, this indicates that the compressibility of CTAB micelles approximates that of CTAB solution at the CMC. This behavior is peculiar to CTAB and was one of the reasons for selecting this surfactant. Relatively large velocity changes were found in solutions of the drug alone. In the concentration range studied, the drug was entirely in monomer form and the observed increase in sound velocity with drug concentration can be attributed to a reduction in solution compressibility due to hydration of the ionic drug. This behavior is typical of ionic species and it has been used as a means of estimating hydration numbers. The presence of CTAB in the drug solutions can be seen to reduce the increases in sound velocity which result from increases in drug concentration. This can only be due to the

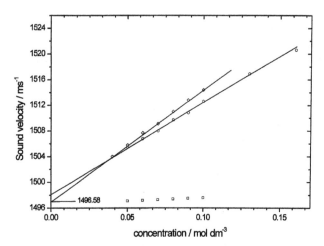

Figure 16 Plots of sound velocity vs concentration for: ($\diamond$) penthianate methobromide; ($\bigcirc$) penthianate methobromide in $0.1 \, \text{mol} \, \text{dm}^{-3}$ CTAB; ($\square$) CTAB. A departure from linearity can be seen for the highest concentration of PMB in CTAB. Although small, this effect is outside the limits of experimental error and indicates that the partition constant, K_p, is not strictly constant over a large concentration range. A similar pair of lines was obtained for the PH–CTAB system.

incorporation of some drug within the CTAB micelles to give rise to a mixed micelle which has the compressibility similar to that of the CTAB micelle alone. If we assume that these mixed micelles have a compressibility that is the same as that of CTAB micelles, then it is a simple matter to estimate how much of the drug is within the micellar phase and how much is in the aqueous phase by the procedure indicated in Fig. 17.

It is unlikely that the incorporation of drug molecules within CTAB micelles will have no effect upon their compressibility and we must consider this method to be a reasonably reliable means of estimating the partitioning of the drug. From the velocity data, it is clear that a linear relationship exists between the concentration of the drug in the aqueous phase m_2 and that in the micellar phase C_2^m. The partition coefficient,

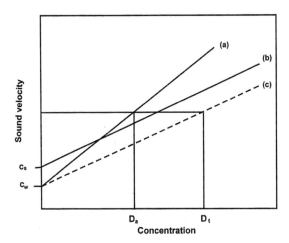

Figure 17 The variation of sound velocity in solution containing the drug alone. This line intercepts the ordinate at C_w, the sound velocity in water. (b) The variation of sound velocity in solutions containing drug in the presence of $0.1\,\text{mol dm}^{-3}$ CTAB. This line intercepts the ordinate at C_s, the sound velocity in $0.1\,\text{mol dm}^{-3}$ CTAB. (c) Is (b) displaced downwards by the amount $C_s - C_w$. (c) Is taken to represent the effect upon the sound velocity of drug which is in the aqueous phase alone, the effect of micellar drug having been allowed for by the displacement mentioned above. When the total drug concentration is D_t, the concentration in the aqueous phase is D_a.

K is defined as shown earlier by Eq. (17). Using the procedure described previously for the alcohol exchange process in the mixed micelles, ΔV the reaction value change, is derived by graphical analysis of Eq. (18) (Fig. 18) and thus R^- can be calculated. From the relationship (19)

$$R^- = k^- C_2^m$$

and the plot shown in Fig. 19, it follows that for PH and PMH, k^- increases with C_2^m in a linear fashion as

$$k^- = k_0(1 + aC_2^m)$$

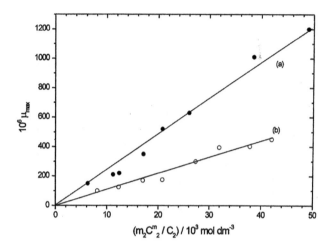

Figure 18 Plots of μ_{max} vs $m_2 C_2^m / C_2$ [Eq. (8)] for: (●) PMB in $0.1\,\mathrm{mol\,dm^{-3}}$ CTAB; (○) PH in $0.1\,\mathrm{mol\,dm^{-3}}$ CTAB. These lines must pass through the origin.

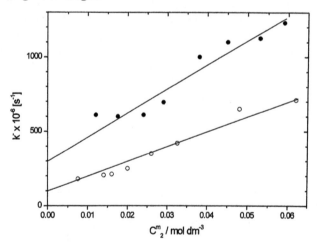

Figure 19 Plots k^- vs C_2^m for: (○) PMB in $0.1\,\mathrm{mol\,dm^{-3}}$ CTAB; (●) PH in $0.1\,\mathrm{mol\,dm^{-3}}$ CTAB.

where k_0 and a are independent of C_2^m.

The increase of k^- with C_2^m again shows that the intrinsic drug release rate from the mixed micelle for PM and PMB increases as the proportion of the drug in the micelles increases, however for CCH, k^- is constant. Finally, we can consider the forward rate constant in the limit $C_2^m \to 0$ by reference to Eq. (23) where

$$k^+ = k_0 Kn$$

where n is the micellar aggregation number. Values of k^+ and k^- for the three drugs PH, PMB, and CCH [61] are given below.

Drug	k^+ $\mathrm{dm^3\,mol^{-1}\,s^{-1}}$	k_0 $\mathrm{s^{-1}}$
PH	2.2×10^9	2.7×10^6
PMH	3.0×10^9	3.7×10^6
CCH	3.6×10^9	4.4×10^6

VI. MIXED UNASSOCIATED POLYMER/ SURFACTANT MIXTURES

Polymers and surfactants are frequently employed together in many industrial and pharmaceutical applications. Mixture of polymers and surfactants in aqueous solution have been utilized in several complex colloidal systems to achieve specific physicochemical properties including emulsification, colloidal stability of flocculation, structuring and suspending properties, and rheology control. As a result, these mixtures have several applications in diverse areas ranging from later paint technology, food products cosmetics as well as crop design formulations. In addition, polymer/surfactant mixtures have also been used as model systems for protein/surfactant interactions. As a result this has prompted a large volume of fundamental studies directed at understanding the equilibrium, kinetic, structural and rheological properties of many different systems. Many of these studies have been documented in several excellent review articles [62–67].

In this section, we shall confine our attention to the interactions between anionic surfactants and nonionic polymers with particular reference to the interactions of SDS to polyvinylpyrrolidone (PVP), polypropylene oxide (PPO) and ethylhydroxyethylcellulose (EHEC). It is now well accepted that when SDS binds to these polymers the surfactant exists [68,69] as micellar type aggregates which are bound to the polymer chain. The kinetic studies in question are focused on the fast relaxation time associated with the monomer/bound micelle equilibrium which can be studied using ultrasonic relaxation. One of the basic prerequisites which is required before any progress can be made at understanding these systems is a reliable binding isotherm. In the past 15 years or so we have shown that this vital thermodynamic information can be obtained using surfactant membrane selective electrodes as follows for the SDS/PVP system.

A. The Binding of SDS to Polyvinylpyrrolidone (PVP)

The EMF data for PVP in the presence of SDS were measured at 25°C and are shown in Fig. 20.

The experiment was carried out in such a way that concentrated SDS solution containing 0.5% w/v PVP is titrated into water and an aqueous solution containing the same amount of polymer. The respective EMFs are then plotted as a function of added SDS for solutions with and without the PVP—the latter being the control experiment; in the electrode data, the CMC of SDS is at the minimum of the EMF plot. In theory, binding of surfactants to the polymer is taking place when the EMF values with and without the polymer are different for each corresponding titration.

As shown in Fig. 20, the EMF data exhibit the characteristic features of a binding process and the two EMFs diverge at an SDS concentration denoted T_1 when SDS starts binding to PVP. The binding proceeds as long as the EMFs of the electrode with and without the polymer are different. When the polymer is fully saturated with bound SDS, no further binding occurs and the two EMFs merge again at an SDS concentration denoted T_2. The corresponding binding isotherm is shown in Fig. 21.

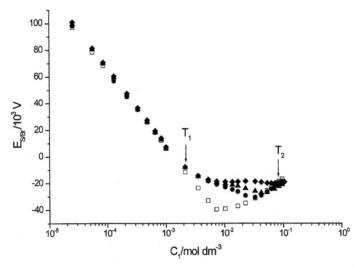

Figure 20 Plot of EMF of SDS electrode as a function of SDS concentration for the PVP ($M_{r,w} = 10\,000$)–SDS system: (□) 0%; (●) 0.5%; (▲) 1% and (◆) 2% PVP.

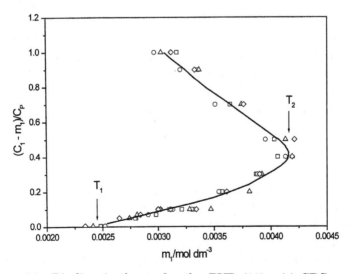

Figure 21 Binding isotherm for the PVP (1% w/v)–SDS system: $M_{r,w}$ of PVP: (□) 10 000; (○) 40 000; (△) 360 000; (◇) 700 000.

The symbols T_1 and T_2 are historical in the sense that they were first introduced by M. N. Jones [62] in his pioneering studies on polymer/surfactant mixtures as characteristic concentration for the state of surfactants. In some cases T_1 is called the "critical aggregation concentration" (CAC) which is synonymous with the formation of bound micellar type aggregates on the polymer. It is now accepted that the bound SDS exists as micellar type aggregates which are considered to grow in size and number during the binding process. The noticeable change in EMF at T_1 as well as the variation of bound SDS as displayed in the binding isotherm of Fig. 21, provide evidence for the cooperative nature of the formation of SDS aggregates during the binding process to PVP. The existence of aggregates have also been confirmed using structural methods such as small angle neutron scattering (SANS) [70]. The ultrasonic relaxation data measured for these systems is confined to SDS concentrations in the binding region. The relaxation is associated with the perturbation of the monomer/polymer bound micelle equilibrium and is characterized by a single relaxation time. The kinetic data in the form of forward (=backward) reaction rate were evaluated at each SDS concentration in the binding region using Eqs (12)–(14) and (ΔV) is assumed to be the same as for micellar SDS. The backward rate R^- is the easiest to handle in the sense that the dissociation of an SDS monomer from the mixed micelle should be proportional to the amount of polymer bound surfactant $(C - m_1)$ which is measured *via* the binding isotherm. Thus

$$R^- = \frac{k^-}{n}(C - m_1) \tag{28}$$

Here we have to take into account n, the aggregation number of the bound micelle. The three polymer/micelle complexes formed by PVP, PPO and EHEC are very different. For example, in PVP (MW 40 000) segments of the polymer chain are attached to the micellar surface and the complex resembles a "bead" type structure with 2–3 SDS micelle bound to each polymer chain as shown in Scheme 4. For the short chain PPO (MW 1000), each surfactant micelle has 2–4 polymer chains bound to

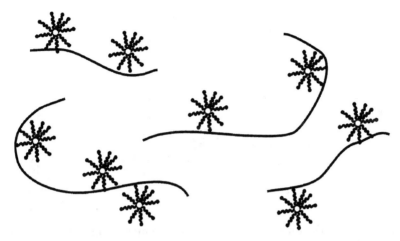

Scheme 4 2–3 SDS micelles bound to each PVP chain.

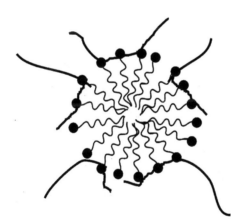

Scheme 5 The structure as "hairy" type micelle with 2–4 PPO chains bound to its surface.

its surface and the structure resembles a "hairy" type micelle as Scheme 5. Finally in EHEC there is strong evidence of an inter-polymer network existing in solution with SDS micelles acting as inter-polymer bridges (Scheme 6).

The rate plots of R^- against $(C - m_1)$ for each polymer are in Figs 22 and 23. Unfortunately, there is very little data

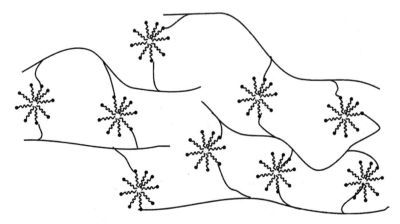

Scheme 6 SDS micelle acting as interpolymer bridges between the EHEC chains.

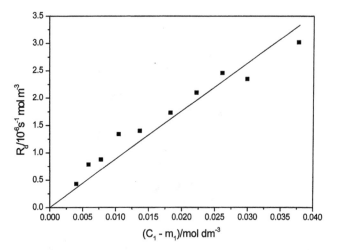

Figure 22 Plot of backward rate (R_d) vs ($C_1 - m_1$) for the PVP (1% w/v, $M_{r,w} = 40\,000$)–SDS system.

available on the aggregation numbers of the bound micelles and the values of k^-/n as a function of SDS concentration are in Fig. 24. The differences clearly relate to the state of the bound micelle and in particular the specific interaction will occur between the polymer and the micelle.

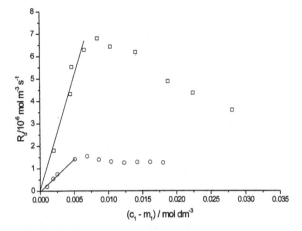

Figure 23 Plot of backward rate (R_d) vs $(C_1 - m_1)$ for the ($\square$) PPO–SDS and the ($\bigcirc$) EHEC–SDS systems.

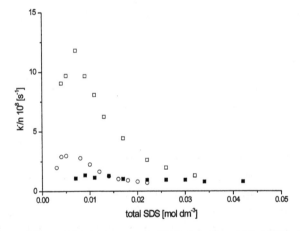

Figure 24 Plot of k^-/n vs total SDS for the ($\blacksquare$) PVP–SDS; ($\square$) PPO–SDS and the ($\bigcirc$) EHEC–SDS systems.

VII. CYCLODEXTRIN INCLUSION COMPLEXES

Inclusion compounds in which the host can admit a guest component into its cavity without any covalent bonds being formed are important in molecular recognition studies and have also found a wide variety of applications [71–74]. The cyclodextrins are known to form several inclusion compounds with substrates varying from hydrophobic to ionic in character. Cyclodextrins are cyclic carbohydrates consisting of 6, 7, or 8 glucose units respectively called α-, β-, and γ-cyclodextrin [73]. They can be regarded as cylinders or more precisely truncated cones with a hydrophilic exterior and hydrophobic interior. The simplest way to describe the inclusion phenomena involving a host and a cyclodextrin as guest is *via* the scheme

$$\text{host} + \text{guest} \rightleftharpoons \text{inclusion complex} \tag{29}$$

For surfactant/cyclodextrin systems, the situation is not simple in the sense that the inclusion mechanism is more complex than (29) and involves the formation of not only $1:1$, but also $1:2$, $2:1$, etc., complexes. Surfactant(s)/cyclodextrin (CD) complexes are formed *via* the stepwise biomolecular mechanism (29)

$$
\begin{aligned}
\text{S} + \text{CD} &\rightleftharpoons \text{S.CD} \\
\text{S.CD} + \text{CD} &\rightleftharpoons \text{S.(C)}_2 \\
\text{S.CD} + \text{S} &\rightleftharpoons \text{S}_2.\text{CD}
\end{aligned} \tag{30}
$$

In order to establish the exact mechanism and also determine the equilibrium constant for each step, a knowledge of the equilibrium concentrations is required. In practice however, this is not a simple matter. For example, a diverse range of different experimental methods have been used to study the complexation between sodium dodecyl sulfate and α-cyclodextrin. It is now reasonably well established that $1:1$ and $1:2$ SDS.CD and SDS(CD)$_2$ complexes are formed [75]. However, estimates of the $1:1$ equilibrium constants from different experiments vary significantly [76]. The reason for this state of affairs is that many of the methods used to study these systems are basically not designed to give reliable estimates of equilibrium concentrations. Indeed in the use of many of these methods, equilibrium concentrations are inferred indirectly from what is essentially measurement of some macromolecular parameters of the system [77]. Recently, we used the SDS selective electrode to study the SDS/α-cyclodextrin system. Although monomer surfactant concentration [S] in mechanism (30) is measured directly in the experiment and the starting concentration of SDS and α-CD are known, the procedure falls under the analytical solution of a cubic equation. We found that the equilibrium constant for step (i) and step (ii) in the mechanism (31)

$$
\begin{align}
\text{SDS} + \text{CD} &\rightleftharpoons \text{SDS.CD} \quad \text{(i)} \\
\text{SDS.CD} + \text{CD} &\rightleftharpoons \text{SDS(CD)}_2 \quad \text{(ii)}
\end{align}
\tag{31}
$$

were respectively 21 000 and 18 000 mol dm^{-3}.

When the measurements were carried out over a wide range of initial SDS and α-CD concentrations, it was found that the uncertainties in these equilibria constants are fairly large. It is therefore not surprising that the use of indirect techniques to study this problem leads to very unreliable data [76]. In theory, these systems are also amenable for kinetic studies using ultrasonic relaxation. In practice however, before embarking on such studies, we need to choose a system in which an unequivocal assignment of the ultrasonic relaxation is possible. In order to facilitate the analysis so that reliable kinetic data is available, a system which only exhibits a 1:1 complex is desirable. Such a system is characterized by a well-defined single relaxation in the ultrasonic experiment. Secondly, in order to confirm the 1:1 stoichiometry of the complex the equilibrium studies must be performed using experiments which measure equilibrium concentrations directly. The systems which satisfied these criteria were respectively pentanol, dicyclomine hydrochloride (DCM), and chlorocyclizone hydrochloride (CCH) forming inclusion compounds with α-cyclodextrin. The equilibrium constant for the 1:1 complexes were measured using head space analysis (Pentanol/α-CD) [78] and a drug membrane selective electrode (DCM and CCH). A chlorocyclizine hydrochloride selective membrane electrode was constructed and monomer chlorocyclizine (CCH) concentrations in the presence of α-cyclodextrin were evaluated as follows. The EMF of the drug electrode relative to the sodium ion reference electrode was measured as a function of increasing

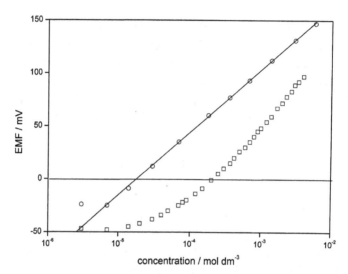

Figure 25 EMF data for the chlorocyclizine (CCH)–selective electrode as a function of [CCH] concentration (○), and in the presence of $0.008\,mol\,dm^{-3}$ α-cyclodextrin (α-CD).

drug concentration until the critical micellar concentration (CMC) of the drug was reached. At concentrations below the CMC, all the chlorocyclizine cations are in monomeric form [3] and in this range, the EMF data yield good Nernstien responses for the electrode as shown in Fig. 25.

Following this, the experiment was repeated by measuring the EMF of the drug electrode in the presence of a constant amount of α-cyclodextrin, again ensuring that the measurements were taken below the CMC of the drug. Typical data are also shown in Fig. 25. The EMF measurements of the electrode were then rechecked against the monomer chlorocyclizine to ensure consistency.

In general a guest ion like the chlorocyclizine cation is expected to form an inclusion compound with the host α-cyclodextrin described by a general equilibrium of the type [4]

$$\text{chlorocyclizine} + \text{cyclodextrin} \rightleftharpoons \text{inclusion compound}$$
$$\quad\;\text{CCH}\qquad\qquad\quad\text{CD}\qquad\qquad\qquad\qquad\text{CCH:CD}$$

In some cases more than one inclusion compound is formed with different stoichiometry, e.g., $1:1$, $1:2$, $2:1$, etc. In the present experiment it is possible, from the EMF data in Fig. 25, to evaluate the monomer drug concentration m_1 at each total concentration of drug for which the measurements have taken place. The next question to address is that concerning the stoichiometry of the inclusion compound. The first step is to assume that chlorocyclizine and α-cyclodextrin form a $1:1$ complex, in which case the binding data from the electrode can be treated using the classical Benesi–Hildebrand plot in the form

$$\frac{1}{\nu} = (1/Km_1) + 1$$

where

$$\nu = \frac{\text{concentration of drug complexed with } \alpha\text{-cyclodextrin}}{\text{total concentration of } \alpha\text{-cyclodextrin}}$$

and K is the complexation constant.

Thus a plot of ν^{-1} against m_1^{-1} should be a straight line with an intercept of 1 for a $1:1$ complex. If the plot is not linear, we recognize that further analysis in terms of more inclusion compounds is required. In the present work, the plot shown in Fig. 26 confirms $1:1$ stoichiometry with $K = 1.2 \times 10^3$ $dm^3\,mol^{-1}$.

Another approach to confirm the $1:1$ stoichiometry is *via* the Scatchard equation in the form

$$\frac{r}{m_1} = K - Kr$$

in which r/m_1 is plotted against r as shown in Fig. 27.

Once the $1:1$ stoichiometry has been established, the ultrasonic data may then be measured. Unfortunately, the number of ultrasonic measurements on α-CD systems were limited partly due to solubility problems and sensitivity. In addition, α-CD did exhibit a small relaxation on its own in solution and we have to compensate for this effect in the

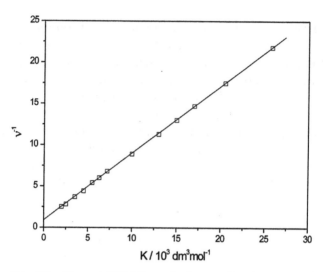

Figure 26 The Benesi–Hildebrand plot confirming a 1 : 1 inclusion compound for CCH and α-CD.

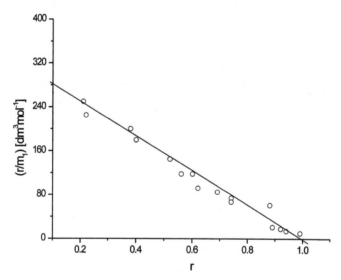

Figure 27 Scatchard plot of α-cyclodextrin–pentanol system.

analysis of the ultrasonic data to yield relaxation parameter for the formation of the $1:1$ complex. The following rate constants were evaluated.

	$k^+ \, dm^3 \, mol^{-1} \, s$	Ref.
Pentanol $+ \alpha$-CD	3.6×10^8	74
CCH $+ \alpha$-CD	1×10^8	57
DCH $+ \alpha$-CD	1.5×10^8	73

The almost constant values of k^+ of close to $10^8 \, dm^3 \, mol^{-1} \, s^{-1}$ is more than a factor of 10 slower than the expected diffusion controlled rate constant. Part of this reason must be due to the fact that in order to form a $1:1$ complex, the drug must approach the α-CD molecule along the axis of the truncated cone structure as shown in Scheme 7.

Finally, we are aware that Nikishawa has also published his results of ultrasonic relaxation for some cyclodextrin inclusion complexes over the range 1–95 MHz. For the α-CD with 1-propanol or ethanol, a single relaxational absorption was observed. The cause of the relaxation was attributed to a perturbation of a chemical equilibrium associated with a complexation reaction between α-CD (host) and alcohols (guest) [79]. In this work, only ultrasonic relaxation data was used in the analysis of the kinetic and equilibrium parameters. In

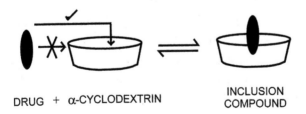

DRUG + α-CYCLODEXTRIN   INCLUSION COMPOUND

Scheme 7 Drug and α-cyclodextrin.

the absence of direct knowledge of equilibrium concentrations, the resulting data cannot be considered as reliable as those described above.

ACKNOWLEDGEMENT

T. T. wishes to thank Salford University and Robert McBride plc. for their support.

REFERENCES

1. M. Eigen, and L. De Maeyer, *Technique of Organic Chemistry* (A. Weissberger, ed.), Vol. VIII, Part 11. John Wiley, New York (1963).

2. S. Claesson (ed.), *Fast Reactions and Primary Processes in Chemical Kinetics*. Interscience (1967).

3. E. F. Caldin, *Fast Reactions in Solution*. Blackwell, Oxford (1964).

4. E. Wyn-Jones (ed.), *Chemical and Biological Applications of Relaxation Spectrometry*. Reidel, Dordrecht (1975).

5. W. J. Gettins, and E. Wyn-Jones (eds) *Techniques and Applications of Fast Reactions in Solution*. Reidel, Dordrecht (1979).

6. G. Czerlinski, *Chemical Relaxation*. Arnold (1966).

7. C. F. Bernasconi, *Relaxation Kinetics*. Academic Press, New York (1976).

8. H. Strehlow, and W. Knoche, *Fundamentals of Chemical Relaxation*, Verlag Chemie, Weinheim (1977).

9. H. J. McSkimin, In W. P. Mason (ed.), *Physical Acoustics*, Vol. 1, Pt. A, p. 272. Academic Press (1964).

10. E. Wyn-Jones, and W. J. Orville-Thomas, In G. M. Burnett and A. M. North (eds), *Transfer and Energy Storage in Molecules*, 11, 265, Wiley-Interscience, London (1969).

11. T. Telgmann, and U. Kaatz, *J. Chem. Phys.* **104**: 4846 (2000) and references quoted therein.

12. J. Gormally, W. J. Gettins, and E. Wyn-Jones, In H. Ratajczak and W. J. Orville-Thomas (eds), *Molecular Interactions*, 2, 143. John Wiley (1980).

13. E. Wyn-Jones and J. Gormally (eds), *Aggregation Processes in Solution*. Elsevier (1983).

14. D. G. Hall, and E. Wyn-Jones, *Journal of Molecular Liquids*, **32**: 63 (1986).

15. R. Zana (ed.), *Surfactant Solutions; New Methods of Investigations*. Dekker, New York (1987), p. 405.

16. J. F. Holzwarth, Ref. 5, p. 47.

17. K. Shinoda (ed.), *Solvent Properties of Surfactant Solutions*, M. Dekker, New York (1976).

18. P. Mukerjee, and K. J. Mysels, *Critical Micelle Concentrations of Aqueous Surfactant Systems*. NSRDS-NBS-35, Washington, USA (1971).

19. R. C. Murray, and G. S. Hartley, *Trans. Faraday. Soc.* **31**: 183 (1935).

20. C. F. Goodene, *Trans. Faraday Soc.* **31**: 197 (1935).

21. D. G. Hall, Ref. 13, Chapter 2, p. 7.

22. E. A. G. Aniansson, Ref. 13, Chapter 3, p. 70.

23. E. A. G. Aniansson, S. N. Wall, M. Algram, N. Hoffmann, H. Kielman, I. Ulbrecht, R. Zana, J. Lang, and C. Tondre, *J. Phys. Chem.* **80**: 905 (1976).

24. E. A. G. Aniansson, and S. N. Wall, *J. Phys. Chem.* **78**: 1042 (1974); **79**: 857 (1975).

25. D. G. Hall, *J. Chem. Soc. Faraday I*, **77**: 1122 (1981).

26. D. G. Hall, *J. Chem. Soc., Faraday II*, **77**: 1973 (1981).

27. D. G. Hall, *Colloids and Surfaces* **4**: 367 (1982).

28. D. Lessner, M. Teubner, and M. Kahlweit, *J. Phys. Chem.* **85**: 1529 (1981).

29. C. Elvingston, and S. Wall, *J. Colloid Interface Sci.* **121**: 414 (1985).

30. W. A. Wan-Bahi, R. Palepu, D. M. Bloor, D. G. Hall, and E. Wyn-Jones, *J. Phys. Chem.* **95**: 6642 (1991).

31. S. G. Cutler, D. G. Hall, and P. Mears, *J. Electroanal. Chem.* **85**: 145 (1977).

32. D. M. Painter, D. G. Hall, and E. Wyn-Jones, *J. Chem. Soc., Faraday Trans. 1* **84**: 773 (1988).

33. N. Takisawa, D. G. Hall, E. Wyn-Jones and (in part), P. Brown, *J. Chem. Soc., Faraday Trans. 1* (1988).

34. H. Gharibi, N. Takisawa, P. Brown, M. A. Thomason, D. M. Painter, D. M. Bloor, D. G. Hall, and E. Wyn-Jones, *J. Chem. Soc., Faraday* **87**: 707 (1991).

35. N. Takisawa, M. Thomason, D. M. Bloor, and E. Wyn-Jones, *J. Colloid and Interface Sci.* **157**: 77 (1993).

36. W. A. Wan-Badhi, T. Lucas, D. M. Bloor, and E. Wyn-Jones, *J. Colloid and Interface Sci.* **169**: 462 (1995).

37. R. Xu, and D. M. Bloor, *Langmuir* **26**: 9555 (2000).

38. R. Palepu, D. G. Hall, and E. Wyn-Jones, *J. Chem. Soc., Faraday Trans.* **86**: 1535 (1990).

39. P. J. Sams, J. E. Rassing, and E. Wyn-Jones, *Chem. Phys. Lett.* **13**: 233 (1972).

40. D. G. Hall, and B. A. Pethica, In M. Schick (ed.), *Non-Ionic Surfactants*, Ch. 16. Marcel Dekker Inc. (1967).

41. J. M. Corkill, J. F. Goodman, T. Walker, and J. Wyer, *Proc. Roy. Soc. A* **312**: 243 (1969).

42. T. Telgmann, and H. Kaatze, *J. Phys. Chem. B* **101**: 7766 (1997).

43. M. Kahlweit, and M. Teubner, *Adv. Colloid and Interface Sci.* **13**: 1 (1980).

44. D. Causon, J. Gettins, J. Gormally, R. Greenwood, N. Natarajan, and E. Wyn-Jones, *J. Chem. Soc. Faraday* **77**: 143 (1981).

45. M. Thomason, D. M. Bloor, and E. Wyn-Jones, *J. Phys. Chem.* **95**: 6017 (1991).

46. P. Jones, G. J. T. Tiddy, and E. Wyn-Jones, *J. Chem. Soc., Faraday Trans. 2* **83**: 2735 (1987).

47. R. Haase, *Thermodynamics and Irreversible Processes*. Addison/ Wesley, Reading, Mass. (1969).

48. D. G. Hall, J. Gormally, and E. Wyn-Jones, *J. Chem. Soc., Faraday. II* **79**: 645 (1983).

49. W. Wan Bahdi, H. Mwakibete, D. M. Bloor, R. Palepu, and E. Wyn-Jones, *J. Chem. Phys.* **96**: 918 (1982).

50. D. N. Rubingh, In K. Mittal (ed.), *Solution Chemistry of Surfactants*. Plenum, New York (1979), Vol. 1, p . 337.

51. P. M. Holland, *Adv. Colloid and Interface Sci.* **26**: 111 (1986).

52. P. J. Sams, J. E. Rassing, and E. Wyn-Jones, *Advances in Molecular Relaxation Processes* **6**: 255 (1975).

53. M. J. Kosta, G. D. Rees, A. Holzwarth, and J. Holzwarth, *Langmuir* **16**: 9035 (2000).

54. E. A. G. Aniansson, Ref. 5, p. 249.

55. R. E. Verrall, *Chem. Soc. Reviews* 135 (1995).

56. J. Gormally, B. Sztuba, E. Wyn-Jones, and D. G. Hall, *J. Chem. Soc., Faraday Trans. 2* **81**: 395 (1985).

57. P. Smith, C. Gould, G. Kelly, D. M. Bloor and E. Wyn-Jones, In W. Knoche and R. Schomater (eds), *Reactions in Compartmentalized Liquids*. Springer-Verlag, Berlin (1989) p. 83.

58. G. Kelly, N. Takasawa, D. M. Bloor, D. G. Hall, and E. Wyn-Jones, *J. Chem. Soc. Faraday Trans. 1* **85**: 4321 (1989).

59. S. Yiv, R. Zana, W. Albuht, and H. Hoffmann, *J. Colloid interface Sci.* **80**: 224 (1981).

60. J. Gormally and S. Sharma, *J. Chem. Soc., Faraday Trans. 1* **82**: 2497 (1986).

61. M. A. Thomason, H. Mwakibete, and E. R. Wyn-Jones, *J. Chem. Soc., Faraday Trans.* **86**: 1511 (1990).

62. E. D. Goddard, In E. D. Goddard and K. P. Ananthapadmanabhan (eds), *Interactions of Surfactants with Polymers and Proteins*. CRC Press, Boca Raton, FL (1993).

63. J. C. Brackman, and J. B. F. N. Engberts, *Chem. Soc. Rev.* **22**: 85 (1993).

64. L. P. Robb, Anionic surfactants, *Surfactant Sci. Ser.* **11**: 109 (1981).

65. J. C. T. Kwak (ed.), *Polymer–Surfactant Systems.* Surfactant Science Series, Vol. 77, Marcel Dekker, New York (1998), pp. 51–68.

66. P. Hansson, and B. Lindmann, *Curr. Opini. Colloid Interface Sci.* **1**: 604 (1996).

67. J. C. T. Kwak (ed.), *Polymer–Surfactant Systems.* Surfactant Science Series, Vol. 77, Marcel Dekker, New York (1998), pp. 183–238.

68. W. A. Wan-Badhi, W. M. Z. Wan-Yunus, D. M. Bloor, D. G. Hall, and E. Wyn-Jones, *J. Chem. Soc., Faraday Trans.* **89**: 2737 (1993).

69. D. M. Bloor, W. M. Z. Wan Yunus, W. A. Wan-Badhi, Y. Li, J. F. Holzwarth, and E. Wyn-Jones, *Langmuir* **11**: 3395 (1995).

70. Y. Li, R. Xu, D. M. Bloor, J. Penfold, J. F. Holzwarth, and E. Wyn-Jones, *Langmuir* **16**: 8677 (2000).

71. Y. Li, R. Xu, S. Courdec, S. M. Ghoreishi, J. Warr, D. M. Bloor, J. F. Holzwarth, and E. Wyn-Jones, *Langmuir* (submitted).

72. W. Waenger, *Angew. Chem. Int. Ed. Eng.* **19**: 344 (1980).

73. M. L. Bender, and M. Koywaga, *Cyclodextrin Chemistry.* Springer Verlag, New York (1987).

74. D. W. Griffiths, and M. L. Bender, *Adv. Catal.* **23**: 209 (1973).

75. W. M. Z. Wan Yunus, J. Taylor, D. M. Bloor, D. G. Hall, and E. Wyn-Jones, *J. Phys. Chem.* (1992).

76. H. Mwakibete, R. Cristantino, D. M. Bloor, E. Wyn-Jones, and J. F. Holzwarth, *Langmuir* **11**: 57 (1995).

77. N. Takisawa, D. G. Hall, E. Wyn-Jones, and P. Brown, *J. Chem. Soc., Faraday Trans. 1* **84**: 3059 (1988).

78. D. G. Hall, D. M. Bloor, K. Tawarah, and E. Wyn-Jones, *J. Chem. Soc., Faraday 1* **82**: 2111 (1986).

79. S. Nishikawa, N. Yokoo, and N. Kuramoto, *J. Phys. Chem.* 4830 (1998).

21

Photocatalysis at Solid/Liquid Interfaces—Photooxidation of Mixed Aqueous Surfactants at the TiO₂/H₂O Interface

H. HIDAKA, M. S. VOHRA, and N. WATANABE
Frontier Research Center for the Global
Environment Protection, Meisei University,
Tokyo, Japan

N. SERPONE
Department of Chemistry and Biochemistry,
Concordia University,
Montreal (Quebec), Canada

SYNOPSIS

A series of anionic (DBS, SDS), non-ionic ($C_{12}E_8$, NPE-9), and cationic (HTAB, BDDAC) surfactant systems were examined alone and in a variety of combinations to assess the influence of co-habitation of two different types in the overall photodegradation of mixed aqueous surfactant solutions. The process was monitored by UV spectroscopy, total organic carbon (TOC) analyses, and surface tension measurements prior to and subsequent to UV illumination of the surfactant solutions in the presence of TiO_2 particles. Results indicate that after about 3 to 4 hr of irradiation, more than *ca.* 90% of the mixed surfactant system was degraded as attested by the TOC remaining in the solution phase. As well, contrary to the stand-alone case, the photodegradation of the cationic surfactants was enhanced in the mixed systems as a result of positive charge neutralization by the other surfactant component. This study provides an overall strategy to minimize pollution caused by intentional or unintentional discharges of surfactants into water bodies.

I. INTRODUCTION

The past few decades have witnessed increased large-scale production and utilization of surfactants resulting in significant global water pollution. For example, approximately 1.1×10^6 tons of different types of surfactants were produced in 1998 in Japan alone; of this amount, the percent production of anionic, nonionic, and cationic surfactants was *ca.* 49%, 43%, and 6%, respectively. Industrial and domestic wastewaters containing these surfactants are usually discharged into natural water systems (e.g., rivers). Environmental problems resulting from such discharges pose serious concerns.

Activated sludge techniques commonly employed in removal of organic aqueous pollutants is not efficient for those nonbiodegradable surfactants that possess a complex chemical structure. Also, many of the surfactants discharged into natural water bodies typically require much longer times for complete biodegradation. Cationic surfactants [1–4] are difficult to degrade because of their sterilizing and antimicrobial properties [5,6]. Nonionic surfactants are also resistant to degradation, in addition to which biodegradation sometimes produces toxic intermediate products, for example nonylphenol, an endocrine disrupter as described by the United States Environmental Protection Agency (EPA) [7,8].

Previously, we investigated the photocatalytic degradation of surfactants using TiO_2 as a semiconductor photocatalyst [9,10], which led to complete mineralization of the surfactant species. Aqueous solutions containing DBS, BDDAC and NPE-9 species can also be degraded using TiO_2 as the photocatalyst under solar exposure [11]. We have also examined the mechanistic details of the photocatalytic degradation of such anionic surfactants as sodium dodecylbenzenesulfonate (DBS; [12,13]) and sodium dodecyl sulfate (SDS; [14]), the cationic surfactants benzyldodecyldimethyl ammonium chloride (BDDAC; [15]) and hexadecyltrimethyl ammonium bromide (HTAB; [16]), and the non-ionic surfactants 4-nonylphenylnanoethoxylate (NPE-9; [17]) and alkyl polyethoxylate [18]. Surfactants with aromatic rings are more easily decomposed than those containing only an alkyl and/or ethoxylate group. Note that although the cationic BDDAC species has an aromatic moiety, its degradation is nonetheless slow because of reduced attachment to the TiO_2 surface. Combination of these two surfactants resulted in very different surface activities [19–21] as described by detergency, foaming power, surface tension, and micelle solubilization, among others. These activities are governed by the functionality of the hydrophilic moiety as well as by the hydrophobic alkyl chain length.

Most of the commercial detergents currently used consist of a mixture of two (or more) different surfactants, softening agents, fluorescent-whitening agents, dyes, perfumes, fillers (e.g., Na_2SO_4), and others. Synergistic effects of mixed surfactant systems generally lead to excellent surface activities. In the present work, we examine the photocatalytic degradation of mixed systems containing different combinations of an aromatic and a nonaromatic surfactant. The following matrices were studied: anionic/cationic systems including DBS/HTAB and SDS/BDDAC, anionic/nonionic systems including DBS/ $C_{12}E_8$, and SDS/NPE-9, and the cationic/nonionic systems included HTAB/NPE-9 and BDDAC/$C_{12}E_8$. The competitive decomposition of the respective surfactants in each matrix was compared in detail.

Structures of Surfactants

II. PHOTOCHEMICAL AND PHYSICAL KINETICS OF PHOTOEXCITATION IN A SEMICONDUCTOR

A TiO$_2$ catalyst (anatase) is a n-type semiconductor having a bandgap of 3.2 eV. When a TiO$_2$ particle is exposed to a UV-light source of wavelengths below 387 nm, electrons present in the valence band are transferred to the conduction band. In other words, the charge separation by UV illumination occurs to generate electron holes (h$^+$) in a valence band and electrons in a conduction band respectively (1). The holes capture the electrons from H$_2$O molecules adsorbed and/or hydroxy groups bound on the TiO$_2$ particle surface [(2) and (3)]. Concominantly, the dissolved oxygen molecules accept the electrons from the conduction band to form $\cdot$O$_2^-$. This radical reacts with protons in aqueous solution to generate $\cdot$OOH radicals [(4) and (5)]. The highly reactive radicals were confirmed by spin-trapping ESR measurements using DMPO reagents.

$$TiO_2 + h\nu \rightarrow TiO_2\{e^- \cdots h^-\} \rightarrow e_{CB}^- + h_{VB}^+ \tag{1}$$

$$\{OH^-\}_{bound} + h^+VB \rightarrow \{^{\bullet}OH\}_{bound} \tag{2}$$

$$\{H_2O\}_{ads} + h^+VB \rightarrow \{^{\bullet}OH\}_{ads} + H^+ \tag{3}$$

$$\{O_2\}_{ads} + e^-CB \rightarrow \left\{^{\bullet}O_2^-\right\}_{ads} \tag{4}$$

$$\left\{^{\bullet}O_2^-\right\}_{ads} + H^+ \rightarrow \{^{\bullet}OOH\}_{ads} \tag{5}$$

The active oxidative radicals of $^{\bullet}$OH or $^{\bullet}$OOH are attached to the organics adsorbed or assisted on the TiO_2 surface, and the organics are photooxidized. In turn, this leaves a positive charge in the valence band referred to as a hole (h^+). In an aqueous system, the TiO_2 surface has OH^- species adsorbed onto specific sites at concentration levels between 5 and 15 OH^-/nm^2 [22]. The hole scavenges an electron from the adsorbed OH^- and produces the strongly oxidative $^{\bullet}$OH radicals, which react with aqueous pollutants to complete mineralization, producing harmless end-products. The involvement of $^{\bullet}$OH radicals in TiO_2 systems was established by electron spin resonance (ESR) methods [23]. Nonetheless, our results do not preclude direct oxidation of aqueous organic pollutants by h^+ species as proposed by Abdullah and coworkers [24], although this possible contribution is likely to be minor, if at all, under our present experimental conditions.

III. PHOTODEGRADATION PROCESS OF SURFACTANT

On the basis of experimental results, we propose a mechanism for the catalytic photodegradation process of anionic DBS surfactant as depicted in Scheme 1.

The rate-influencing steps in the photocatalytic degradation of surfactants in as irradiated TiO_2 system can be classified into three stages; (1) adsorption of the surfactants onto the TiO_2 surface; (2) fast photoinduced steps (hole–electron pair formation, •OH or •OOH radical formation, and/or cation-radical formation); (3) slow dynamic steps (aromatic ring opening, peroxide formation, aldehyde formation, carboxylic acid formation, and finally CO_2 evolution).

IV. EXPERIMENTAL DETAILS

The following reagent grade chemicals were employed: anionic surfactants including sodium dodecylbenzenesulfonate (DBS, $C_{12}H_{25}-C_6H_4-SO_3^- Na^+$) and sodium dodecyl sulfate (SDS, $C_{12}H_{25}-O-SO_3^- Na^+$), both from Tokyo Kasei; cationic surfactants including hexadecyl-trimethyl ammonium bromide (HTAB, $C_{16}H_{33}-N^+(CH_3)_3 Br^-$) from Tokyo Kasei, and benzyl-dodecyldimethylammonium chloride (BDDAC, $C_{12}H_{25}-N^+$ $(CH_2-C_6H_5)(CH_3)_2 Cl^-$) from Wako Pure Chem. Ind. Co. Ltd.; nonionic surfactants 4-nonylphenyl nanoethoxylate (NPE-9, $C_9H_{19}-C_6H_4-O-(CH_2CH_2O)_9-H$) from Miyoshi Oil and Fat Co. Ltd., and dodecyl octaoxyethylene ether ($C_{12}E_8$, $C_{12}H_{25}-O-(CH_2CH_2O)_8-H$) from Nikko Chem, Co. Ltd. The photocatalyst TiO_2 was a kind gift from Degussa (P-25; 87% anatase and 13% rutile).

The analytical instruments include a Shimadzu TOC-5000A total organic carbon analyzer for aqueous TOC determinations, a CBVP-Z automatic surface tensiometer from Kyowa Interface Science Co. Ltd., for surface tension measurements at ambient temperatures, and a JASCO V-570 UV spectrophotometer for quantifying the aromatic surfactant moieties.

A batch-type reactor setup consisting of four reactor vessels was employed using a procedure reported earlier [25]. The TiO$_2$ catalyst (loading 100 mg per 50 mL) was transferred to each of four reactors. In all cases, the total surfactant volume transferred to each was 50 mL. The concentration of all surfactant stock solutions was 0.100 mM. In the studies involving mixed surfactant systems, the respective volumes of each solution was such that the total initial concentration was also 0.100 mM. The aqueous surfactant/TiO$_2$ dispersion was initially sonicated for about 10 min; the reactor was then sealed appropriately and purged with oxygen gas for about 20 min. Subsequently, the suspension in all four reactors was mixed for a few minutes with a magnetic stirrer, following which an equal amount of sample was collected from each reactor using a syringe. These reactors were then transferred to a UV-reactor setup as described previously [14]. As the photodegradation proceeded, samples were collected at different time intervals to quantify the degradation process.

V. RESULTS AND DISCUSSION

A. The DBS/HTAB System

The temporal courses of the degradation of DBS (critical micelle concentration, CMC, 1.2 mM at 25°C), and HTAB (CMC, also 1.2 mM at 25°C) are shown in Figs 1A and 1B, respectively; the mixed DBS/HTAB system is summarized in Fig. 2. The disappearance of the aromatic moiety in the DBS system alone (Fig. 1A) occurs faster compared to the case when DBS is present in the mixed DBS/HTAB system (Fig. 2). Additionally, faster TOC removal is noted for the DBS system (after allowance for a 30-min induction period; Fig. 1A) than either the mixed DBS/HTAB system (Fig. 2) or the HTAB system alone (Fig. 1B). Also, comparing the results of Fig. 1A

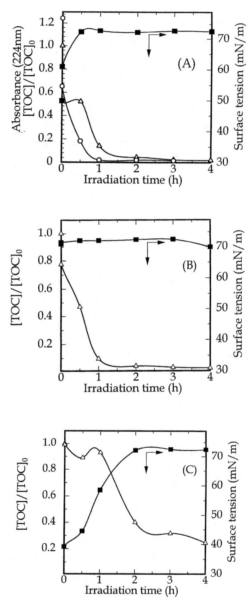

Figure 1 Disappearance of aromatic moiety (○), TOC (△), and surface tension (■) trends resulting during the photodegradation of surfactants alone (total concentration of surfactants, 0.100 mM; TiO$_2$ loading, 100 mg. (A) DBS; (B) SDS; (C) HTAB; (D) BDDAC; (E) C$_{12}$E$_8$; (F) NPE-9.

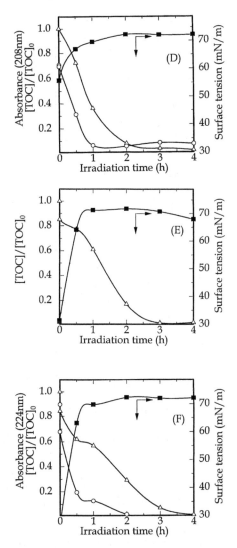

Figure 1 Continued

with those of Fig. 2, both DBS and TOC removal increase with increase in initial DBS concentration. Note that even at smaller concentrations, loss of DBS is lowered significantly in the presence of HTAB; for example, after 2 hr of irradiation about $2.7\,\mu M$ (or 2.7%) of the substrate remained for the DBS (0.100 mM) system alone, whereas for the 0.025 mM

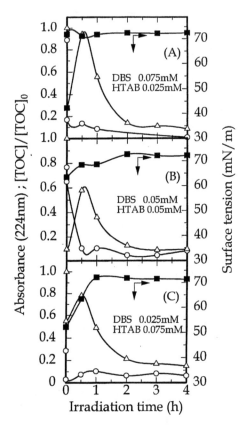

Figure 2 Disappearance of aromatic moiety (○), TOC (△), and surface tension (■) trends resulting during the photodegradation of the mixed DBS/HTAB system using TiO_2 (100 mg) catalyst. (A) DBS (0.075 mM)/HTAB (0.025 mM); (B) DBS (0.050 mM)/HTAB (0.050 mM); (C) DBS (0.025 mM)/HTAB (0.075 mM).

DBS/0.075 mM HTAB mixed system, about 6.25 μM (or 25%) of the substrate remained intact.

The above-mentioned trends are understood in terms of formation of H^+ species from the reaction $h^+ + H_2O \rightarrow$ $^{\bullet}OH + H^+$ which takes place during photocatalysis; this renders the TiO_2 surface positively charged. Since HTAB is a cationic surfactant, electrostatic repulsion between HTAB and TiO_2 particles prevents direct attachment of the former species onto the latter. This in turn causes reduced interactions between

HTAB molecules and the surface bound $^{\bullet}OH$ radicals, eventually resulting in reduced HTAB mineralization. By contrast, DBS is an anionic species so that its adsorption on the positively charged TiO_2 surface will be enhanced; for example, the data summarized in Table 1 indicates that about 50% of DBS is adsorbed on TiO_2 particles prior to irradiation.

We have noted that the cationic HTAB species does not adsorb on the positively charged TiO_2 surface. However, the environmental conditions are different for the mixed DBS/HTAB system (Fig. 2). Significant TOC removal is observed before illumination (Table 1). Such a temporal reduction of aqueous TOC at time zero can be rationalized either by one or both of two possible events: (i) a precipitate is formed between the anionic DBS surfactant and the cationic HTAB molecules, (ii) adsorbed anionic DBS moieties likely form an inner-sphere type complex with the catalyst surface ions, which may in turn attract the cationic HTAB molecules. This results in an accumulation of HTAB close to the TiO_2 particle surface through formation of outer-sphere type complexes. For example, for the case of the "DBS (0.025 mM)/HTAB (0.075 mM)" mixed system (Table 1), about 45% of TOC is initially removed. Since DBS accounts for only 24% (or 5.4 ppm) of the total-initial-TOC concentration and assuming that all of it remains adsorbed, at least 21% of HTAB is initially attached to the TiO_2 surface. However, when only HTAB is present, surface complexation does not occur and no HTAB attachment results (Table 1). This phenomenon is also supported by the experimental observations in Fig. 3 (see below).

For the mixed DBS/HTAB systems (Fig. 2), an increase in TOC is noted at 30 min of irradiation. Although both the DBS and HTAB species in the mixed systems are initially adsorbed to the catalyst surface, their photodegradation intermediate products may not be, thus resulting in their migration to the aqueous bulk phase and causing increased aqueous TOC levels in the aqueous phase. Note that degradation of DBS molecules does occur after 30 min irradiation with about 94% aqueous TOC remaining, of which the DBS accounts only for 16% (Fig. 2A). The reaction intermediates and remaining HTAB molecules account for the rest of the aqueous TOC. It is also

Table 1 Initial values of $[TOC]/[TOC]_0$ and surface tension of aqueous surfactant and mixed surfactant solutions in the absence and presence of TiO_2 particles prior to illumination.

Systems	Concentration (mM)	$[TOC]/[TOC]_0$		Surface tension (mN/m)	
		No TiO$_2$	with TiO$_2$	No TiO$_2$	With TiO$_2$
Surfactants alone					
DBS	0.100	1.0	0.53	50	61
SDS	0.100	1.0	0.77	71	71
HTAB	0.100	1.0	1.0	29	40
BDDAC	0.100	1.0	1.0	29	56
$C_{12}E_8$	0.100	1.0	0.85	31	31
NPE-9	0.100	1.0	0.85	25	27
Mixed surfactants					
DBS/HTAB	0.075/0.025	1.0	0.35	43	71
	0.050/0.050	1.0	0.10	25	63
	0.025/0.075	1.0	0.55	22	52
SDS/BDDAC	0.075/0.025	1.0	0.40	71	71
	0.050/0.050	1.0	0.10	40	67
	0.025/0.075	1.0	0.59	17	46
DBS/$C_{12}E_8$	0.075/0.025	1.0	0.30	36	58
	0.050/0.050	1.0	0.30	30	47
	0.025/0.075	1.0	0.50	32	17
SDS/NPE-9	0.075/0.025	1.0	0.47	33	59
	0.050/0.050	1.0	0.42	26	45
	0.025/0.075	1.0	0.50	25	37
HTAB/NPE-9	0.075/0.025	1.0	1.0	33	35
	0.050/0.050	1.0	1.0	29	33
	0.025/0.075	1.0	0.83	26	29
BDDAC/$C_{12}E_8$	0.075/0.025	1.0	0.95	36	40
	0.050/0.050	1.0	0.95	37	37
	0.025/0.075	1.0	1.0	31	32

relevant to note that the level of TOC remaining at the 30 min period increases with increasing initial DBS content in the mixture (Figs 2A and 2C). However, no such increase in aqueous TOC is observed for the DBS system alone in spite of a 50% initial adsorption (Fig. 1A). This indicates that for the

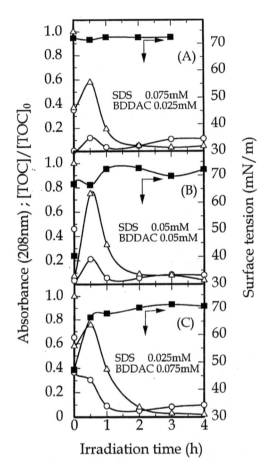

Figure 3 Disappearance of aromatic moiety (○), TOC (△), and surface tension (■) trends resulting during the photodegradation of the mixed SDS/BDDAC system using TiO_2 (100 mg) catalyst. (A) SDS (0.075 mM)/BDDAC (0.025 mM); (B) SDS (0.050 mM)/BDDAC (0.050 mM); (C) SDS (0.025 mM)/BDDAC (0.075 mM).

mixed DBS/HTAB system the surface charge characteristics of the mixed micelle are different when compared to the DBS system alone.

The initial surface tension of aqueous DBS (before irradiation and with TiO_2 present; Table 1) was 61 mN/m. After 30 min of irradiation, the surface tension increased to *ca.* 72 mN/m. For aqueous HTAB alone, the initial surface tension of the solution [before irradiation and in the presence of TiO_2 (Table 1)] was 40 mN/m, which increased to 45 mN/m after 30 min of irradiation (Fig. 1C). With further irradiation, the surface tension gradually reached the value of pure water. (i.e., 72.2 mN/m) after 2 hr. Thus, HTAB requires longer irradiation times for complete mineralization as inferred by both surface tension and TOC data (Fig. 2).

B. The SDS/BDDAC System

Figure 3 shows the photocatalytic degradation trend for the SDS/BDDAC mixed surfactant system at a total initial concentration of 0.100 mM. The CMC of both surfactants is about 8 mM. The surface tension of the aqueous BDDAC solution in the absence and presence of TiO_2 (before irradiation) was, respectively, 29 and 56 mN/m (Table 1). At 30 min of irradiation, the surface tension reached 67 mN/m and after 1 hr reaction time it was 70 mN/m (Fig. 1D). A similar trend was noted when BDDAC was greater than SDS in the mixed system (Fig. 3C). Note that the initial surface tension also increased with an increase in the initial SDS concentration in the mixed system (Table 1); by contrast, for the SDS system alone the surface tension remains at 71 mN/m, both with and without TiO_2 (Table 1).

The initial TOC data for the mixed systems (before irradiation, Table 1) show noticeable substrate removal. This suggests formation of SDS/BDDAC. As well, the data infer attachment of SDS/BDDAC species similar to the DBS-induced HTAB attachment on the TiO_2 surface. This may lead to an initial decrease in the TOC. Attachment of BDDAC to the TiO_2 surface is also consistent with the increase in BDDAC concentration at 30 min of irradiation (see Fig. 3).

At 30 min irradiation, a significant increase in aqueous TOC also results for the mixed SDS/BDDAC system (Fig. 3); we attribute this to detachment of reaction intermediates and unreacted BDDAC species from the TiO_2 surface by analogy with observations for the mixed DBS/HTAB system (Fig. 2). As the reaction proceeded, more than 90% TOC was removed after 2 hr of irradiation. Comparison between the TOC results at

30 min irradiation time at different initial concentration of BDDAC shows that the quantity of TOC removed decreases with an increase in the initial BDDAC content. This results from a reduced adsorption of cationic BDDAC molecules on the TiO$_2$ surface.

The extent of degradation of the aromatic BDDAC species was assessed by monitoring the loss of UV absorption (Figs 1D and 3). For the BDDAC alone, the aromatic moiety disappeared within 1 hr of irradiation (Fig. 1D). Note that the UV absorbance for the mixed SDS/BDDAC system (Figs 3A and 3B) show an increase at 30 min of irradiation, followed by a decrease in UV absorption with further irradiation to 4 hr. We infer that most of the BDDAC species were successfully degraded at the longer illumination time, and that any residual UV absorption is probably caused by some remaining reaction intermediate(s).

C. The DBS/C$_{12}$E$_8$ System

The degradation results for the anionic DBS and nonionic C$_{12}$E$_8$ (CMC, 0.07 mM at 25°C) surfactants are illustrated in Figs 1A and 1E, respectively; the mixed DBS/C$_{12}$E$_8$ system is summarized in Fig. 4. For the C$_{12}$E$_8$ system alone, the initial surface tension (before and after TiO$_2$ addition, but before irradiation, Table 1) was 31 mN/m. After 30 min of irradiation, the surface tension reached 64 mN/m (Fig. 1E) and with further irradiation increased to 72 mN/m. Moreover, the surface tension data at 30 min irradiation are 72, 72, 72, and 69 mN/m, respectively, for [C$_{12}$E$_8$] = 0.100 mM, 0.075 mM, 0.050 mM, and 0.025 mM (Figs 1A and 4).

Hence, at the higher C$_{12}$E$_8$ concentrations, more reaction time is needed to attain a surface tension of 72 mN/m. This indicates that the C$_{12}$E$_8$ molecules are more resistant to photocatalytic degradation than are the DBS species. TOC trends are consistent with this assertion.

As seen for the previous systems (Figs 1–3), a significant increase in aqueous TOC results at 30 min irradiation time for the mixed DBS/C$_{12}$E$_8$ system (Figs 4A–C). No such increase results for DBS or C$_{12}$E$_8$ system alone (Figs 1A and 1E,

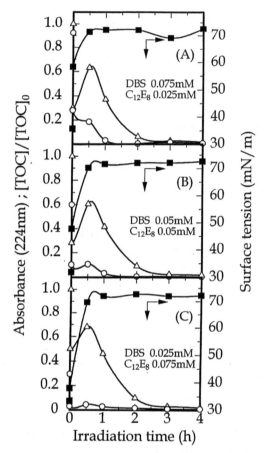

Figure 4 Disappearance of aromatic moiety (○), TOC (△), and surface tension (■) trends resulting during the photodegradation of the mixed DBS/$C_{12}E_8$ system using TiO_2 (100 mg) catalyst. (A) DBS (0.075 mM)/$C_{12}E_8$ (0.025 mM); (B) DBS (0.050 mM)/$C_{12}E_8$ (0.050 mM); (C) DBS (0.025 mM)/$C_{12}E_8$ (0.075 mM).

respectively). However, the UV absorption features at 30 min show significant DBS loss. This indicates that the increase in aqueous TOC at 30 min results either from dissolution of reaction intermediates (which may not be prone to adsorption on the TiO_2 surface) or desorption of $C_{12}E_8$ species. It should be noted, however, that contrary to the results for the HTAB and BDDAC surfactant system alone, some initial adsorption of

$C_{12}E_8$ species on the TiO_2 surface does occur (*ca.* 15%, Fig. 1E). The most negatively charged function on the nonionic surfactant molecules initiated their adsorption on the positively charged TiO_2 particles [25]. As the photoreaction proceeds and thus with the TiO_2 surface charge characteristics changing, the attached nonionic surfactant molecules and reaction intermediates detach and migrate to the aqueous phase. As well, the TOC remaining after 1 hr irradiation increases with increase in the initial $C_{12}E_8$ concentration (Figs 1A, 1E, and 4). Apparently, the nonionic $C_{12}E_8$ species require a longer irradiation time for complete mineralization.

D. The SDS/NPE-9 System

The photodegradation of the anionic SDS and non-ionic NPE-9 (CMC, 0.075 mM at 25°C) surfactant systems is depicted in Figs 1B, 1F, and 5. The decomposition of both SDS and NPE-9 occur concurrently. Note that the initial surface tension before illumination and at 30 min illumination decreases slightly with increase in the initial NPE-9 concentration (Table 1 and Figs 5B and 5C).

A similar trend was noted for the DBS/$C_{12}E_8$ system (Fig. 4 and Table 1). There is also a rapid increase in surface tension from 0 to 30 min (Figs 1B, 1F, and 5) followed by a gradual increase. For example, for the NPE-9 system alone (Fig. 1F) the initial surface tension for pure and TiO_2 mixed NPE-9 solutions (before irradiation) were 25 and 27 mN/m, respectively (Table 1). However, after 30 min of degradation, the surface tension reached 63 mN/m (Fig. 1F). With further irradiation, the surface tension reached 70 and 72 mN/m after 1 and 2 hr of irradiation, respectively.

The initial TOC results (before irradiation) indicate 15% attachment at the TiO_2 surface for the NPE-9 system, and more than 50% for the mixed SDS/NPE-9 systems (Table 1). TOC increased at 30 min of irradiation (see also Fig. 5). The temporal changes in TOC increased slightly with increasing initial SDS concentration (SDS is more prone to adsorption on the TiO_2 surface, before illumination because of its anionic nature). However a fraction of NPE-9 would also adsorb.

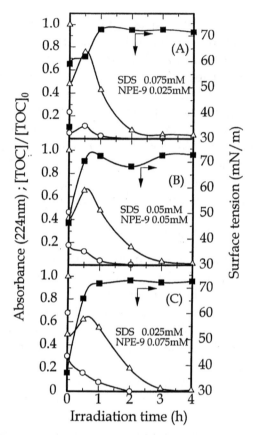

Figure 5 Disappearance of aromatic moiety (○), TOC (△), and surface tension (■) trends resulting during the photodegradation of the mixed SDS/NPE-9 system using TiO_2 (100 mg) catalyst. (A) SDS (0.075 mM)/NPE-9 (0.025 mM); (B) SDS (0.050 mM)/NPE-9 (0.050 mM); (C) SDS (0.025 mM)/NPE-9 (0.075 mM).

As noted earlier, reaction intermediate species may not remain attached to the TiO_2 surface because of different charge properties. Their desorption from the TiO_2 surface causes an increase in aqueous TOC at 30 min irradiation. Nevertheless, complete mineralization does result as evidenced by the TOC results at longer irradiation times. Longer irradiation times to achieve the same degree of TOC removal are required for higher initial NPE-9 concentrations.

E. The HTAB/NPE-9 System

Figures 1C, 1F, and 6 summarize the photodegradation of the cationic HTAB and nonionic NPE-9 surfactants. For systems with greater HTAB quantities (Figs 1C, 6A, and 6B), an initial lag (up to 30 min) in surface tension increase is quite prominent. Such a trend is not observed when the concentration of NPE-9 is greater than that of HTAB and for NPE-9 alone (Figs 6C and 1F, respectively).

Additionally, the TOC results of Table 1 indicate no initial adsorption (before irradiation) of the surfactant molecules at the TiO_2 particle surface. These results are different from those of DBS/HTAB and SDS/NPE-9 systems. The photodegraded solution becomes acidic because of proton formation during •OH radical production (see above). As the HTAB molecules bear a positive charge, they will be repelled by the positive TiO_2 surface. Hence, a much smaller change in surface tension is noticed, especially for systems of greater HTAB concentrations. For example, the results of Fig. 6A show a surface tension of 35 mN/m before irradiation; after 30 min irradiation the surface tension is only 37 mN/m. It is also important to note that for the HTAB system alone (Fig. 1C) a more gradual increase in surface tension results even after 30 min irradiation, whereas results of Figs 6A and 6B show a more pronounced increase.

The TOC results also show an opposite trend compared to the anionic/nonionic systems (Figs 1, 4, and 5), i.e., the percent TOC removal at a specific irradiation time increases with increase in initial nonionic surfactant concentration. At 1 hr irradiation, the TOCs are 86, 78.5, and 70% for each component ratio (Figs 6A, 6B, and 6C, respectively). The results from Figs 5A–C show TOC remaining after 1 hr irradiation to be 43, 48, and 55%, respectively. Consequently, the cationic surfactant slows down the overall mineralization process. In addition, the quantity of TOC removed is greater for the anionic/nonionic system. From these observations, anionic surfactants actually enhance the mineralization of nonionic surfactants.

An initial increase in aqueous TOC at 30 min irradiation time, as noted for the previous systems (Figs 1–5) is not observed in Fig. 6. This observation and the above results infer

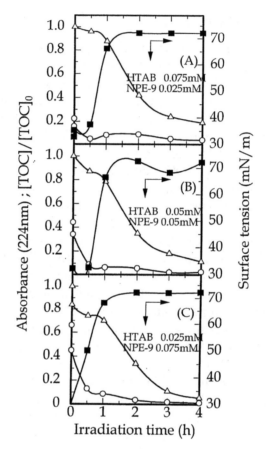

Figure 6 Disappearance of aromatic moiety (○), TOC (△), and surface tension (■) trends resulting during the photodegradation of the mixed HTAB/NPE-9 system using TiO_2 (100 mg) catalyst. (A) HTAB (0.075 mM)/NPE-9 (0.025 mM); (B) HTAB (0.050 mM)/ NPE-9 (0.050 mM); (C) HTAB (0.025 mM)/NPE-9 (0.075 mM).

that the details of the degradative pathway for the photodegradation of cationic/nonionic surfactant systems are different from those of anionic/nonionic systems.

F. The BDDAC/$C_{12}E_8$ System

The photodegradation of the cationic BDDAC/nonionic $C_{12}E_8$ system is depicted in Fig. 7. Although general trends are similar

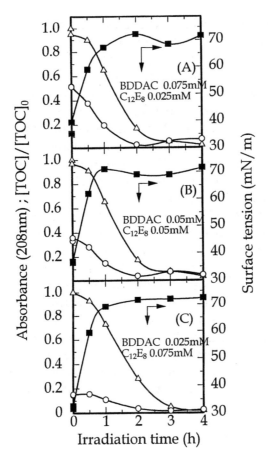

Figure 7 Disappearance of aromatic moiety (○), TOC (△), and surface tension (■) trends resulting during the photodegradation of the mixed BDDAC/$C_{12}E_8$ system using TiO$_2$ (100 mg) catalyst. (A)BDDAC (0.075 mM)/$C_{12}E_8$ (0.025 mM); (B) BDDAC (0.050 mM)/$C_{12}E_8$ (0.050 mM); (C) BDDAC (0.025 mM)/$C_{12}E_8$ (0.075 mM).

to those reported in Fig. 6, two major differences exist. First, the initial lag in surface tension increase (up to 30 min) noted for the HTAB/NPE-9 system does not occur; rather, a sharp increase in surface tension results within the first 30 min of irradiation followed by a gradual increase to 72 mN/m. Second, comparatively higher TOC removal rates are observed for the BDDAC/$C_{12}E_8$ system.

Similar to the HTAB/NPE-9 system, no significant initial surfactant adsorption occurs on the TiO_2 surface (Table 1). As noted earlier, the positively charged TiO_2 surface repels cationic surfactant moieties. However, adsorption of both anionic and nonionic surfactants occurs on the TiO_2 surface (Table 1) so that some adsorption of the nonionic surfactant $C_{12}E_8$ was also expected but not observed. This shows that attachment of nonionic surfactants in anionic/nonionic systems is somehow influenced by the anionic species; if that were not the case, and nonionic surfactant moieties did adsorb to the TiO_2 surface, then similar observations should have been obtained for the cationic/nonionic systems. These trends also support the notion that adsorption of cationic species on the TiO_2 surface occur in the presence of an anionic moiety (Table 1). In summary, increased attachment of both cationic and nonionic surfactant functions on the TiO_2 surface does occur in the presence of an anionic surfactant such as DBS.

Complete mineralization of both BDDAC and $C_{12}E_8$ occurs as indicated by the TOC results at long irradiation times (>3 hr; Fig. 7). The smaller increase in UV absorption, similar to the SDS/BDDAC systems, noted at longer irradiation times is likely due to some unknown reaction intermediate(s).

VI. CONCLUSIONS

The present studies examined the synergistic effects resulting during competitive (and simultaneous) TiO_2-assisted photo-catalytic degradation of two surfactant types. Usually, the $^{\bullet}OH$ radicals attack an electrophilic function in the surfactant molecule. For example, the benzene moiety having a high density of $\pi-\pi^*$ electrons is degraded more rapidly than the alkyl chain portion.

The $^{\bullet}OH$ radicals and H^+ species are equally formed during the photodegradation process from oxidation of H_2O. The H^+ species chemisorb on the TiO_2 surface, and in turn give it an overall positive charge. This results in enhanced adsorption of anionic surfactants on the TiO_2 surface, which in turn accelerates surfactant degradation. By contrast, it would be difficult for a cationic surfactant to attach itself to the positively charged catalyst surface because of electrostatic repulsion, causing lower degradation rates for these surfactants. Also in mixed 1:1 anionic/cationic systems (e.g., DBS/HTAB), complexes (e.g., DBS–HTAB) are generated that precipitate as insoluble species. This also causes an overall reduced mineralization rate in mixed anionic/cationic surfactant systems. Similarly, the degradation rates were also low for the cationic/nonionic systems in comparison with the anionic/nonionic systems. Great adsorption of nonionic surfactant species onto TiO_2 was noted using anionic surfactants as compared to cationic surfactants. These results re-affirm the notion that attachment of surfactant molecules to the TiO_2 particle surface plays an important role in the overall mineralization of pollutants.

ACKNOWLEDGMENTS

We are grateful to the Japanese Ministry of Education, Culture, Sports, Science, and Technology (Grant-in-Aid for Scientific Research No. 14540544 to H.H.), to the Japan Society for the Promotion of Science (JSPS Fellowship to M.S.V.), and to the Natural Sciences and Engineering Research Council of Canada (No. A5443 to N.S.) for generous financial support.

REFERENCES

1. R. J. Larson, and R. D. Vashon, *Develop. Ind. Microbiol.* **24**: 425 (1983).

2. R. J. Shimp, and R. J. Larson, *Ecotoxicol. Environ. Safety* **34**: 85 (1996).

3. M. T. Garcia, I. Ribosa, T. Guindulain, J. Sanchez-Leal, and J. Vives-Rego, *Environ. Pollution* **111**: 169 (2001).

4. R. S. Boethling, *Water Res.* **18**: 1061 (1984).

5. E. Tomlinson, M. R. Brown, and S. S. Davis, *J. Med. Chem.* **20**: 1277 (1977).

6. M. H. Angele, *Seifen Öle Fette Wachese* **104**: 433 (1978).

7. E. Stephanou, and W. Giger, *Environ. Sci. Technol.* **16**: 800 (1981).

8. W. Giger, E. Stephanou, and C. Schaffner, *Chemosphere* **10**: 1253 (1981).

9. H. Hidaka, and J. Zhao, *Colloids and Surfaces* **67**: 165 (1992).

10. E. Pelizzetti, C. Minero, H. Hidaka, and N. Serpone, *Photocatalytic Purification and Treatment of Water and Air* (D.F. Ollis and H. Al-Ekabi, eds), Elsevier, Amsterdam (1993), p. 261.

11. H. Hidaka, S. Yamada, S. Suenaga, H. Kubota, N. Serpone, E. Pelizzetti, and M. Gratzel, *J. Photochem. Photobiol. A: Chem.* **47**: 103 (1989).

12. H. Hidaka, H. Kubota, M. Gratzel, N. Serpone, and E. Pelizzetti, *Nouv. J. Chim.* **9**: 67 (1985).

13. H. Hidaka, J. Zhao, E. Pelizzetti, and N. Serpone, *J. Phys. Chem.* **96**: 2226 (1992).

14. H. Hidaka, H. Kubota, M. Gratzel, E. Pelizzetti, and N. Serpone, *J. Photochem. Photobiol.* **35**: 219 (1986).

15. H. Hidaka, Y. Fujita, K. Ihara, S. Yamada, K. Suzuki, N. Serpone, and E. Pelizzetti, *J. Jpn. Oil Chem. Soc.* **36**: 836 (1987).

16. H. Hidaka, S. Yamada, S. Suenaga, J. Zhao, N. Serpone, and E. Pelizzetti, *J. Mol. Catal.* **59**: 279 (1990).

17. H. Hidaka, K. Ihara, Y. Fujita, S. Yamada, E. Pelizzetti, and N. Serpone, *J. Photochem. Photobiol. A: Chem.* **42**: 375 (1988).

18. H. Hidaka, J. Zhao, K. Kitamura, K. Nohara, N. Serpone, and E. Pelizzetti, *J. Photochem. Photobiol. A: Chem.* **64**: 103 (1992).

19. H. Hidaka, S. Yoshizawa, M. Takai, and M. Moriya, *J. Jpn. Oil. Chem. Soc.* **31**: 489 (1982).

20. M. Takai, H. Hidaka, S. Ishikawa, M. Takada, and M. Moriya, *J. Am. Oil Chem. Soc.* **55**: 382 (1980).

21. M. Takai, H. Hidaka, S. Ishikawa, M. Takada, and M. Moriya, *J. Am. Oil Chem. Soc.* **57**: 183 (1980).

22. C. S. Turchi, and D. F. Ollis, *J. Catal.* **122**: 178 (1990).

23. C. D. Jaeger, and A. J. Bard, *J. Phys. Chem.* **83**: 3146 (1979).

24. M. Abdullah, K. L. Low, and R. W. Matthews, *J. Phys. Chem.* **94**: 6820 (1990).

25. J. Zhao, H. Hidaka, A. Takamura, E. Pelizzetti, and N. Serpone, *Langmuir* **9**: 1646 (1993).

Index

Milton Keynes UK
Ingram Content Group UK Ltd.
UKHW020001071024
449327UK00031B/2610

9 780367 578138